INDUSTRIAL ORGANIC CHEMICALS

OTHER BOOKS BY THE AUTHORS

The Phosphatides, by Harold A. Wittcoff, Reinhold, New York, 1950.

The Chemical Economy by Bryan G. Reuben and Michael L. Burstall, Longman, London, 1973.

Industrial Organic Chemicals in Perspective, Part 1: Raw Materials and Manufacture, Part 2: Technology, Formulation, and Use, by Harold A. Wittcoff and Bryan G. Reuben, Wiley, New York, 1980.

Industrial Organic Chemistry, an ACS tape course, by Harold A. Wittcoff, ACS, Washington DC, 1984.

The Pharmaceutical Industry—Chemistry and Concepts, an ACS tape course, by Harold A. Wittcoff and Bryan G. Reuben, ACS, Washington DC, 1987.

The Cost of "Non-Europe" in the Pharmaceutical Industry, Research in the Cost of "Non-Europe, Basic Findings Vol. 15, by Michael L. Burstall and Bryan G. Reuben, Commission of European Communities, Luxembourg, 1988.

Pharmaceutical Chemicals in Perspective, by Harold A. Wittcoff and Bryan G. Reuben, Wiley, New York, 1990.

Cost Containment in the European Pharmaceutical Market, by Michael L. Burstall and Bryan G. Reuben, Marketletter, London, 1992.

Implications of the European Community's Proposed Policy for Self-Sufficiency in Plasma and Plasma Products, by Bryan G. Reuben and Ian Senior, Marketletter, London, 1993.

INDUSTRIAL ORGANIC CHEMICALS

HAROLD A. WITTCOFF

Scientific Adviser, Chem. Systems Inc., Vice President of Corporate
Research, General Mills, Inc. (retired)

BRYAN G. REUBEN

Professor of Chemical Technology, South Bank University, London;
REMIT Consultants, London

A Wiley-Interscience Publication

JOHN WILEY & SONS, INC.

New York / Chichester / Brisbane / Toronto / Singapore

Library of Congress Cataloging in Publication Data:
Wittcoff, Harold A.
   Industrial organic chemicals/Harold A. Wittcoff, Bryan G.
Reuben.
      p.   cm.
   Rev. ed. of: Industrial organic chemicals in perspective. c1980.
   "A Wiley-Interscience publication."
   Includes bibliographical references and index.
   ISBN 0-471-54036-6 (cloth: alk. paper)
   1. Organic compounds—Industrial applications.   I. Reuben, Bryan G.
II. Wittcoff, Harold A. Industrial organic chemicals in perspective.
III. Title.
TP247.W59   1996
661.8—dc20   95-35580

Printed in the United States of America

10 9 8 7 6 5 4 3

To
Anthony Jacob, Bessie, David, Debbie,
Michelle, Ralph, Ted, and Virginia.

PREFACE

In the early 1970s, one of us (BGR) wrote a book celebrating the rapid growth of the adolescent chemical industry. The organic chemicals industry at the time was growing at four times the rate of the economy. It was indicated nonetheless that "trees do not grow to the sky." In 1980, in another book, we both declared the industry to be middle-aged with slow or zero growth. In this totally revised and expanded version of our earlier book, we reflect that the industry, at any rate in the developed world, is showing many of the illnesses of late middle-age.

The problems have arisen first from the undisciplined building of excess capacity with consequent fierce competition and low prices. Second, the entry of numerous developing countries into the industry has exacerbated the situation (Section 1.3.3), and third, there has been much stricter government legislation (Section 1.3.7). There is massive worldwide restructuring and continual shifting of commodity chemical manufacturing to areas other than the United States, Western Europe, and Japan. The Middle East and Southeast Asia are the principal new players in the game. Perhaps this trend will continue and the present developed world will in the future confine itself to the manufacture of specialties, but the economic and political forces at work are more complex than that. We hope to be able to discuss their resolution in another edition in about 10 years' time.

Meanwhile, some things have not changed. The organic chemicals industry is still based on seven basic raw materials all deriving from petroleum and natural gas. The wisdom of teaching about the chemical industry on the basis of these seven building blocks has been confirmed by the fact that, since the publication of our first book, one of us (HAW) has delivered by invitation 200 courses in 27 countries on the fundamentals of the industry based on this pattern. Most of these courses are for industrial personnel, but academia has not been neglected.

Furthermore, some changes have been positive. For example, there have been exciting new processes such as the development of metallocene catalysts (Section 15.3.12). Section 4.6.1 describes new methyl methacrylate processes that give a potentially cheaper product, that do not produce ecologically undesirable ammonium hydrogen sulfate byproduct or (in another process) that eliminate the use of dangerous hydrogen cyanide.

In this book, our main objective is still to present the technology of the organic chemicals industry as an organized body of knowledge, so that both the neophyte and the experienced practitioner can see the broad picture. Nonetheless, we have expanded its scope to include not only new processes but many apparently less important reactions that are significant because they give rise to the more profitable specialty chemicals. The lesser volume chemicals have been clearly delineated as such, and the reader who wishes to see the industry on the basis of its large tonnage products can omit these sections.

We hope this book will be useful both to college students who have studied organic chemistry and to graduates and industrial chemists who work in or are interested in the chemical industry. Even though much of the chemistry has remained the same, the change in the way the industry looks at its problems provides ample justification for our offering this volume as a fresh perspective on industrial organic chemicals.

Tarrytown, New York   HAROLD A. WITTCOFF
London, England   BRYAN G. REUBEN

ACKNOWLEDGMENTS

We gratefully acknowledge the help of Professor Maurice Kreevoy, who reviewed the chapter on catalysis, checked the mechanisms, and offered many helpful suggestions. Ms. Teresa Castracan typed much of the book cheerfully and efficiently. We are particularly grateful to Chem. Systems Inc. whose numerous reports on virtually all phases of the chemical industry provided detailed information on reaction conditions for many of the processes we have described.

CONTENTS

UNITS AND CONVERSION FACTORS

WEIGHT

thousand pounds	metric tons (tonnes or thousand kg)	long tons	short tons
1	0.4536	0.4464	0.5000
2.2046	1	0.9842	1.1023
2.2400	1.0160	1	1.1200
2.0000	0.9072	0.8929	1

VOLUME

liters $(10^{-3}\,m^3)$	cubic feet	US gallons	Imperial gallons
1	0.03532	0.2642	0.2200
28.32	1	7.481	6.229
3.785	0.1337	1	0.8327
4.546	0.1605	1.201	1

PRESSURE

atmospheres	bar	torr (mm Hg)	psi	kg cm^{-2}
1	1.01325	760	14.696	1.033
0.9869	1	750.06	14.504	1.020
0.001316	0.001333	1	0.01934	0.00136
0.06805	0.06895	51.715	1	0.0703
0.968	0.980	735.3	14.225	1

(1 bar = 10^5 pascal or newtons per square meter)

TEMPERATURE

Expressed as °C (= degrees Centigrade or degrees Celsius)
Degrees Fahrenheit (°F) = 1.8(°C) + 32
(°C) = 0.556(°F − 32)
Degrees Kelvin (°K) = (°C) + 273.15
0°K = − 273.15°C = − 459.7°F

HEAT

kilojoules	kilocalories	British thermal units $(10^{-5}$ therms$)$
1	0.239	0.948
4.184	1	3.968
1.054	0.252	1

One tonne of oil is equivalent to 3.97×10^7 Btu
.01 teracalories
.042 terajoules

1.5 tonnes of coal (typical calorific value)
3 tonnes of lignite (typical calorific value)

0.805 tonnes of LNG
1111 m^3 of natural gas
39200 cubic feet of natural gas

12000 kWh of electricity

One cubic foot of natural gas = 1000 Btu
One m^3 of natural gas = 9000 kcal = 37600 kJ
One kWh = 3412 Btu = 860 kcal

NOBLE METALS

Noble metals—gold, silver, platinum, palladium, rhodium etc.—are traded in
ounces. These are not the familiar avoirdupois ounces (= 28.35 g) but troy or
apothecary ounces (= 31.15 g). One ounce troy = 1.097 ounces avoirdupois.
1 pound troy = 12 ounces troy; 1 pound avoirdupois = 14.58 ounces troy. 1000
ounces troy = 31.15 kg = 0.03115 tonnes.

SPECIAL UNITS IN THE CHEMICAL INDUSTRY

PETROLEUM AND REFINERY PRODUCTS

Crude oil and some refinery products are traded in barrels (bbl) of 42 US gallons (gal) (= 35 Imperial gal). As the gallon is a unit of volume, the weight of a barrel depends on the density of the product. Approximate conversion factors follow:

1000 lb = 3.32 bbl crude oil, 3.83 bbl gasoline, 3.54 bbl kerosene,
        3.40 bbl gas oil, and 3.04 bbl fuel oil.
1 tonne = 7.33 bbl crude oil, 8.45 bbl gasoline, 7.80 bbl kerosene,
        7.50 bbl gas oil, and 6.70 bbl fuel oil.

Liquefied petroleum gases are sold by the US gallon or by the tonne. One tonne contains 521 gallons of propane, 453 gallons of n-butane or 469 gallons of isobutane. LPG can be mainly propane, mainly butane, or a "mixed" cargo, in which case intermediate conversion factors based on composition must be applied.

GASES

Natural gas is measured in standard cubic feet (scf) at 1 atmosphere (atm) and 60°F or in cubic meters (m³) at 1 atm and 0°C. $1 \text{ m}^3 = 37.33$ scf; $1 \text{ scf} = 0.0268 \text{ m}^3$. Thermal units (heat liberated when a volume of gas is burned) are sometimes used. Calorific values depend on the composition of the gas but are usually 900–1000 Btu scf^{-1}. Accordingly 1 therm $= 10^5$ Btu $= 100$–110 scf.

Other gases are also measured in scf and m³. If the molecular weight of a gas is M, then 10^6 scf of the gas weigh 2.635 M thousand pounds. For example, 10^6 scf of hydrogen weigh 5.312 thousand pounds and of oxygen 84.32 thousand pounds. Similarly, 1000 m³ of a gas weigh 0.0446 M tonnes. 1000 m³ of hydrogen weigh 0.0900 tonnes and of oxygen 1.427 tonnes.

COAL TAR PRODUCTS

Coal tar and materials traditionally derived from it such as benzene, toluene, and xylenes are sometimes measured in thousands of US gallons. One thousand US gallons of benzene at 20°C weigh 7320 lb, of toluene 7210 lb, of o-xylene 7300 lb, of m-xylene 7161 lb, and of p-xylene 7134 lb.

ETHANOL

Ethanol is measured in mixed volume and concentration units. One Imperial gallon (1.201 US gal) of 100% ethanol contains the same amount of ethanol as 1.75 proof gal, and concentration is measured in degrees proof. The specific gravity of ethanol is 0.79, hence 1 proof gal contains 4.5 lb (2.04 kg) ethanol. It is also the alcoholic equivalent of a US liquid gal at 15°C containing 50% ethanol by volume. The metric units are hectoliters (= 100 liters), and concentration is expressed in degrees Gay-Lussac. Ninety-five degrees Gay-Lussac represents 95% by volume at 15°C. One hectoliter = 22 Imperial gal = 26.4 US gal. One hectoliter of 100% ethanol weighs 174 lb (79 kg). The US liquid gallon (as above) is identical with the Queen Anne wine gallon. A US tax gallon for spirits of 100 proof or over is equivalent to the proof gallon; for spirits of less than 100 proof to the wine gallon.

INDUSTRIAL ORGANIC CHEMICALS

CHAPTER 0

HOW TO USE INDUSTRIAL
ORGANIC CHEMICALS

Industrial Organic Chemicals is an updated, expanded and completely rewritten version of Volume 1 of a two-volume set, *Industrial Organic Chemicals in Perspective*, published in 1980. Volume I of the set described where industrial organic chemicals came from; volume II described how they were used. Over a decade later, chemicals still come from the same sources, but there are many new processes to be described. The application of chemicals has changed much less, and a revision of volume II is not planned at present.

0.1   WHY THIS BOOK WAS WRITTEN AND HOW IT IS STRUCTURED

The petrochemical industry provides well over 90% by tonnage of all organic chemicals. It grew rapidly in the 1950s and 1960s. Many new processes and products were introduced. Large economies of scale proved possible. The prices of chemicals and polymers dropped so that they could compete with traditional materials. Cheerfully colored plastic housewares, highly functional packaging, and easy care garments of synthetic fibers were no longer exciting new technology but had become an accepted and routine part of modern life.

By the 1970s growth was leveling off. The first and second oil shocks increased the price of crude oil, and hence of its downstream products. Economies of scale ran out. The industry had matured. As its technology became better known, developing countries started their own petrochemical industries, competing with the developed countries, and depressing profitability. Furthermore, the impact of the industry on the environment became evident.

In the 1980s and early 1990s, new products were no longer the name of the game, in part because the 1960s and 1970s had provided an arsenal of them to attack new applications. Also, the industry became subject to strict government monitoring. Expensive toxicity testing was required before a new compound could be introduced (Section 13.7).

Rather than developing bigger, better plants to manufacture novel chemicals, the industry became concerned with lessening pollution, improving processes, and developing specialty chemical formulations and niche products that could be sold at higher profit margins. Research and development became highly process oriented, in part to find less polluting processes, and in part to combat maturity and gain an edge over competition with money-saving technology. Examples are given in the preface and throughout the book.

Chapter 1 shows how the chemical industry fits into the overall economy, and then defines the industry in terms of its characteristics.

Chapter 2 describes where organic chemicals come from, and then shows how the major sources, petroleum and natural gas, provide seven basic chemicals or chemical groups from which most petrochemicals are made. The basic building blocks comprise olefins—ethylene, propylene, and the C_4 olefins (butadiene, isobutene, 1- and 2-butenes); aromatics—benzene, toluene, and the xylenes (ortho, meta, and para); and one alkane, methane. Chapter 2 explains how the olefins derive primarily from steam cracking and secondarily from catalytic cracking, and how the aromatics derives primarily from catalytic reforming in the United States and from reforming and steam cracking in Europe. Methane occurs as such in natural gas. This chapter emphasizes the important interface between the refinery and the petrochemical industry.

Chapters 3 and 4 describe the chemistry of ethylene and propylene. They are the most important of the seven building blocks and are treated accordingly.

Chapters 5 and 6 deal with the C_4 and C_5 olefins. The C_5 compounds and their derivatives are low in volume and are not included in the seven basic building blocks. They are nonetheless an important source of isoprene and of the precursor for the octane improver *tert*-amyl methyl ether (TAME).

Chapters 7–9 describe the chemistry of the aromatics: benzene, toluene, and the xylenes. Benzene has been overshadowed by ethylene and propylene since the 1960s but is still the third most important of the building blocks.

Chapter 10 describes the chemistry of methane, a relatively unreactive molecule, which nonetheless is the source of synthesis gas for ammonia and methanol manufacture. Acetylene is discussed here, since it may be made from methane. Whereas it was very important 50 years ago, its significance has been steadily decreased by newer chemistry based on ethylene and propylene.

Chapter 11 is devoted to the small amount of industrial chemistry based on alkanes other than methane.

Chapters 12–14 deal with nonpetroleum sources of chemicals: coal, fats and oils, and carbohydrates. The chemical industry in the nineteenth and early twentieth centuries was based on chemicals derived from coal tar or coke oven distillate. Today, this is a specialty area and our major interest in coal focuses on its conversion to synthesis gas. This would be the first stage in building a coal-based chemical industry should petroleum and natural gas become depleted. The chemistry of fats and oils is reflected in the surfactant area and in numerous speciality performance products. Carbohydrate-based chemicals are also largely specialties.

Since over one-half of all organic chemicals manufactured end up in polymers, Chapter 15 is devoted to polymerization processes and polymer properties. Chapter 16 deals with the all-important subject of catalysis without which there would hardly be a chemical industry.

It is these new processes that provided the incentive for this new volume, but we have also expanded its scope to include many apparently less important reactions that are significant because they give rise to the more profitable specialty chemicals. We hope it will be useful both to college students who have studied organic chemistry and to graduates and industrial chemists who work in or are interested in one of the most remarkable industries of the twentieth century.

We intend each chapter to be self-sufficient, hence there is inevitably a degree of repetition. We have tried to minimize this by extensive cross-referencing and hope the reader will be tolerant of what remains.

0.2 STANDARD INDUSTRIAL CLASSIFICATION

The United States Government provides statistics on all branches of industry, dividing them according to the Standard Industrial Classification (SIC). Each major segment of the economy is classified under a number between 1 and 99 (Table 1.1). Manufacturing industries are classified under numbers 20–39 and the chemical and allied products industry falls within this category at 28. Statistics for segments and subsegments of the industry are provided under three and four digit numbers. Thus 282 is *Plastic Materials and Synthetics.* 2821 Is *Plastic Materials and Resins*, 2822 is *Synthetic Rubber*, 2823 is *Cellulosic Man-made Fibers*, and 2824 is *Organic Fibers, Non-Cellulosic.* We have relied heavily on these figures for our book, although it is never possible to obtain up-to-date figures. Thus the material published in 1994 contains information for 1992. Statistics from other sources are often more up-to-date but are less authoritative (Section 0.4.5).

The industries that form the chemical and allied products industries are shown in Table 1.2. Although at times one might wish for even more detailed information, the SIC provides a wealth of it. Other countries do not have comparable data bases; many have SICs, but none provides so much information. The classifications in other countries rarely correspond to those in the United States or to each other and analysts wishing to tackle official statistics outside the United States should be aware of the pitfalls.

0.3 UNITS AND NOMENCLATURE

The widespread adoption of the SI system of units (Système Internationale d'Unités) based on the meter, the kilogram, and the second has worsened rather than improved the plethora of units used in the chemical industry. Three kinds of tons are in common use—the short ton (2000 lb), the metric ton or tonne (1000 kg or 2204.5 lb), and the long ton (2240 lb). United States statistics are usually given in millions of pounds, which are at least unambiguous, and we give all our figures either in these units or in tonnes. In addition, we try to quote figures in the units actually used by industry—petroleum is measured in barrels, benzene in gallons, mixed xylenes in gallons, and (incredibly) p-xylene in pounds—and to give conversions into better known units. A table of conversion factors is given on the end papers.

Similarly, in naming chemicals, we tend to use the names conventional in industry rather than the more academic nomenclature of the International Union of Pure and Applied Chemistry (IUPAC). Thus we write hydrogen not dihydrogen; ethylene, acetylene, and acetic acid; not ethene, ethyne, and ethanoic acid.

Industry makes no effort to use consistent nomenclature. Ethene and propene are universally known as ethylene and propylene and would scarcely be recognized by their IUPAC names. The C_4 olefins, however, are frequently referred to as butenes rather than butylenes and we have followed this style. We use trivial names where industry does so. Thus we refer to $C_6H_5CH(CH_3)_2$ as cumene, the name by which it is bought and sold, rather than the more informative names of isopropylbenzene, 2-phenylpropane, or (1-methylethyl)benzene. The term ethanal would be likely to be misread or misheard in industry as ethanol, and the compound is known as acetaldehyde. So important is trivial nomenclature that the pharmaceutical industry could not exist without it.

We regret the lack of consistency that the use of trivial nomenclature entails, but we feel it best serves our aim of introducing the student to chemical industry practice.

0.4   GENERAL BIBLIOGRAPHY

In many ways the greatest service that a book like this can provide is to introduce the student to the industrial chemical literature. We follow each chapter with an annotated bibliography that lists some of the standard literature on the subject of the chapter, cites the sources of much of our own information, and adds occasional notes to matters discussed in the chapter. We largely confine ourselves to material published after 1975 but have included occasional "classics." References to earlier work may be found in *Kirk–Othmer* and other encyclopedias, and in B. G. Reuben and M. L. Burstall, *The Chemical Economy*, Longman, London, 1974.

0.4.1   Encyclopedias

The most important single reference work is *Kirk–Othmer's Encyclopedia of Chemical Technology* (22 volumes plus one supplementary volume and index) 3rd ed., R. E. Kirk and D. F. Othmer, Interscience, New York, 1978–1984.

A fourth edition edited by J. I. Kroschwitz and M. Howe-Grant started in 1991 and 12 volumes had been issued through 1994. *Kirk–Othmer* provides comprehensive and well-referenced coverage of almost every aspect of industrial chemistry. The later volumes of the 3rd ed. are inevitably dated but provide information not readily available from other sources. If a subject is not treated in the new edition, it is always worth consulting the older one.

The Encyclopedia of Polymer Science and Engineering, J. I. Kroschwitz, Ed. (17 volumes plus supplement and an index volume) Interscience, New York, 1985–1989, provides comprehensive coverage of polymer chemistry. It is well referenced but weak on technology. The 1st ed. comprised 15 volumes and was published between 1964 and 1971. As with *The Encyclopedia of Chemical Technology*, the earlier edition still contains valuable material.

The Encyclopedia of Chemical Processing and Design, J. J. McKetta, Ed., Dekker, New York, has a chemical engineering orientation. Forty-nine volumes of this ambitious work had been published by mid-1994 and cover subjects classified under A–S. Thus it was about 85% complete. As publication started in 1976, it is perhaps inevitable that the approach is inconsistent. Individual articles are worthwhile but the content is unpredictable.

A welcome edition to the chemical literature is *Ullmann's Encyclopedia of Industrial Chemistry*, W. Gerhartz, Ed. VCH, Weinheim. This is an ambitious expansion of the early works of Ullmann published in the early 1900s. It is divided into two sets. The A volumes contain alphabetically arranged articles, whereas the B volumes include what is termed "basic knowledge." The publication was started in 1985 and by 1994 Volumes Al–A24 had been published. Twenty-eight are scheduled. In the B series, eight are scheduled and six had been published by 1994.

An ambitious undertaking is the *Dictionary of Scientific and Technical Terms*, S. P. Parker, Ed. The 4th ed. was published by McGraw-Hill in 1989. A newer venture is R. D. Ashford, *Dictionary of Industrial Chemical Properties, Production and Uses*, R. D. Ashford, Wavelength, London, 1994.

Some consulting companies publish reports on a continuing basis that contain a wealth of up-to-date information on chemistry, engineering, and markets of numerous industrial chemicals. These are, however, quite expensive and are usually found only in industrial libraries, the subscriber agreeing to keep the information confidential. One such program is entitled *Process Engineering and Research Planning*, Chem. Systems, Inc., Tarrytown, New York, 10591, which covers in depth the chemistry, engineering, and market data for many of the basic petrochemicals, as well as important specialty chemicals. A less in-depth compendium but one that covers a greater breadth of subjects is the *Chemical Economics Handbook*, Stanford Research Institute, Menlo Park, CA.

0.4.2 Books

Before the spectacular growth of the chemical industry after World War II, three classic books appeared that encompassed much of what was done at that time.

These books have been repeatedly revised and updated and, although they seem old-fashioned in some ways, they are certainly worthy of mention. The oldest is *Riegel's Handbook of Industrial Chemistry*, 9th ed., J. A. Kent, Ed., Van Nostrand-Reinhold, New York, 1992. Riegel first appeared in 1928 and is now a multiauthor survey of the chemical and allied products industry. *The Chemical Process Industries*, R. N. Shreve and N. Basta, 6th ed., McGraw-Hill, New York, 1994, was first published in 1945 and covers many of the process industries such as cement and glass as well as the mainstream chemical industry. *Faith Keyes and Clark's Industrial Chemicals* first appeared in 1950. The fourth and most recent edition was revised by F. A. Lowenheim and M. K. Moran and was published by Wiley-Interscience in 1975, and is now out of print. It provides details of manufacture and markets for the 140 most important chemicals in the United States and is important because of its interdisciplinary approach, which has never been repeated.

Another volume of note is *Chemistry and the Economy*, M. Harris and M. Tischler, Eds., American Chemical Society, Washington DC, 1973. Although old, it provides an excellent overview of the impact of chemical technology on the economy. It describes how industrial chemistry interfaces with numerous industries, gives insight into the makeup of the chemical industry, and provides numerous historical facts that tend to humanize the industry. The book is good reading for students as well as for practicing chemists and engineers.

In the 1970s interest increased in the teaching of industrial chemistry in colleges and universities. There is still a long way to go to impress the academic community with the important role that industrial chemistry has played not only in the application of chemical technology but also in the development of new knowledge. Nonetheless, a number of important books were published.

One of the first books on industrial organic chemistry was *Industrial Organic Chemistry*, J. K. Stille, Prentice Hall, Englewood Cliffs N J, 1968. This small volume, now out of print, contains a wealth of material about the industry as it existed prior to 1970 and is written from the interesting perspective of an academic organic chemist.

The Chemical Economy, B. G. Reuben and M. L. Burstall, Longman, London, 1974 is a guide to the technology and economics of the chemical industry and provides an overview of the industry emphasizing organic chemicals. It is biased somewhat toward European practice and contains annotated bibliographies.

Basic Organic Chemistry V: Industrial Products, J. M. Tedder, A. Nechvatal, and A. H. Jubb, Wiley, Chichester, UK, 1975, is the fifth volume of a series on organic chemistry but the title is somewhat misleading as the book can stand by itself as a textbook on industrial organic chemistry. It comprises a multiauthor survey oriented towards chemistry rather than technology and toward British practice. Insufficient references are given. The book is a mine of information for the specialized reader and was reprinted but not revised recently.

Principles of Industrial Chemistry, C. A. Clausen III and G. Mattson, Wiley-Interscience, New York, 1978, is aimed at chemists and provides an enthusiastic introduction to chemical process principles, process development, and various commercial aspects of the chemical industry.

Chemicals from Petroleum, 4th ed., A. L. Waddams, John Murray, London, 1978, was a pioneering account of petrochemicals and provides an insight into the relationship of the refinery to the chemical industry.

Among the important books dealing specifically with the organic chemicals industry is *An Introduction to Industrial Organic Chemistry*, 2nd ed., P. Wiseman, Halsted Press, New York, 1979. It is well organized and well written and is oriented toward the pure chemistry that provides a base for technology. The same author provided *Petrochemicals*, Ellis Horwood, Chichester, UK, 1986.

The Structure of the Chemical Processing Industries, J. Wei, T. W. F. Russell, and M. W. Swartzlander, McGraw-Hill, New York, 1979, is similar in structure to *The Chemical Economy* (*loc. cit.*) but deals primarily with the economic structure of the industry and much less with technology. More recent is *Dynamics of the US Chemical Industry*, S. Greenbaum, Kendall Hunt, New York, 1994.

At the end of the 1970s appeared our two-volume work, *Industrial Organic Chemicals in Perspective, Part I: Raw Materials and Manufacture*, and *Part II: Technology, Formulation and Use*, H. A. Wittcoff and B. G. Reuben, Wiley, New York, 1980. Part I concentrated on the production of organic chemicals from seven major building blocks, while Part II dealt with the downstream sectors of the chemical industry detailing the chemistry that was involved in the use of chemicals for plastics, fibers, elastomers, surface coatings, adhesives, surface active agents, pharmaceuticals, solvents, lubricating oils, plasticizers, agrochemicals, food chemicals, and dyes and pigments. Both volumes were reprinted by Krieger, FL, in 1990.

The 1980s and early 1990s saw some new editions but relatively few new books on industrial organic chemistry. *Guide to the Chemical Industry: R & D, Marketing and Employment*, W. S. Emerson, Wiley-Interscience, New York, 1983, reprinted Krieger, FL, 1991, provides useful insights into how the industry functions but the technical part contains several unfortunate errors.

Chemicals are discussed from the point of view of the consumer in an interesting and original book, *Chemistry in the Market-Place*, 4th ed., B. Selinger, Harcourt-Brace, 1990. Selinger is a pioneer of the Australian consumer movement and chairs a committee on toxic waste disposal. He describes the formulation of many domestic products together with the reasons for the various additives and the theory behind them.

An English translation of *Industrial Organic Chemistry*, 2nd ed., K. Weissermehl and H. J. Arpe, VCH, Weinheim appeared in 1993. Like the 1976 first edition, it is beautifully laid out, easy to follow, and concentrates on the upstream large tonnage processes.

One of the most significant new books is *Organic Building Blocks of the Chemical Industry*, H. Szmant, Wiley, New York, 1989. This volume contains a wealth of information even about small tonnage chemicals and emphasizes recent practice. It lists the prices at the time of publication of all the chemicals it mentions and illustrates vividly the adding of value as chemicals further and further downstream of the oil refinery are produced.

Survey of Industrial Chemistry, 2nd ed., P. J. Chenier, VCH, New York, 1992, covers some inorganics as well as organics and spends some time on economic aspects. *Handbook of Chemical Production Processes*, R. A. Meyers, Ed., McGraw-Hill, New York, 1986, describes lucidly forty industrial processes, divided between organics, inorganics, and polymers.

C. A. Heaton has edited two books of note: *An Introduction to Industrial Chemistry*, 2nd ed., Blackie, London 1991, and *The Chemical Industry*, 2nd ed., Blackie, London, 1993. The UK Society of Chemical Industry published *The Chemical Industry*, D. H. Sharp and T. F. West, Eds., Ellis Horwood, London, 1981, to celebrate their centenary. It contains "a glance back and a look ahead at problems, opportunities, and resources..." *Industrial Chemistry, Vol. 1*, E. Stocchi, Ellis Horwood, London, 1990, appeared to be the start of a series but was largely devoted to inorganic chemicals and no further volumes have yet appeared. The most recent book we have located is *Industrial Chemicals, Their Characteristics and Development*, G. Agam, Elsevier, Amsterdam, The Netherlands, 1994. It deals entertainingly with the things that academic chemists in general do not know about—formulations, specifications, standards, assays, scale-up, safety, patents, and so on. Crucial environmental problems are discussed in *Waste Management in the Chemical and Petrochemical Industries*, Fontes, Elsevier, Amsterdam, 1994.

A historical perspective of the chemical industry that is interesting for anyone who is engaged in it, as well as to academics who want insight into how basic chemistry is translated into technology, is *Petrochemicals: The Rise of an Industry*, P. H. Spitz, Wiley, New York, 1988. This book can be highly recommended. Other recent histories include *The History of the International Chemical Industry*, F. Attalion, University of Pennsylvania Press, PA, 1991, and *Milestones in 150 years of the Chemical Industry*, P. J. T. Morris, Ed., Royal Society of Chemistry, London, 1991.

0.4.3 Journals

A serious student of the chemical industry must read the trade press. New products and processes, changes in the structure and prospects of the industry, take-overs and trades as well as economic trends can all be followed.

A selection of news magazines for English-speaking readers includes *Chemical and Engineering News* (weekly, ACS, Washington, DC); *Hydrocarbon Processing* (monthly, Gulf Publishing, Houston, TX); *Chemical Week* (weekly, Chemical Week Associates, New York; there are European and US editions); *European Chemical News* (weekly, IPC International Press, London); *Chemistry and Industry*, (fortnightly, Society of Chemical Industry, London); *Manufacturing Chemist* (weekly, Morgan Grampian, London); and *Chemical Marketing Reporter* (weekly, Schnell Publishing, New York). *European Chemical News* provides United States and European prices of the major bulk chemicals. A sister publication, *Asian Chemical News*, was started in 1994. *Chemical Marketing Reporter* carries a comprehensive list of US prices of almost all widely sold chemicals.

CHEMTECH (monthly, ACS, Washington, DC) is an ideas magazine rather than a news magazine. It aims to be conceptual and at the same time to humanize chemistry. It does both admirably.

0.4.4 Patents

Patents are a device whereby the government grants inventors the sole right to exploit their inventions for a period of 17 years in the United States, 20 years in the European Community, and similar periods in other countries. In return, the inventors disclose details of their inventions in their patent specifications. Recent legislation in the United States extends the life of a pharmaceutical patent to 22 years under certain circumstances and similar patent term restoration has been enacted in Europe. As a result of the GATT (General Agreement on Tariffs and Trade) negotiations, the United States is shortly to extend all 17-year US patents to 20 years.

Patents lie at the heart of a developed society. It is difficult to see how innovation could take place if innovators were not rewarded for their efforts. "I knew that a country without a patent office ... was just a crab," said Mark Twain, "and couldn't travel any way but sideways or backwards." Meanwhile, the patent literature has grown exponentially. In the United States, it took about 200 years to amass 4 million patents, the 4 millionth having been issued in 1976. It took only 15 years to accumulate 1 million more patents and US Patent 5 000 000 was issued on 19 March 1991 to L. O. Ingram et al. It described the use of modern biotechnology to produce one of the oldest synthetic organic chemicals—ethanol.

Patent specifications are a major source of technical information. They often disclose information at a much earlier date than the scientific literature; sometimes they are the only source of such information. Negative results often appear in patents but not in scientific journals, and knowledge of what has been tried without success may save the working scientist much time.

Academic scientists shun patents because the introductions and claims are written in legal jargon with long convoluted sentences. Librarians shun them because they are published as individual items and are difficult to collect and bind. They have, however, one overwhelming advantage. They are classified by subject and can be subscribed to in this way, a copy of a US patent costing $3.00.

Patent applications are numbered consecutively as they are received by the US Patent Office (US serial number) and, when the patent is granted, it is assigned another number (US patent number). Other patent offices do the same.

Brief accounts of patents appear in the chemical trade literature. *Chemical Abstracts* publishes a numerical patent index that lists each patent number together with its corresponding *Chemical Abstracts* abstract number, country of origin, and serial number. It also provides a worldwide list of major patent offices and their addresses. *Chemisches Zentralblatt* (Akademie Verlag, Berlin) offers a similar service together with a guide to its use (*Chemisches Zentralblatt: das System*). Derwent Publications Ltd, (Rochdale House, 128 Theobald's Road, London WCIX 8RL, England) publishes analyses and abridgements of patents from every country classified by subject, and provides monthly bulletins, for example, *Organic Patents Bulletin* and *Pharmaceutical Patents Bulletin*. Derwent has contributed greatly to making patent literature available.

The *Official Gazette*, copies of patents, coupon books (a convenient way to pay for copies), listings of patents by subject, copies of foreign patents, and much other information may be obtained from the Commissioner of Patents and Trademarks, Washington, DC 20231. Although many official and commercial organizations exist to help the student of the patent literature, a thorough search can be conducted only at the National Patent Library, Washington, DC.

In the United Kingdom, the equivalent of the *Official Gazette* is the *Official Journal* (*Patents*), and it and other information are available from the Patent Office, State House, 66–71 High Holborn, London WC1R 4TP. A thorough search can be carried out at the British Library (Science Reference Section), 25 Southampton Buildings, London WC2A 1AW. It is still widely known by its former name, The Patent Office Library.

Information on subject codes and many other aids to patent searching may be found in *Kirk–Othmer* (Section 0.4.1) and in *Patents—A Source of Technical Information*, Central Office of Information, HMSO, London, 1975. Highly praised for its clarity and sound advice is *A Better Mousetrap: A Guide for Inventors*, 2nd ed. P. Bissel and G. Barker, Wordbase, Halifax, West Yorkshire, UK.

Access to patents has been simplified greatly by computerized searching of patent databases of which Derwent and *Chemical Abstracts* are the most important. The London Patent Office is itself on line. Although use of these data bases requires skill, the user is rewarded by the access these bases provide a vast amount of information. It is said that the Japanese have been able to accomplish a great deal in the chemical industry because of their skill in reading and interpreting patents. As the industry becomes more and more competitive, it is important to follow trends, to know what other companies are doing, and to avoid duplication. The patent literature can contribute more than any other source to this kind of knowledge.

0.4.5 Statistics

Students of the commercial side of the chemical industry will require access to statistics of production and consumption. Comprehensive US statistics are published annually by the US International Trade Commission as *Synthetic Organic Chemicals: United States Production and Sales*. Data for 1992 were published in the 67th annual edition in February 1994. Buying these has been a problem for many years, especially to those living outside the United States and, starting with the 1993 data, the report will be sold by the Government Printing Office.

Figures for the major chemicals plus information about companies, employment, and related topics are published more rapidly in *Chemical and Engineering News* at the end of June or the beginning of July of the subsequent year. Thus the data for 1993 were published in the 4 July 1994 edition.

On a worldwide basis, The United Kingdom Chemical Industries Association (CIA has a quite different meaning in the United Kingdom) publishes statistical reviews, notably *Basic International Chemical Industry Statistics 1963–1992*, London, 1993, and *Chemical Industry Main Markets, 1989–2000*, London 1990. The former provides charts and tables for Western Europe, the United States and Japan.

The Chemical Industry: Annual Review is compiled by the Economic Commission for Europe and published by the United Nations. The 1992 edition appeared in November 1993. Up to 1988, it was published as the *Annual Review of the Chemical Industry*. Then there is *The Chemical Industry: Europe*, M-G Information Services, Benn, London, 1994.

Detailed statistics may be obtained rather belatedly from Government sources in most countries. In the United Kingdom, disaggregated figures appear relatively quickly in the *Business Monitor*, HMSO, London, and are summarized in, for example, *Business Monitor, Report on the Census of Production, 1991*, Summary Volume, HMSO, London.

In Europe, too, many useful data and some comments are published by the industry association CEFIC (Centre européen des fédérations de l'industrie chimique) and by their subsidiary APPE, the Association of Petrochemicals Producers in Europe, which produces an annual *Activity Review* of exceptional interest.

0.4.6 CD-ROM and On-line Databases

An early book on the complex topic of retrieving chemical information is *Chemical Information*, Wolman, Y., Wiley, New York, 1983. Meanwhile, information gathering has been transformed since the first version of this book by the availability of CD-ROM and on-line databases, not to mention the Internet. It is not only that huge amounts of information can be packed onto a single CD-ROM disc but also that the searching procedure is faster and less user-hostile than in the early days.

Access to on-line databases is obtained through vendors who maintain systems for dialling into them. Typical of the scores of vendors is STN International whose United States address is c/o Chemicals Abstracts Service, 2540 Olentangy River Road, PO Box 3012, Columbus, Ohio 43210-0012. In the UK, access to scientific databases, especially those of the Institute of Scientific Information (ISI) is through the Bath Information and Data services (BIDS), University of Bath, Claverton Down, Bath BA2 7AY.

Vendors of databases are listed in *Gale Directory of Databases, Vol. I: On-line Databases*, Ed. Marcacio, K.Y., Gale Research Inc., Detroit, 1994.

Chemical information is widely available. Sources of on-line patent information are listed in Section 0.4.4. On-line technical information about the chemical industry may be obtained from *Chemical Abstracts* and *Compendex*. Market information is gathered in databases such as *Prompt*. The Royal Society of Chemistry, Thomas Graham House, Science Park, Milton Road, Cambridge, CB4 4WF, UK, produces a *Chemical Business Newsbase*, (originating in 1985) which is international but biased towards Western Europe. *Uncover* is a database of information from multi-disciplinary journals since 1983, which provides tables of contents. Articles can then be ordered directly or by fax.

CD-ROM databases include *Compendex Plus* (originating in 1985), which provides an engineering viewpoint, and *Ei Chemdisc* (originating in 1984) which takes more of an applied chemical engineering view. Both are published by Dialog. There is a general business database, *ABI/Inform*, which abstracts chemical business journals such as *Chemical and Engineering News*, *Chemical Marketing Reporter*, *Chemical Week*, and *Modern Plastics*. The coverage is largely American, but there is an invaluable extension called *Business Periodicals Ondisc*, which provides the full text of many articles on the main database.

WIPO, the World Intellectual Property Organization, based in Geneva, produces many CD-ROM and on-line publications that concern the state-of-play of patents throughout the world. This will become more and more important as more countries come into line with the GATT (General Agreement of Trade and Tariffs) regulations.

BEST (Building Expertise in Science and technology, Longman Cartermill, Technology Centre, St Andrews, Fife, Scotland KY16 9EA) is another typical database, which maintains records of scientists and engineers and what they are doing. It comes from the researchers themselves, from industry and from grant awarding bodies. It is available to industry and, in a subsidized form, to academic institutions. It is widely used as a repository of technical information to supplement what one might get out of a patent search.

The Internet is a virtual world on its own and surfing it is a skill that has to be learned. It carries little systematic information but we have, for example, picked up valuable information about the uses of blood products, when engaged in a project on fractionation of blood plasma. Meanwhile, the Royal Society of Chemistry launched a world wide web server in 1995, and access to industrial chemical information is likely to improve rapidly. Books on the Internet are obsolete almost before publication but three reasonably recent ones are P. K. McBride, *The Internet Made Simple*, Made Simple Books, 1995; J. J. Manger, *The World Wide Web, Mosaic and More*, London: McGraw Hill, 1995, and J. H. and M. V. Ellsworth, *The Internet Business Book*, New York: Wiley, 1994.

CHAPTER 1

THE CHEMICAL INDUSTRY

The United States, the countries of Western Europe, and Japan are the most complex societies that have ever existed. Division of labor has been carried to the point where most people perform highly specialized tasks and rely on many others to provide them with the goods and services they need. In return for these goods and services they provide their output to satisfy the needs of others. All men are brothers in a material sense, just as they should be in a moral sense.

The various segments of the economy are interrelated in a complex way. For example, manufacturing industry draws heavily on the output of the mining sector by buying iron ore from which to make steel. In turn it may convert that steel to machinery to sell back to the mining industry who will use it in mining operations.

1.1 THE NATIONAL ECONOMY

The interdependence of a society's activities may be seen more clearly if its economy is divided into specific industries or groups of industries. This is normally done according to the standard industrial classification (SIC) of the US Bureau of the Census, in which each industry is allocated a code number. Table 1.1 shows the main sectors of a developed economy. Each contributes "value added." This is defined as the value of shipments less cost of materials, supplies, containers, fuels, purchased electricity and contract work plus net change in finished goods, and work-in-progress inventory and value added in merchandising activities of manufacturing establishments. It is thus the value added to all the inanimate inputs into an industry by the people working in it. The total value added throughout the economy is the gross domestic product

TABLE 1.1   Main Sectors of a Developed Economy

The Economy	SIC Classification
Agriculture, forestry, and fisheries	(01–09)
Mining	(10–14)
Construction	(15–17)
Manufacturing	(20–39)
Transportation and communication and electric, gas, and sanitary services	(40–49)
Wholesale trade	(50–51)
Retail trade	(52–59)
Finance, insurance, and real estate	(60–67)
Other services, including medicine, education, social services, and entertainment	(70–97)
Government and Government enterprises	(98)
Nonclassifiable establishments	(99)

Breakdown of Manufacturing Category SIC 20–39

SIC Code	Group	Value Added, 1991[a] ($ billion)
20	Food	145.3
21	Tobacco products	24.5
22	Textile mill products	26.9
23	Apparel and other textile products	33.4
24	Lumber and wood products	27.0
25	Furniture and fixtures	20.7
26	Paper and allied products	58.3
27	Printing and publishing	103.8
28	**Chemical and allied products**	**154.8**
29	Petroleum and coal products	24.0[b]
30	Rubber and miscellaneous plastic products	50.3
31	Leather and leather products	4.3
32	Stone, clay and glass products	31.8
33	Primary metal industries	46.6
34	Fabricated metal products	76.7
35	Industrial machinery and equipment	124.2
36	Electronic and other electric equipment	106.7
37	Transportation equipment	152.0
38	Instruments and related products	82.5
39	Miscellaneous manufacturing industries	20.0
	Total	1314.8

[a] The 1991 statistics were the latest available (US Bureau of the Census, Anual Survey of Manufacturers, 1993) when this book was published. However, the relative positions of the various groups remain constant over long periods.

[b] Contrast the value added of the petroleum industry ($24.0 billion) with that of the chemical and allied products industry ($154.8 billion). The chemical industry's value added is 5.6 times as great despite the fact that its sales are less than twice as great. It is this greater value added that motivated the petroleum companies to develop chemical divisions.

TABLE 1.2   Chemical and Allied Products Industry (SIC 28) 1990

SIC	Number	Industry	Value of product Shipments ($ million)
281		**Industrial inorganic chemicals**	**26,690.8**
	2812	Alkalies and chlorine	2,709.8
	2813	Industrial gases	3,058.1
	2816	Inorganic pigments (except carbon black)	3,203.9
	2818	Industrial organic chemicals n.e.c.[a]	17,719.0
	2819	Industrial inorganic chemicals n.e.c.[a]	4,657.0
282		**Plastic materials and synthetics**	**48,419.8**
	2821	Plastics materials and resins	31,325.8
	2822	Synthetic elastomers	4,210.3
	2823	Cellulosic synthetic fibers	1,456.7
	2824	Organic fibers, noncellulosics	11,427.1
283		**Drugs**	**53,719.7**
	2833	Medicinals and botanicals	4,919.4
	2834	Pharmaceutical preparations	4,419.4
	2835	Diagnostic aids	2,462.2
	2836	Biological producers, except diagnostics	2,155.8
284		**Soap, cleaners, and toilet goods**	**41,437.9**
	2841	Soap and other detergents	15,373.4
	2842	Polishes and sanitation goods	5,847.9
	2843	Surface-active agents	3,168.3
	2844	Toilet preparations	17,048.4
	2851	Paint and allied products	14,238.7
286		**Industrial organic chemicals**	**65,695.5**
	2861	Gum and wood chemicals	6,042.9
	2865	Cyclic intermediates and crudes	10,892.7
	2869	Industrial organic chemicals n.e.c.[a]	54,160.0
287		**Agricultural chemicals**	**18,307.4**
	2873	Nitrogenous fertilizers	3,113.4
	2874	Phosphatic fertilizers	4,636.2
	2875	Mixtures of fertilizers	2,018.8
	2879	Agricultural chemicals n.e.c.[a]	8,538.0
289		**Miscellaneous chemical products**	**19,674.0**
	2891	Adhesives and sealants	5,485.1
	2892	Explosives	1,324.6
	2893	Printing ink	2,754.4
	2895	Carbon black	691.9
	2899	Chemical preparations n.e.c.[a]	9,418.0

Source: United States Bureau of the Census, Annual Survey of Manufacturers, 1992.

[a] Not elsewhere classified—n.e.c.

(GDP), the sum of wealth produced by the nation, in this case about $6.4 trillion for the United States in 1993. Within each sector are subsectors designated by a three number code and subsubsectors, each with a four number code. This is demonstrated in Table 1.2.

The chemical industry falls within the manufacturing classification, SIC 20-39. Table 1.2 also shows the various categories of this classification and the value added each contributes. The manufacturing sector in 1991 contributed about $1.3 trillion of value added, which was 23% of that year's GDP of $5.7 trillion. This underscores the point that manufacturing, the traditional means for creating wealth, is no longer the major part of our national economy.

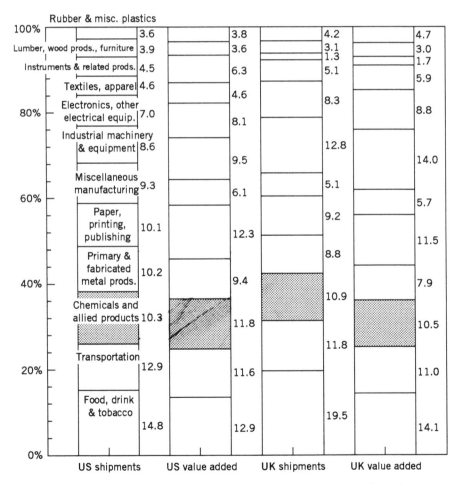

Figure 1.1 Rank of chemical and allied products industry among manufacturing industries in the United States and United Kingdom in 1991.

The chemical industry provides the largest amount of value added ($154.8 billion) although the transportation equipment ($152 billion) and food ($145.3 billion) are close.

The food industry, as is the case in all developed economies, has the largest total revenue ($387.6 billion) in the United States, with the transportation equipment industry second ($364 billion), and the chemical and allied products industry third ($292.3 billion). This is shown in Figure 1.1.

1.2   SIZE OF THE CHEMICAL INDUSTRY

The division of shipments is shown in Table 1.2 and illustrated in Figure 1.2, which also shows the division by value added. The industry grossed about

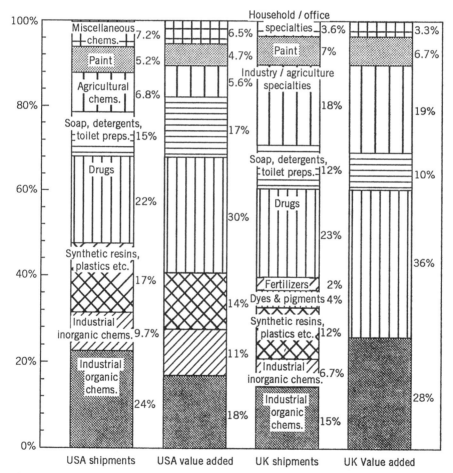

Figure 1.2   Subdivision of the chemical and allied products industry in the United Kingdom and the United States by value of shipments and value added.

TABLE 1.3   Projected Growth of the Chemical and Allied Products Industry in the United States ($ billion)

	Dollar Volume 1992	Projected Dollar Volume 2002 (1992 constant $)
Chemical industry	125.0	157.0
Allied products industry	175.0	236.0
Total	300.0	393.0

$300 billion in the United States in 1992 and $340 billion in 1994. The chemical and allied products industry is subdivided in the standard industrial classification into many component industries. Broadly speaking, the chemical industry (SIC 281) isolates or synthesizes chemicals, whereas the allied products industries (SIC 282-289) modify, formulate, and package products based on those chemicals. Indicated in Table 1.3 is our projection, assuming no major economic disruptions, for the size of the industry by 2002 in constant dollars, that is dollars with 1992 buying power. If inflation were taken into consideration the projections would be higher.

Comparison of the shipments and value added diagrams shows that so-called "fine chemicals," such as ingredients for pharmaceuticals and pesticides, dyestuffs, and food additives, make a larger contribution to the chemical industry's value added than they do its shipments. They tend to be high-priced products with specialized markets, and their manufacture is less capital and more labor intensive than the manufacture of the run-of-the-mill general chemicals. Their importance to the chemical industry is best represented by the value added figure which, for example, emphasizes, the importance of the pharmaceutical sector.

1.3   CHARACTERISTICS OF THE CHEMICAL INDUSTRY

The chemical and allied products industry has certain well-defined characteristics that govern its attitudes and its performance. These are listed in Table 1.4, and we shall discuss them in turn.

1.3.1   Maturity

Maturity, highly prized in an individual, is feared in an industry. When we wrote about the chemical industry in 1980 (see notes) we stressed its growth. Between 1954 and 1974 the US chemical industry grew at a rate of 8.5–9%. Between 1964 and 1974 the Japanese industry grew at a rate of 11.7%. Western Europe during that period enjoyed a 9.7% growth. In 1976 the US industry, coming out of a recession, grew 17%, although this included 5–8% of inflation. We pointed out

TABLE 1.4   Characteristics of the Chemical Industry

1. Maturity and its consequences
2. Participation in international trade
3. Competition from the developing countries
4. Capital intensity and economies of scale
5. Criticality and pervasiveness
6. Freedom of market entry
7. Strong regulation
8. High research and development expenses
9. Dislocations

that inevitably growth must lessen and that the industry would eventually grow at the rate of the economy as a whole. We pointed out also that government regulations relative to pollution, worker safety, and ecology generally would take its toll on profitability. By the early 1990s all of this had come to pass. No company could operate in the chemical industry unless it understood fully all the ramifications of maturity, which expresses itself in overcapacity, intense competition, low prices, and low profitability. Ultimately, it leads to restructuring. All of these things have happened in the chemical industry particularly with commodity chemicals, the ones we are mainly concerned with in this volume.

Maturity occurs because of market saturation, wide diffusion of technology and low barriers to entry to the industry. Engineering companies are eager to build turnkey plants and train clients to operate them. Maturity is hastened by offshore competition, and this is discussed below.

For the years 1987–1994 maturity is demonstrated in another way in Table 1.5, which shows the sales in the chemical industries in the United States, Western Europe, and Japan in current dollars and in constant 1987 dollars. Although the value in current dollars increased appreciably, the corresponding value in constant dollars increased annually at only 2.7% in the United States and 1.9% in Japan. The Western European industry contracted at 1.9% per year.

Restructuring is the inevitable result of overcapacity and is well demonstrated with ethylene. By the late 1970s the capacity for ethylene manufacture far exceeded the demand. Restructuring started in 1986 and involved first reduction in personnel. The chemical industry in the United States employs about 1.15 million people. This number was cut by 7–8% by attrition, early retirement, and redundancy. Second and most important was reduction in productive capacity to bring it in line with demand.

Figure 1.3 shows what happened relative to United States and West European demand versus capacity for ethylene (see note). Capacity in 1981 in the United States was about 40 million lbs. By 1986 this had decreased to about 36 million lbs. This 10% cut in capacity coupled with growth in ethylene usage brought demand and production into line. The growth can be attributed first of all to stockpiling of downstream products—ethylene itself cannot be stockpiled—since companies tend to amass materials when a shortage appears

TABLE 1.5 Sales of the United States, Western European, and Japanese Chemical Industries

	Sales (in millions of dollars)							
	1987	1988	1989	1990	1991	1992	1993	1994
United States								
Current	214,640	240,500	256,000	260,000	289,000	295,000	312,000	334,000
1987 Constant Dollars	214,640	231,577	227,020	230,568	245,477	243,582	251,098	257,982
Western Europe								
Current	302,220	318,000	325,000	335,000	336,000	338,000	340,000	341,000
1987 Constant Dollars	302,220	306,202	299,455	297,078	285,398	279,087	273,632	263,388
Japan[a]								
Current	136,992	193,018	190,630	197,000	201,000	202,180	201,000	203,000
1987 Constant Dollars	136,992	185,857	175,646	174,700	170,729	166,940	161,765	156,797

US Profitability (Before Taxes)

Year	Billion Dollars	% of Sales
1989	24.5	9.6
1991	21.0	7.2
1992	23.5	8.0
1993	26.0	8.3

[a] Does not include pharmaceutical business.

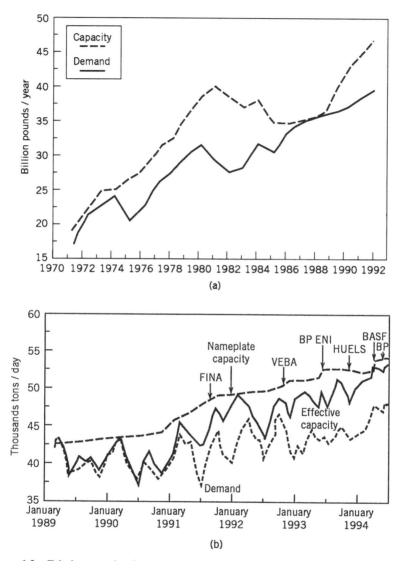

Figure 1.3 Ethylene production capacity *vs.* demand: (a) United States 1970–1992; and (b) Western Europe 1989–1994. Note the capacity-demand balance at the end of 1988 followed by the planning of additional capacity in Western Europe, which came on stream in the early 1990s and created massive overcapacity. Nameplate and effective capacities are shown together with the companies bringing new crackers on stream.

imminent, as it does when capacity is being eliminated. But also, when business improves, companies are more prone to invest in new applications and to expand old ones. This reflected itself in growth for the polyethylenes, which in turn resulted in growth of ethylene. Thus, the period encompassing 1987 and

1988 was probably the most profitable the chemical industry in the United States has ever enjoyed, the value of shipments between 1987 and 1989 in real terms increasing 8.6%. But the honeymoon was over by 1989 for the industry started to debottleneck existing plants and to build new capacity. By 1991 there was again appreciable overcapacity and need for further restructuring. Western European capacity and demand were similarly in balance in 1989/90 but the construction of several new crackers opened up a large gap by 1994.

Compared with other industries, the chronic overcapacity of the chemical industry is related to its being capital intensive with high fixed and low variable costs. There is a large gap between the break-even price of products (when price equals total cost) and the price at which it is rational to shut down a plant (when price equals variable cost). Furthermore, because of economies of scale, it is far more economic to run a plant at full capacity even if it means cutting prices. Finally, third world countries see a chemical industry as an acceptable way to industrialize, and many of them have access to cheap indigenous supplies of oil and natural gas. Thus the chemical industry is always inclined to "shoot itself in the foot" (see note).

Another aspect of restructuring is realignment of business segments. As an example, Union Carbide had a small, and presumably barely profitable, polyether polyol (Section 4.7.1) business. Carbide sold this business to ARCO, a major manufacturer of the propylene oxide from which the polyols are made. ARCO not only gained more captive use for its propylene oxide but increased its polyether polyol business well above the critical mass.

A switch of much greater impact is the trade in 1992 in which Du Pont took on ICI's European nylon business and ICI received in return Du Pont's methyl methacrylate business. Each company will increase its impact in major business areas. And still another example is the combination proposed in 1994 of Montedison's and Shell's polyolefin businesses. These companies, the first and second largest polypropylene producers, when combined, will be five times larger than the current third largest producer, Amoco.

1.3.2   Participation in International Trade

The chemical industry in the United States always had two advantages, which made its products cheap not only at home but in the international arena. First, there was ample natural gas, which provided ethane and propane for steam cracking (Section 2.2.1), and it generally was cheaper to crack gas than the liquids naphtha and gas oil. Second, the United States has ample supplies of propylene, not only because it is produced in steam cracking, but because the huge capacity for catalytic cracking (more than the rest of the world combined) required by the gasoline industry means that the few percent of propylene produced in this reaction translates itself into billions of pounds of product (Section 2.2.2).

Chemical net trade in the United States as compared with Western Europe and Japan is shown in Table 1.6. Several points are important. Western Europe exports considerably more than the United States because companies like BASF, Hoechst, Bayer, ICI, Rhone Poulenc, and ENI are all highly export oriented. The countries in which these companies are located use about 25–40% of their production and the rest must be exported in order for the plants to operate efficiently. Naturally, this is not the most profitable way to proceed, for exports are frequently the least profitable part of the business. The American companies have preferred to assure profits, tenuous those these sometimes might be, by satisfying the local market and then assigning incremental production to world trade. Japan, having based its chemical industry on relatively expensive imported naphtha and struggling with an expensive yen, cannot compete in the world market, although in the early 1990s it has enjoyed sales to neighboring countries with expanding economies, primarily China.

The American chemical industry exports products to more than 185 countries. In 1991 it exported $19 billion more than it imported. Table 1.6 indicates that improvement in the balance of trade started in the late 1980s. These favorable figures should continue at least until 1996 or 1997. In favor of such a projection is the relatively cheap US dollar coupled with reliable manufacturing and trading practices. The value of chemical exports in the early 1990s was equivalent to the dollars spent for importing crude oil. Meanwhile, because of the demand for oxygenates in gasoline, from 1995 the United States will import about 9 billion lb of methyl tertiary butyl ether (MTBE) and probably one-half of the methanol to make MTBE domestically.

Between 1980 and 1991, the US chemical industry had trade surpluses totalling $140 billion. Ten percent of US exports are chemicals. Seldom noted is the fact that the US chemical industry's international investments provide appreciable income from the net earnings of foreign subsidiaries and from the licensing of US technology to those subsidiaries as well as to other companies. Such earnings amounted to $5.1 billion in 1989 as compared with $3.6 billion in 1960. Between 1985 and 1990 they totalled $21 billion.

TABLE 1.6 Chemical Net Trade for the United States, Western Europe, and Japan ($ billion)

	1980	1983	1986	1987	1988	1989	1990	1991	1992	1993	1994 (est)
United States	11.8	10.8	7.8	10.2	12	16.2	17	19	17.1	17.5	17
Western Europe[a]	18.4	n.a[b]	23.9	28.5	27.9	25.8	25	25	24.5	24	24
Japan[c]	1	(0.1)	(0.3)	(0.2)	(1.0)	(1.0)	(0.2)	0.1	0.1	0.1	0.1

[a] Figures from Western Europe do not include intra-West European trade.
[b] Not applicable = n.a.
[c] Figures in parentheses represent net imports.

1.3.3   Competition from Developing Countries

But what about the future? Natural gas has been discovered in many places in the world and of course many countries have petroleum. Many of them are eager to enter the chemical business because it promises greater value added then is possible when the gas or oil is used for energy. Thus an awesome list of countries have built or are building chemical industries. The United States, Western Europe, and Japan have long-standing industries. Newcomers include Saudi Arabia and other Gulf states, Canada, Mexico, Venezuela, Brazil, Argentina, and other Latin American countries including Trinidad and Chile, the former member of the USSR and other Eastern European countries, and East Asian countries including Taiwan, Korea, China, Thailand, Indonesia, Malaysia, the Phillipines, and Singapore. How seriously some of these countries regard the chemical industry is indicated by the statistic that in Taiwan the chemical industry accounts for 30% of manufacturing as opposed to 10% in the United States and Western Europe. Many of these countries enter the chemical business to provide for their own needs. Taiwan and Thailand indicate that they will not be major exporters because they can consume locally most of their production. Korea, Saudi Arabia and Canada and most other countries will, however, become formidable competitors in the international trade arena. Indeed the impact of Canada and Saudi Arabia has already been felt, as Figure 1.4, which relates solely to ethylene trade, indicates.

Trade in ethylene derivatives by exporting countries from 1978–2005 (estimated) is shown in Figure 1.4. Japan and Western Europe have shrunk while the Middle East and Canada have gained. Japan lost its stake in trade of ethylene derivatives very early and indeed this was true of virtually all Japanese chemicals because of expensive raw material and a strong currency. The trade they enjoyed in 1993 was due largely to the needs of mainland China, where shipments from Japan have lower freight costs. Western Europe does not derive much of its huge chemical exports from ethylene derivatives, and much of what it enjoyed formerly has been taken over by Canada and Saudi Arabia. The United States still maintains a role and it is believed that this will continue through the year 2005, particularly if gas continues to be available for steam cracking (Section 2.2.1).

The impact a developing country can make is exemplified by the declining production of petrochemical ethanol in the United States. The data are shown in Table 1.7. Admittedly, the use of ethanol as a chemical feedstock is declining, and corn-based ethanol for fuel is being subsidized. Nonetheless, Saudi Arabia's cheap ethylene can be converted to ethanol and imported into the United States at a cost that makes its domestic manufacture uneconomic. It is predicted that ethanol manufacture in the United States will come to an end by the year 2000.

As already indicated, numerous other countries have entered the chemical business, but Saudi Arabia is impressive for two reasons. First of all, it has 25% of the world's oil reserves, and second it has a reasonable supply of so-called associated gas, the gas that comes out of the ground when oil is pumped. Since

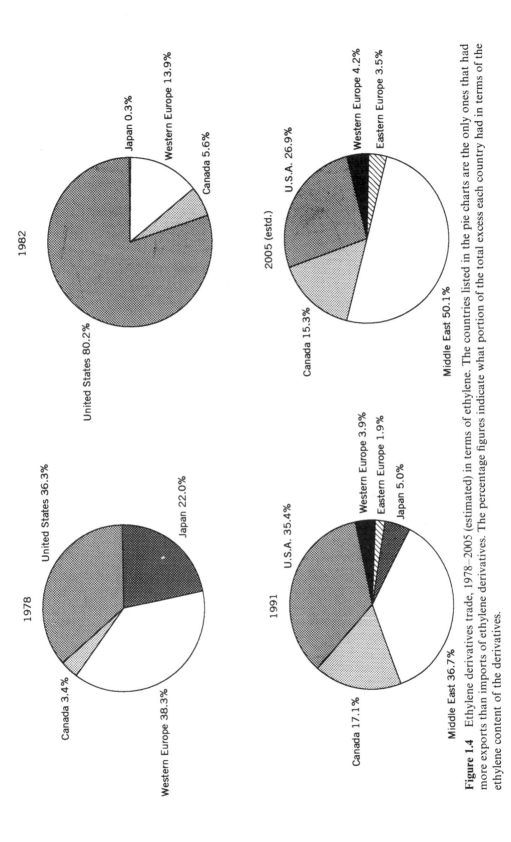

Figure 1.4 Ethylene derivatives trade, 1978–2005 (estimated) in terms of ethylene. The countries listed in the pie charts are the only ones that had more exports than imports of ethylene derivatives. The percentage figures indicate what portion of the total excess each country had in terms of the ethylene content of the derivatives.

TABLE 1.7 Decline in Production of Petrochemical Ethanol in the United States

	1982	1985	1988	1900	2000 (projected)
Production (million lb)	1023	573	500	132	0

there is no alternative use for this gas in Saudi Arabia, it has zero value at the wellhead, which reflects itself in favorable economics when the ethane it contains is cracked to ethylene.

Since Saudi Arabia entered the chemical business cracking only ethane, they could participate only in ethylene derivatives trade. By 1995, however, another cracker was on stream for the cracking of naphtha, and this provided Saudi with the basic chemicals they needed to become a full-fledged competitor in the world's chemical arena.

1.3.4 Capital Intensity and Economies of Scale

The chemical industry is capital intensive. It produces huge quantities of homogeneous materials, frequently liquids or gases, which can be manufactured, processed, and shipped most economically on a large scale. This was less so through the nineteenth century until World War II. The early chemical industry used more general purpose equipment and operated batch processes that required little capital investment but had high labor costs. Typical of such processes were the Leblanc route to sodium carbonate and the benezensulfonate route to phenol (Section 4.6).

The petroleum refining industry was the first to convert to continuous operation on a large scale. The engineering developed for the petroleum industry was applied to the chemical industry after World War II. Plant sizes escalated as dramatic economies of scale became possible. The capacity of a typical ethylene cracker rose from 70 million lb/year in 1951 to 1 billion lb in 1972. In the early 1990s plants with 1.5 billion lb/year capacity were built. Currently, there are few batch processes of any size in operation for commodity chemicals, and substantial economies of scale have become a characteristic of the modern petrochemical industry.

Economies of scale arise not only from improved technology but also from purely geometric factors. The capacity of a great deal of chemical equipment (e.g., storage tanks and distillation columns) varies with its volume, that is the cube of its linear dimensions. The cost, on the other hand, is the cost of a surface to enclose the volume and varies with the square of the linear dimensions. Consequently, cost is proportional to (capacity)$^{2/3}$. This is called the square-cube law. It does not apply to all equipment. The capacity of a heat exchanger depends on its surface area so cost is proportional to (capacity)^1 and there are few economies of scale. Control systems are not affected by capacity at all, so cost is proportional to (capacity)^0 and economies are infinite. It is claimed that for a modern petrochemical plant overall, cost is proportional to (capacity)$^{0.6}$.

The size and complexity of a modern chemical plant demand high capital investment. Although other industries invest more capital per dollar of sales, the chemical industry has the highest investment of current capital. That means that the chemical industry invests more each year than do such other capital intensive industries as mining, where equipment once bought remains in service for many years.

Capital intensity has a number of corollaries. The return on capital is relatively low. Because high capital investment reduces the labor force required, labor productivity (i.e., value added per employee) is high. Salaries contribute relatively little to costs (of the order of 2.0%), and employers need worry less about pay increases than in labor intensive industries such as food or apparel. Consequently, labor relations are unusually good.

The assets of a company are the estimated value of the plant, land, and other capital goods it owns. Such ratios as assets per employee, sales per dollar assets, and sales per employee are measures both of the capital and labor intensity of an industry.

Generally, the petroleum refining industry has both the highest assets and the highest sales per employee. The chemical industry shows lower figures, but still ranks high. The food and clothing industries are usually at the low end of the scale. Service industries with few assets and many employees, such as consulting companies, will show both low assets and low sales per employee.

The move to specialty chemicals has altered these perceptions as far as the overall chemical industry is concerned, although not for the sectors described in this book. Small, high-value, low-tonnage chemicals are frequently made by batch processes in computer controlled equipment. Such equipment brings some of the advantages of continuous processing to batch processes.

1.3.5 Criticality and Pervasiveness

A chemical industry is critical to the economy of a developed country. In the nineteenth and first half of the twentieth century, a nation's industrial development could be gauged from its production of sulfuric acid, the grandfather of economic indicators. Today one uses ethylene production as a yardstick of industrial sophistication. An advanced economy cannot exist without a chemical industry; neither can a chemical industry exist without an advanced economy to support it and to provide the educated labor force it requires.

The chemical industry is not replaceable. There is no other industry that could fulfill its function. It is pervasive and reflected in all goods and services. Not only is the chemical industry here to stay, but also it is a dynamic and innovative industry that has grown rapidly and on which the world will continue to rely in the future. Many of the problems concerning pollution and energy have been detected and monitored by chemical methods, and chemistry is playing a part in their solutions.

1.3.6   Freedom of Market Entry

Another characteristic of the chemical industry is freedom of market entry. Anyone who wants to manufacture bulk chemicals may do so by buying so-called "turnkey" plants from chemical engineering contracting companies. Such companies have processes for preparation of virtually any common chemical and will build a plant guaranteed to operate for anyone who wishes to invest the money. This was the way that many of the petroleum companies gained entry to the petrochemical business and also the way that many of the developing countries are laying the foundations of their own chemical industries.

The one requirement is large amounts of capital. To enter the basic chemical business requires a bare minimum of $1 billion. Large sums of money may be required for purposes other than capital investment. In the pharmaceutical industry, large sums are required for development, which in the early 1990s was of the order of $200 million per drug. The detergent industry, on the other hand, requires money for massive amounts of advertising. These two industries underscore the importance of large cash flows to support research and development, and merchandising. Thus entry into the chemicals market is free in the sense in which economists use the word, but the expense is such that only governments, oil companies, and other giant enterprises can find the necessary capital.

What has been said about capital applies primarily to very large volume basic chemicals such as the seven basic chemicals or chemical groups and their first-line derivatives. Beyond that there may be other barriers to entry such as lack of necessary technology or reluctance on the part of a patent holder to license technology. An example is Du Pont's process for making hexamethylenediamine by the hydrocyanation of butadiene (Section 5.1.3). This is probably the preferred process, but it is not available for license. On the other hand, the Dutch company AKZO was able to enter the Aramid (Du Pont's Kevlar) business, presumably by finding loopholes in the Du Pont patents. Kevlar (Section 9.3.3) is used for fibers that are stronger, weight for weight, than steel and Du Pont believes that its future will be profitable. But it will be a future to be shared with others since the patents were apparently not invincible.

Downstream operations may provide a barrier to entry. Thus the manufacturers of poly(methyl methacrylate) convert their product to acrylic sheets, which are then sold to molders. Potential manufacturers must decide whether they want to gain the expertise that participation and marketing the sheet requires. Indeed, will the company's culture allow participation in a business so far removed from basic chemical manufacture.

Low price of the product, and correspondingly low profitability, may deter entry. Furfural provides a classic example. When introduced many years ago, it was priced so low that it never attracted competition until well into the product's life cycle, and then major competition came not from US companies, but from China. Companies with products whose patents are about to expire often use this technique to discourage other manufacturers. Monsanto used it successfully with one of its herbicides that came out of patent. It was able to manufacture it cheaply in a depreciated plant by an optimized process. It was not worthwhile for other companies to invest the fresh capital in order to compete. And, to Monsanto's surprise, the market increased as farmers used more of this cheap herbicide to control weeds in place of the more cumbersome processes of plowing or covering the ground with plastic. Union Carbide similarly exploited the technique with its pesticide carbaryl. By pricing it low, they successfully avoided serious competition after the patent had expired.

1.3.7 Strong Regulation

The chemical industry is one of the most highly regulated of all industries. The regulations are intended to protect and improve the worker's and the nation's health, safety, and environment. The Chemical Manufacturers' Association (CMA) has documented industry's vigorous response to the need for pollution abatement.

Of all the regulations, the stringent requirements of the Clean Air Act will have the most far-reaching economic impact on the industry, with a cost projected at $25 billion/year. A brief description of nine of the acts that affect the chemical industry is given in Table 1.8.

TABLE 1.8 Legislation Affecting the Chemical Industry

1. The Emergency Planning and Community—Right-to-Know Act	Reporting of production, handling and storage of hazardous materials
2. Clean Air Act	41 pollutants must be controlled by 1995; 148 more by 2003. Cost to industry estimated at $25 billion per year
3. Toxic Substances Control Act	Requires premanufacture notification to EPA
4. Resource Conservation and Recovery Act	Clean up of hazarous and nonhazardous waste sites. Cost to industry $9–60 billion in decade 1990–2000
5. Superfund	Clean up of hazardous sites. Mostly funded by taxes in industry.
6. Clean Water Act	Ensures high quality water
7. Safe Drinking Water Act	Sets standards for 83 chemicals in water
8. Chemical Diversion and Trafficking Act	Prevents use of chemicals to make illegal drugs
9. Occupational Safety and Health Act (OSHA)	Defines hazards in an attempt to prevent industrial accidents. Defines permissible exposure limits for 600 hazardous chemicals.

The situation with asbestos provides an example of where regulation seems to have got out of hand. Table 1.9 compares the expenditure in the United States for the removal of asbestos from public buildings with the annual budget of the National Institute of Health (NIH). It is almost unbelievable that more is spent on the former than on the latter. Also in Table 1.9 are estimates of the harm that asbestos exposure can do to a school child as compared to other causes of death. The chance of death from asbestos for a school child is tiny. In assessing the risk from a material, one needs to know both how carcinogenic it is and what an individual's chances are of exposure to the material. Thus the US Food and Drug Administration (FDA) has announced that short-fiber asbestos and dioxins, though harmful, provide little danger to the general population. An article by Ames (see notes) places the problem of carcinogenicity of chemicals in perspective. Asbestos is harmful only to those who are exposed continually to it in the workplace.

TABLE 1.9   Expenditures on Asbestos Removal and Comparison of Incidence of Death from Asbestos and from other Hazards

Asbestos Abatement Costs versus National Institute of Health (NIH) Budget

Year	NIH Budget	Asbestos Abatement
1989	$7.2 billion	$ 5.2 billion
1990	$7.6 billion	$ 7.0 billion
1991	$8.1 billion	$ 8.3 billion
1992	$8.7 billion	$ 9.8 billion
1993	$9.2 billion	$11.5 billion

Asbestos Risks Compared with other Hazards

Cause of death	Annual Death Rate per Million at Risk
Asbestos exposure in schools	0.005–0.093
Whooping cough vaccination (1970–1980)	1–6
Aircraft accidents (1979)	6
High school football (1970–1980)	10
Drowning (ages 5–14)	27
Motor vehicle accidents, pedestrian (ages 5–14)	32
Home accidents (ages 1–14)	60
Long-term smoking	1200

1.3.8 High Research and Development Expenditures

The chemical industry is research intensive. It hires many graduates—over 15% of all scientists and engineers in the United States—and most of them work in research and development laboratories (see note). Chemical R & D is compared with total R & D in the United States in Table 1.10. The chemical industry's proportion of the total almost doubled from 1970 to 1992. Research expenditures of some of the top chemical and pharmaceutical companies are shown in Table 1.11. Pharmaceutical companies like Merck and chemical companies with major pharmaceutical subsidiaries such Ciba–Geigy, Bayer, and Monsanto spend more on R & D than do the mainline chemical companies. At the opposite extreme is Exxon, which is mostly involved in commodity chemicals and spends a relatively small 2.1%. Monsanto has made an important transition from a commodity to a specialty company, having sold off businesses that yielded $4 billion of commodity sales. Monsanto's R & D expenditures at 8.0% of sales is high because it has both a specialty and pharmaceutical orientation. Actually, 3–4% of sales is considered normal for companies that do not have a pharmaceutical arm and are only marginally involved in specialty chemicals. Research-based pharmaceutical companies with few other interests spend 10–15% of sales on research. True specialty chemical companies have budgets about one-half of those of the pharmaceutical companies.

How is the R & D budget spent? Research is a risky and expensive business. Finding the conditions that maximize the cost-effectiveness of an R & D budget preoccupies many pharmaceutical managers. Unfortunately, it is not a science, and success in the laboratory often depends on serendipity. Should a company rely on discoveries emerging from the interests of its researchers, or should it try to satisfy the pull of the marketplace? "Technology push" was the initial approach. Research and development received a boost from government tax incentives during and immediately after World War II. Much of this research related to finding new materials for which uses could be created. Thus the period between 1940 and 1965 was a time of great discovery. Nylon, other synthetic fibers, the polymers that provide plastics, elastomers, coatings, and adhesives are examples. Chemicals were also discovered with numerous applications that made industry generally more efficient, such as corrosion inhibitors, electronics chemicals, and food antioxidants.

In the mid-1960s, however, the concept changed to "demand pull." What problems are there in the marketplace that require technical solutions? Market research to answer such questions became a discipline and, for the past 25 years, the industry has talked of "market orientation." Examples of technology push include television, sulfonamides, and lasers. Examples of demand pull include hard water compatible detergents, jumbo jets, and automobiles with low-exhaust emission. A catalytic cracking catalyst that gives increased amounts of isobutene (Section 2.2.2) is an obvious example of the result of a market-oriented research project, isobutene being required for the production of methyl *tert*-butyl ether for unleaded gasoline. Both kinds of research should be part of

TABLE 1.10 Research and Development Expenditures in the United States ($ billion current)

	1970	1980	1983	1984	1985	1986	1987	1988	1989	1990	1991	1992	Annual Growth Rates 1970–1980	1980–1990	1991–1992
Total R & D for all manufacturing industries	17.36	42.69	61.93	69.90	77.53	80.38	84.31	89.78	93.57	95.33	89.51[a]	91.21	9.4%	8.4%	3.1%
Chemical and allied products	1.72	4.64	7.19	7.93	8.54	8.84	9.64	10.77	11.47	12.34	14.65	16.71	10.1%	10.3%	14.1%
Chemical R & D as a percent of the total	9.9	10.9	11.6	11.3	11.0	11.0	11.4	12.0	12.3	12.9	16.6	18.3			

Source: In part from *The Chemical Industry Statistical Handbook 1992*, Chemical Manufacturers Association, 2501 M Street, N.W., Washington, DC 20037.

[a] Decrease due to cuts in government spending.

TABLE 1.11   **Research Expenditures of Selected Companies, 1993 ($ billion)**

Company	Sales (predominantly but not exclusively chemical)	R & D Expenditure	R & D as Percent of Sales
Hoechst	26.5	1.75	6.6
BASF	23.3	1.11	4.7
Bayer	23.6	1.82	7.7
ICI	12.5	0.26	2.1
Dow	18.6	1.26	6.8
Rhone-Poulenc	13.6	1.08	7.9
DuPont	20.9	1.09	5.2
Solvay	6.8	0.33	4.9
Zeneca	6.6	0.77	11.7
Exxon	8.6	1.86	2.2
Mitsubishi Kasei	10.4	0.52	0.5
Monsanto	7.9	0.63	8.0
Merck	10.5	1.17	11.1
Union carbide	4.6	0.14	3.1
Ciba-Geigy	15.3	1.48	9.7

Source: Adapted from *Chemical Insight* No. 542, September 1994.

any large company's game plan, although there has been a marked trend to deemphasize the "blue-skies" research that leads to truly novel discoveries.

A major area for research in today's world is monitoring and reducing pollution. One-fifth of new capital expenditure in the 1990s was for pollution abatement and control; approximately the same amount of the R & D budget of a large company is likely to be spent on ecologically oriented projects. Thus, first-generation research was "blue-skies" research. It required little participation by management, and researchers were generally regarded as a group of people difficult to communicate with. It was only when a project reached the development and marketing stages that management was required. Du Pont, General Motors, and IBM are examples of companies that made discoveries and brought them to the marketplace. Second generation research involved going to the marketplace to find out what was needed and applying accounting processes to the monitoring of R & D projects. This demanded strong participation by the marketing branch, but still little participation by top management.

Today there is a third generation of research managers who recognize that research should be a part of the organization not apart from it. Research and development should figure in corporate objectives and should take its direction from these objectives in exactly the same way as any other business function. It should thus help the organization to achieve its overall goal. Indeed, it should even help to set goals by managing technology as opposed only to inventing and applying it.

Thus the R & D department has to determine which technologies may be developed internally, which may be obtained through licensing, and which may be obtained through strategic alliances. This is very much like a "make-or-buy" decision in manufacturing. It was obviously better for many companies to license BPs ammoxidation technology (Section 4.4) than to try to work out an acrylonitrile process on their own. Similarly, when Himont wanted to develop highly sophisticated catalysts for propylene polymerization, it joined forces with Mitsui, for both companies had strong backgrounds in catalyst development, and jointly they could bring these to bear on the objective. Du Pont decided to develop a superior maleic acid process, since this was an important raw material for a new process they invented to prepare Spandex, their most profitable product, in the late 1980s and early 1990s. Having discovered a new and, what appeared to them, superior route (Section 5.4) they joined forces with Monsanto for the development of the process, Monsanto being the largest maleic anhydride producer with many years of experience. (Monsanto subsequently withdrew, because it planned to shed the maleic anhydride business, which it did by selling it to Huntsman).

A sensible R & D strategy avoids duplication, but a large amount of duplication takes place in the world's research laboratories. The patent literature discloses 25 processes for the manufacture of 1,4-butanediol and a similar number for propylene oxide. At least 15 companies have worked on the homologation of methanol to higher alcohols, a process that none has commercialized. There are many other examples.

There is duplication in the competition between companies in the specialty and performance polymers business. Not only is there intercompany competition to sell the same polymer but there is also much interpolymer competition as different polymers vie for use in a given application. In 1993 there were at least 51 companies manufacturing engineering thermoplastic polymers, that is, sophisticated polymers to replace metal. Sixty percent of the business was enjoyed by only five companies: General Electric, Du Pont, Bayer, Hoechst Celanese, and BASF. Another eight companies had 19% of the business; these were Allied, Monsanto, Asahi Chemical, Mitsubishi Gas Chemical, Tiejin, Mitsubishi Chemical, DSM, and Toray. Subsequently, the two Mitsubishi companies combined forces. The remaining 21% of the business was the province of 37 companies, none of which had achieved the critical mass required for profitability in the engineering polymer business. Understandably, this kind of operation leads to a great deal of waste.

Industrial R & D expenditures in the United States have increased impressively between 1970 and 1991 in terms of current dollars as shown in Table 1.10. During the 1980s chemical R & D expenditures were a fairly constant proportion of the total dollars spent. By the mid-1990s, the rate of growth of chemical R & D spending has slowed in constant dollars, but considerably less than the slowing of total R & D spending. Decreased military R & D is largely responsible for this.

Industry allocates three-quarters of its R & D expenditure to development and one-quarter to applied research. Basic research is almost nonexistent. There has always been an academic argument that holds that basic research is the province of the university. This may well be so, but in any laboratory there may be need for theory that is nonexistent but necessary for the solution of a problem. An industry may require such basic research in order to fulfill its objectives.

Applied research is usually described as the type of R & D that leads to new uses for existing products or new products that fill needs in the marketplace. The development of a new polymer that absorbs many times its weight of water for use in disposable diapers is a good example of applied research as is the extension of the use of that material to other areas such as agriculture. Although applied research accounts for about 25% of today's research expenditures, very little effort is being expended on new products. One reason for this is that the maturity of the industry does not offer so many opportunities for new products. Also, the many regulations that govern the chemical industry, particularly the Toxic Substances Control Act (Section 1.3.7) require extensive and expensive testing before a product can be test marketed. Usually, the risk is deemed greater than the potential benefit. A major exception is found in the pharmaceutical and agricultural chemical businesses. In the early 1990s it cost approximately $200 million to develop a new pharmaceutical. The pharmaceutical companies are willing to make such expenditures, however, because of the lure of "blockbuster" drugs whose annual sales may reach $1 billion with concomitant profits.

The bulk of R & D dollars is spent on development. This includes work on new and improved processes, finding new uses for existing products, solving ecological problems, and pursuing the analytical activities on which a modern laboratory depends.

1.3.9 Dislocations

An important concept in today's chemical industry is the ever-present possibility for dislocations. This is particularly important for planners who, all too often, find their scenarios askew because of a dislocation. Dislocations are defined as events over which a given company has no control but which markedly affect that company's business. In planning, one cannot forecast what a dislocation might be. Indeed, if it could be forecast, it would not be a dislocation. But what must be anticipated in planning is that there will be dislocations either for good or ill.

A few examples illustrate the point. The advent of unleaded gasoline made lead tetraalkyls obsolete in the United States. The major manufacturer of these compounds was Ethyl Corp. with a reputed $90 million of profit. That figure rapidly declined to $20 million and would have been lower had it not been for export sales. Obviously, Ethyl was a victim of a dislocation. This motivated Ethyl to use its skills to expand into a variety of specialized and semicommodity businesses, which allowed them to recoup their profits. Thus they became a large supplier of the bulk pharmaceutical, ibuprofen, for sale to packagers who converted it into a consumer item. Their synthesis involved organometallic chemistry developed for the unrelated area of α-olefin production.

The same unleaded gasoline dislocation proved to be a windfall for ARCO. ARCO's two-for-one process for the manufacture of propylene oxide and tert-butanol (Section 4.7) made available to them large quantities of the latter for which, at the time the plant went on stream, there was very little use. The need for octane improvers in unleaded gasoline soon provided a market for it. Dehydration to isobutene and reaction with methanol gave MTBE, and ARCO in the early 1990s was the world's largest supplier of it.

A third example also relates to unleaded gasoline. At least one petroleum company announced that it would achieve the desired octane number by removing lead and increasing the aromatics content of their gasoline. A few years later, the Clean Air Act specified that the aromatics content of gasoline must be decreased from about 35 to 25%. Thus the Clean Air Act provided a second dislocation that negated that company's reaction to the earlier dislocation provided by unleaded gasoline.

A fourth example: Phillips Petroleum Company, never used the metathesis reaction (Section 2.2.9) to convert propylene into more expensive ethylene and 2-butene which, in turn, may be dehydrogenated to butadiene. One might assume that it did not opt to carry out this interesting chemistry because of the widely held belief in the 1970s that within a 15-year period declining US gas supplies would make naphtha and gas oil the major steam cracking feeds. Accordingly, large quantities of butadiene would become available. The United States has always imported butadiene from Europe because insufficient quantities were produced by the cracking of gas. Actually, naphtha and gas oil never became major feedstocks in the United States (Section 2.2.1) because of another dislocation provided by Saudi Arabia's entry into the chemical business. Saudi Arabia decided to use only the ethane in their associated gas, making large quantities of liquefied petroleum gas (LPG) (a propane/butane mixture) available at low world prices. The United States now uses LPG and has not found it necessary to switch to liquid feeds. Thus the United States still imports butadiene, a situation that metathesis might have helped to avoid.

Dislocations frequently result from advances in technology. The producers of propylene oxide by the chlorohydrin route suffered a serious dislocation when Arco announced its new process via tert-butyl hydroperoxide (Section 4.7). Every manufacturer except one went out of business. Similarly, Monsanto's acetic acid process using methanol and CO closed down every producer of acetic acid who used acetaldehyde as a starting material.

The answer to dislocations is the concept of robustness. A robust process is one that can accommodate a variety of dislocations. For example, some companies, uncertain of their feedstock supply, built steam crackers that could operate on gaseous or liquid feedstocks. Their plants cost more than a single feedstock plant would have done, but were sufficiently robust to withstand dislocations in feedstock supply. The fact that the petrochemical business in the United States in 1993 remained relatively profitable is due largely to such flexible crackers. Western Europe suffered because only 5 of its 52 crackers are flexible.

Finally, the chemical business is dynamic. It is affected not only by what it does itself, such as creating new technology, but by what others do around it. Modern managers keep abreast as much as possible with what the rest of the world is doing that might affect their business.

1.4 THE TOP CHEMICAL COMPANIES

Table 1.12 provides data about the 30 leading chemical companies in the world in 1993 (see note). Practically all these companies have nonchemical activities, although the figures for the most part reflect only chemical sales. Chemical arms of petroleum companies include, Shell, Exxon, and Elf Atochem. Thus Exxon is the fourth largest chemical company in the United States, even though its chemical sales are only 9.2% of the total for the company. Table 1.12 shows that the three largest chemical companies in the world are the three arms of the pre-World War II German company, I.G. Farbenindustrie. This giant company was divided into three companies after the war. Their profitability, however, is small judged on percent of sales, and a major reason for this is that they are highly export-oriented (Section 1.3.2). In 1993, ICI demerged its pharmaceuticals, agrochemicals, and specialty businesses to form a new company, Zeneca. Zeneca's profitability is much higher than ICIs because it inherited speciality sectors while ICI remained holding the commodity chemical sectors.

Du Pont and Dow occupy the fourth and fifth places in Table 1.12, respectively. Dow is an efficient company with highly integrated processes and the highest sales per employee in the chemical business. Dow also has a pharmaceutical arm, and Du Pont is known for supporting its marketing operations with strong technical service. In the tenth position is the largest Japanese chemical company, Mitsubishi Kagaku, formed from a merger between Mitsubishi Kasei, who occupied fourteenth position in 1991, and Mitsubishi Petrochemical. Japanese companies are relatively small and their profitability is the lowest in the industry. Sumitomo in the seventeenth position shows only slightly better results than Mitsubishi, Merck and Glaxo in positions 9 and 15 are stellar performers with 20.6 and 23.0% margins on sales. They are the first or second largest pharmaceutical companies, and their profitability is the result of several "blockbuster" drugs including the anticholesteremic drug, Mevacor (Merck) and the antiulcer drug Zantac (Glaxo). For seven years until 1992, Merck was voted America's most admired company in a Fortune magazine poll. Both companies had considerably greater dollar profits than the companies ranked ahead of them in the table. Other companies with high profits as percent of sales include Sandoz, Roche, Pfizer and Zeneca, all of which count heavily on pharmaceuticals. State-owned companies such as Enichem are among the least

TABLE 1.12 Sales and Profitability of Major Chemical Companies Worldwide, 1993

Company	Rank 1993	Country[a]	Sales ($ billion)	Profit as % of Sales[a]
Hoechst	1	D	26.52	1.6
Bayer	2	D	23.62	3.3
BASF	3	D	23.36	2.1
DuPont	4	US	20.90	0.5
Dow	5	US	18.06	3.5
CIBA	6	CH	15.25	7.9
Rhone-Poulenc	7	F	13.64	1.2
ICI	8	UK	12.47	0.4
Merck	9	US	10.50	20.6
Mitsubishi Kagaku	10	J	10.39	Loss
Sandoz	11	CH	10.17	11.3
Roche	12	CH	9.64	17.2
Shell	13	UK/NL	9.32	Loss
Exxon	14	US	8.64	4.1
Glaxo	15	UK	8.48	23.0
AKZO	16	NL	8.50	3.3
Sumitomo Chemical	17	J	8.39	1.4
Elf Atochem	18	F	8.31	n.a.
Monsanto	19	US	7.90	6.3
Pfizer	20	US	7.48	8.8
Solvay	21	B	6.76	n.a.
Zeneca	22	UK	6.57	9.9
Eli Lilly	23	US	6.45	4.6
ENiChem	24	I	6.21	Loss
Huls	25	D	5.85	Loss
l'Air Liquide	26	F	5.14	7.3
Mitsubishi Kasi	27	J	See Mitsubishi Kagaku	
General Electric	28	US	4.99	n.a.
BOC	29	UK	4.84	6.3
Union Carbide	30	US	4.64	1.3

[a] D = Germany, CH = Switzerland, US = United States, UK = United Kingdom, F = France, I = Italy, J = Japan, B = Belgium, NL = Netherlands, n.a. = not available.

profitable, in part because inefficient operations are often retained simply to provide employment.

Although the chemical industry is proliferating around the world, the largest companies are still concentrated in the United States, Western Europe, and Japan. In Table 1.12 nine countries are represented, notably Germany, United Kingdom, United States, Netherlands, Switzerland, Italy, France, Japan, and Belgium. Future lists may well include companies from the Middle East and Southeast Asia.

It appears that Merck has the highest value added per employee, a ratio more widely known as productivity. That such a ranking is achieved by a company in the relatively labor-intensive pharmaceuticals business rather than in the capital-intensive bulk chemicals business is an indication of the poor margins earned by bulk chemicals producers.

The US chemical companies with chemical sales of $1 billion or more, of which in 1993 there were 50, are listed in Table 1.13. Pharmaceutical companies

TABLE 1.13   United States Chemical Companies with Sales of at Least $1 billion in 1993

Company	Chemical Sales as % of Total	Profit as % of Sales[a]
Sales of above $10 billion		
Du Pont	42.1	4.8
Dow Chemical	69.3	6.2
Exxon	9.2	6.4
Sales of $4–7 billion		
Hoechst Celanese	92.0	8.8
Monsanto	71.5	12.9
General Electric	8.3	16.5
Union Carbide	100.0	7.6
Occidental Petroleum	47.6	4.5
BASF	77.6	n.a.
Sales of $3–4 billion		
Eastman Chemical	100.0	12.2
Shell Oil	18.6	8.5
Amoco	13.2	8.5
ICI Americas	100.0	5.3
Mobil	5.5	0.7
Rohm & Haas	100.0	8.2
Arco Chemical	100.0	13.3
Miles	47.3	n.a.
Sales of $2–3 billion		
Air Products	87.3	14.3
W.R. Grace	65.7	13.5
Allied-Signal	23.6	11.1
Chevron	7.5	8.3
Ashland Oil	25.4	4.2
Ciba-Geigy	55.2	n.a.
Praxair	100.0	14.0
Phillips Petroleum	18.8	4.2
Akzo	100.0	8.8
Rhone-Poulenc	100.0	n.a.
Dow Corning	100.0	11.5

TABLE 1.13 (Continued)

Company	Chemical Sales as % of Total	Profit as % of Sales[a]
Sales of $1–2 billion		
Ethyl	100.0	8.6
Huntsman Chemical	100.0	n.a.
Hercules	66.5	13.7
FMC	48.9	10.7
National Starch	100.0	15.9
Great Lakes Chemical	100.0	23.2
Elf Atochem	100.0	n.a.
Lubrizol	100.0	12.5
American Cyanamid	33.0	n.a.
Nalco Chemical	100.0	18.9
Lyondell Petrochemical	34.4	4.3
Morton International	55.2	10.6
Witco	57.5	8.8
Solvay America	100.0	n.a.
Cabot	73.8	12.5
International Flavors	100.0	26.2
PPG Industries	20.0	12.0
Olin	46.1	n.a.
Texaco	3.2	n.a.
BP America	6.6	1.8
Henkel	100.0	n.a.
Cytec	100.0	n.a.

[a] not available = n.a.

are not included in this list unless they are part of a chemical company. Chemical arms of oil companies include Exxon, Occidental, Mobil, Amoco, Shell, Chevron, ARCO, Ashland, Phillips, Lyondell, and Texaco. United States subsidiaries of foreign chemical companies include Hoechst Celanese, BASF, Shell, ICI Americas, Miles (formerly Mobay), Rhone-Poulenc, Ciba–Geigy, AKZO, Elf Atochem, and Solvay. The demerger of ICI has already been mentioned. American Cyanamid also demerged in 1993–1994. The pharmaceutical, pesticide, and specialty sectors retained the original name and the commodity sectors took the name Cytec.

1.5 THE TOP CHEMICALS

Table 1.14 lists the 50 most important chemicals by volume manufactured in the United States in 1993. We have also listed the production figures for 1977, which

TABLE 1.14   Top 50 Highest Volume Chemicals in the United States for 1993

Rank			Production (billion lbs)		Growth 1977–1991
1993	1977	Chemical	1993	1977	(% year)
1	1	Sulfuric acid	80.31	68.80	1.0
2	6	Nitrogen	65.29	24.04	6.4
3	4	Oxygen	46.52	31.86	2.4
4	5	Ethylene	41.25	24.65	3.3
5	2	Lime	36.80	37.78	− 0.2
6	3	Ammonia	34.50	32.35	0.4
7	8	Sodium hydroxide	25.71	21.00	1.3
8	7	Chlorine	24.06	21.30	0.8
9		Methyl tert-butyl ether	24.05		
10	10	Phosphoric acid	23.04	15.60	2.5
11	13	Propylene	22.40	12.56	3.7
12	9	Sodium carbonate	19.80	15.97	1.4
13	15	Ethylene dichloride	17.95	10.48	3.4
14	11	Nitric acid	17.07	14.77	0.9
15	12	Ammonium nitrate	16.79	13.97	1.2
16	16	Urea	15.66	8.99	3.5
17	23	Vinyl chloride	13.75	5.81	5.5
18	14	Benzene	12.32	11.25	0.6
19	18	Ethylbenzene	11.76	7.3	3.0
20	26	Carbon dioxide	10.69	4.45	5.6
21	20	Methanol	10.54	6.46	3.1
22	19	Styrene	10.07	6.82	2.5
23	25	Terephthalic acid (includes dimethyl terephthalate)	7.84	5.0	2.8
24	21	Formaldehyde (37% by weight)	7.61	6.08	1.4
25	22	Xylenes (mixed)	6.84	6.05	0.8
26	24	Hydrochloric acid	6.45	5.13	1.4
27	17	Toluene	6.38	7.73	− 1.2
28	32	p-Xylene	5.76	3.02	4.1
29	27	Ethylene oxide	5.68	4.42	1.6
30	30	Ethylene glycol	5.23	3.47	2.6
31	28	Ammonium sulfate	4.80	3.82	1.4
32	33	Cumene	4.49	2.64	3.4
33	37	Phenol	3.72	2.38	2.8
34	34	Acetic Acid	3.66	2.58	2.2
35		Potash (as K_2O)	3.31		
36	41	Propylene oxide	3.30	1.90	3.5
37	29	Carbon black	3.22	3.48	− 0.5
38	31	Butadiene	3.09	3.19	− 0.2
39	45	Vinyl acetate	2.83	1.60	3.6

TABLE 1.14   (Continued)

Rank			Production (billion lbs)		Growth 1977–1991
1993	1977	Chemical	1993	1977	(% year)
40	48	Titanium dioxide	2.56	1.36	4.0
41	44	Acrylonitrile	2.51	1.64	2.7
42	40	Acetone	2.46	2.14	0.9
43	38	Aluminum sulfate	2.23	2.32	− 0.3
44	39	Cyclohexane	2.00	2.34	− 1.0
45	46	Sodium silicate	1.97	1.56	1.5
46	43	Adipic acid	1.56	1.85	− 1.1
47	35	Sodium sulfate	1.44	2.51	− 3.4
48	36	Calcium chloride	1.40	2.42	− 3.4
49		Caprolactam	1.36		
50	49	n-Butanol	1.33		
		Total organics	**257.4**		
		Total inorganics	**427.9**		
		Grand total	**685.3**		**2.3**

we gave originally in our book *Industrial Organic Chemicals in Perspective,* and the rate of growth over the intervening years. The rank order would be more or less the same in any developed country.

Sulfuric acid heads the list by a large margin as befits its position as an economic indicator, although its maturity means that its growth has been slow. Though it has many applications, about 45% is used for phosphate and ammonium sulfate fertilizers. Of the first 15 chemicals only four—ethylene, methyl *tert*-butyl ether, propylene, and ethylene dichloride—are organic. Six are associated with the fertilizer industry—sulfuric acid, nitrogen, ammonia, phosphoric acid, nitric acid, and ammonium nitrate. Oxygen is used by the steel industry and for welding. Sodium carbonate is important in the glass industry. Most of these chemicals are also used to make organic chemicals, but their main markets lie elsewhere. Chlorine has a number of uses including the bleaching of paper, as a disinfectant, and as a component of organic compounds, most important of which is vinyl chloride, whose precursor is ethylene dichloride. Many chlorine compounds, however, are now considered ecologically undesirable, as is the use of chlorine for bleaching paper and disinfecting swimming pools. Its use between 1990 and 1991 decreased by 1 billion lbs and its growth rate since 1977 was only 0.8%/year.

The two most important organic chemicals, ethylene and propylene, occupy positions 4 and 11. Benzene, the third most important building block, occupies position number 18 with production of 12.32 million lbs. The majority of remaining chemicals in the top 50 are organic, and these form the backbone of the so-called heavy organic chemical industry. Heavy organics are defined as large volume commodity chemicals such as ethylene and propylene as opposed to specialty chemicals such as dyes and pharmaceuticals. Some of the chemicals have only one very large use. For example, the major use for ethylene dichloride (No. 13) is to make vinyl chloride (No. 17). The major use for ethylbenzene (No. 19) is to make styrene (No. 22). p-Xylene (No. 28) is converted primarily into terephthalic acid (No. 23). Cumene (No. 32) is converted to phenol (No. 33) and acetone (No. 42). Cyclohexane (No. 44) is used primarily to make adipic acid (No. 46) and caprolactam (No. 49).

Many of the top 50 chemicals are monomers for polymers, including ethylene, propylene, vinyl chloride, styrene, terephthalic acid, formaldehyde, ethylene oxide, ethylene glycol, phenol, butadiene, propylene oxide, acrylonitrile, vinyl acetate, adipic acid, and caprolactam.

Comparison of the 1993 data with the 1977 data shows the maturity of the chemical industry. Only four chemicals have dropped out of the list since 1977, notably isopropanol, acetic anhydride, sodium tripolyphosphate, and petrochemical ethanol (Nos. 42, 47, 49, and 50, respectively, in 1977). Isopropanol has been hit by the reduction in use of acetone as a solvent as a result of legislation on air pollution (see note), and Exxon has said it is quitting the isopropanol → acetone business. Acetone demand can now be satisfied by the coproduct material from the cumene-phenol process (Section 4.5). Acetic anhydride has suffered as a result of the decline of the cellulosics industry (Section 14.3). Sodium tripolyphosphate has diminished in importance with the marketing of low-phosphate detergents and regulations banning of its use in some states and limiting it in others. Synthetic ethanol has been affected both by imports from Saudi Arabia and the subsidies to fermentation ethanol from corn.

These products were near the bottom of the top 50 anyway, so their dropping out need be little more than a small fluctuation in demand. What products have replaced them? The most significant is methyl tert-butyl ether for unleaded gasoline, which we have mentioned repeatedly, and which now occupies position 9 in the rankings. The other newcomers are the fertilizer, potassium chloride (No. 35), which is mined rather than manufactured, caprolactam (No. 49) and n-butanol. The appearance of caprolactam indicates a slight shift in the United States from nylon 6,6 to nylon 6 (Sections 7.2 and 7.3).

Not only has the composition of the top 50 remained largely unchanged, but the rank order has varied less than might be expected. Chemicals produced at this level usually grow at about the rate of the gross national product, say 2 or 3% per year. The top 50 itself has grown at 2.3% per year. Some of the exceptions are noteworthy. Ethylene and propylene have grown faster than the GNP particularly in the late 1980s and early 1990s because of the growth of high and linear low-density polyethylene, α-olefins, and polypropylene. Butadiene has shown negative growth because its major use is to make styrene–butadiene rubber (Section 5.1) for tires. Because cars in the United States are smaller, less rubber is required for tires. The spare tire in automobiles has been practically eliminated, and radial tires last longer and require less synthetic and more natural rubber. There is an embarrassing excess of butadiene in Western Europe, where it is produced in much larger quantities than in the United States, because liquids are steam cracked there as opposed to gas in the United States (Section 2.2.1). Solvent applications for chemicals have decreased generally because of the more stringent requirements to prevent air pollution. Solvents in the top 50 include benzene, methanol, toluene, mixed xylenes, and acetone. In contrast to the others, methanol has grown at a reasonable rate because of its use in methyl *tert*-butyl ether.

NOTES AND REFERENCES

United States Chemical Industry Statistical Handbook contains general information about the chemical industry including sales, volumes, pollution and environmental problems, and trends. It is published by the Manufacturing Chemists Association, 1825 Connecticut Ave., Washington, DC 20009. Information is also published annually in *Fortune, Chemical and Engineering News, Chemical Week,* and *Chemical Insight* (Quadrant House, The Quadrant, Sutton, Surrey SM2 5AS). A statistical analysis of the US chemical industry is contained in *Kline Guide to the Chemical industry*, Kline Industrial Marketing Guide (Charles H. Kline & Co., Inc., Fairfield, NJ 07006). The guide is updated at intervals.

Few books are published on nonscientific aspects of the chemical industry because geographers, economists, and so on, do not understand the chemistry. It is therefore worth welcoming K. R. Payne, *The International Petrochemical Industry: Evolution and Location,* Blackwell, Oxford, UK, 1991.

Section 1.1   Table 1.1 is based on data in the US Bureau of the Census, *Annual Survey of Manufactures,* 1975. Figures 1.1 and 1.2 are based on US Bureau of the Census figures and the *Report on the Census of Production 1991,* UK Business Monitor, HMSO, London.

Section 1.3   The question of why the chemical industry continually builds excess capacity is discussed in more detail in *When Markets Quake,* J. L. Bower, Harvard Business School Press, Boston, MA, 1986. Bower, who is a Professor at the School, does not arrive at a solution to the problem of "feast and famine" beyond the two obvious and unpopular concepts of cartelization and strict government control.

Section 1.3.1   For comments on the growth of the chemical industry prior to 1980 see *Industrial Organic Chemicals in Perspective, Part One: Raw Materials and Manufacture,* H. A. Wittcoff and B. G. Reuben, Wiley, New York, 1980, pp. 20–21.

Data for Western European overcapacity are given in the *APPE Activity Review 1993–1994,* available from CEFIC, Avenue E. Van Nieuwenhuyse, 4-Box 2, B-1160 Brussels, Belgium.

Section 1.3.4   Economies of scale are discussed in more detail in "Economies of Scale or Diminishing Returns," B. G. Reuben, *Process Eng.,* November, 1974, p. 100.

Section 1.3.7   Figure 1.8 is taken from the *US Chemical Industry Statistical Handbook 1992, loc. cit.* The data in Table 1.9 on the risks posed by asbestos are adapted from an article in *Science,* 19 January 1990. Data on the chemical industry and its contribution to pollution abatement can be obtained from the Chemical Manufacturers' Association, Government Relations Department, Washington, DC. The CMA indicated in a bulletin issued in 1992 that pollution from the chemical industry had decreased 60% and that the industry in 1990 had spent $3.8 billion on pollution abatement. This was five times the 1975 level. Cost for pollution abatement in the year 2005 in 1990 constant dollars is estimated to be $11 billion, mostly as a result of the Clean Air Act. In the early 1990s about 20% of all capital investment of the industry was directed towards pollution abatement and pollution control.

Bruce N. Ames' article on cancer risks and how tests for carcinogens should be interpreted will be found in *CHEMTECH* **19** 591 (1989). He points out that modern sophisticated analytical techniques can detect carcinogens in natural as well as synthetic substances and amazingly that they are ubiquitous in the former.

Section 1.3.8   The effect of government regulations on research and development expenditures has been described for General Motors in an article by P. F. Chenea, *Res. Manag.*, March 22, 1977.

About 65% of all chemists and chemical engineers in industry are employed by companies who are not in the chemical industry but require the services of chemists. Polaroid, 3M, Bell Telephone, and IBM are examples. "Technology push" inventions are quite rare and tend to be solutions in search of a problem. See *Invention and Economic Growth*, J. Schmookler, Harvard University Press, Boston, 1966.

Section 1.4   Tables 1.12 and 1.13 are based on *Fortune's* Top 500 for 1993.

The order of sales of companies from several countries can change with currency value fluctuations. For this reason, sales are not an aboslute indication of size. Also, it is sometimes difficult to separate chemical from nonchemical operations. Thus order of magnitude becomes more important than actual ranking. Actually, the 1993 listing is unusual since BASF is generally considered to be the largest chemical company in the world.

The listing of the top chemical companies in the world is adapted from *Chemical Insight*.

Section 1.5   The highest volume chemicals are listed yearly in *Chem. Eng. News.* For 1993 they will be found in the issue for July 4, 1994, pp. 30–36. This issue contains a wealth of information about the chemical industry worldwide.

The legislation on air pollution mentioned on p. 43 was reversed in 1995 when this book was going to press.

CHAPTER 2

CHEMICALS FROM NATURAL GAS AND PETROLEUM

Where do industrial organic chemicals come from? Table 2.1 provides a guide. Natural gas and petroleum are the main sources. From them come seven chemical building blocks on which a vast organic chemical industry is based. These are ethylene, propylene, the C_4 olefins (butenes and butadiene), benzene, toluene, the xylenes, and methane. The olefins—ethylene, propylene, and butadiene and the butenes—are derived from both natural gas and petroleum.

TABLE 2.1   Sources of Organic Chemicals

Natural Gas and Petroleum	
Ethylene (Chapter 3)	Benzene (Chapter 7)
Propylene (Chapter 4)	Toluene (Chapter 8)
Butenes and butadiene (Chapter 5)	Xylenes (Chapter 9)
Methane (Chapter 10)	
Coal (Chapter 12)	
Synthesis gas (CO and H_2)	
Acetylene[a]	
Carbohydrates (Chapter 14)	*Fats and oils* (Chapter 13)
Cellulose	Fatty acids and derivatives
Starch	
Sugars	
"Gums"	*Miscellaneous natural products*

[a] Although coal is the traditional source of acetylene, much of it is currently produced either from methane or as a byproduct in ethylene production (Section 10.3).

The aromatics, benzene, toluene, and xylenes, are derived from petroleum and to a very small extent from coal. Methane comes from natural gas.

Whether natural gas fractions or petroleum is used for olefins varies throughout the world depending on the availability of natural gas and the demand for gasoline. Both light and heavy naphthas (Table 2.2) are petroleum fractions that

TABLE 2.2   Crude Oil Distillation

Fraction	Boiling Point Range	Comments
1. Gases		
Methane (65–90%) ethane, propane, butane	Below 20°C	Similar to natural gas. Useful for fuel and chemicals. Also obtained from catalytic cracking and catalytic reforming. Much of it flared because of cost of recovery. *n*-Butane, however, is practically always recovered.
2. Naphtha		
Light naphtha (C_5, C_6 hydrocarbons)	70–140°C	Naphtha is predominantly C_5–C_9 aliphatic and cycloaliphatic compounds. May contain some aromatics. Base for gasoline. Useful for both fuel and chemicals. Light naphtha now considered undesirable in gasoline because catalytic reforming yields benzene, which is toxic and has a relatively low-octane number.
Heavy naphtha (C_7–C_9 hydrocarbons)	140–200°C	
3. Atmospheric gas oil		
Kerosene	175–275°C	Contains C_9–C_{16} compounds useful for jet, tractor, and heating fuel.
Diesel fuel	200–370°C	Contains C_{15}–C_{25} compounds, mostly linear. Useful for diesel and heating fuels. Gas oil is catalytically cracked to naphtha and may be steam cracked to olefins.
4. Heavy fractions		
Lubricating oil	Above 370°C	Used for lubrication
Residual or heavy fuel oil		Used for boiler fuel. Vacuum distillation gives vacuum gas oil for catalytic cracking.
Asphalt or "resid"		Used for paving, coating and structural applications.

can be cracked to make olefins. They can also be used for gasoline. In the United States demand for gasoline is higher than for other petroleum fractions. Consequently, the price of naphtha has traditionally been high and the chemical industry has preferred to extract ethane and propane for cracking to olefins from what has hitherto been abundant natural gas. Even with the decrease in natural gas reserves, the shift to liquids for cracking in the United States has been slowed by the availability of cheap LPG (see note).

In Western Europe and Japan the demand for gasoline is lower because on a per capita basis there are half as many cars as in the United States, and they are smaller and are driven shorter distances. Thus in Western Europe more naphtha is produced than is required for gasoline. Also, natural gas is less abundant even with the North Sea discoveries and contains less ethane. Consequently, naphtha has been the important European raw material for the manufacture of olefins. This is equally true in Japan, which lacks resources of either natural gas or oil.

Figure 2.1 illustrates the difference in demand for refinery products in the United States, Western Europe, and Japan. For the past 12 years, gasoline has represented 42% of the US oil barrel, up from 40% in the late 1960s, so that usage has remained more or less steady in spite of a rise in the number of vehicles. In Western Europe, the proportion of the barrel going into gasoline has climbed from 18% in 1967 to 26% in 1993, reflecting the replacement of fuel oil by natural gas, which did not come on stream on a large scale until the 1970s.

What was once a straightforward raw material supply situation has been complicated by the discovery of natural gas in many parts of the world in the 1970s. Dramatic discoveries in Siberia mean that the CIS possesses a large share of the world's natural gas reserves. Gas fields have also been discovered in

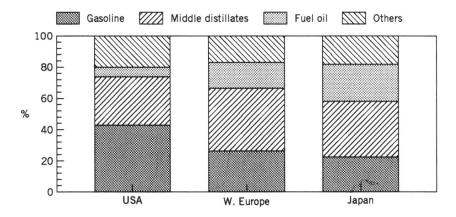

Middle distillates = kerosenes and gas and diesel oils.
Fuel oil includes bunker crude for ships.
Others = refinery gas, LPG, solvents, petroleum coke, lubricants, bitumen, wax, refinery fuel, and losses.

Figure 2.1   Oil consumption by product group, 1993.

Canada, New Zealand, the Arabian peninsula, Thailand, Malaysia, Trinidad, and North Africa. In the Middle East, particularly Saudi Arabia, large quantities of associated gas (the gas that accompanies oil deposits as opposed to that occurring in separate gas fields) are available. In many of these countries, particularly Canada and Saudi Arabia, this gas has become the basis for new chemical industries. The economic consequences of these developments will be discussed further in Section 2.2.1.1.

Probably 90% by weight of the organic chemicals the world uses comes from petroleum and natural gas, and we therefore devote considerable space to them. In addition, we consider what might happen if natural gas and petroleum supplies are exhausted in the next 50–60 years. The reserves-to-current production ratio in 1994 was 42 years for oil and 66 years for gas, but these figures have *risen* from 31 and 41 in 1967. Nonetheless, it has been authoritatively predicted that the United States will never again bring to the surface as much natural gas as they did in 1973. There is a need for strategies for the future. The glut of oil and gas in the 1980s and early 1990s clouds the inevitability of the development of future shortages.

A less important current source of chemicals is coal. Coal was historically important, and much of the progress in the chemical industry until World War II was motivated by the availability of coal. Indeed the famous English chemist, W. H. Perkin, could claim to have founded the organic chemical industry in 1865 when, while trying ineptly to synthesize quinine, he obtained a dye with a mauve color from coal tar intermediates. So important was Perkin's dye that its color gave its name to a period of history known in literature as the "mauve" decade.

The decline of coal coincided with the rise of petrochemicals. However, reserves of coal are much greater than those of oil. If petroleum becomes scarce, will coal come into its own again? Is it a realistic part of an alternative strategy? The so-called C_1 chemistry (Sections 10.5 and 10.6) that developed in the 1960s and 1970s when petroleum shortages loomed large assumes that it will, if the industry is willing to provide the tremendous capital investment a switchover will require.

The third and final source of organic chemicals is the group of naturally occurring, renewable materials of which triglycerides (fats and oils) and carbohydrates are the most important. Although they account at present for only a few percent by weight of the products of the chemical industry, there has been much discussion of how they might be used to replace nonrenewable fossil materials (oil, coal, or natural gas) if the latter ran out. We review some of this chemistry too.

The third group includes more obscure natural products that make contributions to specialized segments of the chemical industry. Examples of such materials are sterols, alkaloids, phosphatides, rosin, shellac, and gum arabic. Their contribution to the chemical industry in terms of value or weight is small but they are irreplaceable for products such as certain pharmaceuticals, flavors, and fragrances.

2.1   PETROLEUM DISTILLATION

To provide an idea of how petroleum is used as a source of chemicals we must describe what happens in a petroleum refinery. Consider a simple refinery (Table 2.2 and Fig. 2.2) in which crude oil, a sticky, viscous liquid with an unpleasant odor, is separated by distillation into various fractions.

The first most volatile fraction consists of methane and higher alkanes through C_4 and is similar to natural gas. These are dissolved in the petroleum. The methane and ethane can be separated from the higher alkanes, primarily propane and the butanes. The methane–ethane mixture is called "lean gas" from which the ethane can be separated if required. The C_3/C_4 mixture (LPG), may be used as a petrochemical feedstock (Section 2.4) or a fuel. Butane is also separated for use in gasoline (Section 2.2.2) and to a lesser extent as a raw material for chemicals (Section 5.4).

In the past, the high cost of compressing or liquefying and shipping these refinery gases has dictated that most of them be flared. As the price of natural gas increases, however, shipping of methane in refrigerated tankers becomes economic. Alternatively, methane may be converted to methanol, which is

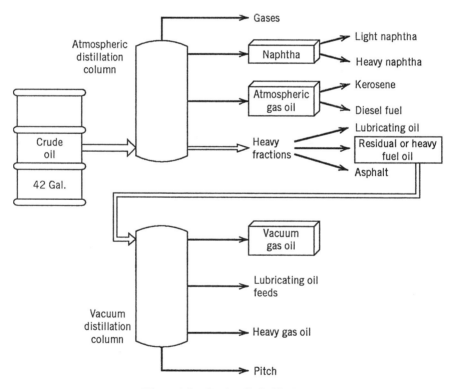

Figure 2.2   Crude oil distillation.

a useful organic chemical building block (Section 10.5) and is more easily shipped. Even today, however, most refinery gas is still flared even in the United States.

The second fraction comprises a combination of light naphtha, or straight run gasoline, and heavy naphtha, and is of particular importance to the chemical industry. The term naphtha is not well defined, but the material steam cracked (Section 2.2.1) for chemicals generally distills in a range between 70 and 200°C and contains C_5–C_9 hydrocarbons. Naphtha contains aliphatic as well as cyclo-aliphatic materials such as cyclohexane, methylcyclohexane, and dimethyl-cyclohexane. Smaller amounts of C_9+ compounds such as polymethylated cycloalkanes, and polynuclear compounds such as methyldecahydronaphtha-lene may also be present. Like the lower alkanes, naphtha may be steam cracked to low molecular weight olefins. Its conversion by a process known as catalytic reforming into benzene, toluene, and xylenes (BTX) (Section 2.2.3) is, in the United States, its main chemical use. Catalytic reforming is also a source of aromatics worldwide, although in Western Europe benzene and toluene are mainly derived from pyrolysis gasoline, an aromatics fraction that results from steam cracking of naphtha or gas oil (Section 2.2.1).

Light naphtha was at one time used directly as gasoline; hence its alternate name, "straight run gasoline." It contains a large proportion of straight-chain hydrocarbons (n-alkanes), and these resist oxidation much more than bran-ched-chain hydrocarbons (isoalkanes), some of which contain tertiary carbon atoms. Consequently, straight run gasoline has poor ignition characteristics and a low-octane number of about 60. It is of little use in gasoline for modern high compression-ratio automobile engines, and its properties are even worse if it is unleaded. Isomerization (Section 2.2.7) of its components to branched com-pounds increases its octane number and thus its utility. Chemically, however, its significance is like that of naphtha, for it can be cracked to low molecular weight olefins. It does not perform well in catalytic reforming, giving large amounts of cracked products and small amounts of benzene. Benzene is no longer a wel-come constituent of gasoline because of its toxicity and relatively low octane number. The situation is summarized in Figure 2.3.

As noted above, international practice has differed. The United States has preferred to crack ethane and propane from natural gas, while the rest of the world has cracked naphtha. The shortage of natural gas in the United States projected in the 1970s led to an increased interest in liquid feedstock cracking. The preferred liquid feedstock is naphtha, which has traditionally been pre-empted in the United States for gasoline manufacture. Accordingly, when a natural gas shortage loomed, gas oil steam cracking was developed. This was not considered when natural gas was plentiful because cracking of gas oil to olefins is accompanied by tar and coke formation. The prediction of a shortage, however, motivated techniques for ameliorating this latter problem and it is now possible to crack gas oil as well as naphtha. The industry has been loathe to do so because it is usually more economic to crack ethane and propane. They are easier to handle, provide fewer coproducts, and less coke.

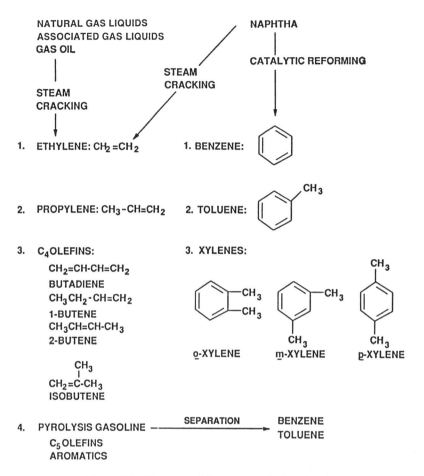

Figure 2.3   Steam cracking and catalytic reforming.

In the 1980s and 1990s, the switch to liquid feedstocks lost momentum for three reasons. First, US natural gas production was maintained. Although reserves are being depleted—the reserves/production ratio was down to 8.8 years in 1993—production was down only 5% from the 1973 peak. Second, natural gas discoveries in Canada meant that cheap natural gas could be imported and is now just under 10% of consumption. Third, Saudi Arabia decided to base its chemical business on the cracking of ethane only, making LPG available on the world market (see note to Section 2). Thus, as indicated in Table 2.3, the percent of gaseous feed cracked in the United States has decreased much less than predicted, and it is believed that it will not fall below 70% until the year 2000.

Because of availability of gas from the North Sea, the percentage of gas cracked in Western Europe doubled by 1995, but even so the predominant

TABLE 2.3   Feedstocks for Ethylene: United States versus Western Europe

Feedstock (% of Total)	1982 US	1982 Western Europe	1985 US	1985 Western Europe	1995–2000 US	1995–2000 Western Europe
Ethane/propane	80	—	79	—	—	—
Ethane/propane/ butane	—	10	—	14	70	20
Naphtha/gas oil	20	90	21	86	30	80

feedstock in Europe will continue to be liquids. Liquefied petroleum gas cannot be used as readily in Western Europe as in the United States because most European crackers are not flexible, accommodating only liquids. This lack of flexibility is in part due to the lack of the expensive infrastructure required to handle gas. Most crackers in the United States are flexible. If the price of propylene or butadiene goes up, more liquids can be cracked; indeed the optimum ratio of feedstocks is determined hourly by linear programming.

The situation in Japan approximates that in Western Europe except that even less gaseous feed is cracked, and what is cracked comprises almost entirely butane. The liquid feed is entirely naphtha and it is not expected that gas oil will be cracked in the foreseeable future.

A decade ago this discussion of world sources of ethylene would have sufficed. By the mid-1980s, however, the impact of the production of ethylene in other parts of the world, primarily the Middle East and Canada, was being felt. Joining these two giants are Asian countries including Korea, Taiwan, Thailand, Malaysia, Singapore, and Indonesia with of course India and China; South and Central American countries including Mexico, Venezuela, Brazil, Argentina, Chile, and Trinidad, and still others including the Eastern European countries. The availability of ethylene derivatives from third world sources had caused exports from Japan and Western Europe to decrease markedly by the late 1980s (Figure 1.4). By 1995 the new producers will enjoy a much larger share of the world's ethylene derivatives export business. The United States will still participate in trade of ethylene derivatives as long as they are able to crack gas.

To summarize, the two reactions in Figure 2.3—steam cracking and catalytic reforming—are the basis for much of the world's petrochemical production valued in 1995 at about $650 billion. The three main raw materials are ethylene, propylene, and benzene, while the C_4 olefins, methane, toluene, and the xylenes are important but to a lesser degree. Methane is an important source of the fertilizer, ammonia, as well as of organic chemicals, primarily methanol. Its most important reaction is the formation of synthesis gas, from which ammonia and methanol are made, and naphtha may also be used for this (Section 10.4) although it seldom is.

Look back to Figure 2.2. Kerosene is a fuel for tractors, jet aircraft, and for domestic heating and has some applications as a solvent. Gas oil is further refined into diesel fuel and light fuel oil of low viscosity for domestic use. Its use as feed for cracking units for olefin production has already been mentioned. Both the kerosene and gas oil fractions may be catalytically cracked to gasoline range materials (Section 2.2.2). Actually, the term gas oil is applied to two types of material, both useful for catalytic cracking. One is so-called atmospheric gas oil which, as its name indicates, is produced by atmospheric pressure distillation. The other is vacuum gas oil, which results from the vacuum distillation of so-called residual oil. It has a much higher boiling range of 430–530°C.

Residual oil (Fig. 2.2) boils above 350°C. It contains the less volatile hydrocarbons together with asphalts and other tars. Most of this is sold cheaply as a high-viscosity heavy fuel oil (bunker oil), which must be burned with the aid of special atomizers. It is used chiefly on ships and in industrial furnaces.

A proportion of the residual oil is vacuum distilled at 0.07 bar to give, in addition to gas oil as mentioned above, fuel oil (bp <350°C), wax distillate (350–560°C), and cylinder stock (>560°C). The cylinder stock is separated into asphalts and a hydrocarbon oil by solvent extraction with liquid propane in which asphalts are insoluble. The oil is blended with the wax distillate, and the blend is mixed with toluene and methyl ethyl ketone and cooled to − 5°C to precipitate "slack wax," which is filtered off. The dewaxed oils are purified by countercurrent extraction with such solvents as furfural, which remove heavy aromatics and other undesirable constituents. The oils are then decolorized with fuller's earth or bauxite and are blended to give lubricants.

Part of the vacuum distillate and the "slack wax" can be further purified to give paraffin and microcrystalline waxes used for candles and the impregnation of paper. The petroleum industry is constantly trying to find methods by which the less valuable higher fractions from petroleum distillation can be turned into gasoline or petrochemicals. Table 2.4 gives an indication of the values of the various fractions relative to the cost of crude oil.

TABLE 2.4   Value of Various Oil Fractions

Product	Typical Ratio of Value of Product to Cost of Crude Oil in the United States[a]
LPG	1.1–1.4
Motor gasoline	1.4–1.5
Naphtha	1.2–1.3
Gas oil	1.15–1.3
Jet kerosene	1.3–1.4
Vacuum gas oil	0.95–1.05
Heavy fuel oil	0.6–0.7

[a] European figures are similar but gas oil prices tend towards the top of the range and naphtha/motor gasoline towards the bottom.

2.2   PETROLEUM REFINING REACTIONS

The production of chemical feedstocks from petroleum is inextricably associated with the production of gasoline and other fuels. Sometimes these two industries compete for raw materials, and sometimes they complement one another. The chemical industry is a junior partner because it consumes only about 6% of refinery output in the United States and rarely more than 8% anywhere else. On the other hand, because it produces premium products, it can compete with other consumers in buying those raw materials it needs. The chemical industry can compete successfully because the petrochemicals and the industrial and consumer products made from them represent very high "value added" as compared to the value of the starting materials. In order to understand this competition and the operation of a petroleum refinery we must examine the chemical processes that follow the physical process of distillation.

We have already noted that straight run gasoline has too low an octane number for high compression-ratio engines. A major objective of a petroleum refinery is to raise this number. It is achieved by way of the reactions summarized in Table 2.5, most of which either modify a petroleum fraction or provide the raw material for another reaction that will give compounds with an improved octane number. In a modern refinery these reactions as well as distillation take place under computer control, which varies conditions and output according to the ever-changing demands of the market and the composition of the feedstock, which in turn may vary from day to day and storage tank to storage tank.

2.2.1   Steam Cracking

Steam cracking, derived from the thermal cracking process introduced as early as 1912, involves the use of heat, but no catalyst. From naphtha feedstocks it yields mainly C_2, C_3, and C_4 olefins and an aromatic fraction called pyrolysis gasoline. Because the olefins are not useful in gasoline, it has long since been superseded for gasoline production by catalytic cracking. It is, however, the mainstay of the petrochemical industry. The insignificance of chemicals' production compared with that of gasoline is illustrated by the fact that only nine percent of total US cracking capacity is steam cracking. The rest is catalytic cracking.

There is little similarity between the old thermal cracking and modern steam cracking processes. The obsolete process used a heavy feedstock and a relatively low temperature and high pressure to maximize gasoline production and minimize gas formation. The new process uses light liquid or gaseous feedstocks (Section 2.1), a high temperature (650–900°C), and a low pressure to maximize the yield of low molecular weight gases. Since it is inconvenient to operate a plant below atmospheric pressure because a small leak could lead to the formation of explosive hydrocarbon–air mixtures, the partial pressure of reactants is reduced by addition of steam as an inert diluent. The steam also serves (Section 10.4) to reduce coke formation in the reactor tubes.

TABLE 2.5   Petroleum Refining Reactions

Cracking

Steam Cracking: Converts n-alkanes, cycloalkanes, and aromatics in oil, or ethane, propane, butane, and higher hydrocarbons in natural gas into ethylene, propylene, butenes, and butadiene. The products are primarily for the chemical industry.

Catalytic Cracking: Produces molecules with 5–12 carbon atoms suitable for gasoline from larger molecules. Facilitates formation of branched chain and aromatic molecules.

Hydrocracking: Upgrades heavy crudes by converting them to more volatile products. Uses catalysts together with hydrogen, which prevents "coke" formation on the catalyst and converts objectionable sulfur, nitrogen, and oxygen compounds to volatile H_2S, NH_3, and H_2O.

Polymerization

Combines low molecular weight olefins into gasoline-range molecules with H_2SO_4 or H_3PO_4 catalysts. Not widely used today, although there has been renewed interest because of the need to enhance octane number in unleaded gasoline. A more appropriate term would be oligomerization, since oligomers, not polymers, are formed.

Alkylation

Combines an olefin with a paraffin (e.g., propylene with isobutane) to give branched chain molecules. Sulfuric acid and hydrogen fluoride are used as catalysts. Very important in achieving high octane number in lead-free gasoline.

Catalytic reforming

Dehydrogenates both straight-chain and cyclic aliphatics to aromatics, primarily BTX over platinum–alumina or rhenium–alumina catalysts. Most widely used refinery reaction in the United States.

Dehydrogenation

Cracking and reforming are basically dehydrogenations. Other dehydrogenations include conversion of ethylbenzene to styrene, butenes to butadiene, and propane to propylene.

Isomerization

Used to convert straight-chain to branched-chain compounds—for example, n-butane to isobutane for alkylation, n-pentane to isopentane, and n-hexane to isohexane in low boiling naphtha and n-butene to isobutene. Other refinery isomerizations include o-xylene and m-xylene to p-xylene.

Coking

Used to remove metals from a refinery steam. Heat in the absence of air "cracks-off" hydrocarbons. The metals stay behind in the "coke."

Hydrotreating

Converts sulfur, nitrogen, and oxygen in petroleum fractions to the gases, H_2S, NH_3, and H_2O. Uses hydrogen from other refining processes such as reforming. Can be applied to heavy feedstocks.

The hydrocarbon feedstock is vaporized, if not gaseous to begin with, mixed with steam, and the mixture passed through tubes about an inch in diameter through a furnace heated by oil and gas burners. The residence time is short—30–100 ms—to minimize coking, which is potentially a major problem since coke and hydrogen are thermodynamically the favored products.

The product gases emerge at about 800°C. As they are to be distilled at low temperatures, they have to be cooled by 900° with minimum waste of heat—a classic chemical engineering problem. Throughout the process, series of heat exchangers conserve the heat used in the early stages and the "coolth" in the later ones. Initially, the gases are quenched to about 400°C in a heat exchanger and the heat used to raise steam to power the turbines of the centrifugal compressors used in the refrigeration process. They are then cooled from 400°C to a little above 100°C in a spray tower cooled by a stream of fuel oil. Further fuel oil condenses. Some of the fuel oil is removed from the bottom of the spray tower as a product and the remainder cooled and recycled. Further cooling to ambient temperature condenses pyrolysis gasoline plus the steam that was originally added to the feedstock, and these are simply separated in a decanter.

The gases are then scrubbed with ethanolamines to remove acid gases such as carbon dioxide and hydrogen sulfide and dried with molecular sieves. The drying is important because ice is highly abrasive and would damage the low-temperature equipment. The remaining gases are compressed to 40 bar and condensed in a cascade of refrigeration units. Cracking processes only became economically viable because of the development of centrifugal compressors that run the refrigeration processes more cheaply and efficiently than reciprocating compressors. They are the greatest contribution of mechanical engineering to the chemical industry. The condensed gases are then distilled in a column called a demethanizer, which carries on top a condenser cooled to $-95°C$ with liquid ethylene. Methane and hydrogen are the top products, and the C_2+ products emerge as a liquid at the bottom.

This bottom stream is distilled again in a deethanizer in which the C_2 components pass overhead and the C_3–C_5 hydrocarbons form the bottom product. The C_2 fraction is hydrogenated to convert small amounts of acetylene to ethylene. The advantage of carrying out the hydrogenation at such a low temperature is that the acetylenic triple bond is selectively hydrogenated. At higher temperatures, double and triple bonds would be indiscriminately reduced. The importance of removing acetylene is that, when polymerized with ethylene, it gives a polymer containing double bonds, which cross-link to give a nonlinear product. The ethylene–ethane mixture is distilled to give an ethylene top product that may be further purified to give polymer grade ethylene, and an ethane bottom product that is recycled.

A depropanizer column separates the C_3 and C_4 streams, and a debutanizer separates the C_4 from C_5 streams. The latter are combined with the pyrolysis gasoline stream from the first column. The C_3 stream is treated in the same way as the C_2 stream to remove propyne.

2.2.1.1 Choice of Feedstock The above basic process may be modified depending on the feedstock. Not surprisingly, the products of steam cracking are affected by this as well as by reaction conditions. Modern crackers operate under severe conditions (high temperatures, low residence times), and in the last quarter century the typical ethylene yield by weight from fairly severe (high temperature) cracking of a naphtha feedstock has been approximately doubled from about 16–31%. As ethylene is the premium product, this is a major advance.

The effect of feedstock is illustrated in Table 2.6. The yield of ethylene decreases and the yield of the coproducts increases as the molecular weight of the feed increases. The economics associated with the use of the different feedstocks depend crucially on the value of the coproducts. Although the cracking of gas has traditionally been the most economical process, there are times when naphtha cracking is preferred because the coproducts are high in price. This is particularly true in Europe where a propylene shortage started to develop in the early 1990s. The United States is isolated from such a shortage by the large amounts of propylene produced in catalytic cracking (Section 2.2.2) Europe and Japan do not have this reservoir because their lower use of gasoline requires fewer catalytic crackers.

Pyrolysis gasoline is an important product obtained from the cracking of liquids and, the heavier the liquid, the more pyrolysis gasoline is produced. It contains aromatics and thus is an important source of benzene and toluene in

TABLE 2.6 Coproduct Yield lb per 100 lb Ethylene

	Ethane[a]	Propane[a]	Butane	Naphtha	Atmospheric Gas Oil	Vacuum Gas Oil
Feedstock required (lb)	120	240	250	320	380	430
Ethylene (yield %)	80	42	38	31	26	23
Propylene	3	50	40	50	55	60
Butadiene	2	5	9	15	17	17
Other C_4 olefins	1	3	17	25	18	18
Pyrolysis gasoline	2	15	18	75	70	65
Benzene	1	5	6	15	23	24
Toluene	neg.	neg.	3	10	12	12
C_5 olefins	neg.	neg.	1	7	8	7
Fuel oil	neg.	2	4	10	70	125
Other[b]	17	65	62	45	50	45

[a] Negligible = neg. [b] Mainly hydrogen and methane.

Europe and Japan. In Europe pyrolysis gasoline provides 52% of total chemical benzene and 42% of toluene. In addition, in Europe 18% of benzene comes from toluene hydrodealkylation (Section 8.1), a portion of which comes from pyrolysis gasoline. Pyrolysis gasoline is a much less important source of these chemicals in the United States where catalytic reforming is more widely practiced.

2.2.1.2 *Economics of Steam Cracking*

Figure 2.4 shows the relative costs of production of ethylene in 1993 from various feedstocks in the United States, Western Europe, Saudi Arabia, South Korea, and Taiwan. The average 1993 prices of ethylene in Western Europe (spot and contract) and the United States (contract) are also shown.

The costs are divided into depreciation, fixed cash costs and variable costs. Depreciation is related to capital cost. In spite of all the engineering ingenuity, the elaborate separation processes are costly to build and operate. Ethane crackers, such as those used in Saudi Arabia (Section 2.1) produce so few byproducts that separation is simple; the same applies to a lesser extent to other light feedstocks. Naphtha and gas oil crackers are more complicated and consequently more expensive. Thus, for the capital cost of a plant to crack various feedstocks, ethane < ethane/propane < light naphtha < full-range

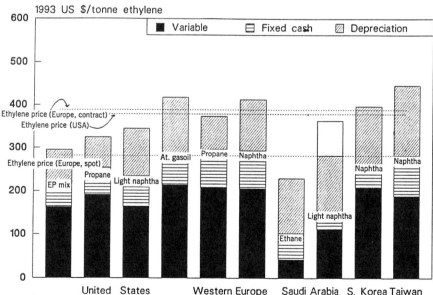

Figure 2.4 Ethylene production costs, 1993.

naphtha < gas oil. Light naphtha in this context usually means "condensate" or "natural gasoline," a cheap C_5+ fraction from certain gas wells. Flexible crackers that can operate on any feedstock are more expensive but involve lower risk. Crackers in developing countries are 10–15% more expensive, in spite of cheap labor, because of the lack of scientific and engineering infrastructure and the need to import many components.

Figure 2.4 does not include interest on finance or corporate overheads. The fixed cash costs comprise direct labor, supervision (see note) maintenance, insurance and local rates, local taxes and interest on working capital. Working capital is the money tied up in stocks of raw materials, product and work in progress, and debts plus the funds needed to pay salaries and so on. These costs are dominated by maintenance charges which, other things being equal, are related to plant size and complexity. Hence, plants for heavier feedstocks carry a further penalty only partly compensated for by income credits.

Meanwhile, the variable costs make up about one-half of all costs and are crucial to the process economics. They comprise raw material costs plus the costs of fuel and other utilities to run the plant minus credit for coproducts. The average costs of the various feedstocks and coproducts in 1993 in dollars per tonne and ¢/lb or ¢/gal are given in Table 2.7. We stress that these represent a single year's prices and that they fluctuate all the time. The table gives some idea of relative prices at a time of European overcapacity, but the reader should consult more recent sources for up-to-date information (see note).

As noted earlier, naphtha is more expensive in the United States than in Europe because of the higher demand for gasoline. The price differential, however, is only about $15/tonne because, at a higher price, it pays producers to ship material westwards across the Atlantic. When gaseous feedstocks are available, they are usually cheaper per tonne than naphtha but do not give so many coproducts. Other things being equal, raw material costs are related to the amount of raw material required to produce a tonne of ethylene. This is higher for naphtha than for gaseous feeds and higher still for gas oil, because they produce more coproducts. Correspondingly, there is more coproduct credit. The coproducts sell at a higher price than the raw materials but at a lower price than ethylene. Depending on the relative prices of the various feedstocks and products, the net raw material price may favor liquid rather than gaseous feedstocks. Nonetheless, even when it exists, this saving is sometimes negated by the higher fixed costs and deprecation.

The flexibility to crack gases or liquids, whichever is more economic, is available to companies who have paid more for their plants. Gaseous feedstocks have traditionally been the more economic. These two factors have accounted for the advantage that the United States, with its historically abundant supplies of natural gas and large supply of refinery propylene, has held in world chemical markets.

The dominance of the United States has been modified by the entry of Saudi Arabia as a major producer. Ethane is produced there at the wellhead from associated gas and is priced at zero plus the logistic cost of separation and

TABLE 2.7 Average Feedstock and Coproduct Prices, 1993[a]

Compound	European Contract Price ($/tonne)	European Spot Price ($/tonne)	United States Contract Price ($/tonne)	European Contract Price (¢/lb)	European Spot Price (¢/lb)	United States Contract Price (¢/lb)
Naphtha	132–155	167	182 (light naphtha 157)		7.6	8.3
LPG	132–155	132–155	132–155	25–32 ¢/gal	25–32 ¢/gal	25–32 ¢/gal
Propane	147		163	27 ¢/gal		31 ¢/gal
Butane	175	175	175	39 ¢/gal	39 ¢/gal	39 ¢/gal
Gas oil		165	175		7.5	7.9
Fuel oil (3.5% sulfur)		63			2.9	
Ethylene	389	282	379	18.3	13.0	21.3
Propylene (polymer grade)	282	266	302	13.0	11.7	13.9
Butadiene	288	267	368	13.1	11.9	17.0
Pyrolysis gasoline		180			8.2	
Benzene	290	293	292	13.1	13.3	13.5
Toluene	244	242	231	11.1	11.0	
Xylenes		249	241		11.3	10.9

[a] Prices are taken from European Chemical News, Platt's European Marketscan or estimated by the authors.

transport, which amounts to about \$27/tonne (50 ¢ per million cubic feet). The Saudi Government wants to penetrate world markets and takes the view that the ethane would have to be flared if it were not separated, hence zero wellhead value is acceptable. Indeed, because flaring would cause pollution, one might even maintain that the wellhead value is negative. On the zero value basis, Saudi ethane-cracking is by far the most economic and sets free for export, quantities of propane plus butane (LPG) not required for fuel (Section 2). On the other hand, the availability of ethane in Saudi Arabia is limited because it comes from associated gas, the quantity of which is linked to production of crude. Subsequent investment has been based increasingly on the use of naphtha and LPG (see note).

Gas cracking is more economic when propylene is not in short supply. The United States has taken full advantage of this, importing small amounts of Saudi LPG in addition to its own supplies of ethane/propane and natural gas liquids. It is believed that cheap LPG will continue through the year 2000. Western Europe cannot take as much advantage of imported LPG because only five of the 52 steam crackers that existed at the end of 1993 can be substantially switched to gaseous feedstocks. European natural gas from the North Sea contains little ethane and propane and only the gas from the northernmost fields provides sufficient C_2+ material to be worth cracking. Hence, the Mossmoran cracker in Scotland is the only dedicated gas cracking unit in Western Europe (see note).

In addition, the infrastructure for handling LPG does not exist in Europe to the extent that it does in the United States. Finally, Europe is dependent on pyrolysis gasoline for over one-half of its benzene and toluene, which gas cracking produces in only small amounts. (The toluene is needed largely for dealkylation to provide more benzene.) Europe's situation is echoed in Japan, Taiwan, and South Korea and is part of the economic structure of their chemical industries. Other Southeast Asian countries are hoping to crack gases (e.g., Thailand, which already has a small plant, and Malaysia with a plant scheduled for late 1995) but as yet these have had little impact on the market. The key countries likely to crack gas in the future, however, are those of the Middle East—Kuwait, Abu Dhabi, and Qatar—all of whom are planning facilities in the next few years following the Saudi example. They will aim to export to Europe. Canada and Venezuela will also crack gas for export to the United States.

Inspection of Figure 2.4 indicates that different countries crack naphtha at more or less similar costs in the region of \$390/tonne, the European contract price in 1993. The exception is Saudi Arabia, which stated that a 30% discount off the lowest export price would be granted on supplies of naphtha and LPG used in their newest petrochemical plant. The extent of the subsidy is indicated by the upper unshaded bar on the Saudi Arabia light naphtha costing.

Figure 2.4 shows that United States ethylene producers are making profits when others are not, having relatively low production costs. Saudi Arabia is even more profitable because of the low value placed on the feedstock in the costing. Ethylene from naphtha, elsewhere in the world, is economically marginal. Substantial overcapacity in 1993 kept prices depressed and forced the sale of surplus product on the spot market. A remarkable sequence of accidents and plant breakdowns in 1994 and early 1995 led to shortages and price rises, but there is general agreement that prices will fall again once repairs have been carried out. Overcapacity seems endemic. Why and how, therefore, do producers based on usually less economic liquid feedstocks keep going? There are four main reasons.

First, many of the crackers are linked into integrated petrochemical complexes. Ethylene is more expensive to ship than to pump over the fence, hence many producers of ethylene derivatives will prefer to draw on their own cracker or a European pipeline to which they are connected rather than to pay for the shipping of ethylene in refrigerated tankers. Nonetheless, the European pipeline network is much shorter than that in the United States. With an effective pipeline network, fewer crackers are needed and, indeed, the United States has fewer crackers than Western Europe.

Second, as the European plants are already built, their owners insist on operating them as long as variable costs are covered. Huge new capacity in Europe, which was planned in the 1980s, came on stream at the start of the 1990s. The plants already operating did not provide sufficient cash flow to justify fresh investment, but the new plants were built on a tide of what currently appears to have been unjustified optimism. The European industry association has tried to encourage reductions in capacity but has been largely unsuccessful. Meanwhile, investment costs in developing countries are usually higher than in Europe.

Third, there is the question of tariffs (see note). These are being reduced in the developed world as a result of the last round of the GATT (General Agreement on Trade and Tariffs) negotiations, but they remain. Southeast Asia is permitted to protect indigenous manufacturers. In May 1994, Indonesia produced no ethylene although a naphtha-based plant will be completed in 1995. In spite of this weakness, polyethylene is heavily protected, hence a large plant has been built there. It is arguable that the Southeast Asian chemical industry will have to rely for its profitability on a local market allied to high tariffs.

Fourth, the price differential between naphtha relative to gaseous feedstocks has been narrowing. If the world ethylene market was swamped with material based on ethane, some naphtha-based crackers would have to close. There would then be a propylene shortage and prices would rise, making the naphtha-based route economic once more.

European producers have survived for the above reasons, but there is no doubt that they have suffered from the new producers around the world and from their misjudgment in building new capacity. Western Europe's share of total international trade in ethylene derivatives dropped from 31% in 1980 to 13% in 1990, although there has been a slight improvement since then. Thus the chemical industries in developed countries are under pressure from those in developing countries. This has everything to do with availability and price of feedstocks, location of plant, and tariffs, and nothing to do with the availability of cheap labor, which makes up a tiny proportion of total costs (see note).

2.2.1.3 Mechanism of Cracking   Catalytic cracking involves carbenium ion intermediates, but steam cracking, which is of greater concern to the chemical industry, is a free radical reaction involving the steps illustrated in Figure 2.5(a–h) which assumes that ethane is being cracked. In the initiation step (a) two methyl radicals form. These attach another molecule of ethane (b) to give

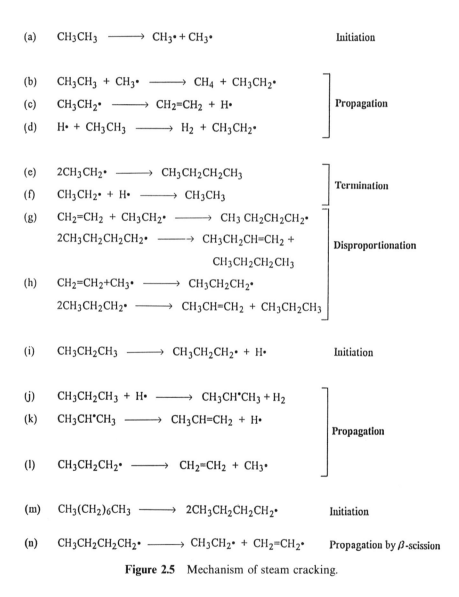

Figure 2.5   Mechanism of steam cracking.

methane plus an ethyl radical. Propagation proceeds as shown in (c) and (d). Thus far only ethylene has been produced and it is indeed the major product of ethane cracking. The termination reaction (e) produces *n*-butane by coupling, and this undergoes cracking to yield small amounts of olefins and propylene. Also, if ethylene reacts with an ethyl radical, C_4 olefins can be produced by disproportionation (g) and if it reacts with a methyl radical (h) propylene may result. Note that a corresponding amount of alkane is produced, which can undergo further cracking.

When propane is cracked Figure 2.5 (i–l), much more propylene is produced. This occurs because propane can be attacked by a hydrogen atom to yield an isopropyl radical (j), which can only convert to propylene. The initial hydrogen atom can come from propane scission (i), which produces a propyl radical. This is in fact a dominant reaction and explains why ethylene is always the major product as in (l). With higher molecular weight feeds, the long chains fragment readily (m), and the preferred reaction of the resulting free radicals is β-scission (n) to give ethylene and another free radical.

2.2.2 Catalytic Cracking

Catalytic cracking, as its name implies, fragments molecules over a catalyst. It also dehydrogenates some of them. Typical heavy naphtha and gas oil feedstocks give a mixture of products consisting mainly of isoalkanes suitable for gasoline together with *n*- and isoolefins, aromatics, and *n*-alkanes from C_3 upwards. Little ethylene but significant amounts of propylene and C_4 olefins are produced. Though important to the gasoline industry, catalytic cracking is not a major route to petrochemicals, except to insure an ample supply of propylene for the United States.

More propylene is produced by catalytic cracking in the United States than by stream cracking, but it occurs in a more dilute stream, and cost after separation is higher. For the preparation of isopropanol (Section 4.6) and cumene (Section 4.5), it is possible to use refinery grade propylene without upgrading. Also, there are legitimate uses for propylene from catalytic cracking in the refinery itself. The two largest are for alkylation (Section 2.2.5) and directly as fuel. In addition, it is mixed in the refinery with LPG and is also used for polymer gasoline (Section 2.2.4). For the C_4 fraction, alkylation with *n*-butenes and isobutene is a major refinery use. Another use is for direct blending of *n*-butane into gasoline, although legislation in the United States now requires that this be eliminated in the summer because volatile butane contributes to the formation of photochemical smog. A major new use for refinery isobutene and isopentene is their reaction with methanol to produce methyl *tert*-butyl and *tert*-amyl ethers (Section 5.2.1).

Catalytic cracking received impetus from World War II because military planes required high-octane gasoline. At that time, naturally occurring aluminosilicates, primarily amorphous montmorillonites, were used as catalysts. In the mid-1960s synthetic zeolites were introduced. These too are aluminosilicates but with a well-defined crystalline structure. Y zeolites, such as a so-called faujasite with a ratio of silicon to aluminum of 1.5:3.0 and a composition of $Na_{56}(AlO_2)_{56}(SiO_2)_{136}$, are used (Section 16.9). Synthetic zeolites are characterized by the presence of microscopic pores of consistent diameter, and chemical reactions take place inside these pores. The sites within the zeolite that create reactive species include Brønsted sites, Lewis sites and even superacid sites. It is these that bring about the formation of carbocations, the intermediates that in catalytic cracking lead to the formation of the branched-chain compounds, prized because they have a high-octane number.

The Y zeolites are modified with rare earths by an ion exchange process. These impart high thermal stability and additional acidity for promotion of chemical reactions. In the early 1990s, catalytic cracking catalysts were announced, which increase the isobutene content of the product so that more will be available for methyl *tert*-butyl ether production.

Catalytic cracking is an excellent example of the important contribution catalysts can make to the improvement of selectivity and yield. In the cracking of gas oil, a conversion of 73% can be obtained with a synthetic zeolite as compared to 64% with a natural silica–alumina.

As indicated in Section 2.1, vacuum gas oil is a major feed for catalytic cracking as is atmospheric gas oil. Virtually any petroleum fraction can be cracked, but vacuum gas oil, derived from residual petroleum fractions, has few other uses. Catalytic cracking takes place at a temperature of about 500°C and at atmospheric pressure or slightly above. The reaction is carried out in a fluidized bed reactor in which the finely divided catalyst is maintained in a fluid state by a stream of steam. Residence time is of the order 1–3 s. The products must be rapidly quenched to prevent their decomposition. Coke deposits on the catalyst reduce its activity. In its fluidized state, the catalyst can be removed continuously from the reactor and is sent to another reactor at 630°C, where the coke is burned off in a stream of air. The regenerated catalyst is then recycled to the fluidized bed. A typical product mix from catalytic cracking is shown in Table 2.8.

TABLE 2.8   Product Mix from Fluidized Bed Catalytic Cracking of Gas Oil

Compound	Approximate Wt.%
Propylene	3.2
n-Butenes and isobutene (no butadiene)	5.8
Isobutane	3.3
Branched C_5+ compounds (gasoline or catalytic cracker naphtha)	36.5
$C_{10}+$ compounds (gas oil)	15.0
Residue and coke	29.0
Ethylene, ethane, propane, *n*-butane, H_2, H_2S	7.2

The search for improved catalysts for catalytic cracking is a vigorous one. The objectives are obvious—to improve yield of aromatics and branched chain compounds and to decrease the yield of olefins (except for isobutene), coke and residue. Zeolite-like materials such as aluminophosphates are reported to offer certain advantages as are zirconium phosphates. These are not true zeolites because they have layered rather than crystalline structures, the sites for catalytic reaction being the space between the layers.

Catalytic cracking is an ionic rather than a free radical reaction but otherwise the mechanism is formally similar to that in Figure 2.5. Why then do the products differ widely? The reason is that, under cracking conditions, free radicals rearrange and fragment at roughly comparable rates whereas carbocations rearrange much more rapidly than they fragment and hence rearrangement to branched chain compounds is the dominant process. This relates to the relative stabilities of free radicals and carbocations. Tertiary free radicals are more stable than secondary are more stable than primary, but the differences are quite small. In contrast, the differences between tertiary, secondary, and primary carbocations are very large, and such ions have a much greater tendency to rearrange. In the later stages of cracking, furthermore, fragmentation with the expulsion of a small free radical is energetically unfavorable but not markedly so, whereas expulsion of a small unstable ion is energetically highly unfavorable. Thus chain transfer to give alkanes is a much more favorable reaction for small ions than fragmentation. Catalytic cracking thus gives only alkanes with at least three carbon atoms.

2.2.3 Catalytic Reforming

Catalytic reforming, a process first commercialized in 1950 by Universal Oil Products, converts aliphatic or cycloaliphatic compounds to aromatics. It is carried out with naphtha (Fig. 2.1), the original intention being to raise the octane number of the gasoline fraction. It has also become the major source of aromatics—BTX—in the United States.

Catalytic reforming comprises three basic reactions: dehydrogenation, isomerization, and hydrogenolysis. Dehydrogenation is exemplified by the dehydrocyclization of a paraffin as shown in Figure 2.6 and by the dehydrogenation of an alicyclic compound to an aromatic one. Isomerization is demonstrated by

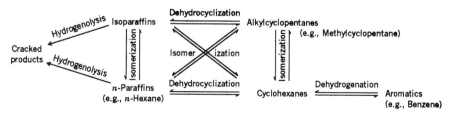

Figure 2.6 Catalytic reforming.

the molecular rearrangement of methylcyclopentane to cyclohexane. Hydrogenolysis is a hydrocracking reaction, which takes place during catalytic reforming, in which heavy molecules are cracked into lighter ones such as propane, isobutane, and n-butane.

Most important are the reactions of dehydrogenation and isomerization. For these a dual function catalyst is required, in this instance platinum or rhenium supported on alumina with a chloride adjuvant that helps to achieve and maintain uniform metal dispersion. The catalyst effects two reactions. In the first, compounds are isomerized or dehydrocyclized (e.g. n-hexane to cyclohexane). In the second the isomerized products are dehydrogenated (e.g. cyclohexane to benzene and methylcyclohexane to toluene). Thus the dual function catalyst precludes an equilibrium state by causing the product of an initial reaction (isomerization) to undergo a second reaction (aromatization) as soon as it is formed. The isomerization step is rate determining. Because it is slower than the aromatization step, there is an opportunity for the undesirable hydrogenolysis or hydrocracking reaction to occur. This can be reduced but not eliminated by lower operating pressures. The reaction is carried out at 400–500°C and 25–35 bar. Gas and hydrogen yield is about 15%, the remaining 85% comprising aromatics and unreacted feed.

Catalytic reforming provides toluene and xylenes for unleaded gasoline. These are preferred to benzene because of their higher octane number and lower toxicity. Accordingly, higher molecular weight naphthas are reformed for unleaded gasoline and a surplus of light naphtha, which yields benzene, started to develop in the mid-1980s. This fraction may be isomerized to branched-chain compounds, which have higher octane numbers and are less polluting than the aromatics. It may also be steam-cracked to produce olefins. It cannot be reformed efficiently to benzene because yield is low owing to the facile formation of cracked products.

A typical catalytic reformate contains BTX in the ratios shown in Table 2.9. Toluene is the major component. The chemical industry, however, requires more benzene than toluene, so there is a mismatch. Little can be done to change the reforming yields since they depend on the composition of the feed. The problem is solved instead by dehydroalkylating toluene to benzene (Section 8.1).

TABLE 2.9 Catalytic Reformate Production Ratio versus US Chemicals Demand

Chemical	Reformate Production Ratio (%)	Chemicals Demand (%)
Benzene	11	56
Toluene	55	10
Xylenes	34	34

The separation of the components of catalytic reformate is complicated. The gas stream from the reformer is cooled, and products containing five or more carbon atoms condense. Hydrogen and the C_1–C_4 alkanes are taken as a top product. The hydrogen is usually used to dealkylate toluene (Section 8.1), and the alkanes are burned as fuel. The liquid product is treated with a solvent that preferentially dissolves aromatic compounds. Diethylene glycol/water, N-methylpyrrolidone/ethylene glycol, and sulfolane (Section 5.1.3.3) are used in various processes. Sulfolane appears to be preferred because it produces an aromatic stream with less than 1% nonaromatics. The aromatics appear in the extract, and the C_5 compounds are left in the raffinate. The solvent is distilled off and recycled, and the benzene, toluene, and mixed xylenes are separated on three fractional distillation columns, leaving a high boiling C_9+ aromatics fraction.

Separation of the mixed xylenes is difficult because their boiling points are so close. The separation process is described in Section 9. As with benzene and toluene, the isomer distribution as shown in Table 2.10 does not match market requirements. Whereas 56% of p-xylene is required, nature provides about 19%. Similarly 4% of m-xylene is needed, whereas the C_8 fraction contains about 43%. The imbalance is corrected by isomerizing unwanted o- and m-xylenes to an equilibrium mixture (Section 9), which must again be separated. A zeolite catalyst ZSM-5 is most often used. It converts ethylbenzene to benzene, although other catalysts isomerize it to the xylene isomers.

An interesting technology developed by Chevron is in situ hydrodealkylation during catalytic reforming, the key being a zeolite catalyst such as ZSM-5 doped with platinum. The reaction is carried out at 540°C, the temperature necessary for the hydrodealkylation of both the toluene and the xylenes. Benzene and methane are the major products. The process was later modified to make benzene and toluene the major products. The zeolite decreases hydrogenolysis to lower alkanes and may even facilitate the conversion of those alkanes, once formed, to aromatics.

TABLE 2.10 Composition (weight %) of C_8 Fractions from Catalytic Reforming and Pyrolysis Gasoline Compared with End-Use Requirements

Component	Catalytic Reforming (%)	Pyrolysis Gasoline (%)	End-Use Requirement (%)
Ethylbenzene	17–22	43–57	
p-xylene	16–20	10–12	56
m-xylene	40–45	23–26	4
o-xylene	17–22	16–19	16
Mixed xylenes (for solvents)			24

The mechanism of benzene formation in catalytic reforming differs from that of toluene and xylene. The latter are obtained by dehydrogenation of alkylcyclohexanes, which in turn result from the dehydrogenation of the corresponding alkanes. Some benzene results similarly from the dehydrogenation of cyclohexane, but also catalytic reforming involves the dehydroisomerization of cyclopentanes, which combines dehydrogenation and carbenium ion reactions. A cyclopentyl cation forms, which undergoes ring opening and the subsequent formation of the cracked products shown in Figure 2.6.

2.2.4   Oligomerization

In the refinery, the oligomerization of olefins is termed polymerization because the term oligomer had not been coined when the process was invented. Low molecular weight olefins produced either by steam or catalytic cracking, but usually by the latter, are dimerized and trimerized to branched chain olefins, suitable for gasoline. They may be used as such or, preferably, hydrogenated. These oligomers are called polymer gasoline or polygas. Their popularity waned with the advent of alkylation (Section 2.2.5). The removal of lead from gasoline has prompted a renewed interest in polygas.

Oligomerization is carried out with propylene, 1- and 2-butenes, and isobutene with a Friedel–Crafts catalyst such as phosphoric, sulfuric or hydrofluoric acids. In the Friedel–Crafts reaction, a low-energy content intermediate species forms readily. Because of its stability, it does not polymerize. Ethylene does not oligomerize in this way because its intermediate species, formed only with difficulty and requiring initiators or catalysts, is highly energetic and polymerizes readily. An advantage of the oligomerization process is that dilute olefinic streams (catalytic cracker off-gases) from the refinery can be used.

The reaction proceeds by a classical carbenium ion mechanism. In the presence of an acid catalyst (HB where B^- is the conjugate base) the olefin takes on a proton from the catalyst to yield the carbenium ion.

$$CH_3CH{=}CH_2 + HB \rightarrow CH_3CH^+CH_3 + B^-$$

This may add to another molecule of olefin in a propagation reaction to yield a new carbenium ion:

$$CH_3CH^+CH_3 + CH_3CH{=}CH_2 \rightarrow (CH_3)_2CHCH_2CH^+CH_3$$

which can be stabilized or terminated by loss of a proton in one of two ways to form a dimer.

$$(CH_3)_2CHCH_2\overset{+}{C}H\text{—}CH_2 \underset{HB}{\overset{B^-}{\rightleftharpoons}} (CH_3)_2CHCH_2CH\text{=}CH_2$$
$$\quad\quad\quad\quad\quad\quad\quad |$$
$$\quad\quad\quad\quad\quad\quad\quad H$$
4-Methyl-1-pentene

$$(CH_3)_2CHCH\text{—}\overset{+}{C}HCH_3 \underset{HB}{\overset{B^-}{\rightleftharpoons}} (CH_3)_2CHCH\text{=}CHCH_3$$
$$\quad\quad\quad\quad |$$
$$\quad\quad\quad\quad H$$
4-Methyl-2-pentene

Repetition of the propagation reaction with another molecule of propylene yields isomeric trimers, and the trimeric carbenium ion may add a fourth molecule of propylene to give a mixture of tetramers. Even pentamers and hexamers may form.

Trimerization

$$CH_3CHCH_2\overset{+}{C}H \quad + \quad CH_3CH\text{=}CH_2 \longrightarrow CH_3CHCH_2CHCH_2\overset{+}{C}HCH_3$$
$$\quad |\quad\quad |$$
$$\quad CH_3 \quad CH_3$$
Dimer

$$\quad\quad\quad\quad\quad\quad\quad\quad\quad CH_3CHCH_2CHCH_2\overset{+}{C}HCH_3$$
$$\quad\quad\quad\quad\quad\quad\quad\quad\quad\quad |\quad\quad\quad |$$
$$\quad\quad\quad\quad\quad\quad\quad\quad\quad CH_3 \quad CH_3$$
I

proton loss →
$$CH_3CHCH_2CHCH_2CH\text{=}CH_2$$
$$\quad\quad |\quad\quad\quad |$$
$$\quad\quad CH_3 \quad CH_3$$
Trimer

I

proton loss →
$$CH_3CHCH_2CHCH\text{=}CH\text{—}CH_3$$
$$\quad\quad |\quad\quad\quad |$$
$$\quad\quad CH_3 \quad CH_3$$
Trimer

Tetramerization

$$\text{I} \quad + \quad CH_3CH\text{=}CH_2 \longrightarrow CH_3CHCH_2CHCH_2CHCH_2\overset{+}{C}H\text{—}CH_3$$
$$\quad\quad\quad\quad\quad\quad\quad\quad\quad |\quad\quad\quad |\quad\quad\quad |$$
$$\quad\quad\quad\quad\quad\quad\quad\quad\quad CH_3 \quad CH_3 \quad CH_3$$
II

proton loss →
$$CH_3CHCH_2CHCH_2CHCH_2CH\text{=}CH_2$$
$$\quad\quad |\quad\quad\quad |\quad\quad\quad |$$
$$\quad\quad CH_3 \quad CH_3 \quad CH_3$$
Tetramer

II

proton loss →
$$CH_3CHCH_2CHCH_2CHCH\text{—}C\text{—}CH_3$$
$$\quad\quad |\quad\quad\quad |\quad\quad\quad |\quad\quad\quad |$$
$$\quad\quad CH_3 \quad CH_3 \quad CH_3 \quad H$$
Tetramer

Because the intermediate species are relatively unreactive, polymers cannot be made in this way. Propylene could not be polymerized until the advent of Ziegler–Natta catalysis (Section 4.1).

The branched C_9 and C_{12} olefins have chemical uses that are discussed later (Section 4.2). Ziegler-type catalysts may be used for the oligomerization in processes such as "Dimersol" (Section 4.2).

2.2.5 Alkylation

Polymerization involving two or three molecules of olefins, was replaced by the more sophisticated alkylation reaction between 1 mol each of an isoparaffin and an olefin. An olefin such as isobutene is used to alkylate a branched-chain hydrocarbon, usually isobutane, in the presence of Friedel–Crafts catalysts such as sulfuric acid or hydrogen fluoride:

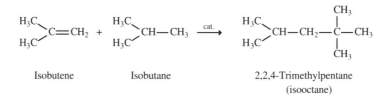

| Isobutene | Isobutane | 2,2,4-Trimethylpentane (isooctane) |

2,3,4- and 2,3,3-trimethylpentanes are also formed. The interaction of propylene with isobutane yields a complex mixture including 38% 2,3-dimethylpentane and 19% 2,4-dimethylpentane. In addition, 25% propane results. The remainder is a mixture of branched chain hydrocarbons with eight carbon atoms. Since the carbenium ion intermediate attacks tertiary carbon atoms, the products are highly branched and have high-octane numbers.

The octane number of a fuel is defined in terms of its knocking characteristics relative to *n*-heptane and isooctane (2,2,4-trimethylpentane), which have been arbitrarily assigned octane numbers of 0 and 100, respectively. The octane number of an unknown fuel is the volume percent of isooctane in a blend with *n*-heptane that has the same knocking characteristics as the unknown fuel in a standard engine. Three kinds of octane numbers are recognized. The first is Research Octane Number (RON), which is the octane number measured under relatively mild test conditions. The Motor Octane number (MON) is determined at higher engine speeds and temperature. A third octane number, RON 100°C, is the RON of the fraction of the gasoline distilling below 100°C. This fraction plays a special role in controlling antiknock performance under conditions of acceleration.

The olefins for alkylation, as for oligomerization, come primarily from the off-gases from catalytic cracking. The isobutane is produced during catalytic cracking as well as in catalytic reforming. It also occurs in the gas fraction from petroleum distillation, and on-purpose isobutane is made by the isomerization of *n*-butane (Section 2.2.8). Hydrogen fluoride is the preferred catalyst in modern refineries because the process can operate in the liquid phase at temperatures in the range of 50°C. With sulfuric acid lower temperatures of 0–10°C are required to prevent oxidation of the olefin. Solid acid catalysts, with their nonpolluting features, are being vigorously explored.

2.2.6 Hydrotreating and Coking

Hydrotreating is a term applied to several refinery processes intended to remove impurities from petroleum and/or to reduce the viscosity of very viscous oils or "bottom of the barrel" fractions. Hydrotreating will become increasingly important when heavier more impurity-laden oils enter the refinery, as they are certain to do since the world's reserves contain a high proportion of heavy oil. Light crude oils have a higher hydrogen-to-carbon ratio than heavy oils. Hydrotreating aims to add hydrogen to heavy oils so that they can be handled in a refinery like the lighter, more desirable materials. Hydrocracking is today's best established technology for adding hydrogen to heavy fuels, particularly heavy gas oils. It can also be applied to the so-called resid fraction, the residue left when practically everything else in the petroleum barrel is removed.

Hydrocracking is a variant of hydrotreating and also of catalytic cracking. A different catalyst is used, and the cracking reactions take place in an environment of hydrogen at 60–100-bar pressure. Because hydrogen is present the catalyst does not "coke" as it does in catalytic cracking; a wider range of feedstocks can be tolerated (e.g. heavy distillate can be used); and objectionable sulfur, nitrogen, and oxygen compounds are converted to hydrogen sulfide, ammonia, and water. The products are paraffins, not olefins, and are fairly low in aromatics and very low in sulfur. They are used for low-sulfur jet fuels and diesel fuels where lack of aromatics and absence of sulfur are desirable.

In addition to hydrocracking, there are several other forms of hydrotreating. The mildest is used to remove sulfur as hydrogen sulfide from gas oil and naphtha with a catalyst comprising cobalt–molybdenum or nickel–molybdenum on alumina at about 300°C and 20–25 bar. Hydrotreating of vacuum gas oil (Section 2.1) is carried out for the same reason but requires a temperature of 350–400°C and a pressure of 35–55 bar. If the feed material is resid, the required temperature may be 450°C and the pressure may vary from 60–120 bar. Hydrotreating is also used for the refining of lubricating oils, primarily to remove waxes by cracking. This is competitive with extraction processes.

Several newer hydrotreating processes are under development. One, known as Dynacracking, developed by the Hydrocarbon Research Institute, combines hydrocracking with the conversion of a portion of the feed to synthesis gas, that is, carbon monoxide and hydrogen (Section 10.4). In this way hydrogen is provided for the process. The so-called Aurabon process, developed by UOP, makes use of metals (e.g., vanadium) and sulfur present in the residual oil as catalysts. It uses less hydrogen than standard hydrocracking and removes metals efficiently. This could be a significant advantage in that a deterrent to the use of hydrotreating is the high cost of hydrogen. It is produced in most refineries but in dilute streams and recovery costs are high. New membrane processes for removing hydrogen from gaseous streams may provide a solution to this problem (see note). The Clean Air Act, once in force, will cause a decrease in hydrogen availability in the refinery because it requires a drastic cut in the aromatic content of gasoline. A major source of hydrogen is the catalytic reforming reaction (Section 2.2.3) used to make aromatics.

Coking is the reverse of hydrotreating, for in this process the hydrogen is rejected. Thus heavy fractions are heated to produce volatiles that are used in the refinery. The coproduct, which is impure carbon, is known as coke. If it has low sulfur and metal content, it can be used to make anodes for aluminum manufacture. Lower quality cokes can be used as fuel. Coke can also be gasified to carbon monoxide and hydrogen (Section 10.4).

Alternate processing involves the extraction of residual oil with light solvents such as propane, butane, or pentane to remove asphalts and to provide a higher grade residual oil for hydrocracking. The asphalts can be used in paving, roofing, as fuels, and for conversion to synthesis gas.

2.2.7 Dehydrogenation

If cracking and reforming are considered dehydrogenations, then dehydrogenation is the most important reaction carried out industrially. In refinery practice, however, the term is restricted to specialized dehydrogenations such as conversion of ethylbenzene to styrene (Section 3.9), or butenes to butadiene (Section 5).

Butenes are dehydrogenated to butadiene, particularly in the United States, where ethane and propane were steam-cracked to produce primarily ethylene and propylene and lesser amounts of the C_4 olefins (Section 2.1 and 2.3.2). Most widely used was Air Products' Houdry process. The dehydrogenation process, however, must compete with imported butadiene, which is often more economical.

Propane dehydrogenation processes have been available for several decades but have been of little interest because of the large amount of propylene produced by catalytic cracking (Section 2.2.3) in the United States. The possibility of a propylene shortage in the late 1980s (Section 2.3.2) has motivated a reexamination of propane dehydrogenation. The Houdry process was chosen for a plant in Belgium, which came on stream in 1992. Universal Oil Products' (UOP) Pacol process was developed for dehydrogenating detergent-range paraffins (from the wax removed from lubricating oil) to internal olefins. This same technology can be used to dehydrogenate propane, and a plant in Thailand went on stream in 1991 after some difficulty. The catalyst in the Air Products process is a chromia-impregnated alumina. The reaction takes place at about 600°C at subatmospheric pressure. Conversion of propane is 65% with an 83% selectivity to the desired product. The UOP catalyst is platinum metal. Still another process developed by Phillips Petroleum makes use of a supported platinum and tin catalyst with promoters at a temperature of about 1000°C and pressures varying from 1–20 bar.

The newest use for dehydrogenation is to convert isobutane to isobutene for further conversion to methyl *tert*-butyl ether (Section 5.2.1).

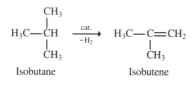

Isobutane   Isobutene

The dehydrogenation goes readily because a labile tertiary hydrogen is involved. Without this reaction, it would be virtually impossible to produce enough methyl *tert*-butyl ether for the unleaded gasoline the United States needs.

2.2.8 Isomerization

Isomerization is used to convert straight- to branched-chain compounds. The term may be applied to cracking and reforming reactions but usually these are excluded. In petroleum refining *n*-butane is isomerized to isobutane for alkylation to supplement isobutane from catalytic cracking and petroleum distillation. Isomerization is also used to increase the octane rating of C_5/C_6 alkanes and to convert a mixture of *m*- and *o*-xylenes and ethylbenzene from which the *p*-xylene has been extracted into a random mixture from which more *p*-xylene can be extracted (Section 2.2.3).

The isomerization of C_5/C_6 alkanes for the gasoline pool is of interest. These are available since light naphtha is now used to a lesser extent in catalytic reforming because of the desire to eliminate benzene from gasoline. Although the C_5/C_6 fraction may be isomerized as such, it has been usual to remove the C_5 fraction by distillation. The *n*-pentane is isomerized to isopentane over a fixed bed platinum catalyst in a hydrogen atmosphere, although hydrogen consumption is low.

The isomerization of *n*-butane to isobutane is effected by an aluminum chloride catalyst in either the liquid or gas phase at temperatures of around 100°C. Isobutane can be readily dehydrogenated to isobutene and the importance of this reaction has already been described (Section 2.2.7).

2.2.9 Metathesis

Metathesis, said to be one of the most important organic chemical reactions discovered since World War II, originated as a refinery reaction at Phillips Petroleum Company in 1964. The objective was to convert cheap propylene into more valuable ethylene and butene. In addition, since the reaction is reversible, it made it possible to interchange between major olefins:

$$2CH_2{=}CHCH_3 \rightleftharpoons CH_2{=}CH_2 + CH_3CH{=}CHCH_3$$

Propylene   Ethylene   2-Butene

Many heterogeneous catalysts are effective at about 500°C, the most important being oxides of tungsten and rhenium, and mixtures of cobalt and molybdenum on silica. Subsequently, it was found that homogeneous catalysts could be used, comprising complexes of molybdenum or tungsten with alkyl aluminum halides. The homogeneous reaction goes at room temperature as opposed to the very high temperature required by the heterogeneous catalyst, demonstrating dramatically an advantage of homogeneous reactions. However, it is apparently not used industrially.

Metathesis reactions can be carried out with most olefins except those with conjugated double bonds or with functional groups near to the double bond. An important application is in the SHOP process (Section 3.2.4).

The original metathesis reaction was used only in Canada by Shawinigan, the process having been instituted commercially only six and a half years after its discovery. This plant, however, is no longer operating because of change in feedstock availability.

Because the metathesis reaction is reversible, its versatility is increased. The potential propylene shortage (Section 2.3.2) has motivated a reaction with the same catalysts in which 2-butene and ethylene are combined to yield propylene.

The shift of equilibrium to the left is achieved by recycling ethylene to give a high ethylene/butene ratio at the reactor inlet. In this way it is possible to obtain nearly 100% conversion of butene. The reaction can be carried out with 1-butene as well as *cis* or *trans* 2-butene provided that another catalyst is added to promote the shift of the double bond from the less thermodynamically stable 1 position to the more stable 2 position.

In 1984, LYONDELL announced plans to build a unit to produce propylene by this reaction, which had been developed by ARCO. Thus a major use for metathesis developed that was opposite from the use that motivated the reaction's invention. The process is also interesting because the 2-butene may result from the dimerization of ethylene, a process also pioneered by Phillips. Nickel catalysts have been widely studied for this dimerization, and one described in a Phillips' patent comprises nickel oxide supported on silica and alumina. A more sophisticated catalyst later described by Phillips is tri-*n*-butylphosphine nickel dichloride mixed with ethyl aluminum dichloride. With this catalyst the reaction goes at 33°C and 12 bar to give an ethylene conversion of 93%. 2-Butene is also available from the C_4 fraction from steam or catalytic cracking.

2.2.9.1 *Metathesis Outside the Refinery* Metathesis has been used for the synthesis of various chemicals outside the refinery. "Dimer acid" is described in Section 13.3, but the other syntheses are described here because the original metathesis process was intended for refinery use.

When applied to cyclic compounds, metathesis yields polymers. Thus cyclopentene and cyclooctene undergo metathesis to yield specialty polymers. The one from cyclopentene is called a pentenomer.

$$n \; \bigpentagon \quad \xrightarrow{\text{cat.}} \quad -\!\!\left[CH_2CH_2CH_2CH\!=\!CH_2\right]_n$$

Cyclopentene   Pentenomer

If there are nonconjugated double bonds in a cyclic structure, a crosslinked polymer forms. Hercules took advantage of this in the development of a reaction injection molding (RIM) compound comprising dicyclopentadiene and a metathesis catalyst which polymerized it to a cross-linked structure (Section 6.2). They have subsequently sold the business.

Ethylene will undergo metathesis with a mixture of diisobutene isomers. The isomer that undergoes metathesis is 2,4,4-trimethyl-2-pentene, whose internal double bond reacts most readily. A catalyst is added to isomerize other diisobutenes with terminal double bonds to the desired pentene. 2,2-Dimethyl-1-butene results together with isobutene. The branched-chain product, known also as neohexene, is a precursor for the preparation of a synthetic musk fragrance.

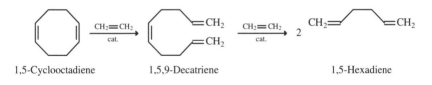

Shell has announced the development of a metathesis route to α,ω-dienes. One such compound, 1,5-hexadiene, results from the reaction of ethylene with 1,5-cyclooctadiene, which in turn comes from the dimerization of butadiene (Section 5.1.3.2).

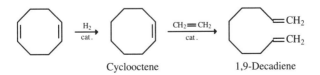

A related product is 1,9-decadiene made by reacting ethylene with cyclooctene, which is made by selective hydrogenation of 1,5-cyclooctadiene.

An isoprene precursor results from the metathesis of isobutene and 2-butene (Section 6.1).

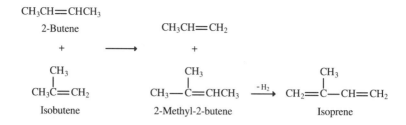

2.2.9.2 *Mechanism of Metathesis*

The mechanism of metathesis is intriguing because it differs from other catalytic transformations of olefins. It can be best explained if one assumes a homogeneous catalyst such as tungsten hexachloride in combination with a metal alkyl such as methyl lithium. As in all such reactions, a metal complex **III** forms first.

$$WCl_6 + CH_3Li \rightarrow CH_3WCl_5 + LiCl$$
$$\textbf{III}$$

This complex reacts with more methyl lithium to form a compound with a carbon–tungsten double bond **IV**. This is a true double bond but **IV** may also be considered as an encumbered carbene, thus reacting like a stabilized carbene.

$$CH_3WCl_5 + CH_3Li \rightarrow CH_2{=}WCl_4 + CH_4 + LiCl$$
$$\textbf{IV}$$

This catalyst brings about the rearrangement of Compound **V**, an olefin with four different R groups by the following propagation steps. The reaction of **IV** and **V** provides **VI**, which in turn dissociates to give **VII** and **VIII**.

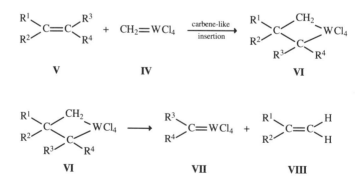

Compound **VII**, which contains alkyl groups R^3 and R^4, reacts with more of the initial olefin to give **IX**. This compound in turn dissociates to give **X** and **XI**. Compound **X** is considered one of the major products of the metathesis since it contains only two of the four groups present in Compound **V**. The other two are resident in the catalyst complex **XI**. Accordingly, **XI** reacts with the initial olefin **V** to give **XII**, which can dissociate just as **IX** did to give **XIII** and **XIV**. **XIII**, counterpart of **X**, contains the other two groups present in the initial olefin **V**. The catalyst complex **XIV** can then continue the reaction just as the complex **XI** did.

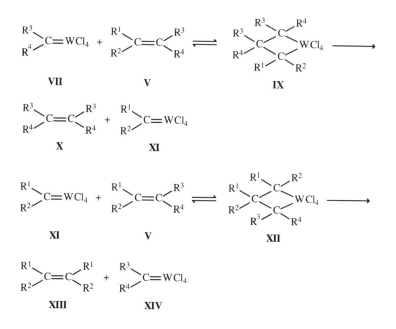

The dissociation reactions are mechanistically the inverse of the carbene-like insertion reactions, and these reactions are also reversible, but such reversions have no observable consequences.

Termination steps are obscure and presumably are brought about by impurities and possibly by dimerization of the catalyst.

2.3   THE REFINERY—A PERSPECTIVE

The petroleum refinery provides feedstocks for the petrochemical industry throughout the world. It underwent basic changes in the 1980s, which it is important for a student of industrial organic chemistry to understand. These changes result from three factors, which will be discussed below. These changes are:

- The prediction of a petroleum and natural gas shortage in the 1970s and the reversal of that prediction in the 1980s when crude oil prices, which rose precipitously in the 1980s, collapsed.

- The entry of new players into the petroleum refining and petrochemical arena, most important of which are Western Canada and Saudi Arabia, but include many countries in the Middle East and Southeast Asia.
- The advent of unleaded gasoline and related environmental pressures.

2.3.1 The Function of the Refinery and the Potential Petroleum Shortage

A major objective of the United States refinery has traditionally been to produce high-octane gasoline. To be sure, the fractions heavier than naphtha (Fig. 2.1) were important for aviation, diesel, boiler and heating fuels, lubricants, and paving materials, but production was driven by the demand for gasoline. Because of need for high octane number, catalytic cracking was developed and always played a far more important role in American refineries than it did in refineries in Europe and Japan that were geared to production of fuel oil, since much less gasoline was needed and cheap coal was not available (Section 2).

The refinery reactions (Sections 2.2.3–2.2.5) of oligomerization, alkylation, and catalytic reforming were all devised to raise octane number, as were lead tetraalkyls and other octane improvers. The many catalytic crackers in the United States produced large quantities of C_3 and C_4 olefins, and it was these that provided feed for the oligomerization and alkylation reactions. They also proved useful in chemical reactions (Sections 2.2.2, 4.5 and 4.6) and in 1991 the refinery provided 48% of the chemical industry's propylene in the United States.

In Europe, the situation is reversed. Less refinery propylene is produced in catalytic crackers (because there are fewer of them) and more is produced from steam cracking (because much more naphtha is cracked). Catalytic cracking in 1991 provided only 11% of the European chemical industry's propylene. Because of Europe's propylene shortage, an effort will be made to obtain more propylene from the refinery in the 1990s, although at best this will amount to only 2–3 billion lbs.

The rise in gasoline prices in the 1970s motivated the development of smaller, lighter cars with lower gasoline consumption. Also diesel fuel became popular. With lower gasoline production, there was lower production of the middle distillate (gas oil fraction, Fig. 2.2) from which diesel fuel comes. Thus the cost of diesel fuel also increased and, in Europe and Japan, there were threatened shortages.

The greater need for diesel fuel motivated the development of a Shell process intended to produce on-purpose diesel fuel (Section 12.2). A Mobil process oligomerized olefins, primarily propylene and butenes, over a zeolite ZSM-5 catalyst, to give a mixture of oligomers, the lower ones constituting gasoline and the higher ones middle distillate. In the early 1990s there were no plans to commercialize this process, but Shell has a Fischer–Tropsch plant in Malaysia to make diesel.

2.3.2 Unleaded Gasoline and the Clean Air Act

The octane number of straight run gasoline has traditionally been raised by catalytic cracking and by addition of lead tetraalkyls plus ethylene dibromide as a lead scavenger (Section 3.11.9). The emission of lead bromide in exhaust gases has since been seen as a toxic hazard and lead additives were banned in many countries.

The advent of unleaded gasoline has affected refinery operations in several ways. Oligomerization units (Section 2.2.4) have been reactivated to obtain branched-chain, high octane number olefins. These had been shut down in the 1960s in favor of alkylation (Section 2.2.5), which also became more important because of unleaded gasoline. Catalytic cracking catalysts were improved so that higher octane products resulted. The production of aromatics by catalytic reforming (Section 2.3) became more important as refiners tried to raise the octane number of gasoline by including higher amounts of aromatics, the most effective being toluene and xylenes. This made it less desirable to reform the low-boiling naphtha that yields benzene, which in turn made low-boiling naphtha available for isomerization to branched chain, higher octane number components (Section 2.2.8) and for cracking to olefins.

Since none of these reactions can satisfy fully the octane needs of most refineries, new high-octane components for gasoline were needed. This was particularly so since the Clean Air Act demanded drastically lower aromatics content (see below). The answer was found in oxygenates, compounds that are primarily the province of the chemical industry.

The best oxygenate appears to be methyl *tert*-butyl ether made by reacting isobutene with methanol over an acid catalyst, usually an acidic ion exchange resin.

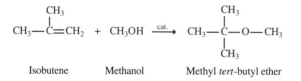

Isobutene	Methanol	Methyl *tert*-butyl ether

Isobutene is available from the refinery as well as from propylene oxide production, where it comes from the facile dehydration of the *tert*-butanol coproduct (Section 4.7). It suddenly became an important chemical, whereas in the past it had always been a cheap byproduct. A process for its production from *n*-butane is described in Section 2.2.8. Estimates indicate that 41 billion lb of methyl *tert*-butyl ether will be required in the United States by 1996 and achievement of this output means that it will be produced on a similar scale to ethylene.

Fermentation ethanol has become an important product in the United States because of government subsidies and is widely used in American gasoline despite high volatility. Ethanol is also used as a motor fuel as such and in combination with hydrocarbon fuels in Brazil. There, however, the alcohol is made by the fermentation of juice from sugar cane, which has a much higher carbohydrate content that does the corn, which yields the starch used as a raw material in the United States. Also, the sugar cane juice is obtained simply by crushing the cane and is thus cheap, whereas starch isolation is a relatively expensive and energy-intensive process. Even in Brazil, however, its use is not economic, even though the energy for distilling the alcohol comes partly from burning bagasse, the sugar cane residue.

The use of fermentation alcohol represents a trend back to renewable natural products for commodity chemicals. All the same, it is difficult to think of ethanol as a renewable resource if it generates less energy as a motor fuel than was spent in its production and purification. The deficit, of course, comes from petroleum and natural gas. The possibility of retrofitting methanol plants so that they will produce mixtures of higher alcohols has also been studied but has not been commercialized. Ethyl *tert*-butyl ether is acceptable in unleaded gasoline and results from the interaction of ethanol and isobutene and so is *tert*-amyl methyl ether from methanol and isopentene.

Methanol, available cheaply from many parts of the world where natural gas is in surplus, has been studied as a neat automotive fuel and as a component in petroleum-based gasoline. Because it separates out in the presence of a small amount of moisture, gasoline containing it must include a cosolvent alcohol. *tert*-Butanol has been used with methanol in a proprietary product known as "Oxinol." It is not, however, as effective as methyl *tert*-butyl ether. Methanol as such is of particular interest for replacing diesel fuel in trucks and buses. However, by 1991 there was no methanol surplus, since it was being consumed in ever larger quantities for manufacture of methyl *tert*-butyl ether.

The final alcohol that has been used in gasoline is isopropanol, which is also an acceptable oxygenate and can be made from refinery propylene. It was marketed in the United Kingdom after World War II under the name "Cleveland Discol."

Although unleaded gasoline provided a "dislocation," the refineries in the United States accommodated to it well. In 1991 a much greater dislocation was provided by the Clean Air Act. Now it was necessary to reduce benzene content of gasoline from about 3 to 1%. In nine areas in the United States where pollution exceeded levels specified by the Environmental Protection Agency (EPA), aromatics (i.e., toluene and xylenes) content had to be reduced from 36 to 25% by 1995. Ironically, this occurred after some refineries had increased their capacity for manufacturing aromatics, which they regarded as a good way to achieve necessary octane number in unleaded gasoline.

There are at least four scenarios for achieving the 1% benzene level. Refineries in the early 1990s were already stripping naphtha to remove the low-boiling C_5–C_6 fraction. More of this can be done to minimize benzene formation in catalytic reforming. Carried to the extreme, a benzene shortage for the chemical industry might result. Even so some benzene will always form in catalytic reforming because of hydrodealkylation (Section 8.1) of toluene and xylenes.

UOP has proposed that the reformate be reacted with propylene in situ to form cumene (Section 4.5), itself an octane improver. A third possibility is to remove the benzene by distillation. This would add about 330 million lb of benzene to the current supply, which would not be disruptive.

A fourth scenario is for the refinery to produce a reformate with as little benzene as possible for gasoline, and another reformate containing benzene for the chemical industry. Although this seems logical, it might be disruptive for the refinery, which is programmed to carry out continuous reactions without interruption. Nonetheless, it is the approach being used in the mid-1990s.

An even greater problem is how to compensate for the octane number lost by such a drastic reduction of aromatics content. The Clean Air Act addresses this by specifying oxygenated and reformulated gasolines containing compounds that contribute 2.7 and 2% oxygen respectively. Methyl *tert*-butyl ether content, to contribute this amount of oxygen, must be 17 and 12.5%. There is not enough isobutene in the world from steam and catalytic cracking and from *tert*-butanol dehydration to make the amount required, hence the importance of the isomerization of *n*-butane to isobutane (Section 2.2.8) and its dehydrogenation to isobutene. Also, at the 17% level, methyl *tert*-butyl ether will contribute more octane number than most US gasoline requires.

Still another problem is the decreased supply of hydrogen that will occur in the already hydrogen-deficient refinery when catalytic reforming is drastically reduced because aromatics content is decreased. Several companies in the early 1990s were considering on-purpose manufacture of hydrogen for refineries.

2.4 SEPARATION OF NATURAL GAS

In addition to methane, most American natural gases contain recoverable amounts of ethane, propane, and higher alkanes. These are extracted as feedstock for steam-gas crackers. At the wellhead the gas is at a high pressure (30–100 bar), and propane and higher alkanes may be absorbed in a high-boiling oil at ambient temperature and subsequently purified by low-temperature fractional distillation. Ethane may be absorbed similarly at $-50°C$. Hydrogen sulfide and carbon dioxide are scrubbed out with aqueous mono- or diethanolamine, and water removed with hygroscopic diethylene glycol.

Methane is the major component of natural gas and may also be "steam reformed" (Section 10.4) to synthesis gas. Even when most C_2+ compounds are removed, some ethane is retained in natural gas to raise its calorific content to the value demanded for heating gas.

Thus natural gas provides the chemical industry with methane primarily for conversion to synthesis gas (Chapter 10), ethane, and propane for steam cracking to olefins, and butane, which is oxidized to maleic anhydride (Section 11.1.3) and isomerized to isobutane (Section 2.2.8).

NOTES AND REFERENCES

Chemicals from natural gas and petroleum occupy much space in the general books on the chemical industry referred to in Section 0.4.2, and we shall not repeat their titles here. The world oil industry, although not a theme of this book, is of importance to anyone concerned with petrochemcials. An old book on the chemistry of petroleum, which has never really been superseded, is B. T. Brooks, C. T. Boord, S. S. Kurtz, and L. Schmerling, *The Chemistry of the Petroleum Hydrocarbons*, 3 vols., Reinhold, New York, 1955. A somewhat different treatment is given in A. A. Petrov, *Petroleum Hydrocarbons: Composition, Structure, Ways of Formation of Various Petroleum Hydrocarbons*, Springer, Berlin, 1987.

The best introduction to the oil industry is *Our Industry—Petroleum*, 5th ed. British Petroleum, London, 1970. At the time of writing, a few copies are still available at a nominal price from the BP Educational Division. Two nontechnical books for general reading are A. Sampson, *The Seven Sisters*, Hodder and Stoughton, London, 1975, which recounts the history of the seven great oil companies, and C. Tugendhat and A. Hamilton, *Oil—The Biggest Industry*, 2nd ed., Eyre Methuen, London, 1975. Those looking for a vivid and entertaining history of the oil industry should read *The Prize*, D. Yergin, Simon and Schuster, London, 1991.

Statistics on world oil production and consumption are published annually in the *BP Statistical Review of the World Oil Industry*, BP, London, a well presented publication either free or at a nominal price. Figure 2.1 in particular was derived from this source. The Shell Petroleum Company also publishes a useful *Information Handbook*, and a *Chemicals Information Handbook*.

The details of gasoline formulation are described in E. G. Hancock, *The Technology of Gasoline Production*, SCI Critical Reports on Applied Chemistry, Blackwell, Oxford, UK, 1985.

Section 2 Liquefied petroleum gas (LPG), (a propane/butane mixture) became available cheaply when Saudi Arabia decided to crack only ethane. Large amounts of a propane–butane mixture were placed on the world market and more was available than energy uses required. Thus it was predicted in the early 1990s that the United States, by feeding LPG to its crackers, will use predominantly gas feedstock well into the next decade.

Section 2.1 The catalytic reforming of low-boiling naphtha, which contains predominantly C_6 compounds, is normally accompanied by cracking and the formation of unacceptably large amounts of gases. New catalysts under development in the late 1980s, primarily large pore type L zeolites doped with various metals, promise to make possible the conversion of light naphtha to benzene for chemical use in high yields. (cf. US Patents 4 539 304–5 March 8, 1984 to Chevron).

Section 2.2.1 Steam cracking, strictly speaking, is not a refinery reaction. Because it derives feedstock from a refinery, however, steam cracking units are usually closely associated with a refinery.

Historical insight into the development of the cracking process is provided in an article in *CHEMTECH*, March 2, 1976, p. 180. The article reproduces William Burton's acceptance address for the 1922 Perkin Medal, which was awarded him for his research on cracking. Another article describing early work on cracking and the development of the petrochemical industry generally has been published by B. Achillaides, *Chem. Ind.*, April, 19, 1975, p. 337. The Burton process for cracking was first, but close on its heels was the Dubbs process. Dubbs was so involved with technology that he named his son Carbon Petroleum and his daughters, Methyl and Ethyl.

Section 2.2.1.1   For descriptions of gas oil cracking, see a series of articles by S. B. Zdonik in *Hydrocarbon Proc.*, September 1975; December 1975; April 1976. Gas oil cracking is also discussed by M. J. Offen, *Hydrocarbon Proc.*, October, 1976, p. 123.

For articles describing the cracking of whole oil see *Eur. Chem. News*, **30**, No. 785 (1977); *Chem. Week*, September 28, 1977, p. 39; December 7, 1977, p. 55.

A discussion of substitute feedstocks by the year 2000 will be found in *Chem. Eng. News*, September 26, 1977, p. 7.

Section 2.2.1.2   Figure 2.4 is based on a costing by Paul Ray of Trichem Consultants, published in "Petrochemical Review" *European Chemical News* "*Chemscope*" September 1994, p. 7, but has been extensively modified by the authors. The costing of a chemical plant is a complicated problem. There are no "correct" answers, only answers that are made under different accounting conventions, which should be stated, and which are useful for different purposes. An important difference is whether or not a costing includes depreciation. If a plant is already built, then it is worth producing so long as the revenue covers the variable costs. Cash costs include the fixed cash costs, described in the text. If the plant is to be viable in the long term, the replacement of worn-out plant must be allowed for. In depreciation costing, this is built into the calculation and what is left at the end is profit. In the cash cost calculation, what is left at the end is cash flow and plant replacement must come eventually out of that flow. Thus cash flow = retained profits plus depreciation.

Paul Ray's costing involved purely cash costs. Depreciation and corporate overheads were ignored. In the figure, we have added depreciation of plant. These costings may be compared with others that include depreciation by Steve Rothman of Chem. Systems Inc, which appeared in *Hydrocarbon Processing*, *Achema '94* Perspective, May 1994. The figures here are still lower and the difference, as far as we can see, is corporate overheads included at 50% of other fixed costs and then depreciated at 10% plus 5% return on capital. If these costs are included, the two costings appear similar. Also, Ray's "net raw materials" corresponds to Rothman's "net raw materials plus utilities." The point is that waste streams on crackers are often used to provide fuel for the cracker furnaces, and byproduct steam is used to drive the refrigeration compressor turbines, so that a byproduct credit can be set against a utilities charge.

Further complications arise because of tariffs and transport charges. For example, if South Korea exports propylene to Taiwan, the freight costs are about \$70/tonne. A costing in Taiwan will include propylene at \$70/tonne more than a similar costing in South Korea, hence ethylene costs from an identical plant will appear about \$35/tonne lower. Similarly a tariff barrier on imported coproducts will raise their value in the importing country and make home-produced ethylene appear cheaper.

There is no easy answer to these problems and the reader should treat all costings with circumspection and demand details of all costing conventions before making judgements.

There is much debate as to whether labor should count as a fixed or variable cost. In the nineteenth century, the foreman at a chemical plant could open the gates at 6 a.m. and sign on exactly the amount of casual labor that he needed for the day. Labor was thus related to the amount of output required and was truly a variable cost. In the chemical industry today, workers are generally highly trained and cannot be recruited or dismissed on a daily basis. Furthermore, a plant operating at 50%

capacity requires exactly the same number of operatives as when it works at 100% capacity. Hence, a modern trend is to treat labor as a fixed cost that is independent of output.

Much of this section draws on the APPE (Association of Petrochemicals Producers in Europe) *Activity Review 1993–1994*, Brussels: CEFIC, 1994. There was a feature on tariffs in *European Chemical News "Chemscope"* in September 1993.

The question of spot and contract prices is complex and depends on availability. Spot prices are higher than contract at times of shortage and lower at times of surplus. Overcapacity has meant that they have been generally lower in recent times although 1994 was an exception. The increase in the price of propylene in the early 1990s resulted because in the mid-1980s steam crackers were shut down to bring the supply of ethylene in line with the demand. This decreased the supply of propylene and its price increased. As indicated earlier, the large catalytic cracking facilities in the United States makes ample supplies of refinery propylene available.

It is not strictly true that Mossmoran is the only gas cracker. There is a small facility operated by Atochem at Lacq in the Pyrenees, France, that has been cracking gas from a small high sulphur deposit of natural gas for over 30 years. It is, however, of little economic significance.

Labor costs vary depending on whether they are direct costs or involve the maintenance staff, security staff, sales staff and so on. The lowest figure we have seen allows labor as 0.5% of total costs, the highest about 1.5%. The BP/ICI olefins 6 cracker at Wilton, Teesside, UK employed fewer than 100 people all pictured in *Cracker*, ICI Schools Liaison Section, Welwyn, 1980.

Section 2.2.1.3   For the mechanism of steam cracking see P. A. Wiseman, *J. Chem. Educ.*, **54**, No. 154 (1977).

Section 2.2.2   The development of zeolites for cracking catalysts is well described in an article in *CHEMTECH* by Plank (April, 1984, p. 243). A thorough review of the topic is to be found in B. W. Wojciechowski, Catalytic Cracking—Catalysis, Chemistry and Kinetics. See also D. de Croocq, *Catalytic Cracking of Heavy Petroleum Fractions*, Inst. Français de Pétrole, Editions Technip, Paris, 1984, and *Advanced Fluid Catalytic Cracking Technology*, K. C. Chuang, G. W. Young, and R. M. Benslay, Eds. AIChemE, 1992.

Section 2.2.3   An excellent description of the development of catalytic reforming has been published by M. J. Sterba and V. Haensel, *Ind. Eng. Chem., Res. Dev.*, **15**, (1976) 2.

The Chevron technology for in situ hydrodealkylation during the catalytic reforming reaction is described in US Patent 4 347 394 (August 31, 1982); US Patent 4 434 311 (February 28, 1984) and in UK Patent Application 2114150A, August 17, 1983.

Section 2.2.5   A review of octane number has been published by J. Benson, *CHEMTECH*, January 1976, p. 16.

Section 2.2.6   Use of hydrocracking on residues from petroleum distillation as well as coal-derived liquids, shale oil, and tar sands is described in an article in *Chem. Week*, February 18, 1976, p. 69.

Membrane processes for hydrogen removal are described by J. Haggin, *Chem. Eng. News.*, June 6 (1988) p. 7. The first industrial hydrogen-separating membranes were Monsanto's Prism separators, made of "skinned" polysulfone and used in the Haber process to recycle hydrogen, so that the wasteful "bleed" to get rid of argon build-up was no longer required. The "skinned" membrane is anisotropic and a thin layer of polymer of small well-defined porosity is supported on a thicker layer of more porous polymer.

Section 2.2.9   The original metathesis reaction was described by its inventors Banks and Bailey in *Ind. Eng. Chem., Prod. Res. Dev.* **3**, 170 (1964). The ARCO development has been reported in *Chemical Marketing Reporter* (November 18, 1985 p. 36) and also in *J. Molecular Catalysis* **8**, 269 (1980). The conversion of 2-butene and ethylene to propylene is also described in US Patent 3 915 897 (October 28, 1975) to Phillips Petroleum Company.

Ethylene dimerization is described in US Patent 4 242 531 (December 30, 1980) to Phillips Petroleum Company. A catalyst which gives primarily 1-butene on ethylene dimerization is discussed in US Patent 4 487 847 (December 11, 1984) to Phillips Petroleum Company.

Section 2.3.1   Interestingly the need for diesel fuel in South Africa decreased markedly because buses for transporting the predominantly black population gave way to jitneys. Whereas the buses used diesel fuel, the jitneys use gasoline.

Section 2.3.2   An article on the affect of unleaded gasoline on the chemical industry has been authored by H. Wittcoff, *J. Chem. Educ.*, **64**, (1987) 773.

The situation regarding lead in gasoline is not consistent throughout the world. At present well over 90% of all cars in the United States run on lead-free gasoline. In continental Western Europe, lead levels have been reduced to about 0.15 g/L.

CHAPTER 3

CHEMICALS AND POLYMERS FROM ETHYLENE

We have described how the petroleum refinery, the steam crackers, and the catalytic reformers provide the seven raw materials on which the petrochemical industry is based. We can now examine the extensive chemistry associated with each of these building blocks. We start with ethylene not only because it has the simplest structure but also because it is the most important in terms of tonnage.

In 1995 the United States produced 45 billion lb of ethylene. This in turn (Section 1.5) was converted to about 180 billion lb of chemicals and polymers, a figure that includes the ethylene itself. Since the total US annual production of organic chemicals and polymers was about 450 billion lb, we can gauge the importance of ethylene as a raw material. By 2000 it is predicted that US ethylene production will be over 54 billion lb. Ethylene, propylene, and benzene production over a 40-year period is shown in Figure 3.1.

About 80% of all the ethylene produced is destined to end up as thermoplastic polymers. Ethylene is itself a monomer for low, high, and linear low-density polyethylenes as well as for ethylene oligomers. It is the raw material for other important monomers including vinyl chloride, vinyl acetate, styrene, and ethylene glycol. Some of these monomers also have nonpolymer uses. Thus ethylene glycol is the main component of antifreeze.

Ethylene chemistry is the most mature in the petrochemical industry. This topic is discussed in Section 3.10. The major chemicals and polymers from ethylene are shown in Figure 3.2. The figure is deceptively simple since many of the reactions are multistep and the intermediates have uses in their own right. Even so, the chart demonstrates a basic tenet of industrial organic chemistry, namely, that most of the tonnage of the industry derives from relatively few materials. The major chemistry of ethylene can be divided into three categories: polymerization and oligomerization; four oxidation reactions practically un-

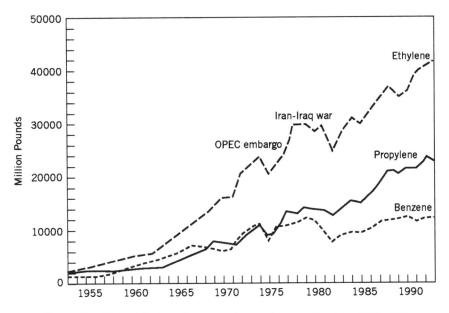

Figure 3.1 United States ethylene and propylene production, 1953–1993.

known in textbook organic chemistry; and two reactions termed "classical" because they are typical of textbook chemistry. Some of the smaller volume chemicals and polymers from ethylene are listed in Figure 3.3 with reference to where they are discussed.

The price of ethylene from 1970 to 1995 is shown in Figure 3.4. For a 14-year period, from 1960 to 1974, the price was almost constant. The OPEC embargo of 1974 caused the price to more than double. Inflation contributed to a steady increase with a precipitous rise in 1980 because of the Iran-Iraq war. Overcapacity, maturity, and weakening oil prices caused the price to decrease in 1985 and 1986. By 1988 price increased markedly, due in part to increased petrochemical demand but more important to industry restructuring, which eliminated marginal suppliers as well as effective producers who felt that capital redeployment would lead to greater profit. Figure 3.4 also shows the price of propylene, which has consistently been about 20% lower than the price of ethylene. Its price relationship to ethylene was maintained until 1986 when the price of the two products became equal. At that point propylene usage was growing faster than that of ethylene. Also, there was a slight shortage of propylene, particularly in Europe, for reasons discussed earlier (Section 2.3.2). The traditional price relationship resumed in the late 1980s and is projected to be maintained during the 1990s.

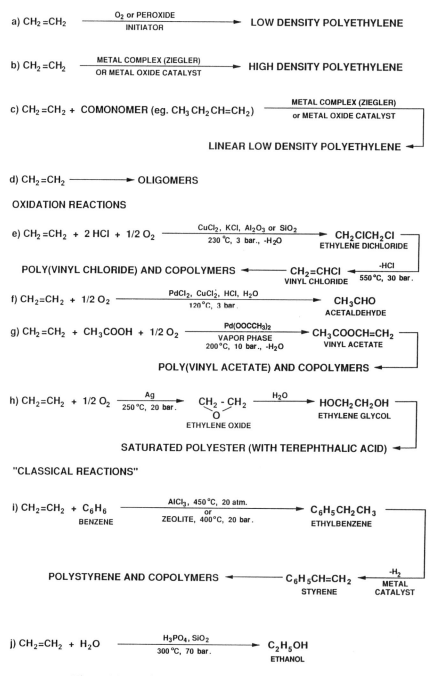

Figure 3.2   Major chemicals and polymers from ethylene.

POLYMERS

VERY HIGH MOLECULAR WEIGHT
POLYETHYLENE (Sec. 3.11.12)
VERY LOW DENSITY POLYETHYLENE (Sec. 3.2.6)
CHLOROSULFONATED POLYETHYLENE (Sec. 3.2.1)
ETHYLENE - VINYL ACETATE (Sec. 3.2.2)
ETHYLENE - VINYL ALCOHOL (Sec. 3.2.2)
ETHYLENE - ETHYL ACRYLATE (Sec. 3.2.2)
ETHYLENE - METHYL METHACRYLATE (Sec. 3.2.2)
ETHYLENE - BUTYL ACRYLATE (Sec. 3.2.2)
ETHYLENE - ACRYLIC ACID IONOMER (Sec. 3.2.3)
POLYETHYLENE - NYLON BLENDS (Sec. 3.2.4)
ETHYLENE - PROPYLENE ELASTOMERS (Sec. 3.2.5)

CHEMICALS

PROPIONALDEHYDE (Sec. 3.11.1)
PROPIONIC ACID (Sec. 3.11.1)
n-PROPANOL   (Sec. 3.11.1)

ETHYL CHLORIDE (Sec. 3.11.2)
ETHYL BROMIDE (Sec. 3.11.2)
CHEMICALS FROM ACETALDEHYDE (Sec. 3.11.3)
METAL COMPLEXES (Sec. 3.11.4)
ETHYLENE DIAMINE AND RELATED
COMPOUNDS (Sec. 3.11.5)
ETHYLAMINES (Sec. 3.11.7)
ETHYLENE IMINE (Sec. 3.11.6.5)
VINYLIDENE CHLORIDE (Sec. 3.11.7)
TRICHLOROETHYLENE (Sec. 3.11.7)
PERCHLOROETHYLENE (Sec. 3.11.7)
CHLOROACETIC ACID (see note)
ETHYLENE GLYCOL OLIGOMERS (Sec. 3.11.6.1)
GLYCOL ETHERS AND ESTERS (Sec. 3.11.6.2)
ETHYLENE CARBONATE (Sec. 3.11.6.3)
AMINOETHYL ALCOHOLS (Sec. 3.11.6.4)
VINYL FLUORIDE (Sec. 3.11.8)
VINYLIDENE FLUORIDE (Sec. 3.11.8)
ETHYLENE DIBROMIDE (Sec. 3.11.9)
VINYL ESTERS AND ETHERS (Sec. 3.11.11)

Figure 3.3 Some lesser volume chemicals and polymers from ethylene.

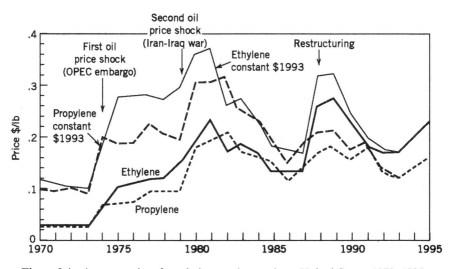

Figure 3.4 Average prices for ethylene and propylene, United States 1970–1995.

Ethylene price is manufacturers' clearing price, not posted price. Propylene price is for chemical grade. Polymer grade is about $(1993) 0.015/lb more expensive.

3.1 ETHYLENE POLYMERS

Polyethylene is cheap and is the largest volume polymer. Consumption in the United States in 1994 approximated 24 billion lb, 45% of which was accounted for by high-density polyethylene (HDPE) and the remaining 55% by low-density and linear low-density polyethylene (LDPE and LLDPE). These are the three important forms of polyethylene. Growth in the 1990s is projected at about 2% higher than that of the gross domestic product.

3.1.1 Discovery of Low- and High-Density Polyethylenes

The discovery of LDPE in 1932 was serendipitous. Fawcett and Gibson at ICI in England were actually studying chemical reactions at high pressures with a view to making synthetic rubber from ethylene and benzaldehyde. The discovery hinged on a leaking autoclave. It had been pumped up to reaction pressure and allowed to stand over the weekend prior to the carrying out of the reaction. Pressure had been lost because of a leak, and more ethylene was added to the reactor which, at this stage, contained traces of oxygen. To the surprise of the chemists, the pressure refused to rise. The reactor turned out to be full of a white powder and the benzaldehyde was intact. After many more experiments, it was realized that oxygen was serving as the initiator.

The conventional wisdom of the 1930s dictated that polyethylene could not be made, because it was not possible to put sufficient energy into the ethylene molecule to achieve the transition state. This is, however, quite possible with pressures in the range of 2000 bar and a temperature of 200°C in the presence of a free radical initiator such as oxygen or various peroxides. The polymerization is carried out in the gas phase, high pressures favoring the propagation reaction, which is in part a function of ethylene concentration. Termination, on the other hand, is independent of ethylene concentration.

The discovery of HDPE was similarly serendipitous by investigators at Standard Oil of Indiana and Phillips Petroleum seeking to oligomerize ethylene to gasoline-size molecules. Whereas LDPE finds its greatest use in the manufacture of film, the stiffer linear high-density material has its major uses in the manufacture of bottles by blow molding and in structural parts by injection molding.

The first linear polyethylene was probably made in 1950 at Standard Oil of Indiana by Zletz, where it was observed that a molybdenum oxide catalyst on silica at mild temperatures and pressures did not give oligomers, but rather high-density polyethylene. Soon thereafter, Banks at Phillips Petroleum obtained similar results with a catalyst comprising chromium oxide supported on silica or alumina. Phillips developed and licensed its process aggressively.

In 1954 Ziegler announced his discovery. He was successfully studying ethylene oligomerization reactions (Section 3.3.2) based on aluminum alkyls. A metal salt impurity in the reaction mixture led to the formation of high molecular weight linear polyethylene. From this evolved the famous Ziegler catalyst, typical of which is a combination of aluminum triethyl with titanium tetrachloride. The Ziegler process attracted great theoretical as well as practical interest, because it can be applied to propylene and to practically any unsaturated compound, unlike the metal oxide processes, which are effective, for practical purposes, only with ethylene. Chromium-based catalysts dominate the production of polyethylene in the United States, although the Ziegler process is used to the extent of 60% in Western Europe. The latest development is the use of metallocenes or single site catalysts. These are described in Section 15.3.12.

3.1.2 Low-Density Polyethylene

Low-density polyethylene may be manufactured batchwise in an autoclave or, more commonly, in a tubular reactor that makes possible continuous processing. Batch-produced product is useful for paper coating, where its highly branched structure is advantageous. The somewhat less branched continuously produced product is useful for film. About 35% of the ethylene is allowed to react in the continuous process, and the remaining 65% is recycled. This is termed 35% conversion per pass and is necessary to eliminate excess branching. With HDPE branching is not a problem and conversion may be 100%. The properties of LDPE and HDPE are listed in Table 3.1.

The branched structure of LDPE profoundly affects its properties. Because the polymer molecules cannot get as close together as they can in HDPE, the crystallinity is of the order of 55% as compared to 85–95% for HDPE. The crystalline melting point, softening point, and tensile strength of LDPE are all a function of the branched structure and are considerably lower than the corresponding values for HDPE. On the other hand, the softer LDPE shows higher elongation at break and higher impact strength than does the rigid

TABLE 3.1 Properties of Polyethylenes

	LDPE	HDPE	LLDPE
Initiator or catalyst	Oxygen or organic catalyst peroxide	Ziegler or Phillips catalyst	Ziegler or Phillips catalyst
Reaction temperature	200–300°C	As low as 60°C	As low as 60°C
Pressure (bar)	1300–2600	1–300	1–300
Structure	Branched	Linear	Linear with short branches
Approximately crystallinity	55%	85–95%	55%
Comonomer	None	None	1-Butene, 1-hexene, or 1-octene
Tensile strength (psi)	1200–2000	3000–5500	2000–2500
Tensile strength (tonnes m^{-2})	850–1400	2100–3900	1400–1800
Elongation at break (%)	500	10–1000	500
Density (g cm^{-3})	0.915–0.925	0.945–0.965	0.915–0.925

HDPE. It is also translucent rather than opaque because of its lower crystallinity. The difference in densities, which characterizes the two polymers, is of the order of 0.3–0.4 g/cm^3. The density of LDPE may be as low as 0.915 g/cm^3 and of HDPE as high as 0.965 g/cm^3.

3.1.3   High-Density Polyethylene

Most HDPE is actually a copolymer containing up to 4% of 1-butene or less commonly 1-hexene. The comonomer is required, particularly when metal oxide catalysts are used, to avoid formation of molecular weights so high that the polymer becomes intractable. The copolymer also has improved low-temperature properties.

The production of HDPE is much less energy intensive than that of LDPE. Reaction temperatures can be as low as 60°C and pressures as low as 1 bar. Nonetheless, temperatures of 130–270°C and pressures of 10–160 bar are used commercially. Conversion per pass approaches 100%.

High-density polyethylene is manufactured in solution, slurry, or fluidized bed processes. In the slurry process, the catalyst is dispersed in a solvent such as hexane, and the ethylene is polymerized batchwise in a series of reactors. The gas phase, fluidized bed process was devised by Union Carbide, BP, and others. Small HDPE particles are fluidized by gaseous ethylene and comonomer (e.g., 1-butene) at 85–105°C and 20 bar. Catalyst is continuously sprayed into the reactor. The ethylene and comonomer copolymerize around the preformed polymer particles. At the same time, the gaseous ethylene removes the heat of reaction.

The initial particles grow to an average diameter of 500 μ over a period of 3–5 h, during which time only about 2–3% of the ethylene polymerizes. The unconverted reactants are recycled. Polyethylene, once prepared, is melted, mixed with stabilizers and other additives, and extruded to form spaghetti-like rods, which are then cut into small pellets. The extrusion is an energy-intensive operation. An objective of the fluidized bed process, not achieved initially, was to obtain the polymer as a powder that could be used as such for molding and extrusion. Further development has apparently made this possible, although the value of the powder is questionable, because its low-bulk density increases shipping costs. Even so, the gas-phase process has proved to be an economical way to prepare both HDPE and LLDPE and has been licensed extensively.

3.1.4   Linear Low-Density Polyethylene

Linear low-density polyethylene is the successful result of a desire to prepare LDPE by the less energy-intensive conditions used for HDPE. High-density polyethylene copolymers with high comonomer content have been known for many years. Their density was less than that of HDPE, their crystallinity was lower, and the properties that depended on crystallinity were altered. Considerable time elapsed before it was recognized that a copolymer of HDPE, in which crystallinity had been reduced to about 55% (the crystallinity of LDPE, see Table 3.1) and its density to about 0.925 g/cm^3 had many of the characteristics of LDPE. Thus a copolymer of ethylene and 6–8% 1-butene resembles LDPE. Like the other polyethylenes, it may be manufactured by solution, slurry, or fluidized bed processes.

The fact that products of this type were known before they were recognized as economically advantageous replacements for LDPE emphasizes the importance not only of discovery but of its recognition.

Linear low-density polyethylene, like LDPE, has branching that inhibits close approach of polymer molecules and decreases crystallinity. The branching in LDPE is irregular and, if the LDPE is prepared by the autoclave process, there are secondary branches on the primary ones. LLDPE has regular branching because of the pendant C$_2$ groups provided by the 1-butene comonomer. This uniformity makes possible closer association of the polymer molecules in the crystalline portion, for which reason LLDPE has a higher tensile strength than LDPE, allowing the use of thinner or lower gauge films.

Its growth in the United States was rapid at first and almost completely at the expense of LDPE. Growth was facilitated because LLDPE could be processed in HDPE equipment. Subsequently, new uses for it were found in stretch wrap film, injection molding, and rotomolding—applications for which LDPE is not suitable. The cost advantage provided for LLDPE manufacture by lower energy use is in large part counterbalanced by the cost of the more expensive monomer, 1-butene, which became one of the fastest growing chemicals of the mid 1980s. Newer processes for LLDPE make use of 1-hexene, 1-octene, and 4-methyl-1-pentene as comonomers.

1-Butene for LLDPE may be obtained either by dimerization (Section 3.3.1) or oligomerization (Sections 3.3.2 and 3.3.3) of ethylene or by isolation from the C$_4$ olefin stream from steam or catalytic cracking (Chapter 5). In fact, practically all of it in the United States and Western Europe is obtained from the last source, which underscores the point that refinery processes usually are more economical than processes in chemical plants. Saudi Arabia makes 1-butene by dimerization of ethylene (Section 3.3.1) since refinery 1-butene is not available. 1-Hexene and 1-octene are obtained solely by ethylene oligomerization and 4-methyl-1-pentene by propylene dimerization (Section 4.2).

3.1.5 Very High Molecular Weight Polyethylene

Very high molecular weight polyethylene with a density of 0.941 g/cm^3 or higher is not used widely because it is difficult to process. It is made under HDPE conditions without comonomer and is used primarily for plastic ropes. One of its interesting newer applications is for the preparation of high-strength polyethylene fibers. Tensile strength in polymeric fibers may be increased by drawing, a process that causes the polymer molecules to crystallize or to align themselves so closely that physical forces of attraction between polymer molecules come into play. However, physical stretching does not cause uncoiling of small portions of molecules that are "tangled" because of folds and crossovers. It has been found that polymer molecules uncoil in very dilute solution if the solvent has high solvating power. Thus a dilute hydrocarbon solution of very high molecular weight polyethylene at 120°C is cooled to give a gel that is extruded into gel-like fibers. The solvent is removed and replaced with dichloromethane, which effects more unraveling. The second solvent is then removed and the dry gel oriented or drawn to provide a fiber whose tensile strength is 10 times as great as that of steel. An obvious disadvantage of polyethylene fibers is their low-melting point as compared with Aramid fibers (Section 9.3.4). On the other hand they absorb less water.

3.2 ETHYLENE COPOLYMERS

3.2.1 Chlorosulfonated Polyethylene

Ethylene is found in many other copolymers in addition to HDPE and LLDPE. All of these are sold in comparatively low volumes. One of the earliest is chlorosulfonated LDPE, known as "Hypalon." It is an elastomer formed by chlorosulfonation of LDPE with sulfur dioxide (SO_2) and chlorine (Cl_2). The SO_2Cl and chlorine groups are inserted into the chain. This is an example of polymer modification by a chemical reaction in which the polymer is one of the reactants. The polymer can be cross-linked through the sulfonyl chloride group (which has destroyed the polymer's crystallinity and made it rubbery) with inorganic oxides such as lead oxide or with di- or polyamines. This elastomer is prized because of its ozone and oxidation resistance. It finds use in gaskets, wire and cable installation, roof coatings, white side-wall tires, and coated fabrics.

3.2.2 Ethylene–Vinyl Acetate

The most important ethylene copolymer after HDPE and LLDPE is ethylene-vinyl acetate (EVA). Its consumption approximated 750 million lb in 1991, which is about 70% of all ethylene copolymers excluding HDPE and LLDPE. Ethylene-vinyl acetate is a random copolymer. Depending on the proportion of vinyl acetate, it may be either hard or rubbery.

The hard form requires less than 50% of vinyl acetate. Like most copolymers, it has a lower density than the homopolymer and is useful in agricultural films because it demonstrates better heat retention, toughness, and greater transparency, trapping more infrared (IR) light than does LDPE. It is also used for adhesives, particularly hot melts where the adhesive is applied as a melt without solvent and bonds almost instantaneously on cooling. It is used as a coating often blended with waxes, polyolefins, and elastomers. An interesting medical application is as a semipermeable film for drugs administered in sustained release dosage forms. A drug such as trinitroglycerin is placed in a reservoir covered with the semipermeable film, which in turn is placed next to the skin.

The drug enters the body since the skin too is semipermeable and makes its way into the bloodstream at a controlled rate via the capillaries.

If more than 50% vinyl acetate is present, a rubbery polymer results, which in the form of a latex is useful for water-based paints along with other homo- or copolymers of vinyl acetate, acrylates, and methacrylates. The ethylene lowers the cost of the polymer and contributes much-prized water resistance.

Related copolymers include ethylene–ethyl acrylate, ethylene–methyl methacrylate, and ethylene–butyl acrylate. Each of these has properties that fit it for specific uses. For example, ethylene–ethyl acrylate is applied mainly as a cable coating. Ethylene–butyl acrylate provides a tough film with excellent low-temperature properties, fitting it for the packaging of frozen foods.

Ethylene–vinyl alcohol is a copolymer prepared by conversion of the ester groups of EVA to alcohol groups by ester interchange with ethanol. Its films are good oxygen barriers and, because of its high polarity, it has good resistance to oils and greases. The copolymer was developed in Japan; United States manufacture started in 1986. Laminates may be made by coextrusion of ethylene–vinyl alcohol copolymer with other polymers such as polyethylene or polypropylene.

3.2.3 Ionomers

Ethylene–acrylic acid and ethylene–methacrylic acid are examples of random copolymers. They are useful as such and also in the preparation of ionomers. Ionomers comprise a class of copolymers of which Du Pont's "Surlyn" is typical. An ionomer contains ionic groups such as pendant carboxyl groups. These react with both divalent and monovalent metal ions such as Zn^{2+} and Na^+ to cross-link the polymer chains by the formation of carboxylates. The monovalent ion is effective because of the formation of ion aggregates through coulombic forces. The hygroscopic aggregates attract water even in a medium as nonpolar as an ethylene-containing polymer. The water serves to stabilize the aggregate and makes possible the reversal of salt formation on heating.

The virtue of an ionomer is that it has some of the stiff, tough properties of a cross-linked resin at room temperature. On heating, the ionic bonds or ion aggregates are disrupted, and the polymer becomes thermoplastic and processible. Major uses for ionomers include ski boots, tough, flexible hosing, and the coating of golfballs, where toughness is a major requirement. Very tough forms of ionomers have been proposed for body panels and bumpers of automobiles.

3.2.4 Copolymer from "Incompatible" Polymer Blends

Polar polymers such as nylon are incompatible with nonpolar polymers such as polyethylene for the same reasons that polar and nonpolar liquids are immiscible. They can, however, be combined in an imaginative process, which uses a third polymer. The technique is exemplified by DuPont's "Selar".

Selar is a blend of nylon and a graft copolymer of ethylene and fumaric acid. This mixture is combined with HDPE, the copolymer serving as a "nail" to hold the incompatible polyethylene and nylon together. Its polar portion, the pendant carboxyl groups, associate with the nylon and may even react chemically with amine end groups. The polyethylene backbone of the copolymer, on the other hand, associates with the polyethylene homopolymer. These physical bonds are sufficiently strong to prevent phase separation and accordingly the nylon, which is the lesser component, "plates out" in the polyethylene matrix when the molten mixture is cooled. The nylon platelets overlap providing a barrier that makes the blend suitable for structures where good oil resistance is required. It is used primarily for the fabrication of automotive fuel tanks. Figure 3.5 demonstrates this important concept for combining incompatible polymers.

3.2.5 Ethylene–Propylene Elastomers

Ethylene–propylene copolymers containing 20–80% ethylene are noncrystalline elastomers, which can be cross-linked with peroxide. Alternatively, a diene may be included such as *trans*-1,4-hexadiene (Section 5.1.2), ethylidene norbornene (Section 5.1.2), or dicyclopentadiene (Section 6.3), each of which contains one active double bond to copolymerize with the ethylene and propylene and one less reactive pendant double bond for cross-linking of the rubber with conventional sulfur-based vulcanizing agents. The polymerization is carried out with a Ziegler catalyst. This elastomer finds specialty uses where long service life is required. Because only enough double bonds are present to permit cross-linking, the polymer is particularly stable and resistant to oxidation.

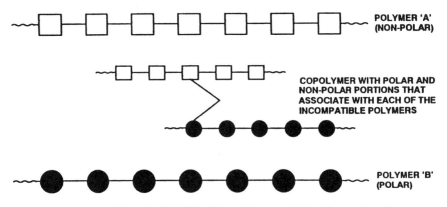

Figure 3.5 A chemical "nail" holds two incompatible polymers together.

3.2.6   Ultra-Low-Density Polyethylene

Ultra-low-density polyethylene is useful in applications normally reserved for ethylene copolymers such as EVA. The designation "ultra-low-density" applies if the density is below 0.915 g/cm^3. A product with a density as low at 0.890 g/cm^3 was announced by Union Carbide in the 1980s. The material is suitable for "noiseless" film of the type required in adult diapers and other applications where low stiffness coupled with reasonable strength is required. The very low density is achieved in part by copolymerization of ethylene with up to 10% of one of the α-olefins used for the preparation of LLDPE.

3.2.7   Photodegradable Copolymers

Photodegradable polymers are one answer to the problem of plastic litter. In the United States only 7% of the contents of waste dumps is plastics. But 40% of the litter along roadsides is plastics and this visibility, abetted by consumer concern, has motivated the development of photodegradable plastics. These were already in use in the early 1990s for yokes for "six-pack" beverage bottles in those states where legislation required it. There is evidence that animals are trapped by the holes in these yokes.

The most advanced product is an alternating copolymer of ethylene and carbon monoxide, $\{CH_2CH_2CO\}_m$ the ketone group contributing the photodegradability. The homogeneous copolymerization is carried out with a catalyst comprising a palladium salt of trifluoroacetic or p-toluenesulfonic acid with a bidentate ligand identified in the patent literature as 2,2′-bipyridine, 1,10-phenanthroline or 1,3-bis(diphenylphosphino)propane.

3.3   OLIGOMERIZATION

Ethylene oligomerizes with difficulty because the free radical intermediate, once formed, is highly energetic and polymerizes rapidly to give high molecular weight polymers. Thus it was not until the advent of Ziegler technology that ethylene oligomerization became feasible. Conversely, ionic intermediates from propylene and butylenes form readily. Since they are much less energetic, they oligomerize, but do not polymerize.

3.3.1   Dimerization

Ethylene may be dimerized either to 1-butene or to the more thermodynamically stable 2-butene. The incentive for the development of dimerization processes was the possibility of dehydrogenating the resulting butenes to butadiene just as refinery butenes are dehydrogenated (Section 2.2.7). Although this was never done commercially with the dimerization products, ethylene dimerization is made use of in ARCO's process for producing propylene from ethylene and 2-butene (Section 2.2.8). Dimerization to 1-butene has been commercialized is Saudi Arabia to obtain this chemical for use in LLDPE since in that country there are no refinery butenes. Another possibility is the production of 1-butene in situ before or during ethylene polymerization to provide the comonomer needed for LLDPE (Section 3.1.4).

Catalysts for ethylene dimerization are primarily combinations of nickel compounds and an alkyl aluminum. This catalyst resulted from a chance observation by Ziegler that nickel from reactor corrosion produced 1-butene rather than ethylene oligomers in the presence of alkyl aluminum compounds. Cobalt salts and titanium complexes are also effective. A catalyst that converts ethylene primarily to 1-butene comprises bis(1,5-cyclooctadiene)nickel, triphenylphosphine, and *p*-toluenesulfonic acid. A catalyst for 1-butene preparation proposed by the Institut Français du Pétrole comprises triethylaluminum with a complex of tetra-*n*-butyl titanate and tetrahydrofuran. A catalyst useful for the formation of 2-butene comprises tri-*n*-butylphosphine nickel dichloride and ethylaluminum dichloride. A simpler catalyst is a mixture of nickel oxide, silica, and alumina.

3.3.2 Ziegler Oligomerization of Ethylene

Related to Ziegler polymerization is the oligomerization of ethylene as shown in Figure 3.6. Triethylaluminum reacts with ethylene to form a trialkylaluminum at 100°C and 100 bar. An aluminum alkoxide results from the interaction of the trialkylaluminum with oxygen. It can be hydrolyzed to straight-chain fatty alcohols with an even number of carbon atoms. Alternatively, the trialkylaluminum may be heated to 280–300°C and 50 bar in the presence of

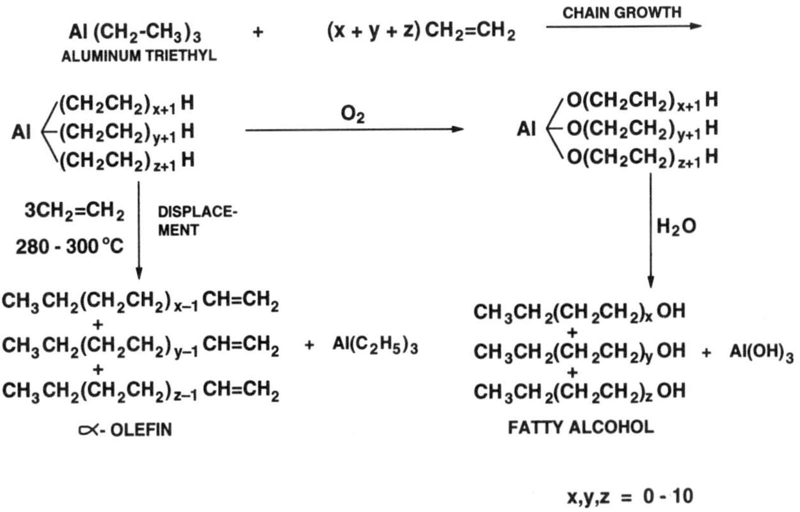

Figure 3.6 Ethylene oligomerization (Ziegler).

ethylene to give linear α-olefins. Both groups of compounds are useful in the manufacture of biodegradable detergents, but the alcohols are much more widely used. The C_{10} α-olefins may in turn be oligomerized to trimers to obtain compositions valuable as synthetic lubricants, and the C_{18} compounds, may be sulfonated to obtain industrial grade surfactants. The C_6 and C_8 olefins are useful as comonomers for LLDPE (Section 3.1.4).

This technology was adapted by several companies. The major problem with the process is an unfavorable molecular weight distribution of the products, which range from C_4 to C_{24}. Conoco (now Vista) produced the alcohols, whereas Ethyl and Gulf (now Standard Oil of California) produced α-olefins. The most desirable materials for surfactants contain 12–14 carbon atoms. A process developed by the Ethyl Corp. recycles α-olefins to the chain growth step so that transalkylation takes place. This gives a higher concentration of C_{12}–C_{18} compounds and fewer of the lower molecular weight materials, although it introduces some branching. The displacement is effected by raising the temperature to 300°C. The alkyl aluminums with chain lengths from 12 to 18 are replaced by the lower molecular weight olefins that form at the high temperature. This provides alkyl aluminums with chain lengths of 4–12 and free C_{12}–C_{18} α-olefins. In other words, the displacement reaction shown in Figure 3.6 is allowed to take place in situ rather than as a separate step, using C_4–C_{12} α-olefins to displace the alkyl chains rather than ethylene. High-pressure favors chain growth since it reduces the number of molecules. Displacement, on the other hand, as indicated above, is favored by high temperature. The effectiveness of the transalkylation reaction is indicated by the fact that C_{12}–C_{14} α-olefin production is increased from 10–18 to 20–35%.

The equations in Figure 3.6 show that the reaction requires stoichiometric quantities of triethylaluminum. However, it is possible to carry it out with catalytic amounts if the conditions are such that ethylene continually displaces the growing chains on the catalyst. Beta-hydrogen abstraction from the growing alkyl chain regenerates an Al–H bond that can start growth of a new alkyl chain. When alcohols are made, stoichiometric amounts of $Al(OH)_3$ are formed. This may be disposed of by converting it to a catalyst support.

3.3.3 Shell Oligomerization of Ethylene

Other processes for the oligomerization of ethylene have been devised. Esso developed one based on a catalyst combination comprising an alkyl aluminum chloride and titanium tetrachloride. The reaction is carried out in an organic solvent and the molecular weight of the α-olefins produced increases as the reaction temperature is raised. Nonpolar solvents also favor higher molecular weights. Shell devised a process, which is integral to their SHOP process (Section 3.3.4), based on a nickel chloride catalyst with ligands comprising diphenylphosphinoacetic acid and triphenylphosphine. Sodium borohydride is present to reduce the nickel salt to a nickel hydride catalyst, and a glycol such as 1,4-butanediol serves as a solvent for the reactants but not the products.

The oligomerization takes place at about 100°C and 40 bar in the presence of excess ethylene. The mixture of α-olefins that forms has chain lengths varying from C_4 to C_{40}. The distribution includes 40.5% of C_{10}–C_{18} α-olefins, 41% C_4–C_8 compounds, and 18.5% of C_{20} and higher materials. The α-olefin distribution is similar to that obtained in a Ziegler process carried out without transalkylation. The mechanism of α-olefin formation, like the one for the alkyl aluminum reaction, involves β-hydrogen abstraction as shown in Section 3.3.2. The product α-olefins precipitate because they are not soluble in the 1,4-butanediol.

3.3.4 The Shell Higher Olefins Process

The Shell higher olefins process (SHOP) process is one of the most ingenious in the chemical industry. Its objective is the preparation of linear or almost linear α-olefins and fatty alcohols with the most suitable chain lengths for surfactants, that is, C_{11}–C_{15}. The natural product—coconut or palm kernel oil (Section 13.6)—provides C_{12} and C_{14} chain lengths, but a carbon atom more or less is also acceptable. The SHOP process comprises a combination of four reactions, each of which was already known, at least in concept, although some required modification. Thus it is an excellent example of that form of creativity that involves the reordering of old knowledge to achieve a new end result.

As shown in Figure 3.7, the first step in the process is the oligomerization of ethylene as described in Section 3.3.3. A spread of chain lengths is obtained with a "bell-shaped" statistical distribution. The processes can be engineered so that the peak of the distribution is at C_{10}–C_{14} and these α-olefins comprise about 30% of the reaction mixture, which means that approximately 70% of unwanted chain lengths is produced.

1. OLIGOMERIZATION

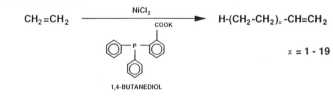

2. FRACTIONATION OF ∝- OLEFINS

    a. C_{10}-C_{14} ∝- OLEFINS

    b. C_4 -C_8 ∝- OLEFINS

    c. C_{16}-C_{40} ∝- OLEFINS

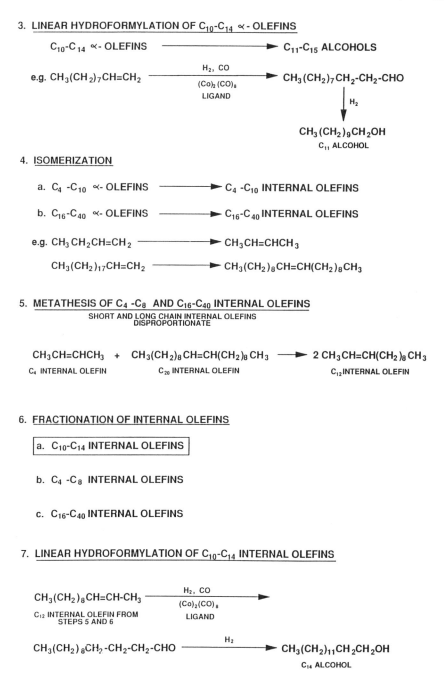

3. <u>LINEAR HYDROFORMYLATION OF C$_{10}$-C$_{14}$ $\propto$- OLEFINS</u>

C$_{10}$-C$_{14}$ $\propto$- OLEFINS $\longrightarrow$ C$_{11}$-C$_{15}$ ALCOHOLS

e.g. $CH_3(CH_2)_7CH=CH_2$ $\xrightarrow[\substack{(Co)_2(CO)_8 \\ LIGAND}]{H_2,\ CO}$ $CH_3(CH_2)_7CH_2\text{-}CH_2\text{-}CHO$

$\downarrow H_2$

$CH_3(CH_2)_9CH_2OH$
C$_{11}$ ALCOHOL

4. <u>ISOMERIZATION</u>

   a. C$_4$ -C$_{10}$ $\propto$- OLEFINS $\longrightarrow$ C$_4$ -C$_{10}$ INTERNAL OLEFINS

   b. C$_{16}$-C$_{40}$ $\propto$- OLEFINS $\longrightarrow$ C$_{16}$-C$_{40}$ INTERNAL OLEFINS

   e.g. $CH_3CH_2CH=CH_2$ $\longrightarrow$ $CH_3CH=CHCH_3$

   $CH_3(CH_2)_{17}CH=CH_2$ $\longrightarrow$ $CH_3(CH_2)_8CH=CH(CH_2)_8CH_3$

5. <u>METATHESIS OF C$_4$ -C$_8$ AND C$_{16}$-C$_{40}$ INTERNAL OLEFINS</u>
   SHORT AND LONG CHAIN INTERNAL OLEFINS
   DISPROPORTIONATE

   $CH_3CH=CHCH_3$ + $CH_3(CH_2)_8CH=CH(CH_2)_8CH_3$ $\longrightarrow$ 2 $CH_3CH=CH(CH_2)_8CH_3$
   C$_4$ INTERNAL OLEFIN   C$_{20}$ INTERNAL OLEFIN   C$_{12}$ INTERNAL OLEFIN

6. <u>FRACTIONATION OF INTERNAL OLEFINS</u>

   | a. C$_{10}$-C$_{14}$ INTERNAL OLEFINS |

   b. C$_4$ -C$_8$ INTERNAL OLEFINS

   c. C$_{16}$-C$_{40}$ INTERNAL OLEFINS

7. <u>LINEAR HYDROFORMYLATION OF C$_{10}$-C$_{14}$ INTERNAL OLEFINS</u>

   $CH_3(CH_2)_8CH=CH\text{-}CH_3$ $\xrightarrow[\substack{(Co)_2(CO)_8 \\ LIGAND}]{H_2,\ CO}$
   C$_{12}$ INTERNAL OLEFIN FROM
   STEPS 5 AND 6

   $CH_3(CH_2)_8CH_2\text{-}CH_2\text{-}CH_2\text{-}CHO$ $\xrightarrow{H_2}$ $CH_3(CH_2)_{11}CH_2CH_2OH$
   C$_{14}$ ALCOHOL

8. <u>REPEAT STEPS 5, 6, AND 7 TO EXTINCTION</u>

Figure 3.7 Shell higher olefins process (SHOP).

From the C_4–C_{40} mixture of α-olefins, C_{10}–C_{14} products are separated by distillation (Step 2, Fig. 3.7). Any other chain length may also be removed by fractionation should a market exist for it. The C_{10}–C_{14} α-olefins are hydroformylated (Step 3, Fig. 3.7) to C_{11}–C_{15} alcohols. Linear hydroformylation normally requires a rhodium catalyst, (Section 4.8), which is best recovered by distilling the product to leave the expensive rhodium in the vessel.

Shell's products have too high molecular weights to allow for rhodium recovery in this way. Perhaps more important, the rhodium catalyst does not shift the double bond to the α-position as is required later in the process. Accordingly, Shell uses a dicobalt octacarbonyl catalyst with ligands such as tributylphosphine. This catalyst has the even more important advantage of making possible the linear hydroformylation of internal olefins, as will be seen later. From the C_{10}–C_{14} α-olefins, linear alcohols result with chain lengths of 11–15 carbon atoms.

In the fourth step (Fig. 3.7) the C_4–C_{10} and the C_{16}–C_{40} olefins are isomerized to internal olefins using heterogeneous catalysts such as magnesium oxide granules. Isomerization takes place at 80–140°C and 4–20 bar. Internal double bonds are distributed randomly throughout the olefin molecules.

The next step (Step 5, Fig. 3.7) involves metathesis (Section 2.2.8) of the short- and long-chain internal olefins to provide new internal olefins with a broad distribution of chain lengths, some of which are in the desired C_{10}–C_{14} range. The example shows a metathesis reaction between a C_4 and a C_{20} internal olefin to yield two molecules of a C_{12} internal olefin. The desired chain lengths are removed by distillation, 10–15% being obtained (Step 6, Fig. 3.7). These internal olefins are again subjected to linear hydroformylation as shown in Step 7, using the above-described cobalt octacarbonyl catalyst, which causes the migration of the double bond to the α position. The intermediate aldehydes, which are produced by hydroformylation, are hydrogenated in situ.

Step 6 provides further quantities of C_4–C_8 and C_{16}–C_{40} internal olefins and these again are subjected to metathesis. Thus Steps 5–7 are repeated to extinction. High yields are obtained in each step.

The process is versatile. Thus C_6 and C_8 olefins may be removed for sale to LLDPE manufacturers (Section 3.1.4) and the C_{10} olefin may be isolated for use in synthetic lubricants. In this case, only the C_4 olefin is used in the metathesis reaction, and if necessary more can be obtained from steam or catalytic cracking (Sections 2.2.1 and 2.2.2). The internal olefins may be sold for the alkylation of benzene (Section 7.4) for surfactants.

3.4 VINYL CHLORIDE

Vinyl chloride is the monomer for poly(vinyl chloride) (PVC). Originally it was made by addition of hydrogen chloride to acetylene

$$CH{\equiv}CH + HCl \rightarrow CH_2{=}CHCl$$

Ethylene-based vinyl chloride was first made by production of ethylene dichloride from ethylene and chlorine which, when heated in contact with pumice or charcoal, yielded vinyl chloride and 1 mol of HCl

$$CH_2=CH_2 + Cl_2 \rightarrow CH_2ClCH_2Cl \rightarrow CH_2CHCl + HCl$$

The byproduct presented a problem, for hydrogen chloride is corrosive and difficult to ship. Unless it can be used on site, its value is small. Disposal presents economic and environmental problems. Furthermore, one-half of the chlorine (produced by electrolysis, which requires expensive electrical energy) was wasted.

One solution was to react the byproduct hydrogen chloride with acetylene in a second plant. In this way, the chlorine was saved but, as the price gap between ethylene and acetylene widened, the process (the so-called integrated chlorine economy) became less attractive. It is still operated in South Africa, but elsewhere a wholly ethylene based process became desirable.

The first idea involved resurrection of a Victorian process. In 1858 Deacon had shown that hydrogen chloride can be oxidized to chlorine by air over bricks soaked in copper chloride. It was the first heterogeneous catalytic process to be operated.

$$4HCl + O_2 \xrightarrow{CuCl_2} 2H_2O + 2Cl_2$$

Shell investigated the reaction and improved the catalyst. The snag was that the hydrogen chloride had to be isolated and oxidized, and the chlorine separated and recycled. Before the process could be instituted, a related process was developed called oxychlorination.

Here the Deacon chemistry has been incorporated into a one-step reaction in which ethylene, hydrogen chloride, and air are passed over a copper chloride–potassium chloride catalyst to give ethylene dichloride. This is mixed with the ethylene dichloride from the chlorination process and cracked to vinyl chloride and byproduct hydrogen chloride, which is returned to the oxychlorination process (Fig. 3.2e). Unless there is a separate source of byproduct hydrogen chloride, the chlorination and oxychlorination plants are integrated. Consequently, one-half of the United States ethylene dichloride is made by chlorination and one-half by oxychlorination. No one, except Du Pont briefly, used the modernized Deacon process; its value lies in the fact that it led to oxychlorination. Oxychlorination is typical of modern petrochemical processes, where the aim is to pass a simple feedstock through a hot tube over an appropriate catalyst, with the desired chemical emerging at the end. Of course, the secret lies in the "appropriate" catalyst.

The direct chlorination of ethylene is generally carried out in the liquid phase with the product ethylene dichloride as the reaction medium and with dissolved $FeCl_3$, $CuCl_2$, or $SbCl_3$ as catalyst at 40–70°C and 4–5 bar.

Oxychlorination is a gas-phase reaction taking place at about 225°C and 2–4 bar with a cupric chloride catalyst supported on alumina or silica together with potassium chloride, whose chloride ion serves as an activator. The cupric chloride is the chlorinating agent. It chlorinates the ethylene and is itself reduced to cuprous chloride. Oxygen regenerates the cuprous chloride by converting it to a double salt of cupric oxide and cupric chloride. The double salt in turn reacts with the hydrogen chloride to give cupric chloride and water.

$$CH_2{=}CH_2 + 2CuCl_2 \rightarrow CH_2ClCH_2Cl + 2CuCl$$

$$4CuCl + O_2 \rightarrow 2CuO \cdot CuCl_2$$

$$CuO \cdot CuCl_2 + 2HCl \rightarrow 2CuCl_2 + H_2O$$

Because the reaction is highly exothermic, a fluidized bed reactor, which has much better heat transfer capability than a fixed bed, may be used. The chemical industry provides many examples of the use of fluidized beds for heat transfer as in acrylonitrile production (Section 4.4) or catalytic cracking (Section 2.2.2).

Vinyl chloride manufacture consumes about 20% of the chlorine produced in the United States. An objective of the chemical industry has been to devise routes to vinyl chloride and other C_2 compounds from ethane rather than ethylene. Several processes have been devised but have not yet been commercialized. ICI has developed a vapor phase oxychlorination of ethane, which yields vinyl chloride, water, and hydrogen chloride. The catalyst comprises metallic silver with salts of manganese or lanthanum impregnated on an offretite zeolite. Although conversions above 95% can be achieved, the selectivity to vinyl chloride is only about 50%. Other chlorinated compounds including ethylene dichloride and ethylene are produced. The latter can be oxychlorinated in a separate operation.

Another research aim for vinyl chloride production is to develop a process in which the overall reaction can be accomplished without isolation of the ethylene dichloride intermediate. The reaction conditions for the two steps, particularly temperatures, differ so widely, however, that satisfactory yields may not prove possible. Another possibility is the intermolecular dehydrogenation of ethylene and hydrogen chloride in the presence of a reactant for the hydrogen produced. This is a thermodynamically feasible but elusive reaction.

The major use for vinyl chloride is the manufacture of polymers or copolymers (Chapter 15). It is also a starting material for vinylidene chloride, trichloroethylene, and tetrachloroethylene (Section 3.11.7). Ethylene dichloride, vinyl chloride's precursor, finds some application as a solvent, but use of most chlorine compounds in chemical processing is decreasing because of their persistence in the body. They are lipophilic and are not destroyed or excreted but are stored in body fat. Vinyl chloride itself is toxic, causing angiosarcoma, a rare type of liver cancer. Accordingly, its concentration in ambient air during manufacture must be strictly limited.

3.5 ACETALDEHYDE

Acetaldehyde is one of the few industrial chemicals whose production has shrunk in the past 15 years. Its decline has been paralleled only by acetylene (Section 10.3) and more recently by US petrochemical ethanol (Section 3.9). In the United States in 1969, 1.65 billion lb of acetaldehyde was manufactured. By the late 1980s this had decreased to an estimated 650 million lb and growth was not foreseen.

Acetaldehyde is manufactured by an ingenious process, the Wacker reaction. Its demise was caused by the discovery of equally ingenious processes for the preparation of the two chemicals for which it served as precursor, *n*-butanol (Section 4.8) and, more important, acetic acid (Section 10.5.2.2).

Acetaldehyde was originally made by the hydration of acetylene over an oxidation–reduction catalyst, mercurous–mercuric sulfate buffered by ferric sulfate. Vinyl alcohol is assumed to form momentarily and rearranges to acetaldehyde at atmospheric pressure and 95°C.

$$CH{\equiv}CH + H_2O \rightarrow [CH_2{=}CHOH] \rightarrow CH_3CHO$$

Acetylene   Vinyl alcohol   Acetaldehyde

Inexpensive ethylene replaced acetylene in the early 1960s and by 1974 only 15% of acetaldehyde was made from acetylene. The remainder was made by the oxidation of ethanol at 450°C and 3 bar with air over a silver gauze catalyst.

$$2CH_3CH_2OH + O_2 \rightarrow 2CH_3CHO + 2H_2O$$

Ethanol   Acetaldehyde

Alternatively, the ethanol may be dehydrogenated over a chromium oxide activated copper catalyst at 270–300°C. This is a more attractive process if a use exists for the byproduct hydrogen.

These routes gave way to the Wacker process described by Parshall as "a triumph of common sense" (see note). It is based on the observation that ethylene is oxidized by palladium chloride to acetaldehyde. As indicated, stoichiometric quantities of palladium chloride are required.

$$CH_2{=}CH_2 + PdCl_2 + H_2O \rightarrow CH_3CHO + Pd^0 + 2HCl$$

Ethylene   Acetaldehyde

In the 1950s, chemists at Wacker Chemie in Germany converted the palladium salt from a stoichiometric to a catalytic component by including cupric chloride, oxygen, and hydrogen chloride in the reaction mixture. Each atom of palladium, when formed, is then oxidized back to palladium chloride. The cuprous chloride is converted back to cupric chloride by oxygen:

$$Pd^0 + 2CuCl_2 \rightarrow PdCl_2 + 2CuCl$$

$$4CuCl + 4HCl + O_2 \rightarrow 4CuCl_2 + 2H_2O$$

The mechanism proposed by Parshall (see notes) is shown in the following cycle:

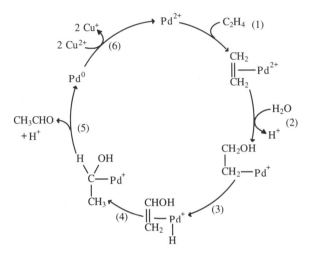

There is general agreement about the formation of the π-complex (step 1). Step (2) is the OH addition. There is controversy as to how this step takes place. Thereafter the mechanism is straightforward, with hydrogen abstraction by the palladium (3) followed by rearrangement (4) and acetaldehyde formation (step 5). Oxidation of the palladium to Pd^{2+} is step 6. The palladium chloride–copper chloride mixture is analogous to the oxychlorination system (Section 3.4).

n-Butanol (Section 4.8) was originally made from acetaldehyde by an aldol condensation (Section 3.11.3). Today it is made from propylene (Section 4.8) by hydroformylation. Acetic acid was made by oxidation of acetaldehyde with either air or oxygen over a manganese or cobalt acetate catalyst at 60°C. The oxidation takes place by a free radical mechanism in which peracetic acid is the intermediate.

$$CH_3CHO + X^\bullet \longrightarrow CH_3\overset{\bullet}{C}{=}O + HX$$
Acetaldehyde

$$CH_3\overset{\bullet}{C}{=}O + O_2 \longrightarrow CH_3C\underset{\overset{\|}{O}}{-}O{-}O^\bullet$$

$$CH_3C\underset{\overset{\|}{O}}{-}O{-}O^\bullet + CH_3CHO \longrightarrow CH_3C\underset{\overset{\|}{O}}{-}OOH + CH_3\overset{\bullet}{C}{=}O$$
Peracetic acid

The peracetic acid in turn reacts preferentially with acetaldehyde to give α-hydroxyethyl peracetate, which decomposes through a cyclic transition state to 2 mol of acetic acid.

$$CH_3C\!-\!O\!-\!OH \;+\; CH_3CHO \longrightarrow \underset{\text{α-Hydroxyethyl peracetate}}{CH_3\overset{O}{\overset{\|}{C}}\!-\!O\!-\!O\!-\!\overset{OH}{\overset{|}{C}}HCH_3} \longrightarrow \underset{\text{Acetic acid}}{2\,CH_3COOH}$$

Cheap naphtha in Europe in the 1950s motivated the development of the primary flash distillate route to acetic acid (Section 10.5.2.2). This is still in use in Europe because it gives valuable byproducts and *n*-butane is oxidized in a related process in the United States. Cheap methanol in the 1970s similarly led to the development of methanol carbonylation (Section 10.5.2.2), a process whose economics are so good that it shut down every US manufacturer using the acetaldehyde route.

Lesser volume processes based on acetaldehyde are described in Section 3.11.3.

3.6 VINYL ACETATE

Like acetaldehyde, vinyl acetate was originally made from acetylene, except that acetic acid rather than water was added across the triple bond. The catalyst comprised zinc acetate on charcoal with sodium acetate and the reaction proceeded at about 200°C.

$$\underset{\text{Acetylene}}{CH\!\equiv\!CH} \;+\; CH_3COOH \rightarrow \underset{\text{Vinyl acetate}}{CH_3COOCH\!=\!CH_2}$$

The modification of the Wacker reaction (Section 3.5) by substitution of acetic acid for water in the presence of potassium acetate yields vinyl acetate. Two processes were developed, one a homogeneous liquid-phase reaction with a palladium chloride–cupric chloride catalyst, and the other a heterogeneous gas-phase reaction with a $PdCl_2$–$CuCl_2$, $PdCl_2$–Al_2O_3 or palladium-on-carbon catalyst onto which a trickle of potassium acetate solution flows. Fixed and fluidized bed processes are in use.

$$CH_2\!=\!CH_2 \;+\; PdCl_2 \;+\; 2\underset{\text{Sodium acetate}}{CH_3COONa} \rightarrow$$

$$\underset{\text{Vinyl acetate}}{CH_3COOCH\!=\!CH_2} \;+\; 2NaCl \;+\; Pd^0 \;+\; CH_3COOH$$

The Pd^0 is regenerated by the $CuCl_2$ in the catalyst analogously to the conversion of Pd^0 to $PdCl_2$ in the acetaldehyde reaction (Section 3.5) except that the Cu^+ ion is converted to Cu^{2+} with air. No HCl is present since, together with the acetic acid, it is formidably corrosive.

The liquid-phase process ran into severe corrosion problems from this combination. These might have been overcome, but there were also serious mass-transfer problems because foaming in the reactor prevented sufficient ethylene from dissolving quickly enough, and the plant was closed.

Although the liquid-phase process never really worked, it is of interest because the water produced in the reaction leads to byproduct acetaldehyde. This can be oxidized to acetic acid, which can be returned to the reactor. The system can be varied so that acetaldehyde is produced in just the right quantity to provide, on oxidation, the correct amount of acetic acid. In most two-for-one reactions (two products, in this instance acetaldehyde and vinyl acetate, from one set of equipment) the ratio of the products cannot be varied [cf. phenol–acetone production (Section 4.5)] and accordingly a market must exist for the products in the ratio in which they are produced. Here the ratio may be varied by controlling the concentration of water. This same flexibility has been demonstrated in a new process that produces both acetic acid and acetic anhydride (Section 10.5.2.3), and in a process for making ethylene glycol and dimethyl carbonate (Section 3.7.1).

The heterogenous gas-phase process is now generally used. The potassium acetate trickle is necessary because the acetate, essential to the catalyst's performance, migrates continually. Byproducts include acetaldehyde, as might be expected, together with methyl and ethyl acetates.

An uncommercialized process for preparing vinyl acetate from methanol is described in Section 10.5.2.4.

Vinyl acetate's major use is for conversion to poly(vinyl acetate), the basis of many adhesives and water-based emulsion paints. Some poly(vinyl acetate) is converted to poly(vinyl alcohol) by saponification or transesterification with ethanol to give ethyl acetate as a coproduct. Poly(vinyl alcohol) is used in Japan in fibers.

3.7   ETHYLENE OXIDE

The most important ethylene-based chemical that is not primarily a polymer precursor is ethylene oxide. It is made (Fig. 3.2h) by direct reaction of ethylene and oxygen over a silver catalyst (Section 16.1.1). The reaction is exothermic, and the simultaneous even more exothermic oxidation of both ethylene and ethylene oxide leads to the byproducts, carbon dioxide and water.

$$CH_2{=}CH_2 + 0.5O_2 \longrightarrow CH_2{-}CH_2$$
$$\diagdown O \diagup$$

Ethylene oxide

$$CH_2{=}CH_2 + 3O_2 \longrightarrow 2CO_2 + 2H_2O$$

Ethylene oxide technology provides an excellent example of yield improvement. High selectivity is important because approximately 75% of the cost of ethylene oxide derives from the raw material, ethylene. Initially, yields with a silver catalyst supported on alumina were of the order of 65–70%. Because of improved catalysts, yields now exceed 82%. Shell Development, Union Carbide, and Halcon/Scientific Design have numerous patents describing improved catalysts. One of Shell's patents discloses the addition of minute amounts of potassium, rubidium, or cesium ions to a silver catalyst to increase yields. Apparently, the concentration of the added ions is critical.

The reaction is carried out at 15 bar and 250°C. A few parts per million of ethylene dichloride is frequently included in the reaction mixture to inhibit oxidation of ethylene to carbon dioxide and water. The ethylene dichloride decomposes to give chlorine atoms, which adsorb on the silver surface on the sites where oxygen would otherwise chemisorb to catalyze the combustion reaction.

The direct addition of oxygen to a double bond, discovered in the 1930s by Union Carbide, provides a striking example of the power of catalysis. The original process for ethylene oxide production, now obsolete, involved addition of hypochlorous acid to ethylene to give ethylene chlorohydrin, which with calcium hydroxide underwent a dehydrohalogenation to provide ethylene oxide and calcium chloride.

$$CH_2{=}CH_2 + HOCl \longrightarrow CH_2OH{-}CH_2Cl$$
Ethylene chlorohydrin

$$2CH_2OH{-}CH_2Cl + Ca(OH)_2 \longrightarrow 2H_2C{-}CH_2 + CaCl_2 + 2H_2O$$
$$\underset{O}{\diagdown\diagup}$$

The process was wasteful of chlorine and, although yields were high, dilute solutions were necessary. The calcium chloride, whose weight was three times greater than that of the product, provided disposal problems. Also, since HOCl forms in an equilibrium reaction $(Cl_2 + H_2O \rightleftharpoons HCl + HOCl)$ ethylene dichloride was a byproduct as was chloroethyl ether, $ClCH_2CH_2OCH_2CH_2Cl$.

About 60% of US ethylene oxide is hydrolyzed to ethylene glycol (Section 3.7.1). Other major uses include reaction with alkylphenols, fatty alcohols (Sections 3.3.2 and 13.6) fatty amides and amines (Section 13.2), and anhydrosorbitol to yield nonionic surfactants. The ethoxylated fatty alcohols are the most important. Copolymers of ethylene oxide and propylene oxide are useful as isocyanate coreactants for urethanes. Ethylene oxide reacts with ammonia to provide aminoethyl alcohols (Section 3.11.6.4), and may be converted to poly(ethylene glycols) (Section 3.11.6.1).

Ethylene oxide reacts with starch and cellulose to provide hydroxyethyl derivatives (Sections 14.2 and 14.3). The former is a more readily dispersible form of starch, which finds application in both food formulation and paper manufacture. The latter is a thickener for water-based paints and a protective colloid for water dispersions.

3.7.1   Ethylene Glycol

Ethylene glycol is prepared either by the acid catalyzed or more often by the uncatalyzed hydration of ethylene oxide (Fig. 3.2h). A huge excess of water is used (18–24 mol of water per mole of ethylene oxide) to prevent the formation of di-, tri-, and higher ethylene glycols. Even with high water/ethylene oxide ratios, mono-, di-, and triglycols are obtained in a weight ratio of about 91:8.6:0.4.

Desired Reaction

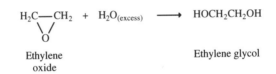

Side Reactions

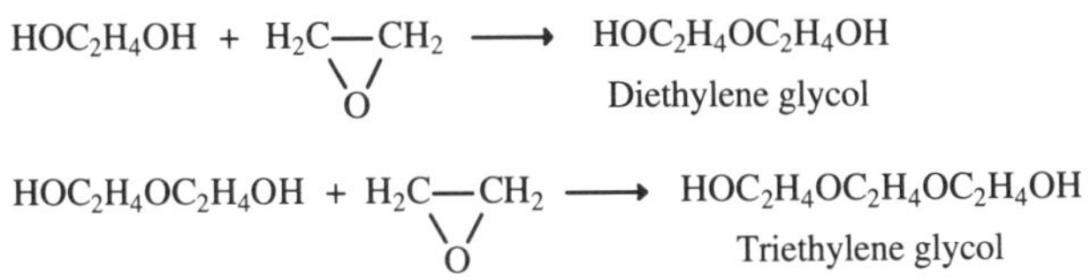

$$HOC_2H_4OH + H_2C{-}CH_2 \longrightarrow HOC_2H_4OC_2H_4OH$$
$$\underset{O}{\diagdown\diagup}$$
Diethylene glycol

$$HOC_2H_4OC_2H_4OH + H_2C{-}CH_2 \longrightarrow HOC_2H_4OC_2H_4OC_2H_4OH$$
$$\underset{O}{\diagdown\diagup}$$
Triethylene glycol

The reaction of ethylene oxide with glycols may form oligomers of ethylene oxide, and they may be prepared in this way "on purpose," if needed, by the side reactions shown.

Purification of the ethylene glycol is complex, requiring repeated fractional distillations to remove the water and to separate the glycol from its oligomers. High purity is necessary for the preparation of polyester resins, one of ethylene glycol's two major uses. Thus, ethylene glycol is reacted with purified terephthalic acid or sometimes with dimethyl terephthalate to produce poly(ethylene terephthalate).

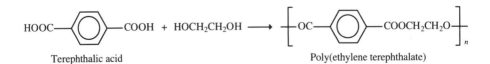

The polymer may be spun into textile fibers; Dacron or Terylene are well-known brand names. It can also be made into strong biaxially oriented films known as "Mylar." More recently it has found an important application as a plastic for the fabrication of soft drink bottles.

Ethylene glycol's other major use, a nonpolymer one, is as antifreeze in automobile radiators. Uses for di- and triethylene glycol are described in Section 3.11.6.1.

A process for converting ethylene oxide to ethylene glycol in high selectivity is described in Union Carbide and Texaco patents. It involves the reaction of ethylene oxide with carbon dioxide at 190°C and 13 bar to give ethylene carbonate (Section 3.11.6.3), which in turn reacts with water at the same pressure but at 170°C to give ethylene glycol and carbon dioxide. The first step is catalyzed by tetraethylammonium bromide and potassium iodide.

$$H_2C\text{---}CH_2 + CO_2 \longrightarrow H_2C\text{---}CH_2 \xrightarrow{H_2O} HOCH_2CH_2OH + CO_2$$

Ethylene
carbonate

The process eliminates large excesses of water as well as the possibility for the formation of di-, tri-, and higher ethylene glycols. Texaco is also believed to carry out a reaction in which the ethylene carbonate is reacted with methanol to give a mixture of ethylene glycol and dimethyl carbonate that can be separated by distillation. The ratio of the two products can be controlled by the concentration of the methanol.

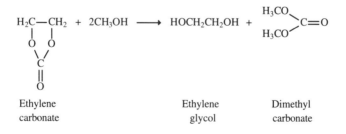

Ethylene   Ethylene   Dimethyl
carbonate   glycol   carbonate

Dimethyl carbonate was originally prepared from methanol and the highly toxic phosgene. The above route eliminates the hazard as does another recently developed route involving the oxidative carbonylation of methanol.

$$2CH_3OH + 0.5O_2 + CO \rightarrow (CH_3O)_2C{=}O + H_2O$$

Methanol   Dimethyl carbonate

The reaction of carbon dioxide with diepoxides leads to carbonate polymers, which have not been commercialized but which are of interest because of their biodegradability.

Still another means for reducing oligomer formation is described in a Carbide patent. The ethylene oxide hydrolysis is carried out with a vanadate or molybdate catalyst attached to an ion exchange resin. Only 2.5 mol of water are used per mole of ethylene oxide. One hundred percent selectivity to ethylene glycol at 100% conversion is claimed. The route has not yet been commercialized, presumably because of the difficulty of removing traces of catalyst from the final product.

3.7.2 Proposed Processes for Ethylene Glycol Production

Even with improved catalysts, yields of ethylene oxide from ethylene epoxidation are only about 82%. Hence, other processes for ethylene glycol are sought. If an economic one were to be found, the market for ethylene oxide would be cut by 50%, since at least that proportion is converted to glycol in Europe and even more in the United States.

In one proposed reaction ethylene, acetic acid, and oxygen combine to give ethylene glycol diacetate. These are the reactants that give vinyl acetate in the Wacker process, but the proportion of acetic acid is greater, and a different catalyst, tellurium oxide with an alkyl halide, is used. Two ethylenic hydrogen atoms instead of one are substituted. This diacetate can be hydrolyzed to ethylene glycol and the acetic acid recycled so that the overall reaction involves only ethylene, air, and water.

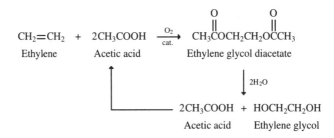

An 800 million lb/year plant was built in the late 1970s but corrosion problems not unlike those associated with the original liquid-phase vinyl acetate plant (Section 3.6) proved its downfall.

Although corrosion resistant materials could undoubtedly be found to withstand the chemistry of this process, there has been no attempt to revive it, but Mitsubishi is using the same chemistry successfully for the manufacture of 1,4-butanediol (Section 5.1) from butadiene. Because of the greater reactivity of the conjugated double bond system, a catalyst comprising palladium and tellurium without halide can be used, so that the environment is less corrosive.

Carbide has pioneered a process for ethylene glycol based on the direct combination of carbon monoxide and hydrogen:

$$2CO + 3H_2 \rightarrow HOCH_2CH_2OH$$

Carbide investigated many catalysts including rhodium carbonyl cluster catalysts, which require temperatures of 240°C and pressures of 1000–3000 bar. In other studies a ruthenium carbonyl complex and an organosilicon compound such as trimethylethoxysilane were used at 400°C and 1000 bar. Reactions under such severe conditions tend to have poor selectivities. In this case selectivity to ethylene glycol is 60–65% and methanol is a major coproduct. Methyl formate, other esters, 1,2-propanediol, glycerol, and water are also formed.

A two-step approach involved condensing carbon monoxide and hydrogen with methanol to methyl oxalate (Section 10.6) followed by hydrogenolysis as in the Du Pont process below. None of these processes has been commercialized.

This poor selectivity led Carbide to join forces with Ube Industries in Japan to develop a process based on the hydrogenolysis of oxalate esters, compounds which Ube had investigated extensively as described in Section 10.6.1. Selectivity is usually improved when reactions are carried out stepwise. Thus in the first step di-*n*-butyl oxalate results, which on hydrogenolysis yields ethylene glycol and butanol for recycling.

$$2\ n\text{-}C_4H_9OH\ +\ 2CO\ +\ 0.5O_2\ \longrightarrow\ n\text{-}C_4H_9OC\!\!\overset{O}{\overset{\|}{-}}\!\!\overset{O}{\overset{\|}{C}}OC_4H_9\text{-}n\ +\ H_2O$$

n-Butanol   Dibutyl oxalate

$$n\text{-}C_4H_9OC\!\!\overset{O}{\overset{\|}{-}}\!\!\overset{O}{\overset{\|}{C}}OC_4H_9\text{-}n\ +\ H_2\ \xrightarrow{\text{cat.}}\ HOCH_2CH_2OH\ +\ 2\ n\text{-}C_4H_9OH$$

Ethylene   *n*-Butanol
glycol

The second step is similar to one used by DuPont until 1968 in which formaldehyde was carbonylated in the presence of water to give glycolic acid at 200°C and 700 bar. The esterification of the acid with methanol provided methyl glycolate which, on hydrogenolysis, yielded ethylene glycol and methanol for recycling,

$$HCHO + CO + H_2O \rightarrow HOCH_2COOH$$
Glycolic acid

$$HOCH_2COOH + CH_3OH \rightarrow HOCH_2COOCH_3 + H_2O$$
Methyl glycolate

$$HOCH_2COOCH_3 + 2H_2 \rightarrow HOCH_2CH_2OH + CH_3OH$$
Ethylene glycol

The carbonylation takes place at 200°C and 70–100 bar at high selectivity, as does the esterification reaction. The hydrogenolysis occurs at 200°C and 30 bar with an appropriate catalyst. The released methanol is recycled so that the overall reaction is

$$2H_2 + HCHO + CO \rightarrow HOCH_2CH_2OH$$

An improvement in this process involves the use of HF in the carbonylation step. Coupled with the development of more active catalysts for the hydrogenolysis (Section 10.3), this could make the Du Pont process attractive should coal-based synthesis gas become a major feedstock.

A Monsanto process for ethylene glycol is based on the hydroformylation of formaldehyde in the presence of a homogeneous rhodium catalyst. The ligand is tri(p-trifluoromethylphenyl)phosphine together with triethylamine. Glycol aldehyde results, which on hydrogenation yields ethylene glycol.

$$HCHO + CO + H_2 \rightarrow HOCH_2CHO \xrightarrow{H_2} HOCH_2CH_2OH$$
Glycol aldehyde

A free radical induced chain reaction between methanol and formaldehyde provides the basis for an interesting process in which free radicals from di-*tert*-butyl peroxide convert methanol to a hydroxymethyl radical. This radical attacks formaldehyde to provide an ethylene glycol radical that attacks another molecule of methanol to provide ethylene glycol and a hydroxymethyl radical for further propagation.

$$CH_3OH + \dot{X} \rightarrow \dot{C}H_2OH + HX$$

$$\dot{C}H_2OH + HCHO \rightarrow HOCH_2CH_2\dot{O}$$

$$HOCH_2CH_2\dot{O} \xrightarrow{CH_3OH} HOCH_2CH_2OH + \dot{C}H_2OH$$
Ethylene glycol

In its current stage of development, however, the chains in the process are too short and many of the hydroxymethyl radicals couple to provide ethylene glycol and terminate the chain reaction. Thus consumption of the peroxide initiator is unacceptably high

$$2\dot{C}H_2OH \rightarrow HOCH_2CH_2OH$$

Another approach involves the electrohydrodimerization of formaldehyde. The problem, as in all electrochemical reactions, is to obtain adequate current efficiencies.

$$2HCHO + 2H^+ + 2e^- \rightarrow HOCH_2CH_2OH$$

3.8 STYRENE

After polyethylene and PVC, the third large tonnage polymer to be made from ethylene is polystyrene. Its monomer, styrene, was used initially not in polystyrene but in a copolymer with butadiene that served as a substitute for natural rubber. Germany had started commercial production of Buna S rubber in 1938 and during World War II was totally dependent on it. With the capture by Japan of the rubber-producing areas of Southeast Asia, the United States too was forced into a crash program for synthetic rubber development. The Buna S-type (GR-S) rubber was the result of an admirable coordinated effort between many academic, industrial, and government laboratories. If either side in the war had lacked synthetic rubber, its war effort might have collapsed.

Synthetic rubbers now account for about two-thirds of world consumption. Styrene-butadiene (SBR) rubber, the descendant of Buna S and GR-S, is the preferred material for automobile tire treads and accounts for some 60% of the world synthetic rubber production, although natural rubber is assuming increased importance. Styrene is widely used in copolymers as well as in homopolymers and rubber-modified styrene polymers.

Styrene provides an example of a "classical" reaction for ethylene. The major process for styrene manufacture involves a Friedel–Crafts reaction between benzene and ethylene to form ethylbenzene. Some ethylbenzene is also produced during catalytic reforming (Section 2.2.3) and can be removed from the C_8 fraction by so-called superfractionation. Dehydrogenation to styrene in the presence of steam and a catalyst provides styrene and hydrogen (Fig. 3.2i). An alternate process for styrene, to be described later (Section 4.7), gives propylene oxide as a coproduct. It is used commercially by ARCO and Shell.

The Friedel–Crafts alkylation was initially carried out at 85–95°C at atmospheric pressure in the liquid phase with aluminum chloride and a small amount of hydrogen chloride. In one process, boron trifluoride is used. The mechanism is well established:

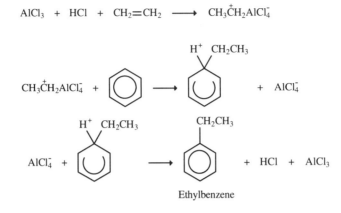

Ethylbenzene

Yields are high, but small amounts of di- and polyethylbenzenes result because the alkylaryl complex is more reactive than the hydrogen aryl complex and can continue to add ethyl groups. They are recycled and transalkylated with excess benzene to give ethylbenzene. The aluminum chloride catalyst originally used is corrosive and hard to dispose of. An improvement was the development of zeolite ZSM-5 as a catalyst in a continuous vapor-phase reaction that can operate at a temperature of 400°C at 18 bar. The higher temperature makes heat recovery easier, and the product is obtained in 99.5% selectivity with 98% ethylene conversion.

The dehydrogenation is endothermic and requires a high temperature (550–600°C) and low pressure. As in steam cracking (Section 2.2.1), superheated steam is added to inhibit coking and reduce the partial pressure of the reactants. The catalyst is highly selective. Its major component, present usually to the extent of more than 50%, comprises the oxides of iron, cobalt, manganese, chromium, or zirconium. If iron oxide is used, another oxide is added as a stabilizer to keep the iron in the ferric state. Alkaline metal oxides, particularly potassium and rubidium oxides, are particularly effective. A third ingredient is a carbon formation inhibitor, which may also be the oxide of potassium or rubidium. An oxide of copper, silver, cadmium, thorium, or vanadium is added as a secondary promoter to increase the effectiveness of the primary active ingredient. A "cement" such as calcium aluminate aids pellet formation. Organic substances such as methylcellulose may be included to introduce porosity when they burn off during the calcining of the catalyst.

The oxidative dehydrogenation of styrene was never considered practical because both ethylbenzene and styrene are oxygen sensitive. In the mid-1980s UOP described a new catalyst, to be used in combination with present catalysts. It preferentially oxidizes the hydrogen produced from the ethylbenzene to water in a separate step, thus not affecting the other components of the reaction mixture. It comprises platinum, tin, and lithium on alumina. Since water formation is exothermic, the heat generated reduces the superheated steam input necessary in the current process. This development had not been commercialized by the mid-1990s.

Proposed routes to styrene include the dehydrogenation of vinylcyclohexene prepared by butadiene dimerization (Section 5.1.2), the conversion of toluene to stilbene followed by metathesis with ethylene (Section 8.4), and the ZSM-5 catalyzed alkylation of toluene with methanol (Section 8.4). The literature contains references to the production of styrene directly from benzene and ethylene in the presence of oxidants for the liberated hydrogen. Only the toluene–methanol route currently holds promise of competing effectively with the dehydrogenation of ethylbenzene.

An obsolete process for styrene production involved oxidation of ethylbenzene to acetophenone and phenylethyl alcohol followed by hydrogenation of the acetophenone to phenylethyl alcohol, which dehydrates easily to styrene.

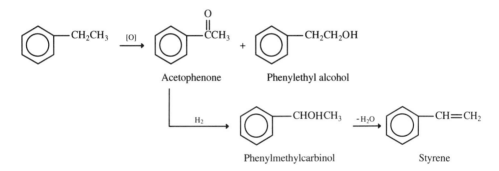

Acetophenone   Phenylethyl alcohol

Phenylmethylcarbinol   Styrene

p-Methylstyrene was at one time proposed as an alternative monomer to styrene. It is made by the ZSM-5-catalyzed alkylation of toluene with ethylene. This zeolite catalyst is para orienting providing high selectivity to *p*-methylbenzene. Dehydrogenation of the ethyl group provides *p*-methylstyrene. Since toluene is cheaper than benzene, *p*-methylstyrene should enjoy an economic advantage. In fact, the initial lower scale of production and escalating toluene prices in the mid- and late-1980s eliminated this advantage and the product was not commercialized.

Styrene's major use is for the manufacture of polystyrene, almost one-half of which is used for disposables such as packaging, and plastic cups and cutlery. The remainder goes into radio and TV cabinets, toys, door liners and trays for refrigerators, housings for small appliances such as clocks, housewares, furniture components, and in construction for window moldings, window shutters, and pipes. The longer lasting applications require so-called high impact polystyrene, made by polymerizing styrene in the presence of polybutadiene elastomer. A fast growing use is for audio and video cassettes and for support materials for magnetic tapes.

The second largest use for styrene is for elastomers including styrene–butadiene rubbers and latexes. A newer application is in so-called thermoplastic rubbers, typical of which is a block polymer of polystyrene–polybutadiene–polystyrene (Section 5.1).

The third largest use is in styrene copolymers, of which the most important is acrylonitrile–butadiene–styrene (ABS), the plastic widely used for telephones and calculator cases. The acrylonitrile provides strength, the butadiene flexibility and impact resistance, and the styrene gloss and hardness. Styrene is also found in styrene–acrylonitrile (SAN) and styrene–maleic anhydride copolymers.

The fourth most important use for styrene is as a reactive solvent for unsaturated polyesters which, after curing, become copolymers of styrene and the polyester.

3.9 ETHANOL

Ethanol was first produced from ethylene in 1930. Ethylene reacted with sulfuric acid to provide ethyl sulfate and diethyl sulfate, both of which were hydrolyzed to ethanol. Sulfuric acid was recycled. Diethyl ether is a byproduct and indeed the reaction can be run selectively to provide it (Section 3.11.10).

$$CH_2=CH_2 + H_2SO_4 \rightarrow CH_3CH_2OSO_3H + (CH_3CH_2O)_2SO_2$$

$$\text{Ethyl sulfate} \qquad \text{Diethyl sulfate}$$

$$CH_3CH_2OSO_3H + H_2O \rightarrow CH_3CH_2OH + H_2SO_4$$

$$(CH_3CH_2O)_2SO_2 + 2H_2O \rightarrow 2CH_3CH_2OH + H_2SO_4$$

The above process was replaced by direct hydration in which water is added to ethylene over a phosphoric acid catalyst on silica or celite.

$$CH_2=CH_2 + H_2O \overset{\text{cat.}}{\rightleftharpoons} CH_3CH_2OH$$

$$\text{Ethanol}$$

Selectivity to ethanol in this process is of the order of 92–98%. However, the reaction is equilibrium controlled, and thus, conversion is as low as 4% per pass although 20% has been reported. This means that large amounts of ethylene must be recycled. Low conversion is also necessary because the reaction is exothermic, so that the high temperatures required to achieve acceptable rates favor the ethylene–water side of the equilibrium.

Typical reaction conditions for the direct hydration are 300°C and 69 bar. The crude ethanol is purified by distillation to produce an azeotrope with water containing 95.6 volume percent ethanol. Absolute or 100% ethanol is usually prepared by azeotropic distillation. Benzene is added and the mixture distilled. The ternary azeotrope that distills carries the benzene and water with it, leaving anhydrous alcohol as the bottom product. Another method uses countercurrent extraction with glycerol or ethylene glycol. The added component hydrogen bonds to the water and allows anhydrous alcohol to be obtained from the top of the column. Less energy-intensive methods for obtaining anhydrous ethanol provide an active area of research. Membrane technology has held promise for at least a decade, but its development is slow.

Until about 1950, ethylene was expensive and was obtained from fermentation ethanol by dehydration—the reverse of the above processes. With the advent of cheap ethylene from steam cracking, the petrochemical route to ethanol became more economical than fermentation. By the early 1970s, scarcely any industrial ethanol was made by fermentation in the United States, although there was a legal requirement in most countries that potable ethanol be made in the traditional way by fermentation.

In the United States in the 1980s this trend was reversed when government subsidies were introduced to facilitate the production of ethanol by fermentation of cornstarch. The product went into automotive fuel known as gasohol. Fermentation ethanol is not economical without subsidies and does not provide a positive energy balance considering the energy required to grow the corn, isolate the starch from it, and distill the resulting ethanol. Since petroleum or natural gas is required to make up the energy deficit, fermentation ethanol cannot be considered a renewable energy source. Fermentation alcohol is discussed in Chapter 14 and Section 2.3.3.

The development of gasohol has motivated the development of other routes to ethanol, the most important of which is methanol homologation, as yet uncommercialized (Section 10.6.1).

Ethanol was once the basis for acetaldehyde and acetic acid production (Section 3.5) and this is still practiced in India and Brazil. In these countries ethylene was also prepared by the dehydration of ethanol as late as the mid-1980s. In addition to gasohol, ethanol has humble uses for ethyl ester formation. Ethyl acetate is made by esterification of acetic acid with ethanol. In some countries, where ethanol is expensive or there is surplus acetaldehyde capacity, ethyl acetate is made by a Tischenko reaction (Section 3.11.3). Ethanol is a solvent for surface coatings, cleaning preparations, and cosmetics. Fermentation ethanol is aerobically fermented to vinegar (dilute acetic acid).

3.10 MAJOR CHEMICALS FROM ETHYLENE: A SUMMARY

Sections 3.1–3.8 have described the "major" chemicals from ethylene, that is, chemicals that are produced in the United States at a level of more than about 1 billion lb/year. These chemicals and their production volumes for 1977 and 1993 are shown in Table 3.2 together with growth rates over this period. The final column gives the quantity of ethylene required to give the product, if it is made directly from ethylene. Thus figures for ethylbenzene and ethylene dichloride are given and not for styrene and vinyl chloride, which are made from them. Only one polymer family, polyethylenes, is included in the table since these are made directly from ethylene. Styrene, vinyl chloride, ethylene glycol, and vinyl

TABLE 3.2 Production of Chemicals from Ethylene (1977–1993)—Summary, United States (million lb)

Polymer or Chemical	1977	1993	Average Annual % Increase 1977–1993	Ethylene Consumption[a]
Ethylene	24,650	41,250	3.3	—
Polyethylenes	8,000	21,956	6.5	21956
Ethylene dichloride	10,480	17,950	3.4	5077
Ethylbenzene	7,300	11,760	3.0	3106
Styrene	6,820	10,070	2.5	—
Vinyl chloride	5,810	13,750	5.5	—
Ethylene oxide	4,420	5,680	1.6	4518
Ethylene glycol	3,470	5,230	2.6	—
Vinyl acetate	1,600	2,830	3.6	921
α-Olefins	820	2,700	7.7	2700
Ethanol[b]	1,300	120	− 14.0	73
Acetic acid	2,580	3,660	2.2	Very small[c]
Acetic anhydride	1,500	d	1.0	Very small[e]
Acetaldehyde[e]	970	300	− 7.0	Very small

[a] This figure is the ethylene required to give the 1993 production. Yields are assumed to be 100% except for ethylene oxide, where 80% is assumed.

[b] Consumption of industrial ethanol was much higher because of imports from Saudi Arabia.

[c] Acetic acid and acetic anhydride now derive from carbonylation processes. Ethylene based products shrank at about 11% per year.

[d] Total acetic anhydride demand for 1987 was estimated at 1.67 billion lb. Growth rate refers to 1977–1987.

[e] Consumption of acetaldehyde in 1974 was 1.0 billion lb. The figure in the third column refers to 1987 and the growth rate to 1977–1987.

acetate are all important monomers for polymers. Indeed this is the most important use for all the chemicals except ethanol and ethylene oxide, which is mainly converted to ethylene glycol, used almost equally in polymers and antifreeze.

Four chemicals—acetic acid, acetic anhydride, acetaldehyde, and ethanol— have lost volume precipitously between 1977 and 1993 insofar as they were based on ethylene. The first three of these are related to the new processes for acetic acid production, especially methanol carbonylation (Section 10.5.2.2), which does not involve ethylene. Acetic acid from ethylene-based acetaldehyde was formerly acetaldehyde's most important use. Acetic anhydride production is dependent on acetic acid and accordingly it too is no longer ethylene based. A new process for acetic anhydride based on methyl acetate carbonylation (Section 10.5.2.3) is also synthesis gas based.

Ethanol was formerly a raw material for acetic acid too, because it was oxidized to acetaldehyde prior to the advent of the Wacker process, but that application had disappeared long before 1977. Ethanol production in the United States has also plummeted (Section 1.5) because of imports from Saudi Arabia. Ethyl chloride production (not shown in Table 3.2 but it was 662 million lbs in 1974) has decreased because its major derivatives, the octane improvers lead tetraethyl and lead tetramethyl, have been phased out of US gasoline. It remains a precursor for silicones.

The other chemicals showing low growth in Table 3.2 are ethylene oxide and its derivative ethylene glycol. One reason for this is low growth in the poly(ethylene terephthalate) fiber market in developed countries. The bottle market, although growing at 8% per year, has only one-third of the market so far and is not making up for the low fiber growth. Ethylene glycol in antifreeze is growing only slowly because sealed radiator formulations last longer, and because there are moves to replace it by propylene glycol which, although less effective as an antifreeze, is not toxic to household pets who might lick up spilled material. Finally, the methyl and ethyl glycol ethers (Section 3.11.6.2) were found to be to teratogenic and were replaced by propylene-oxide-based analogues.

Relative newcomers to the list of major chemicals are α-olefins and fatty alcohols (Section 3.3). Fatty alcohols are increasingly used in surfactants for detergents because of their greater biodegradability. α-Olefins are comonomers in linear low-density and high-density polyethylene production. 1-Decene finds a growing use because its trimer is a synthetic lubricant.

Much of the growth in ethylene consumption came from the polyethylenes, which continue to find new applications. Low-density polyethylene usage is declining as it is being replaced by LLDPE. A market for LDPE will continue because in some applications it is used in combination with LLDPE, and because it is essential for extrusion coating of paper board for milk and juice cartons.

The final column of Table 3.2 shows the amount of ethylene going into each of the end-uses. The requirements of these major end-uses account for 93% of ethylene consumption even on the basis of the assumed 100% yields. That leaves about 7% to go into the myriad lesser volume chemicals, which will be discussed in the remainder of this chapter.

3.11 LESSER VOLUME CHEMICALS FROM ETHYLENE

This section deals with chemicals from ethylene that are produced in volumes of less than 1 billion lb/year in the United States. We shall include the copolymers mentioned in Section 3.2.

3.11.1 Hydroformylation: Propionaldehyde, Propionic Acid, and *n*-Propanol

The oxo or hydroformylation reaction on ethylene produces propionaldehyde, which in turn can be oxidized to propionic acid or reduced to *n*-propanol.

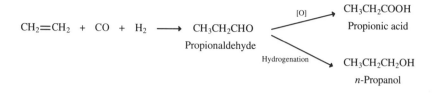

Hydroformylation, which is discussed in Section 4.8, was carried out initially with a dicobalt octacarbonyl, $(Co)_2(CO)_8$, catalyst at temperatures of about 150°C and 250–300 bar. A breakthrough was the discovery that rhodium chloride with ligands such as triphenylphosphine allowed the reaction to take place at temperatures of around 100°C and 10–25 bar. Hydrogenation of the aldehyde to the alcohol takes place with a nickel catalyst at 2–3 bar at about 115°C in the gas phase. In the liquid phase, higher pressures are required.

Propionaldehyde reacts with formaldehyde to provide trimethylolethane.

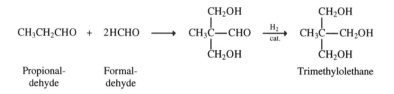

The formaldehyde condenses with the two active hydrogens to give a dimethylolaldehyde. The aldehyde group may then be reduced by another mole of formaldehyde, which is itself oxidized to formic acid. Alternatively, the dimethylolaldehyde may be isolated and reduced catalytically. The same chemistry is used to synthesize pentaerythritol (Section 3.11.3). Trimethylolethane is used for alkyd resins and urethanes (Section 7.3.1).

n-Propanol reacts with ammonia to form mono-, di- and tripropylamines:

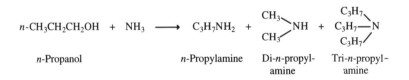

One of n-propanol's important uses is as a solvent, particularly for flexographic inks used with natural rubber rolls that are attacked by more powerful solvents. It also finds small use in the production of propyl esters such as propyl acetate.

Propionic acid may be made by the oxidation of propionaldehyde. The liquid-phase reaction is catalyzed by a cobalt salt at 100°C and 6–7 bar. An alternative process involves the carbonylation of ethylene by a so-called Reppe reaction, which is carried out with a nickel carbonyl catalyst in the liquid phase at 200–250 bar and 300°C. Nickel carbonyl is poisonous and in this reaction it is formed in situ.

$$CH_2=CH_2 + CO + H_2O \rightarrow CH_3CH_2COOH$$

Propionic acid

A low-pressure process for carbonylation of ethylene in the presence of water, which has not been commercialized, makes use of a nickel–molybdenum carbonyl catalyst with iodide and phosphoric acid promoters at about 200°C and 10–30 bar.

The route to acetic acid by oxidation of butane or naphtha (Section 11.2.1) gives propionic acid as a byproduct. In the United Kingdom, such a plant is kept in operation because of the value of the byproducts. A proposed propionic acid synthesis involves the homologation of acetic acid with CO and H_2 at 220°C and 270 bar. Butyric and valeric acids form as minor coproducts. The reaction is homogeneous and is catalyzed by ruthenium or rhodium with ligands such as acetylacetonate, with an iodide promoter or cocatalyst.

$$CH_3COOH + CO + 2H_2 \rightarrow C_2H_5COOH + H_2O$$

The major use of propionic acid is as a feed and grain preservative. An important use, particularly for the calcium or sodium salts, is as a food preservative especially in baked goods. A minor use is for the preparation of cellulose propionate via its derivative propionic anhydride.

3.11.2 Ethyl Halides

Ethyl chloride and ethyl bromide can be made by the addition of the corresponding hydrogen halides to ethylene. The reaction between ethylene and hydrogen chloride takes place either in the liquid or the gas phase at moderate temperatures and pressures with halide catalysts such as $AlCl_3$. A second process involves the chlorination of ethane, which, unlike methane (Section 10.2), gives 80% ethyl chloride before the dichloro compound starts to form. The chlorination of ethane generates ethyl chloride and hydrogen chloride, which can then be added to ethylene to provide more ethyl chloride in an integrated process.

$$C_2H_6 + Cl_2 \rightarrow C_2H_5Cl + HCl$$

$$CH_2=CH_2 + HCl \rightarrow C_2H_5Cl$$

The chlorination reaction takes place at 400°C. Since chlorine will not add to an olefin (Section 4.10.1) at this high temperature, the ethylene and ethane may be in the same reaction mixture, the chlorination of the ethane taking place in the presence of the essentially inert ethylene.

A route to ethyl chloride from ethanol and hydrogen chloride is obsolete because ethanol is more expensive than either ethane or ethylene.

Ethyl chloride consumption in the United States in 1975 was approximately 660 million lbs. By 1991 this had dropped by 80%. Its major use was in the manufacture of tetraethyl lead antiknock agents for gasoline that were almost completely phased out from the United States by 1991. Ethyl chloride is still used to etherify cellulose to ethyl cellulose. Small quantities are consumed as refrigerants and local anesthetics. Ethyl bromide is a low volume specialty chemical which, like its chlorine analog, can be made either by the addition of HBr to ethylene or by the bromination of ethane.

3.11.3 Acetaldehyde Chemistry

Acetaldehyde has been discussed previously (Section 3.5) as a derivative of ethylene whose consumption has declined because its major uses have disappeared. There are a number of lower volume chemicals still requiring acetaldehyde. Acetaldehyde with formaldehyde in the presence of alkali yields pentaerythritol:

$$CH_3CHO \ + \ 3HCHO \ \xrightarrow{cat.} \ HOCH_2 \overset{\displaystyle CH_2OH}{\underset{\displaystyle CH_2OH}{-\overset{|}{\underset{|}{C}}-}} CHO \ \xrightarrow[cat.]{H_2} \ HOCH_2 \overset{\displaystyle CH_2OH}{\underset{\displaystyle CH_2OH}{-\overset{|}{\underset{|}{C}}-}} CH_2OH$$

| Acetal-
dehyde | Formal-
dehyde | | Pentaerythritol |

In pentaerythritol, as in trimethylolethane formation, the aldehyde group of the trimethylolacetaldehyde may be reduced by 1 mol of formaldehyde, which oxidizes to 1 mol of formic acid in a crossed Cannizzaro reaction. Alternatively, the aldehyde may be reduced catalytically. Calcium hydroxide is frequently used as a catalyst. Well over 90% of pentaerythritol production goes into alkyd resins. The tetranitrate is an explosive and is also a vasodilator for the treatment of *angina pectoris*.

$$HOCH_2 \overset{\displaystyle CH_2OH}{\underset{\displaystyle CH_2OH}{-\overset{|}{\underset{|}{C}}-}} CH_2OH \ \xrightarrow{4HNO_3} \ O_2NOCH_2 \overset{\displaystyle CH_2ONO_2}{\underset{\displaystyle CH_2ONO_2}{-\overset{|}{\underset{|}{C}}-}} CH_2ONO_2 \ + \ 4H_2O$$

Pentaerythritol tetranitrate

Acetaldehyde undergoes the aldol condensation with an alkaline catalyst. Mild hydrogenation to avoid dehydration leads to 1,3-butanediol. This is of historical interest because on double dehydration it yields butadiene. This reaction was important during World War II when the monomer was needed with styrene for synthetic rubber.

$$2CH_3CHO \longrightarrow CH_3CHOHCH_2CHO \xrightarrow[\text{cat.}]{H_2} CH_3CHOHCH_2CH_2OH \xrightarrow{-2H_2O}$$

Acetaldol   1,3-Butanediol

$$CH_2{=}CH-CH{=}CH_2$$

Butadiene

Dehydration of acetaldol in the presence of acetic acid gives crotonaldehyde, which can be oxidized to crotonic acid. The oxidation is a mild one, carried out at room temperature and slightly above atmospheric pressure. Crotonic acid is a specialty monomer for use in copolymers where pendant carboxyl groups are required.

$$CH_3CHOHCH_2CHO \xrightarrow{-H_2O} CH_3-CH{=}CH-CHO \xrightarrow{[O]} CH_3-CH{=}CHCOOH$$

Crotonaldehyde   Crotonic acid

Mild hydrogenation of crotonaldehyde yields *n*-butyraldehyde. More extensive hydrogenation attacks both the double bond and the aldehyde group to give *n*-butanol. Today, the route is obsolete and *n*-butanol comes from the hydroformylation of propylene (Section 4.8).

$$CH_2CH{=}CH-CHO \xrightarrow[\text{cat.}]{H_2} C_3H_7CHO \xrightarrow[\text{cat.}]{H_2} C_3H_7CH_2OH$$

Crotonaldehyde   *n*-Butyraldehyde   *n*-Butanol

Crotonaldehyde, in a curious reaction with ethanol, gives butadiene and acetaldehyde.

$$CH_3CH{=}CHCHO + CH_3CH_2OH \rightarrow$$

Crotonaldehyde   Ethanol

$$CH_2{=}CHCH{=}CH_2 + CH_3CHO + H_2O$$

Butadiene   Acetaldehyde

The latter can be recycled and converted to more crotonaldehyde. This synthesis of butadiene is still in use in India and in the People's Republic of China, although it will no doubt be phased out as petrochemical C_4 fractions become available.

Sorbic acid, 2,4-hexadienoic acid, in which both double bonds are *trans*, was originally made by two processes, one of which involves an aldol condensation between acetaldehyde and crotonaldehyde. The resulting sorbic aldehyde was then mildly oxidized to sorbic acid.

$$CH_3CH{=}CHCHO \ + \ CH_3CHO \xrightarrow{-H_2O} \ CH_3CH{=}CH{-}CH{=}CH{-}CHO \xrightarrow{[O]}$$

Crotonaldehyde   Acetal-dehyde   Sorbaldehyde

$$CH_3CH{=}CH{-}CH{=}CHCOOH$$

Sorbic acid

Selectivities are low since several aldol condensations can take place. In today's more sophisticated process, crotonaldehyde is reacted with ketene to give a β-lactone intermediate, which is hydrolyzed and dehydrated to sorbic acid.

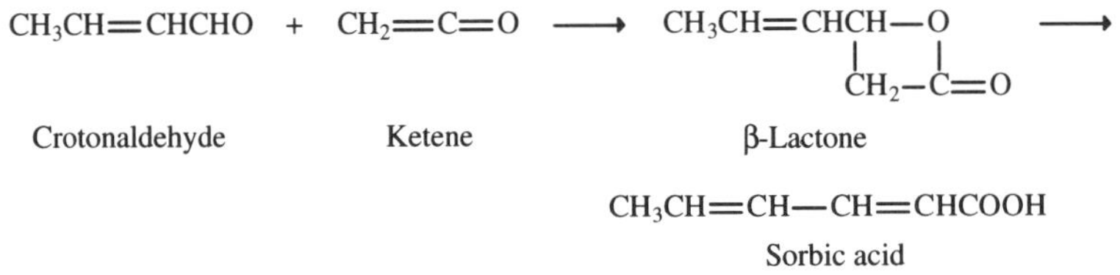

$$CH_3CH{=}CHCHO \ + \ CH_2{=}C{=}O \longrightarrow CH_3CH{=}CHCH{-}O \longrightarrow$$

$$\underset{CH_2{-}C{=}O}{|\qquad|}$$

Crotonaldehyde   Ketene β-Lactone

$$CH_3CH{=}CH{-}CH{=}CHCOOH$$

Sorbic acid

It is also possible to form an intermediate polyester by the self-polymerization of the lactone. It too will hydrolyze to sorbic acid. Ketene results from the pyrolysis of either acetone or acetic acid (Section 10.5.2.3).

Sorbic acid and its calcium and potassium salts (Section 3.11.1) are used in the same way as calcium propionate as preservatives for foods, especially baked goods.

Acetaldehyde is a source of ethyl acetate by way of the Tishchenko reaction (Section 3.9). The catalyst is aluminium ethoxide.

$$2CH_3CHO \xrightarrow{cat.} CH_3COOCH_2CH_3$$

Ethyl acetate

Pyridine and some of its derivatives are still isolated from coal tar. Synthetic pyridines from acetaldehyde have achieved importance. Ammonium acetate catalyzes the reaction of acetaldehyde and ammonia to 2-methyl-5-ethyl-pyridine by way of aldol condensations between four molecules of the aldehyde, followed by ring closure with ammonia. Triple dehydration brings about aromatization. The reaction takes place at high temperatures and pressures.

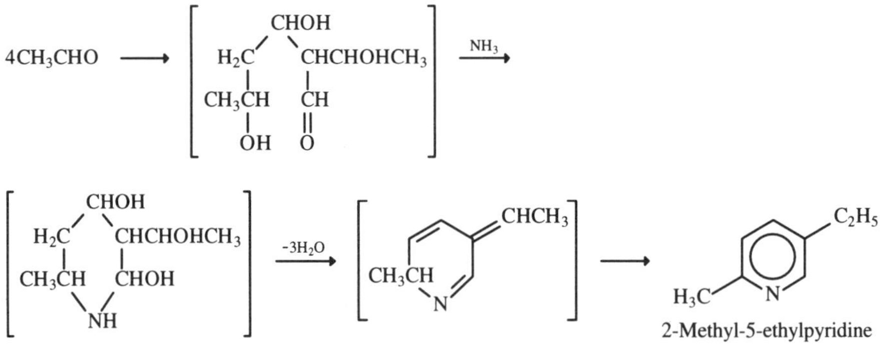

2-Methyl-5-ethylpyridine

Oxidation of 2-methyl-5-ethylpyridine with nitric acid gives nicotinic acid by way of the decarboxylation of the intermediate dicarboxylic acid.

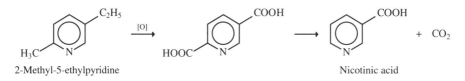

2-Methyl-5-ethylpyridine   Nicotinic acid

Nicotinic acid and its amide are both termed vitamin B6. They lower lipid levels in the blood and have a long established but more doubtful pharmaceutical use as peripheral vasodilators.

The reaction of acetaldehyde and formaldehyde in the presence of ammonia and a catalyst yields a mixture of pyridine and 3-picoline. Acetaldehyde and ammonia in a 3:1 ratio, when passed over various dehydration/dehydrogenation catalysts (e.g., PbO or CuO on alumina; ThO_2 on ZnO; or CdO on silica–alumina) give equimolar quantities of 2- and 4-picolines. 2-Picoline is a component of a coccidiostat; 4-picoline is a precursor of the antituberculosis drug isoniazid. Pyridines generally are of value as intermediates for herbicides, pesticides, and pharmaceuticals.

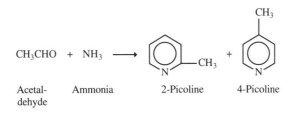

    Acetal-   Ammonia   2-Picoline   4-Picoline
    dehyde

The chlorination of acetaldehyde yields trichloroacetaldehyde or chloral. In practice ethanol is chlorinated and is oxidized to the aldehyde in situ. In the presence of the hydrogen chloride that forms, the trichloroacetaldehyde reacts with more ethanol to give the acetal.

$$CH_3CH_2OH + 4Cl_2 \longrightarrow 5HCl + [CCl_3CHO] \xrightarrow{2C_2H_5OH} CCl_3CH\begin{smallmatrix}OC_2H_5\\OC_2H_5\end{smallmatrix} + H_2O$$

Diethylacetal of
trichloroacetaldehyde

The reaction of the acetal with sulfuric acid provides the desired trichloracetaldehyde and ethyl hydrogen sulfate.

$$CCl_3CH\begin{smallmatrix}OC_2H_5\\OC_2H_5\end{smallmatrix} + 2H_2SO_4 \xrightarrow{H_2O} CCl_3CHO + 2C_2H_5OSO_3H + 2H_2O$$

Trichloro-
acetaldehyde

The hydrate of chloral is stable and in the past has been useful as a soporific known in the vernacular as "knock-out" drops. The major use for trichloracetaldehyde has been in the manufacture of the insecticide 1,1,1-trichloro-2,2-bis (p-chlorophenyl)ethane (DDT) (Section 7.6).

Peracetic acid (more appropriately peroxyacetic acid) is an intermediate in the oxidation of acetaldehyde to acetic acid (Section 3.5). If the oxidation is carried out at temperatures no higher than 40°C and at about 35 bar, it can be isolated.

$$CH_3CHO + O_2 \longrightarrow CH_3C-OOH$$
$$\underset{O}{\overset{\|}{}}$$

Peracetic acid

The alternative route to peracetic acid is the reaction of acetic acid with expensive hydrogen peroxide. Peracetic acid is used for several industrially important epoxidation reactions including the epoxidation of soybean oil (Section 13.7), α-olefins, and certain diunsaturates such as vinylcyclohexene (butadiene dimer), which give reactive intermediates for the formation of materials resembling epoxy resins.

3.11.4 Metal Complexes

Triethylaluminum is important for the production of Ziegler alcohols and α-olefins (Section 3.3.2). Its combination with titanium salts provides the Ziegler–Natta catalyst (Sections 3.1 and 4.1). Triisobutylaluminum is an alternative. The process for preparing triethylaluminum is more complex than that for its higher analogs. Triisobutylaluminum results from the interaction of isobutene with hydrogen and aluminum at 150°C and 200 bar.

$$3CH_3CCH_3 + 1.5H_2 + Al \longrightarrow \left[\underset{CH_3}{\overset{CH_3CHCH_2}{\underset{|}{}}} \right]_3 Al$$

Isobutene   Triisobutylaluminum

Triethylaluminum, on the other hand, cannot be made directly from ethylene, hydrogen, and aluminum because so high a temperature is required that the triethylaluminum formed reacts with more ethylene to give a mixture of higher alkyl aluminums. Thus it is necessary to react preformed triethylaluminum with aluminum powder and hydrogen to produce diethylaluminum hydride, which then reacts with more ethylene to yield more triethylaluminum than was used initially. Alternatively, both steps can be combined by simultaneous addition of hydrogen and ethylene to a mixture of triethylaluminum and aluminum powder.

$$4Al(C_2H_5)_3 + 2Al + 3H_2 \xrightarrow{120°C} 6AlH(C_2H_5)_2$$

Triethyl-   Diethylaluminium
aluminium   hydride

$$AlH(C_2H_5)_2 + C_2H_4 \xrightarrow{70°C} Al(C_2H_5)_3$$

The exchange reaction between triisobutylaluminum and ethylene also gives triethylaluminum. Isobutene is recycled.

$$
\underset{\text{Triisobutyl-}\atop\text{aluminium}}{\text{Al(CH}_2\overset{\text{CH}_3}{\overset{|}{\text{CH}}}\text{CH}_3)_3} + 3\text{CH}_2{=}\text{CH}_2 \longrightarrow \text{Al(C}_2\text{H}_5)_3 + \underset{\text{Isobutene}}{3\text{CH}_3{-}\overset{\text{CH}_3}{\overset{\|}{\text{C}}}{-}\text{CH}_3}
$$

3.11.5 Ethylene Diamine and Related Compounds

The reaction of ethylene dichloride (Section 3.3) with ammonia provides ethylene diamine and higher homologs including diethylene triamine, triethylene tetramine, and tetraethylene pentamine. The compounds form as hydrochlorides and are transformed to free bases with sodium hydroxide. The structures of the bases are shown in the equation. The reaction is carried out in the liquid phase with molar excesses of ammonia as great as 30:1 to minimize the formation of higher homologs. If the latter are required, less ammonia is used.

$$
\text{ClCH}_2\text{CH}_2\text{Cl} + 2\text{NH}_3 \longrightarrow \underset{\substack{\text{Ethylene diamine}\\\text{dihydrochloride}}}{\text{NH}_2\text{CH}_2\text{CH}_2\text{NH}_2 \cdot 2\text{HCl}}
$$

$$
+ \underset{\text{Diethylene triamine}}{\text{NH}_2\text{CH}_2\text{CH}_2\text{NHCH}_2\text{CH}_2\text{NH}_2} + \underset{\text{Triethylene tetramine}}{\text{NH}_2\text{CH}_2\text{CH}_2\text{NHCH}_2\text{CH}_2\text{NHCH}_2\text{CH}_2\text{NH}_2}
$$

$$
+ \underset{\text{Tetraethylene pentamine}}{\text{NH}_2\text{CH}_2\text{CH}_2\text{NHCH}_2\text{CH}_2\text{NHCH}_2\text{CH}_2\text{NHCH}_2\text{CH}_2\text{NH}_2}
$$

The above reaction gives vinyl chloride and organic-contaminated ammonium chloride as byproducts by the dehydrohalogenation of ethylene dichloride:

$$
\underset{\substack{\text{Ethylene}\\\text{dichloride}}}{\text{ClCH}_2\text{CH}_2\text{Cl}} + \text{NH}_3 \rightarrow \underset{\substack{\text{Vinyl}\\\text{chloride}}}{\text{CH}_2{=}\text{CHCl}} + \text{NH}_4\text{Cl}
$$

These present environmental problems. Accordingly, processes have been devised for the production of ethyleneamines from ethanolamines (Section 3.11.6.4) and ammonia. Piperazine, aminoethylpiperazine, and hydroxyethylpiperazine form as byproducts.

$$
\underset{\text{Ethanolamine}}{\text{HOC}_2\text{H}_4\text{NH}_2} + \text{H}_2 + \text{NH}_3 \longrightarrow \text{Di- and polyamines shown in the above equation plus}
$$

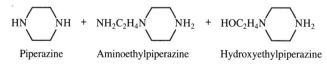

Piperazine   Aminoethylpiperazine   Hydroxyethylpiperazine

The reaction, a reductive amination, is carried out at high pressure in the vapor phase in the presence of hydrogen, with an amination catalyst such as nickel, cobalt, or copper. These polyamines are used as cross-linking agents for epoxy resins (Section 4.10.1) and for the preparation of low molecular weight fatty polyamide resins by reaction with so-called dimer acids (Section 13.3).

The reaction of ethylene diamine with carbon disulfide and sodium hydroxide gives the sodium salt of a bisdithiocarbamate, an important fungicide, "Dithane."

$$NH_2C_2H_4NH_2 \ + \ 2CS_2 \ + \ 2NaOH \ \longrightarrow \ Na\overset{\overset{S}{\|}}{C}SNHC_2H_4NH\overset{\overset{S}{\|}}{C}Na \ + \ 2H_2O$$

Ethylene diamine	Carbon disulfide		Dithane	

The reaction of ethylene diamine with formaldehyde and sodium cyanide in alcohol solution gives the sodium salt of ethylenediaminetetraacetic acid.

$$H_2NCH_2CH_2NH_2 \ + \ 4HCHO \ + \ 4NaCN \ + \ 4H_2O \ \xrightarrow[80°C]{NaOH}$$

Formaldehyde   Sodium cyanide

$$(NaOOCCH_2)_2NCH_2CH_2N(CH_2COONa)_2 \ + \ 4NH_3$$

Disodium salt of ethylenediaminetetraacetic acid

An alternative route involves the reaction of ethylene diamine with sodium chloroacetate. The product is a chelating agent both for alkaline earth and heavy metal ions.

Ethylene diamine reacts with ethylene oxide to provide hydroxyethylamino-ethylamine, a textile finishing agent and an epoxy resin curing agent.

$$NH_2C_2H_4NH_2 \ + \ H_2C\overset{O}{\overset{/\ \backslash}{\longrightarrow}}CH_2 \ \longrightarrow \ HOC_2H_4NHC_2H_4NH_2$$

Ethylene oxide   Hydroxyethylamino-
                             ethylamine

Piperazine dihydrochloride forms from the condensation of 2 mols of ethylene diamine dihydrochloride. Reaction of piperazine with ethylene dichloride yields triethylenediamine or DABCO, the acronym of its chemical name 1,4-diaza-bicyclo-(2,2,2)-octane. Triethylene diamine is an important catalyst for the interaction of isocyanates with polyols in urethane technology.

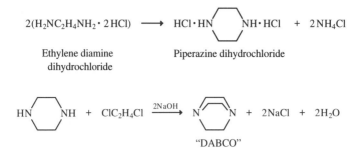

3.11.6 Ethylene Oxide and Ethylene Glycol Derivatives

3.11.6.1 Oligomers Ethylene oxide oligomers—diethylene glycol and tri-ethylene glycol—form during the hydrolysis of ethylene oxide to ethylene glycol (Section 3.8.1). If more of the oligomers are desired, less water is used for the hydrolysis, but normally more than enough byproduct diethylene glycol is available. Additional triethylene glycol is produced by the reaction of ethylene oxide with ethylene glycol or diethylene glycol (Section 3.7.1).

Di- and triethylene glycols are used in the manufacture of unsaturated polyesters (Section 9.3.4) to which they impart resilience and toughness. The ether linkages, however, increase water sensitivity. They are also used as solvents in polyurethane formulations and as the basis for textile chemicals, which serve as lubricants, softeners, finishers, and dye assistants. They have been used for the extraction of aromatic hydrocarbons from aliphatic–aromatic mixtures such as catalytic reformate (Section 2.2.3) but sulfolane (Section 5.1.4) has largely replaced them. The glycols are useful for dehydration of natural gas, in the manufacture of plasticizers and surfactants, as tobacco humectants, and at a concentration no greater than 5% as antifreeze components. Triethylene glycol serves as a coalescing agent in water base paints to fuse solid particles of the vehicle into a film. The oligomeric polyethylene glycols are water soluble, waxy solids useful in cosmetic formulations and as ion coordination catalysts.

It is possible to obtain high molecular weight polyethylene glycols by the polymerization of ethylene oxide with an iron catalyst. These have not found extensive use. One application is in cosmetics. Another is to reduce the viscosity of water by decreasing its hydrogen bonding. Lower viscosity water enables easier flow through pipes particularly in fire fighting systems.

3.11.6.2 Glycol Ethers and Esters Ethylene oxide reacts with alcohols to provide glycol ethers, the most important of which was ethylene glycol mono-ethyl ether (Cellosolve, Oxitol) the reaction product of ethanol and ethylene oxide. Oligomers, for the most part of little value, also form despite the use of large excesses of ethanol.

$$H_2C\overset{\diagdown}{\underset{O}{\diagup}}CH_2 \ + \ C_2H_5OH \ \longrightarrow \ C_2H_5OCH_2CH_2OH$$

Ethanol   Ethylene glycol monoethyl ether

The glycol ethers may be esterified with acetic acid to give ethylene glycol ether acetates or further etherified by conversion of the hydroxyl to a sodium salt, which in turn will react with methyl chloride to give 1-methoxy-2-ethoxyethane.

$$C_2H_5OCH_2CH_2OH \ \xrightarrow[-H_2O]{NaOH} \ C_2H_5OCH_2CH_2ONa \ \xrightarrow[\substack{Methyl \\ chloride \\ -NaCl}]{CH_3Cl} \ C_2H_5OCH_2CH_2OCH_3$$

1-Methoxy-2-ethoxyethane

Other commercial glycol ethers are based on methanol and butanols. The range is large; most of the commercially important compounds are shown in Table 3.3. The total figures include propylene glycol ethers and esters.

The ethylene glycol methyl and ethyl ethers and their acetates, however, have been found to harm the reproductive and development process in mammals. Accordingly, very low emission levels have been specified. To some extent they are being replaced by related materials either with longer alkyl chains or based on propylene oxide (Section 4.7), which may not be quite as good solvents because of their branched structure.

The glycol ethers and esters are largely used as solvents for protective coatings and in smaller volumes as solvents for printing inks, liquid cleaners, dyestuffs, and cosmetics. They are jet fuel deicing agents. Ethylene glycol monobutyl ether is particularly useful in water-borne coatings. Ethylene glycol dimethyl ether is used as an aprotic solvent (Section 4.4). A synthesis that decreases raw material costs by eliminating the need for sodium hydroxide and chlorine involves formation of the formal of the ether, which on hydrogenolysis yields one mol of the dimethyl ether and one mole of the starting material for recycle.

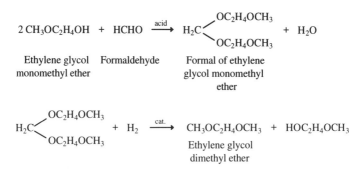

2 $CH_3OC_2H_4OH$ + HCHO   Ethylene glycol monomethyl ether   Formaldehyde   Formal of ethylene glycol monomethyl ether

3.11.6.3 Ethylene Carbonate

Ethylene carbonate can be prepared by reaction of ethylene oxide and carbon dioxide with basic catalysts such as sodium hydroxide, and tertiary amines or quaternary ammonium compounds. The reaction is one of the few apart from the Kolbe synthesis, methanol formation (Section 10.4.2), and urea production in which carbon dioxide participates. The product is a high boiling, aprotic solvent useful in the synthesis of rubber chemicals and textile agents. Its use as an intermediate in ethylene glycol and dimethyl carbonate formation has been described (Section 3.7.1).

$$H_2C\text{---}CH_2 + CO_2 \longrightarrow H_2C\text{---}CH_2$$

Ethylene
oxide

Ethylene
carbonate

3.10.6.4 *Aminoethyl Alcohols (Ethanolamines) and Derivatives* The inter-
action of ethylene oxide with ammonia provides ethanolamine (often known as monoethanolamine), diethanolamine, and triethanolamine. The reaction is carried out with aqueous ammonia at about 100°C and 100 bar. If a high selectivity to ethanolamine is desired, excess ammonia is used. The formation of the di- and tri-products is favored because the rate of reaction of ammonia with mono-ethanolamine is greater than with ethylene oxide.

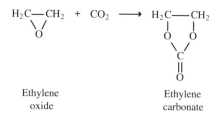

Ethanolamine   Diethanolamine   Triethanolamine

The reaction of fatty acid methyl esters such as methyl laurate with ethanolamines provides fatty ethanolamides widely used as foam stabilizers in detergent formulations.

$$CH_3(CH_2)_{10}COOCH_3 + NH_2C_2H_4OH \xrightarrow{\text{cat.}} CH_3(CH_2)_{10}COONHC_2H_4OH + CH_3OH$$

Methyl laurate   Ethanolamine   N-Hydroxyethyllauramide

Ethanolamines are used for the removal of acid gases from gaseous streams. Monoethanolamine absorbs carbon dioxide and hydrogen sulfide but diethanolamine is required to absorb carbonyl sulphide. Being weakly basic, the ethanolamines form loose compounds with acid gases, but these are decomposed by steam stripping and the ethanolamines regenerated for recycle.

Of the three ethanolamines, triethanolamine enjoys about 40% of total demand, its major use being in the formation of a quaternary salt with fatty acids to provide a surfactant useful in dry cleaning and cosmetic formulations.

TABLE 3.3 Commercially Important Ethylene Glycol Ethers and Esters

Trivial name	Chemical Name	Formula	US Production 1991 (tonnes)
Ethylene glycol ethers (including all polyhydric alcohol ethers)			1,010,280
Diethylene glycol		$HOCH_2CH_2OCH_2CH_2OH$	221,185
Triethylene glycol		$HOCH_2CH_2OCH_2CH_2OCH_2CH_2OH$	53,302
Tetraethylene glycol		$HOCH_2CH_2OCH_2CH_2OCH_2CH_2OCH_2CH_2OH$	12,782
Polyethylene glycol		$HO(CH_2CH_2O)_nH$	61,519
Ethylene glycol monomethyl ether	2-Methoxyethanol	$HOCH_2CH_2OCH_3$	n.a.[a]
Ethylene glycol monoethyl ether	2-Ethoxyethanol	$HOCH_2CH_2OCH_2CH_3$	31,886
Ethylene glycol mono-n-butyl ether	2-n-Butoxyethanol	$HOCH_2CH_2OCH_2CH_2CH_2CH_3$	156,437
Ethylene glycol dimethyl ether	1,2-Dimethoxyethane	$CH_3OCH_2CH_2OCH_3$	n.a.
Diethylene glycol monomethyl ether	2-(2-Methoxyethoxy)ethanol	$HOCH_2CH_2OCH_2CH_2OCH_3$	19,959
Diethylene glycol monoethyl ether	2-(2-Ethoxyethoxy)ethanol	$HOCH_2CH_2OCH_2CH_2OCH_2CH_3$	12,797
Diethylene glycol mono-n-butyl ether	2-(2-n-Butoxyethoxy)ethanol	$HOCH_2CH_2OCH_2CH_2OCH_2CH_2CH_2CH_3$	109,070
Diethylene glycol dimethyl ether	1,1'-Oxybis(2-methoxyethane)	$CH_3OCH_2CH_2OCH_2CH_2OCH_3$	n.a.
Diethylene glycol di-n-butyl ether	1,1'-Oxybis(2-n-butoxyethane)	$CH_3(CH_2)_3OC_2H_4OC_2H_4O(CH_2)_3CH_3$	n.a.

Triethylene glycol monomethyl ether	2-[2-(2-Methoxyethoxy)-ethoxy]ethanol	$HOCH_2CH_2OCH_2CH_2OCH_2CH_2OCH_3$	11,959
Triethylene glycol mono-*n*-butyl ether	2-[2-(2-*n*-butoxyethoxy)-ethoxy]ethanol	$HOCH_2CH_2OCH_2CH_2OCH_2CH_2O(CH_2)_3CH_3$	12,250
Ethylene glycol esters (including all polyhydric alcohol ethers)			133,005
Ethylene glycol monomethyl ether acetate	2-Methoxyethyl acetate	$CH_3COOCH_2CH_2OCH_3$	n.a.
Ethylene glycol monoethyl ether acetate	2-Ethoxyethyl acetate	$CH_3COOCH_2CH_2OCH_2CH_3$	n.a.
Ethylene glycol monobutyl ether acetate	2-Butoxyethyl acetate	$CH_3COOCH_2CH_2OCH_2CH_2CH_2CH_3$	6,173
Glycols			
Diethylene glycol		$HOCH_2CH_2OCH_2CH_2OH$	221,185

Source: Synthetic Organic Chemicals, US Production and sales 1991, USITC Pub. 2607, Washington DC, 1993. More recent editons of this publication do not include such detailed information.

[a] Not available = n.a.

$$
\underset{\substack{\text{Triethanolamine}}}{(HOC_2H_4)_3N} + \underset{\substack{\text{Stearic acid}}}{C_{17}H_{35}COOH} \longrightarrow \underset{\substack{\textit{tris}\text{-Hydroxyethylammonium} \\ \text{stearate}}}{C_{17}H_{35}COO^-\overset{H}{\overset{|}{N}}{}^+(C_2H_4OH)_3}
$$

Morpholine results from the dehydration of diethanolamine with sulfuric acid, a sulfate forming, which on treatment with alkali provides the free base.

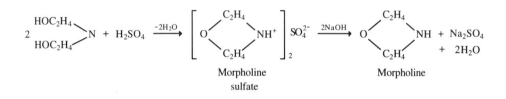

$$
2\underset{\substack{HOC_2H_4}}{\overset{HOC_2H_4}{>}}N + H_2SO_4 \xrightarrow{-2H_2O} \left[\underset{\substack{\text{Morpholine} \\ \text{sulfate}}}{O\overset{C_2H_4}{\underset{C_2H_4}{<}}NH^+}\right]_2 SO_4^{2-} \xrightarrow{2NaOH} \underset{\substack{\text{Morpholine}}}{O\overset{C_2H_4}{\underset{C_2H_4}{<}}NH} + Na_2SO_4 + 2H_2O
$$

3.11.6.5 Ethyleneimine
Ethyleneimine can be made either from monoethanolamine or ethylene dichloride. The aminoalcohol is esterified with sulfuric acid and the resulting compound treated with a stoichiometric amount of sodium hydroxide to generate the imine.

$$
\underset{\substack{\text{Monoethanolamine}}}{H_2NCH_2CH_2OH} + H_2SO_4 \xrightarrow{-H_2O} \underset{\substack{\text{Ethanolamine sulfate}}}{H_2NCH_2CH_2OSO_3H} \xrightarrow{NaOH} \underset{\substack{\text{Ethyleneimine}}}{H_2C\!\!-\!\!CH_2} + H_2SO_4
$$

Ethylene dichloride with ammonia in the presence of lime also yields ethylene imine.

$$
\underset{\substack{\text{Ethylene} \\ \text{dichloride}}}{ClCH_2CH_2Cl} + 2NH_3 \xrightarrow{CaO} \underset{\substack{\text{Ethyleneimine} \\ \text{hydrochloride}}}{H_2C\!\!-\!\!CH_2} + NH_4Cl
$$

There is a large body of chemistry associated with ethyleneimine, most of which is beyond the scope of this volume. Most important is its polymerization to poly(ethyleneimine), a cationic flocculent useful for the purification of waste waters, particularly water from paper manufacture. Reaction with phosphoryl trichloride, $POCl_3$, provides *tris*-(1-aziridinyl phosphine oxide) a reactive intermediate for use in imparting fire resistance and creaseproofing for textiles. Ethyleneimine and some of its derivatives are toxic and ecologically harmful.

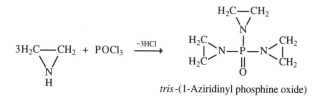

tris-(1-Aziridinyl phosphine oxide)

3.11.6.6 Ethylene Glycol Derivatives

The gas-phase oxidation of ethylene glycol at 300°C yields glyoxal. Catalysts are silver or copper inhibited with halide additives to prevent total oxidation of the glycol. Alternatively, acetaldehyde can be oxidized with nitric acid less selectively to give glyoxal and a number of byproducts including glyoxylic acid, OHCCOOH.

$$HOCH_2CH_2OH \xrightarrow[\text{Nitric acid}]{\text{cat. or}} OHCCHO$$

Ethylene glycol   Glyoxal

This is the major product when the oxidation is carried out at a higher temperature. It was at one time important in the synthesis of ethylene glycol (Section 3.7.1). Since anhydrous glyoxal polymerizes readily, it is sold as a hydrate. In textile finishing it functions as a shrinkproofing agent by crosslinking the cellulose chains of cotton through the formation of cyclic acetals with adjacent hydroxyl groups.

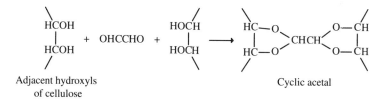

Adjacent hydroxyls
of cellulose   Cyclic acetal

Its ability to react with cellulose also means that it lends wet strength to paper and increases its absorbency. It insolubilizes starch and protein and has been used as a reducing agent in the process of silvering mirrors. It reacts with amino groups and hence insolubilizes casein and animal glues. It is used to immobilize enzymes by bonding both to them and a substrate such as glass. Glyoxal mixed with formaldehyde is the basis for embalming fluid.

Glyoxal and glyoxylic acid are both used in the synthesis of 4-hydroxyphenylglycine, the side-chain precursor in the manufacture of ampicillin. In one process—a modified Mannich reaction—phenol in aqueous ammonia reacts with glyoxylic acid to give 4-hydroxyphenylglycine, presumably via the intermediate p-hydroxymandelic acid.

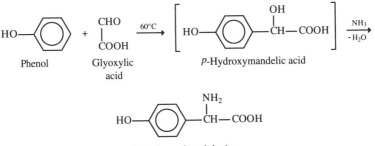

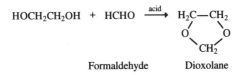

4-Hydroxyphenylglycine

In the second method, phenol, glyoxal, and urea react to give a hydroxyphenyl-hydantoin that is hydrolyzed by water to 4-hydroxyphenylglycine.

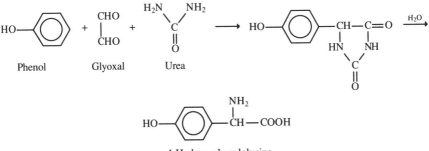

4-Hydroxyphenylglycine

Dioxolane is related to ethylene carbonate (3.11.6.3) and is the acetal from the acid-catalyzed reaction of ethylene glycol and formaldehyde.

$$HOCH_2CH_2OH + HCHO \xrightarrow{acid} H_2C-CH_2$$

Formaldehyde   Dioxolane

It is a comonomer in the formation of polyacetal polymers for which the major monomer is formaldehyde. It is also a strong aprotic solvent. Being an acetal, it is unstable in strong acid.

Another ethylene glycol derivative is 1,4-dioxane, which results from the dehydration of ethylene glycol with dilute sulfuric acid. Presumably, diethylene glycol forms as an intermediate and cyclizes to dioxane, since diethylene glycol may also be used as the starting material.

$$2HOCH_2CH_2OH \xrightarrow[acid]{-H_2O} HOCH_2CH_2OCH_2CH_2OH \xrightarrow[acid]{-H_2O}$$

Diethylene glycol   Dioxane

Ethylene oxide, ethylene chlorohydrin, or 2,2-dichlorodiethyl ether (prepared by treating ethylene chlorohydrin with sulfuric acid) all condense to form 1,4-dioxane. Dioxane's major use is as a solvent, but care must be exercised because it readily forms explosive peroxides.

Ethylene glycol can be nitrated to form ethylene glycol dinitrate, useful as a freeze inhibitor for explosives.

$$HOCH_2CH_2OH + 2HNO_3 \rightarrow O_2NOCH_2CH_2ONO_2$$

Ethylene glycol dinitrate

3.11.7 Vinyl Chloride and Ethylene Dichloride Derivatives

Vinyl chloride is the basis for other chlorinated compounds. Vinylidene chloride (1,1-dichloroethylene) results from the chlorination of vinyl chloride followed by dehydrohalogenation. The intermediate 1,1,2-trichloroethane can also be made by chlorinating ethylene dichloride:

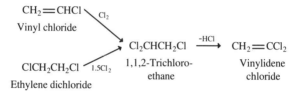

Vinylidene chloride is the basis for "Saran," which is a copolymer of 85% vinylidene chloride with 15% vinyl chloride. The polymer has high-tensile strength, which makes possible the formation of thin films whose electrostatic properties cause it to cling and assume the shape of the products, primarily food, which it protects. Poly(vinylidene chloride) films are clear because they are biaxially oriented, that is, they are stretched in two directions simultaneously. The polymer is also used as a coating for nitrocellulose and polypropylene films to impart air and water impermeability.

The direct chlorination of ethylene dichloride gives a mixture of trichloro-ethylene and tetrachloroethylene in addition to 1,1,2-trichloroethane, mentioned above.

$$ClCH_2CH_2Cl + Cl_2 \rightarrow Cl_2C{=}CHCl + Cl_2C{=}CCl_2 + HCl$$

Trichloro-   Tetrachloroethylene   ethylene

To avoid waste of the hydrogen chloride byproduct, oxychlorination can be used as it is in vinyl chloride manufacture (Section 3.4). The mechanism of the process, however, is somewhat different since it makes use of ethylene dichloride and chlorine in the presence of oxygen with a cupric chloride catalyst. In the vinyl chloride process, cupric chloride is used as a source of chlorine. In this process, chlorine, oxygen, and ethylene dichloride react to provide tri- and tetrachloroethylene and water. The hydrogen chloride that forms presumably oxidizes immediately to chlorine for further chlorination.

The traditional synthesis of tri- and tetrachloroethylene started with acetylene. This route is still used in Europe.

$$CH{\equiv}CH + 2Cl_2 \rightarrow CHCl_2CHCl_2 \xrightarrow{-HCl} Cl_2C{=}CHCl \xrightarrow{Cl_2}$$

Trichloro-
ethylene

$$CHCl_2CCl_3 \xrightarrow{-HCl} Cl_2C{=}CCl_2$$

Trichloro-
ethylene

Trichloroethylene is used as a solvent, a degreasing agent in metal working, and in dry cleaning. Tetrachloroethylene, also called perchloroethylene, is used to extract fats and oils. Both solvents, however, are in jeopardy because of toxicity and ecological problems and consumption is rapidly decreasing in the 1990s.

3.11.8   Vinyl Fluoride and Vinylidene Fluoride

Vinyl fluoride can be prepared either from acetylene (Section 10.9) or from vinyl chloride. In the latter process, vinyl chloride is treated with hydrogen fluoride with a mercury salt or metallic fluoride catalyst to give an addition product, which on dehydrohalogenation yields vinyl fluoride.

$$CH{\equiv}CH + HF \rightarrow CH_2{=}CHF$$

Vinyl fluoride

$$CH_2{=}CHCl + HF \xrightarrow{cat.} CH_3CHClF \xrightarrow{-HCl} CH_2{=}CHF$$

Vinyl fluoride

The noncatalyzed addition of hydrogen fluoride to vinyl chloride proceeds at mild temperatures and pressures. Dehydrochlorination at 500–600°C with copper powder, copper–nickel catalyst, or in special steel tubes yields vinyl fluoride. A byproduct, 1,1-difluoroethane, can be converted to additional vinyl fluoride by dehydrofluorination. Vinyl fluoride's major use is for free radical polymerization to poly(vinyl fluoride) with peroxide initiation. Films of this polymer have excellent resistance to weathering, ultraviolet (UV) light, chemicals, and solvents.

Vinylidene fluoride results form the dehydrochlorination of 1,1,1-chlorodifluoroethane.

$$ClF_2CCH_3 \xrightarrow{-HCl} CF_2{=}CH_2$$

Chlorodifluoro-   Vinylidene
ethane   fluoride

Either alkali or heat is used. The precursor results from the addition of two moles of HF to acetylene followed by dehydrofluorination.

$$CH{\equiv}CH \xrightarrow{\ HF\ } [CH_2{=}CHF] \xrightarrow{\ HF\ } CH_3CHF_2 \xrightarrow{\ -HF\ } CH_2{=}CHF$$

Vinylidene fluoride undergoes free radical polymerization to poly(vinylidene fluoride) which, unlike polytetrafluoroethylene (Teflon), may be molded by conventional techniques. It is useful for the preparation of high temperature wire insulation and heat-shrinkable tubing. As an emulsion it is used to coat metal for building panels.

Tetrafluoroethylene is not produced from ethylene but from methyl chloride and is described in Section 10.2.

3.11.9   Ethylene Dibromide

Ethylene dibromide results from direct addition of bromine to ethylene.

$$CH_2{=}CH_2 + Br_2 \rightarrow CH_2BrCH_2Br$$

This reaction at one time provided the major outlet for bromine. Ethylene dibromide was used as a lead scavenger in leaded gasoline. The tetraethyl lead that served as an octane improver would otherwise have been oxidized to nonvolatile lead oxide, which would have accumulated in the engine. Addition of ethylene dibromide meant that lead bromide was formed, which was volatile enough to be swept out of the cylinders. This application became obsolete in the United States in the early 1990s. Ethylene dibromide is also a fumigant for soil and grain, but its use was prohibited in the United States in 1984 because of toxicity problems. Declining consumption has meant that a phase-transfer catalytic route to ethylene dibromide was never commercialized (Section 16.10)

Dehydrobromination of ethylene dibromide yields vinyl bromide, used in small quantities in the formulation of fire retardants for carpet textiles.

$$CH_2BrCH_2Br \xrightarrow{\ cat.\ } CH_2{=}CHBr + HBr$$
$$\text{Vinyl}$$
$$\text{bromide}$$

Reaction of ethylene dibromide with sodium cyanide gives succinonitrile, an aprotic solvent.

$$CH_2BrCH_2Br + 2NaCN \rightarrow CH_2CNCH_2CN + 2HBr$$

A better synthesis, if hydrogen cyanide is available, involves the addition of that compound to the active double bond in acrylonitrile. Triethylamine serves as catalyst.

$$CH_2=CHCN + HCN \rightarrow CH_2CNCH_2CN$$
    Acrylonitrile

Hydrogenation of succinonitrile gives 1,4-diaminobutane:

$$CH_2CNCH_2CN \xrightarrow{\text{H}_2/\text{cat.}} H_2NCH_2CH_2CH_2CH_2NH_2$$
                                    1,4-Diaminobutane

1,4-Diaminobutane reacts with adipic acid to give nylon 4,6, a development product that came on the market in the mid 1980s. Nylon 4,6 has more amide groups that nylon 6,6 (Section 7.2.1.1) per equivalent of molecular weight. Accordingly, there is more opportunity for hydrogen bonding, which leads to greater tensile strength and higher melting point. On the other hand, the greater concentration of amide groups provides greater moisture sensitivity.

3.11.10   Ethanol Derivatives

Diethyl ether is prepared from ethanol by reaction with sulfuric acid (Section 3.8) to form ethyl hydrogen sulfate, which is transformed to diethyl ether by heating with more alcohol at 140–150°C, a temperature slightly below that at which it decomposes to ethylene. Diethyl ether is an important laboratory solvent and was at one time used as an anesthetic.

$$C_2H_2OH + H_2SO_4 \xrightarrow{-\text{H}_2\text{O}} C_2H_5OSO_3H \xrightarrow{\text{C}_2\text{H}_5\text{OH}} C_2H_5OC_2H_5 + H_2SO_4$$
   Ethanol   Ethyl hydrogen   Diethyl ether
                                         sulfate

Ethylamines result when ethanol and ammonia are passed over a dehydration catalyst such as alumina or a mixture of silica and alumina. Amines and water are removed and unreacted ammonia and ethanol recycled. The ratio of ammonia to alcohol is varied from 2:1 to 6:1 depending on the mix of amines required.

$$C_2H_5OH + NH_3 \xrightarrow{\text{cat.}} C_2H_5NH_2 + (C_2H_5)_2NH + (C_2H_5)_3N$$
                            Ethylamine   Diethylamine   Triethylamine

The ethylamines are used at a much lower level than the methylamines. Ethylamine finds its largest use in the preparation of herbicides by reaction with cyanuric chloride and isopropylamine. An ethylaminoisopropylamino-substituted chlorotriazine results, which is the herbicide atrazine.

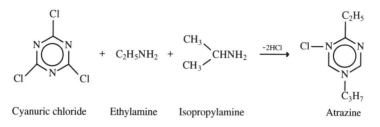

Cyanuric chloride　Ethylamine　Isopropylamine　　　　Atrazine

Diethylamine with carbon disulfide and sodium hydroxide gives tetraethyl thiuram disulfide or disulfiram:

$$2(C_2H_5)_2NH + CS_2 \xrightarrow{NaOH} (C_2H_5)_2N\overset{\overset{\displaystyle S}{\|}}{C}-S-S-\overset{\overset{\displaystyle S}{\|}}{C}N(C_2H_5)_2$$

Disulfiram

This compound is a rubber cure accelerator but is better known as "Antabuse," a pharmaceutical for the treatment of alcoholism.

Finally, triethylamine forms a salt with phosphoric acid, which is a corrosion inhibitor.

3.11.11　Vinyl Esters and Ethers

One synthesis for ethyl acetate involves the direct addition of acetic acid to ethylene.

$$CH_2=CH_2 + CH_3COOH \xrightarrow{cat.} CH_3COOC_2H_5$$

Ethyl acetate

Analogously, vinyl esters of higher acids are prepared by their addition to acetylene. Vinyl ethers result from the interaction of alcohols and acetylene.

$$CH\equiv CH + RCOOH \xrightarrow{cat.} RCOOCH=CH_2$$

Vinyl ester

$$CH\equiv CH + ROH \xrightarrow{cat.} ROCH=CH_2$$

Vinyl ether

Vinyl esters may also be prepared by an exchange reaction in which an acid higher than acetic reacts with vinyl acetate in the presence of a catalyst. The vinyl acetate process is a transesterification and is catalyzed by mercuric acetate with sulfuric acid, or palladium chloride with lithium chloride.

$$CH_3COOCH=CH_2 + RCOOH \xrightarrow{cat.} RCOOCH=CH_2 + CH_3COOH$$

Vinyl acetate　　　　　　　　　　　Vinyl ester　　　　Acetic acid

R = C_3 or higher

Vinyl esters of fatty acids are used primarily as comonomers to impart flexibility to polymers based on vinyl chloride.

NOTES AND REFERENCES

Ethylene is central to the petrochemical industry, hence discussion of it features in all the books on the industry listed in the general bibliography. The classic work is *Ethylene and its Industrial Derivatives*, S. A. Miller, Ed., Benn, London, 1969 but is does not appear to have been either updated or superseded, except, possibly by *Ethylene: Keystone to the Petrochemical Industry*, L. Kniel, Dekker, New York, 1980. A relatively simple book from the University of York (UK) Science Education Group is *Ethene: Industry's most important Chemical*, M. F. Uttley, 1986.

The latest thinking about ethylene technology and economics is to be found in *Ethylene Producers Conference Proceedings, Volumes* 1 & 2, *AIChemE*, 1992, 1994.

Section 3 Chloroacetic acid is made by liquid-phase chlorination of acetic acid. In the United States and Western Europe, some acetic acid is made from hydrocarbon oxidation (Section 111.2.1), but much of it comes from the carbonylation of methanol (Section 10.5.2.2). In this sense, chloroacetic acid is not an ethylene derivative. However, some acetic acid in Europe is still made by oxidation of acetaldehyde and, in this sense, chloroacetic acid is an ethylene derivative.

Section 3.1.5 High Strength polyethylene fibers are the subject of Allied Corporation's US Patent 4 413 110 (November 1, 1983).

Section 3.2.2 Ethylene–vinyl alcohol copolymer resins with particularly good barrier properties are described in *Chem. Week*, April 17, 1985, p. 13.

Section 3.2.6 Carbide's development of ultra-low-density polyethylene is described in European Patent Application 0 120 503 (October 3, 1984).

Section 3.2.7 Many patents granted to Shell Research describe ethylene–carbon monoxide copolymers. Typical is US Patent 4 880 909 (November 14, 1989).

Section 3.3.1 The catalyst for 2-butene formation, tri-*n*-butylphosphine nickel dichloride mixed with ethylaluminum dichloride, is described in US Patent 4 242 531 (December 30, 1980) issued to Phillips Petroleum Company.

The nickel oxide catalyst for 2-butene formation is described in US Patent 3 341 620 (September 12, 1967) issued to Phillips Petroleum Company.

A catalyst for preparing 1-butene is described in US Patent 4 487 847 (December 11, 1984) issued to Phillips Petroleum Company.

The Institut Français du Pétrole catalyst for 1-butene preparation is claimed in European Patent Application 0135 441 A1 (March 27, 1985).

Section 3.3.3 The Shell oligomerization process is described in US Patent 4 020 121 (April 26, 1977) issued to Shell Oil.

Section 3.3.4 The isomerization step in the SHOP process is described by R. A. Nieuwenhuis, *Petrole et Techniques*, No. 268, January, 1980, p. 46. The SHOP process has been described in detail in *J. Chem. Educ.* **65**, 605 (1988) by B. G. Reuben and H. A. Wittcoff.

Section 3.4 The ICI developments are described in British Patent Applications 2 095 242A and 2 095 245A (September 29, 1982). Previous work in this area is described in a Lummus patent [US Patent 3 557 229 (January 19, 1971)].

Section 3.5   Parshall's statement about the Wacker reaction is found in his excellent book *Homogeneous Catalysis*, Wiley, New York, 1980, p. 102.

Acetaldehyde's demise has been discussed by H. A. Wittcoff, *J. Chem. Educ.*, **60**, 1045 (1983).

Section 3.7   Patents describing improved catalysis for ethylene oxide production include European Patent Application 0 076 504 (April 10, 1982) to Union Carbide, British Patent 1 512 625 (June 1, 1978) to Shell, US Patent 4 125 480 (November 14, 1978) to Shell and West German Offen. 2 809 835 (September 14, 1978) to Halcon.

Section 3.7.1   Union Carbide's process for converting ethylene oxide to ethylene glycol via the carbonate is described in five British Patents 2 011 400; 2 011 401; 2 011 402; 2 010 685, 2 010 694.

The conversion of ethylene oxide to ethylene glycol in high selectivity with a vanadate or molybdate catalyst is described in International Patent Application WO85/04393 (October 10, 1985) to Union Carbide.

Section 3.7.2   Typical of the many patents describing Carbide's process for ethylene glycol via the direct combination of CO and H_2 is US Patent 4 360 600 (November 23, 1982). The Ube modification is described in many patents, typical of which are US Patent 4 229 589 (October 21, 1980), US Patent 4 229 591 (October 21, 1980), and Japanese Patent 8 242 656 (March 10, 1982).

The free radical-induced chain reaction between methanol and formaldehyde has been described by M. Oyama, *J. Org. Chem.*, **30**, 2429 (1965). It is further developed in US Patent 4 337 371 (June 29, 1982) to Redox Technologies.

The electrohydrodimerization technology is described in US Patents 4 478 694 (October 23, 1984) and 4 270 992 (June 2, 1981).

Catalysts that effect hydrogenolysis at mild conditions are described in International Patents WO86/03189 (June 5, 1986) and WO82/03854 (November 11, 1982).

The process for preparing ethylene glycol from synthesis gas has been discussed by Dombek, *J. Chem. Educ.*, **63**, 210 (1986).

Section 3.8   UOPs catalyst, which makes possible the oxidative dehydrogenation of styrene, is the subject of US Patent 4 565 898 (January 21, 1986).

The use of ZSM-5 for ethylbenzene production is described in US Patent 3 751 506 (August 7, 1973) to Mobil Oil Company and in US Patent 4 107 224 (August 15, 1978) to Mobil.

Catalyst formulation for ethylbenzene dehydrogenation is disclosed in numerous patents typical of which are US Patents 2 414 585 (January 21, 1947) to Shell, US Patent 2 461 147 (February 8, 1949) to Shell, US Patent 2 603 610 (July 15, 1952) to Dow, US Patent 3 205 179 (September 7, 1965) to Dow, US Patent 4 143 083 (March 6, 1979) to Shell, and World Patent WO8 300 687 (August 17, 1981) to Shell.

Work on the production of styrene by the coupling of benzene and ethylene is described in two publications of R. S. Shire of Phillips Petroleum Co.: *J. Chem Soc, Chem. Commun.*, 1510 (1971) and *J. Catalysis*, **26**, 112 (1972). This oxidative coupling with a Pd^{2+} catalyst takes place with oxygen at a total benzene plus oxygen pressure of 20 bar at 80°C. Selectivity is 95% but conversion is only 2.3% based on ethylene.

Section 3.9   The use of fermentation alcohol as an automotive fuel has been discussed in *Chem. Eng. News* November 5, 1984, p. 25. Although Brazil's program aimed at replacing petroleum-based fuel for automobiles with ethanol was successful technically, its economics were poor in light of the low petroleum costs of the late 1980s. Collar, the president elected in 1989, campaigned on a platform that included phase-out of the alcohol program. In fact, market interactions helped him, for the world price of sugar increased and many refiners chose to convert sugar cane juice to sugar rather than alcohol. By so doing, they created an alcohol fuel shortage, which caused Brazil to import alcohol to fuel its 1 million cars that could not operate on petroleum-based fuel. The Iraqi invasion of Kuwait caused a stunning reversal of both Collar's campaign platform and the manufacture of sugar because it caused the price of crude petroleum to increase to a point where the economics of Brazil's alcohol fuel program became more reasonable.

Section 3.10   Most of the figures for Table 3.2 are taken from *Chem. Eng. News*, July 4, 1994, and Wittcoff and Reuben, Part I, *op. cit.* The published ethanol figure is probably in error and has been replaced with one from data bases maintained by Chem Systems Inc., Tarrytown, NY.

Section 3.11.1   The homologation of acetic acid has been described by J. F. Knifton, *J. Chem. Soc, Chem. Commun.* 41 (1981), and *Hydrocarbon Processing*, December 1981, p. 113.

The low-pressure process for ethylene carbonylation is described in British Patent Application 2 099 430 (December 8, 1982) issued to Halcon.

Section 3.11.6.2   The toxicity of glycol ethers and their derivatives is discussed in *Chem. Week*, June 4, 1986, p. 10.

Section 3.11.6.6   Patents involving the production of hydroxyphenylglycine from glyoxal and glyoxylic acid include British Patents 1 576 678 to Tanabe Seiyaku, 2 012 756 to Ajinomoto, US Patent 4 175 206 to Tanabe Seiyaku and European Patents 24181, 8514; Japan Patent 7 941 876; German Patent 21 515 210 all to Beechams.

Section 3.11.9   The addition of HCN to acrylonitrile is described in Canadian Patent 1 135 72 (November 16, 1982) to Stamicarbon BV.

CHAPTER 4

CHEMICALS AND POLYMERS FROM PROPYLENE

After ethylene the most important olefin is propylene. In 1993 the United States produced about 36 billion lb of propylene, about 34% as a coproduct of ethylene production in the thermal cracking of ethane, propane, and higher alkanes (Section 2.2.1) and 66% as a byproduct of catalytic cracking and of other refinery processes (Section 2.2.2). About 23 billion lb was used for chemicals. Much of the remainder was reacted with isobutane to give alkylates for gasoline (Section 2.2.5) or, to a lesser extent, oligomerized to polygas (Section 2.2.4). In this respect propylene differs from ethylene, which has no nonchemical or fuel uses.

Figure 4.1 shows the ratio in which the three major olefins were used in 1991 in the United States for chemical applications and indicates also the proportion

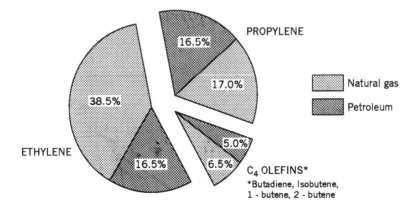

Ratio of usage and preparation coming from natural gas and petroleum

Figure 4.1 Source of C_2, C_3, and C_4 olefins: United States early 1990s.

of each that comes from natural gas and petroleum. Ethylene claims 55% of the total consumption of olefins, propylene 33.5%, and the butenes and butadiene 11.5%.

Most natural gas contains about 2.5 times as much ethane as propane. On the other hand, catalytic cracking (Section 2.2.2) of higher petroleum fractions, which gives negligible amounts of ethylene (see note), is a large although dilute source of propylene. The steam cracking of ethane, propane, butane, naphtha, or gas oil inevitably gives propylene as one of the coproducts, although the cracking of ethane yields so little it is usually not isolated (see note). The amount depends on the feed (the higher the molecular weight the more is produced) and on severity of cracking. For naphtha and gas oil, it varies from 0.9 to 0.4 moles of propylene per mol of ethylene as the cracking becomes more severe. A typical figure is about 0.5. This means that high severity (high-temperature) cracking leads to a product rich in ethylene. Low-severity cracking increases propylene content.

For many years propylene was ethylene's ugly sister, a byproduct that was sold at a fraction above its fuel value or else was burned. This availability stimulated chemists' creativity, and many ingenious chemical uses have been developed. Propylene's price is usually lower than that of ethylene, although shortages in Europe have at times erased this differential. By the mid-1990s propylene's price was again lower than ethylene's (see Fig. 3.4).

The main polymers and chemicals made from propylene are shown in Figure 4.2. The three types of reactions that ethylene undergoes—polymerization and oligomerization, oxidation, and "classical" reactions—apply here also. The unconventional chemistry in which one of ethylene's hydrogen atoms is replaced by

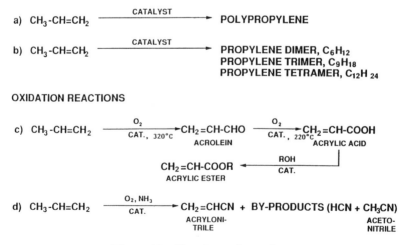

Figure 4.2 Chemistry of propylene.

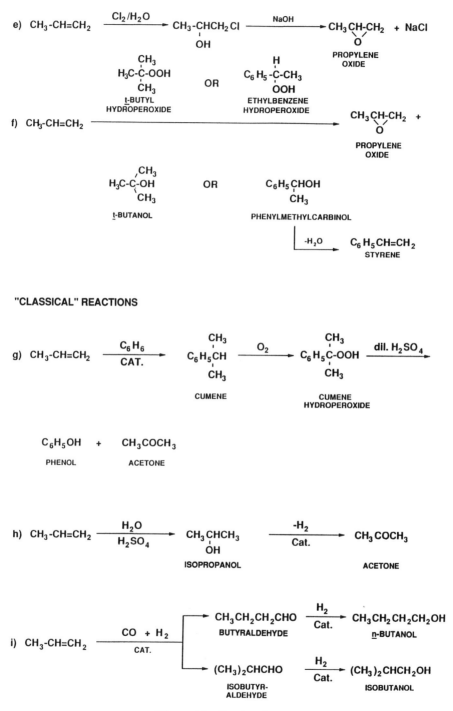

Figure 4.2   Continued

chlorine, hydroxyl, and acetoxy to give vinyl chloride, acetaldehyde, and vinyl acetate is not repeated with propylene. Instead, propylene undergoes some unconventional chemistry of its own based on the stability of the vinyl moiety and the reactivity of the methyl group containing allylic hydrogens. This active methyl can be oxidized directly to acrolein, acrylic acid, and with ammonia to acrylonitrile.

4.1 PROPYLENE POLYMERS AND COPOLYMERS

Polypropylene is the biggest consumer of propylene. It is made with Ziegler–Natta catalysts (Section 3.1.1), typical of which is one developed by Himont and Mitsui Petrochemical based on titanium tetrachloride with a Lewis base complex and a support of magnesium or manganese chloride. The Lewis base blocks nonstereospecific sites on the catalyst. A cocatalyst is used, which is a combination of an alkylaluminum such as triethylaluminum with a Lewis base.

Polypropylene development provides a superb example of process improvement. As indicated in Section 1.5, polypropylene can exist in two crystalline forms, isotactic and syndiotactic, as well as in the amorphous or atactic form. Initially 1.2 lb of monomer were consumed per pound of isotactic polymer produced. Today only about 2% excess monomer is required. Today's fourth generation catalyst will produce 25,000 lb of polymer per pound of catalyst as compared to 1000–2000 lb initially. This very high yield means that deashing or catalyst removal is not necessary. Similarly, an isotactic polymer level of 97% is achieved, which makes unnecessary the removal of atactic material. The catalyst controls molecular weight distribution, important where a narrow weight distribution is required. Finally, modern catalysts can produce polypropylene in spherical form eliminating the need for extrusion into ropes and chopping into pellets. It is significant that the catalyst can control the polymer's morphology. Spherical particles are produced because they replicate the shape of the catalyst particles but have a diameter many times greater. Thus the catalyst serves as a template. The catalyst diameter may range from 300 μ to 4 mm; the polymer molecules can be 40 times larger. The simplification of a polypropylene plant made possible by a fourth generation catalyst as compared to the plant required by a first generation catalyst is shown in Figure 4.3. The technology that eliminates extrusion was in use in only one plant by the mid-1990s, and appears to require further development.

With a Ziegler catalyst, polymerization takes place at 40–80°C. Polypropylene has been produced by slurry, solution, bulk, and vapor-phase or fluid bed processes. In the slurry process, the catalyst is dispersed in an inert diluent such as hexane or heptane. The isotactic polymer that forms is insoluble in the hydrocarbon but the atactic material is soluble. Thus an in situ separation is effected.

In the solution process, the isotactic polymer is soluble in the solvent. This process has fallen into disuse because of the expense of solvent losses and the

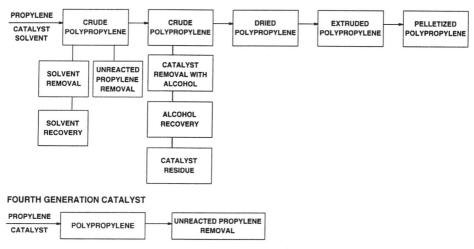

Figure 4.3 The fourth generation catalyst simplifies the polypropylene plant.

difficulty of removing all the solvent from the polymer. The bulk process makes use of pressures that maintain the propylene in liquid form. Atactic product is insoluble in liquid propylene, whereas the stereospecific polymer is soluble. The fluid bed process resembles one originally developed for polyethylene production (Section 3.1.3). This technology is now considered the most desirable, particularly since removal of atactic polymer and catalyst residues is no longer necessary.

There are fewer copolymers of propylene that there are of ethylene. Ethylene–propylene elastomers have already been mentioned (Section 3.2.5). Also important is a block copolymer, comprising a polypropylene block and an ethylene–propylene copolymer block. This is made by first polymerizing propylene and then adding ethylene and more propylene, with the same catalyst as in the first step. The fresh material forms a random copolymer block that grafts onto the original polypropylene block. The copolymer block plasticizes the polypropylene, improving its impact strength, which is poor at low temperatures. About 40% of polypropylene production is the copolymer. The homopolymer is used largely for fibers, although it may be molded if low-temperature resistance is not required.

Random copolymers of propylene with ethylene or 1-butene have poorer low-temperature impact strengths because they are highly crystalline and therefore stiff. The block copolymer has lower crystallinity and greater flexibility, which gives it better low temperature properties but lower clarity.

Copolymers of propylene with olefins other than ethylene and with monomers such as vinyl chloride and styrene are being explored as possible replacements for expensive resins such as acrylonitrile–butadiene–styrene (ABS) (Section 5.1.1) and nylon.

Polypropylene's major use is for injection molded products such as housings and parts for small and large domestic appliances, furniture, and office equipment. In automotive applications it is used for battery cases, bumpers, interior trim, and air ducts. Injection molded packaging includes tubs for margarine, medicine bottles, and syringes. Reusable household food containers are made from copolymers. Polypropylene has a sufficiently high modulus (stiffness) and sufficiently low extensibility to be used for textile fibers. It is found in clothing, particularly for sports, and as filament yarn in carpeting, upholstery, and automobile seats. It is used for carpet backing. Polypropylene cloth can replace canvas in luggage and shoes. Thick mats made from polypropylene are useful in stabilizing soil. They are finding increasing use in road building to replace the rocks or aggregate normally combined with asphalt. Polypropylene film and sheet have greater strength and better high temperature properties than those from polyethylene, and oriented film can be used for cigarette and food packaging and as a shrink wrap. Polypropylene fiber is useful as an asbestos replacement when high temperature resistance is not required.

4.2 OLIGOMERIZATION

The oligomerization of propylene to highly branched dimers, trimers, and tetramers (Fig. 4.2b), which are olefins, takes place in the presence of Friedel–Crafts catalysts such as sulfuric or phosphoric acids. A gasoline fraction (polygas) results (Section 2.2.4) from which the highly branched dimers, trimers, and tetramers can be isolated by fractional distillation, if desired. Zeolites and pillared clays (Sections 13.3 and 16.9) have been investigated as catalysts for oligomerization but have not found commercial application. The feed for oligomerization is usually catalytic cracker off-gases and includes not only propylene but the butenes, which are preferred because they form more highly branched compounds with higher octane numbers. This technology practically fell into disuse with the development of alkylation (Section 2.2.5). It has been revived because the oligomers are useful in unleaded gasoline.

A newer process, called "Dimersol," developed by the Institut Français du Pétrole, produces branched oligomers from propylene or propylene–ethylene blends using a nickel complex catalyst activated by an organometallic compound such as tri-*n*-propylaluminum. With *n*-butenes alone, or mixed with propylene, more nearly linear molecules result for petrochemical use such as conversion to oxo alcohols (Section 4.8) for plasticizers.

The important oligomers for chemical use are branched dodecene (propylene tetramer) and branched nonene (propylene trimer), octene (butenes dimer), heptene (dimer of propylene and butenes), diisobutene (dimer of isobutene), and hexene (propylene dimer). Propylene tetramer formerly was used for the alkylation of benzene to make a branched-chain dodecylbenzene that was subsequently sulfonated to provide a nonbiodegradable surfactant. This application is now insignificant in the United States as more biodegradable detergents have taken over.

Phenol may be alkylated with propylene trimer or tetramer to give nonylphenol and dodecylphenol, respectively. Both are used in lubricating oil additives. They react with ethylene oxide to give nonionic detergents but consumption of these is also declining because they biodegrade very slowly.

Propylene trimer and tetramer and butenes dimer undergo the oxo reaction (Section 4.8) and the aldehydes can be hydrogenated to isodecanol, tridecanol, and nonanol, respectively. Their major use is as their phthalates, which are poly(vinyl chloride) (PVC) plasticizers.

2-Methyl-1-pentene is a dimer of propylene produced with the aid of a tri-*n*-propylaluminum catalyst. It is the basis for an isoprene synthesis that was never commercialized (Section 6.2). Another dimer, 4-methyl-1-pentene, was developed by ICI as a monomer for a specialty polymer, which despite its crystallinity, is transparent, because the refractive indexes of the crystalline and noncrystalline portions are the same. Therefore light is not reflected at crystal boundaries. The catalyst for the dimerization is a combination of sodium and potassium carbonate. The process was sold to Mitsui who manufacture the polymer. The monomer has a new lease on life because of its use as a comonomer in LLDPE (Section 3.1.4).

4.3 ACRYLIC ACID

At the beginning of this chapter, we mentioned that propylene underwent some unconventional chemistry based on the stability of its vinyl moiety and the reactivity of its methyl group. This methyl group may be oxidized to a carboxyl group in the same way as the methyl group of toluene may be oxidized to give benzoic acid. Similarly, the methyl groups of both toluene and propylene can be chlorinated and subjected to ammoxidation. The mechanisms that cause these reactions to take place with the two molecules are different and perhaps this is the reason why the similarity was not recognized earlier. The chlorination is discussed in Section 4.10.1; the ammoxidation reaction in Section 4.4; and the simple oxidation is described here.

The oxidation of propylene gives acrolein first and then acrylic acid. There are several other routes to acrylic acid and its esters shown in Fig. 4.4, all but one of which has, at one time or another, been commercial.

Acrylonitrile hydrolysis (c) is an obvious route. It is economic if small amounts of acrylic acid are required and is still used by some acrylonitrile producers to make acrylic acid for captive use. Another route involves the reaction of hydrogen cyanide with ethylene oxide to give ethylene cyanohydrin, which may be simultaneously dehydrated and hydrolyzed with sulfuric acid [Reaction d(1)]. This process was last used in 1971. A third method involves the treatment of acetaldehyde with hydrocyanic acid to give a different cyano-

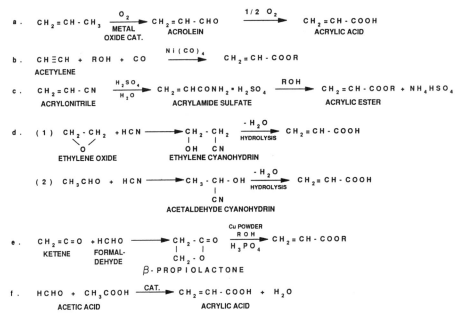

Figure 4.4 Acrylic acid: possible production methods.

hydrin, which may be similarly dehydrated and hydrolyzed [Reaction d(2)]. This process too is obsolete.

Reaction e in Figure 4.4 starts with ketene from the pyrolysis of acetic acid (Section 10.5.2.3). It reacts with formaldehyde to give β-propiolactone which, on reaction with an alcohol, gives an acrylate ester directly. The high toxicity and the carcinogenicity of the lactone made this route unattractive.

Reaction b in Figure 4.4 is a relic of the great days of acetylene chemistry. Acetylene is reacted with an alcohol and carbon monoxide in the presence of a nickel carbonyl catalyst to give an alkyl acrylate directly. The process, now obsolete, accounted for almost one-half of the total acrylic acid and ester production in the United States in 1976. It was still used in the United States in the early 1980s, and was one of the last acetylene-based reactions to give way to modern chemistry. A few small depreciated units continued to operate in Europe despite the economic advantage of propylene oxidation.

Reaction f in Figure 4.4 shows a proposed process in which formaldehyde reacts with acetic acid to give acrylic acid, presumably by way of a hydroxy-methyl intermediate. The process, although not commercialized, is of interest because it provides a route to acrylic acid from synthesis gas (Section 10.4) since acetic acid is made from carbon monoxide and methanol, which in turn is made from synthesis gas. Similarly, formaldehyde is made by the oxidation of meth-anol. The vapor-phase reaction is carried out with a vanadium orthophosphate catalyst. Conversion is quantitative at approximately 60% selectivity. Formaldehyde is used in a similar way in a new process for methyl methacrylate (Section 4.6.1).

Propylene oxidation (Reaction a) is now the preferred route. The two-step process yields acrolein as an intermediate, that can be isolated if desired (Section 4.10.4). Optimum catalyst and temperatures are different for each of the two steps, the first step requiring a much higher temperature than the second. Molybdenum-based catalysts predominate and a typical catalyst for the first stage of the two step conversion is $Fe_4BiW_2Mo_{10}Si_{1.35}K_{0.6}$. With this catalyst at 320°C, propylene conversion is 97% and acrolein selectivity is 93%. In addition, however, there is a 6% selectivity to acrylic acid making the useful selectivity 99%. These results are typical of what can be achieved with modern sophisticated catalysts. In the second stage, a catalyst may be used comprising $Mo_{12}V_{4.6}Cu_{2.2}Cr_6W_{2.4}$. The catalyst is supported on alumina and at 220°C a 98% molar yield of acrylic acid is obtained.

Figure 4.5 shows a proposed mechanism for the conversion of propylene to acrolein. The catalyst (**I**), a bismuth molybdate, abstracts a hydrogen from propylene (**II**) and π bonds the resulting radical to give (**III**). The formation of a carbon–oxygen bond to give (**IV**) and a hydride shift as in (**V**) leads to the formation of acrolein (**VI**) and a reduced form of the catalyst (**VII**). The catalyst is regenerated by reaction with oxygen as shown at the bottom of the figure.

Most acrylic acid is converted to an important range of esters—methyl, ethyl, butyl, and 2-ethylhexyl acrylates. These are polymerized usually with comonomers including methyl methacrylate and/or vinyl acetate. The polymers, in the form of emulsions, are used in water-based paints and this market accounts for 30–35% of the use of all acrylates. They are also used in solvent-based coatings, as copolymers with methacrylates, for product finishes such as thermosetting automotive topcoats. For this application, the acrylates are combined with melamine (Section 10.4.1). Acrylic coatings are prized for their decorative quality and their film durability, particularly out of doors, where their resistance to ultraviolet (UV) light is of benefit. Acrylic emulsions are also found in textile chemicals and adhesives, which are the second and third largest uses, as well as in paper coatings, binders for non-woven fabrics, polishes, and leather coatings. Thermoplastic acrylate coatings have been much prized for automobile topcoats. They contain much more solvent than the thermosets, however, and have been phased out because of the contribution of the solvents to air pollution.

A two-package coating based on hydroxyl-containing acrylic copolymers (made with hydroxyethyl acrylate as one of the monomers) may be cross-linked with diisocyanates to give a hard thermoset polymer, which cures at room or slightly elevated temperatures. It is the preferred vehicle for coatings for automobile refinishing.

A new and growing application for acrylic monomers is in UV-cured coatings and printing inks. Pentaerythritol tetraacrylate and oligomers of it are typical. They are applied to a substrate such as plywood for subsequent polymerization into a film by UV light. The films must be clear, since pigments absorb UV rays,

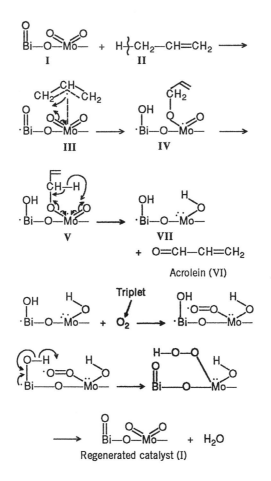

Figure 4.5 Mechanism of acrolein formation.

and can be applied only to a horizontal surface. Printing inks based on acrylic monomer and oligomers can be cured by UV light because their low-volume pigmentation does not seriously interfere with curing. Heavily pigmented coatings can be cured by electron beams, but the equipment is expensive and the process requires safety precautions.

Two new applications for acrylic acid were developed in the mid-1980s, which motivated expansion of acrylic acid production. One is the use of polyacrylic acid (prepared by the peroxide-induced free radical polymerization of acrylic acid) or an acrylic acid–maleic anhydride copolymer as a builder in detergents to replace phosphates for chelating of calcium and magnesium ions. They serve also as dispersing aids and inhibit soil redeposition.

The second use is as a water-absorbing agent for application in disposable diapers and personal care products. The polymers are termed superabsorbent because they can absorb over a thousand times their weight of water. A typical formulation comprises a copolymer of acrylic acid, sodium acrylate, and a small quantity of a cross-linking agent such as trimethylolpropane triacrylate. Cross-linking is necessary in order to impart salt stability to the gels. Another type of water absorber, irreverently termed "superslurper," is made from a starch graft polymer in which acrylonitrile is grafted onto starch to give long chains. Thereafter the nitrile groups are partially hydrolyzed to amide and carboxyl groups. Here too a small amount of a cross-linking agent—a polyfunctional acrylate or methacrylate—must be used to achieve stability of the gel in the presence of salts. Starch graft polymers may also be prepared with acrylamide and acrylic acid rather than acrylonitrile.

4.4 ACRYLONITRILE

Closely related to the direct oxidation of propylene to acrolein and acrylic acid is oxidation in the presence of ammonia to give acrylonitrile (Fig. 4.2d). It is the second largest volume derivative of propylene. The process, developed by Sohio in the United States and by Distillers Company in Great Britain, is known as ammoxidation. It excited much attention when discovered in the late 1950s, for there was no previous example of the formation of a C–N bond in this way.

In the early 1960s ammoxidation displaced various processes for acrylonitrile production such as the addition of hydrogen cyanide to acetylene, the interaction of ethylene oxide with hydrogen cyanide followed by dehydration, and the reaction of propylene with nitric oxide. A bismuth phosphomolybdate catalyst was initially used in the United States. This was replaced by uranium bismuthate based on depleted uranium, which in turn has been superseded by proprietary catalysts used in the vapor phase that have increased the yield of acrylonitrile and boosted reactor throughput.

The catalysts are described in an extensive patent literature. Most of them reduce byproduct acetonitrile production almost to zero but byproduct hydrogen cyanide is still produced. A typical catalyst contains silica-supported oxides of selenium, iron, and tellurium. One such catalyst gives 100% conversion of propylene with 86.9% selectivity to acrylonitrile and 2.1% selectivity to hydrogen cyanide. It makes operation possible at the relatively low temperature of 320°C with a contact time of 2–5 seconds. It is interesting that this catalyst does not contain bismuth, molybdenum, or phosphorus. Still another catalyst, which requires a higher temperature of 420°C, comprises iron, antimony, molybdenum, vanadium, tellurium, and copper, supported on silica. It gives an overall acrylonitrile yield of 85% and a hydrogen cyanide yield of 3% with no acetonitrile. The ammoxidation reaction is highly exothermic and is carried out in a fluidized bed to ensure effective heat exchange and temperature control.

The fact that the methyl group of propylene can be oxidized preferentially to give acrolein indicates that not all of the hydrogen atoms of propylene are equivalent. The methyl hydrogen atoms are allylic, and hence reactive; the allyl radical $CH_2{=\!=}CH{=\!=}CH_2$, on the other hand, is relatively stable (see note). It is the recognition of this difference that makes the oxidation and ammoxidation processes feasible.

The mechanism for the formation of acrylonitrile from propylene is similar to the one previously proposed for acrolein formation (Fig. 4.5) and is shown in Figure 4.6. The mechanism is illustrated for the original Sohio catalyst, an

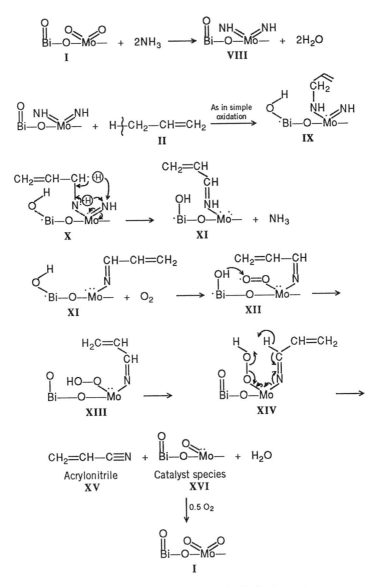

Figure 4.6 Mechanism of acrylonitrile formation.

oxidized bismuth molybdenum species, which reacts with ammonia to give the imminium compound (**VIII**). This bonds with an allyl radical resulting from the abstraction of a hydrogen from propylene (**II**) to give (**IX**). As indicated in (**X**) a double hydride shift occurs to liberate 1 mol of ammonia and to give the species (**XI**). This is oxidized to give (**XII**), which forms (**XIII**) by a hydride shift. Compound (**XIII**) undergoes the transformation shown in (**XIV**) to give the desired product, acrylonitrile (**XV**), and the catalyst species (**XVI**), which with oxygen regenerates the initial catalyst species (**I**).

In the early years of the process, there was the problem of the acetonitrile and hydrogen cyanide byproducts. The hydrogen cyanide was used by some companies for methyl methacrylate (Section 4.6.1), by Du Pont for hexamethylene diamine (Section 5.1.1), and for sodium cyanide and chelating agents, but only small uses were found for the acetonitrile. One suggestion was reaction with methane to give acrylonitrile:

$$CH_3CN + CH_4 + O_2 \xrightarrow{\text{cat.}} CH_2{=}CHCN + 2H_2O$$

The catalyst comprises potassium or calcium halides on quartz. Molybdenum and bismuth oxides individually or combined on quartz or other supports are also proposed. The process was never commercialized because the amount of acetonitrile byproduct was reduced by catalyst development.

Some acetonitrile is used as an aprotic solvent, particularly for the separation of butadiene from other C_4 olefins (Chapter 5). Modern catalysts are so efficient in decreasing acetonitrile production that in the early-1990s there was actually a shortage of acetonitrile for use as an aprotic solvent.

On the horizon is a process for the ammoxidation of propane rather than propylene (Section 11.2.1). This is formally analogous to the oxidation of butane to maleic anhydride (Section 5.4) with the simultaneous introduction of a double bond and functional groups into a saturated hydrocarbon. BP America has scheduled the commercialization of this process for the late-1990s.

4.4.1 Uses of Acrylonitrile

Acrylonitrile is used primarily to make polymers. Its most important application is for the production of polyacrylonitrile for textile fibers. Because polyacrylonitrile is not thermally stable, it must be spun from solution in the same way as cellulose acetate. The problem is that acrylonitrile polymerizes to a very high molecular weight, hence the polymer is only slightly soluble in ordinary solvents, despite the fact that it is thermoplastic. Spinning was made possible by the discovery of so-called aprotic solvents such as N,N-dimethylformamide (DMF) and N,N-dimethylacetamide. Polyacrylonitrile is also soluble in aqueous solutions of inorganic salts and acids such as sodium thiocyanate solution and, to a lesser degree, nitric acid and zinc chloride. These are used competitively with the aprotic solvents. The problem of spinning polyacrylonitrile thus gave rise to the science of aprotic solvents, which find application today in extractive distillation and in the synthesis of aromatic polymers whose monomers are insoluble in more usual solvents.

Acrylic fibers are also made more tractable by inclusion of 10–15% of comonomers such as methyl acrylate of vinyl acetate. p-Vinylbenzenesulfonic acid is also added to make the fiber dyable with basic dyes. Familiar trade names of acrylic fiber are Orlon, Acrilan, and Courtelle.

Another important copolymer made from acrylonitrile is ABS resin (Section 3.8). Acrylonitrile–butadiene–styrene is the most widely used engineering (i.e., metal replacing) plastic and US consumption in 1995 reached 3000 million lb. It is a two-phase polymer system, with the elastomeric butadiene–acrylonitrile copolymer dispersed in a rigid styrene–acrylonitrile matrix. Because light reflects at the phase boundaries, it is opaque. Butadiene–acrylonitrile rubber was developed in Germany prior to World War II and is still used today. The acrylonitrile imparts valuable oil and abrasion resistance.

High acrylonitrile-containing polymers have good barrier properties to carbon dioxide and oxygen. They are clear and have good extensibility, which makes them useful for the fabrication of bottles. Monsanto developed such a resin based on a copolymer of acrylonitrile and styrene. The development was banned from food packaging by the Food and Drug Administration (FDA) because of the slight possibility that some undetectable levels of free acrylonitrile might remain in the resin. This was proved to be unlikely, but Monsanto never reestablished the project and the market for bottles for carbonated beverages was taken over by poly(ethylene terephthalate). Styrene–acrylonitrile remains a useful copolymer but sells only on the scale of about 100 million lb/year. Other acrylonitrile formulations contain styrene, methyl methacrylate, and styrene–butadiene rubber. Barrier resins based on acrylonitrile are available for nonfood packaging.

The electrohydrodimerization of acrylonitrile provides a significant route to adiponitrile, which in turn can be hydrogenated to hexamethylene diamine for nylon 66 production. It has partly been displaced by the hydrocyanation of butadiene (Section 5.1.2). What is presumed to be a catalytic route for the dimerization of acrylonitrile was developed by ICI and was scheduled for commercialization in the early 1990s but ICI instead left the nylon business.

Acrylonitrile production grew rather slowly in the period 1980–1993, and domestic United States demand is expected to be more or less constant until the end of the decade. The market for acrylic fibers shrank from 671 million lb in 1984 to 433 million lb in 1993 because of fashion changes. Acrylonitrile consumption has only been maintained by growth in ABS resins, hexamethylene diamine and acrylamide.

4.5 CUMENE AND CUMENE HYDROPEROXIDE

Cumene (isopropylbenzene) is the precursor of phenol and acetone. It results from the acid-catalyzed alkylation of benzene with propylene as shown in Figure 4.2g. It is a continuous reaction similar to ethylbenzene production (Section 3.8) where benzene is alkylated with ethylene. Both liquid- and gas-phase alkylations are possible. Almost any Friedel–Crafts catalyst can be used. Gas-phase alkylation is the most popular with aluminum chloride, phosphoric acid, or boron trifluoride catalyst at 200–350°C and 10–15 bar. Solid acid catalysts, such as the zeolites ZSM-5, which are more ecologically acceptable, were introduced in the early 1990s.

The molar ratio of benzene to propylene is 8–10:1. Either chemical grade (92 wt%) or refinery grade propylene is suitable. The latter is a dilute propylene stream (20–25 wt%) from catalytic cracking (Section 2.2.2) also containing propane. Cumene yields are 96–97% based on benzene and 91–92% based on propylene. Selectivity to cumene is 92–94% based on propylene and an excess of benzene is used to depress the formation of di- and triisopropylbenzene as well as a small amount of *n*-propylbenzene.

The reaction was first carried out in refinery polymerization units. Cumene was an important octane improver in aviation gasoline in World War II (Section 2.2.4). Its production was fostered by the fact that polygas units became available when that product was replaced by alkylate (Section 2.2.5). Conversely, oligomerization can be carried out in units built specifically for cumene production.

Virtually all cumene is converted to phenol and acetone. This is the classical two-for-one reaction in which two products are manufactured in one plant. Phenol will be discussed in detail in Section 7.1.

Cumene reacts with oxygen from the air in the presence of aqueous alkali at 130°C and pH 6–8 to give cumene hydroperoxide at a concentration of about 25 wt% (Fig. 4.2g). An emulsifying agent is often added to facilitate contact between the cumene and aqueous phases. Soda ash is added to maintain the pH. Metallic catalysts such as cobalt, copper, or manganese with promoters, may be used, but usually are not because they facilitate oxidation of the cumene itself to carbon dioxide.

As might be expected, the reaction is sensitive to acids such as formic and acetic, which decompose the cumene hydroperoxide. These are only slowly neutralized by the alkali present, which is in the aqueous rather than the organic phase, but they are also volatilized in the oxygen stream. Other possible impurities in the cumene such as sulfur compounds, phenol, or aniline may also attack the hydroperoxide.

Surplus cumene is distilled off for recycle. The cumene hydroperoxide is cleaved to phenol and acetone without further purification by treatment with 10% sulfuric acid at about 50°C in the liquid phase. The mechanism for the decomposition is shown in Figure 4.7. Byproducts are dimethylphenylcarbinol, α-methylstyrene, and acetophenone, as well as diacetone alcohol, which forms

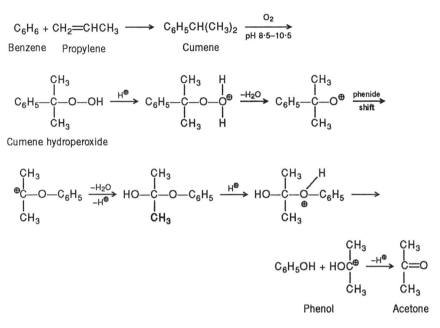

Figure 4.7   Mechanism of the cumene–phenol process.

from the self-condensation of acetone. The stoichiometry demands that 0.6 tons of acetone are formed for every ton of phenol.

4.6   ACETONE AND ISOPROPANOL

The cumene–phenol process is the major source of acetone. This two-for-one reaction suffers from the same problem as all such reactions in that it is seldom that the marketplace requires both products in the ratio in which they are produced. In this case, there has traditionally been a shortfall of acetone. It would be uneconomic to satisfy acetone demand by the accumulation of unsalable phenol. Hence, an alternative route to acetone was required.

The alternative route is shown in Figure 4.2h. Absorption of propylene in concentrated sulfuric acid gives isopropyl sulfate, which is hydrolyzed to iso-propanol. This has a number of uses, primarily as a solvent, but some of it is air-oxidized or dehydrogenated to acetone.

$$CH_3CHOHCH_3 \rightarrow CH_3COCH_3 + H_2 \text{ (dehydrogenation)}$$

$$CH_3CHOHCH_3 + 0.5O_2 \rightarrow CH_3COCH_3 + H_2O \text{ (oxidative dehydrogenation)}$$

The dehydrogenation process is preferred and employs a zinc oxide catalyst at 300–400°C and 3 bar or a copper or brass catalyst at 500°C. Conversions per pass are over 90% and an overall yield of 96% results. The oxidative dehydrogenation takes place in the presence of air at 400–600°C over a silver or copper catalyst. Low pressures, a temperature of 550°C, and a zinc oxide catalyst are employed. The dehydrogenation is endothermic and lack of a use for the byproduct hydrogen may make the oxidative route, which is exothermic, worthwhile.

During the 1980s, acetone from isopropanol accounted on average for about 8% of acetone production in the United States. By 1990 this had fallen to 4.5%. It was projected to disappear because more phenol would be needed for bisphenol A (Section 7.1.2) and less acetone for solvent use because of the Clean Air Act. Shell has gone out of the business and Exxon is scheduled to leave in 1996.

A variant of the Wacker process (Section 3.5) is used in Japan to obtain acetone from propylene. Propylene reacts with oxygen in the presence of catalytic amounts of palladium chloride and a stoichiometric amount of cupric chloride, which is reduced to cuprous chloride. The latter is then reoxidized to cupric chloride in a separate reactor. In the original Wacker process, both the $CuCl_2$ and $PdCl_2$ were used in catalytic amounts, the CuCl being reoxidized to $CuCl_2$ in situ.

A Shell process, not in use, involves the oxidation of isopropanol to acetone and hydrogen peroxide in a two-for-one process.

$$CH_3CHOHCH_3 + O_2 \rightarrow CH_3COCH_3 + H_2O_2$$

Isopropanol   Acetone   Hydrogen peroxide

The three major chemical (i.e., nonsolvent) uses for acetone are for the preparation of methyl methacrylate (Section 4.6.1), methyl isobutyl ketone (Section 4.6.2) and bisphenol A (Section 7.1.2).

4.6.1 Methyl Methacrylate

The classical route to methyl methacrylate starts with the condensation of acetone and hydrocyanic acid to give acetone cyanohydrin, as shown in Figure 4.8 Reaction 1. Hydrolysis of the nitrile to an amide group followed by esterification provides methyl methacrylate. The conversion of acetone cyanohydrin to methacrylamide sulfate via acetone cyanohydrin sulfate is carried out with 98% sulfuric acid at temperatures as high as 140°C. The amide reacts with methanol at 80°C to yield methyl methacrylate. Overall molar yield based on acetone has been raised in recent years form 75 to 92%, an indication of the intense process development to which old processes are frequently subjected.

A major problem with this process is the production of about 1.5 tonnes of ammonium bisulfate byproduct per tonne of methyl methacrylate by the reactions shown in Figure 4.8 Reaction 1. This byproduct problem was never

1. CONVENTIONAL ROUTE

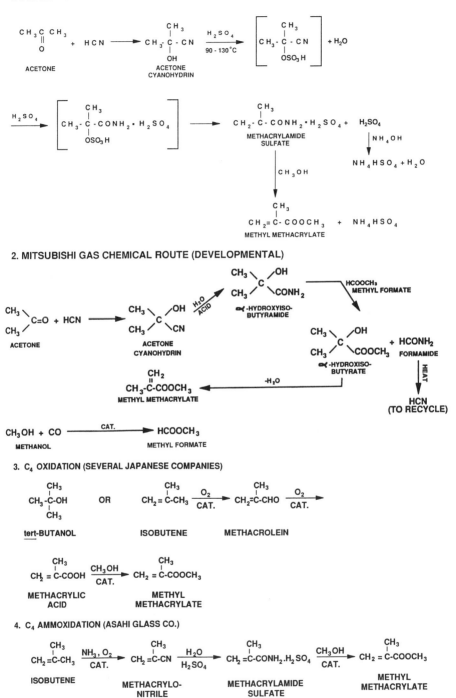

2. MITSUBISHI GAS CHEMICAL ROUTE (DEVELOPMENTAL)

3. C₄ OXIDATION (SEVERAL JAPANESE COMPANIES)

4. C₄ AMMOXIDATION (ASAHI GLASS CO.)

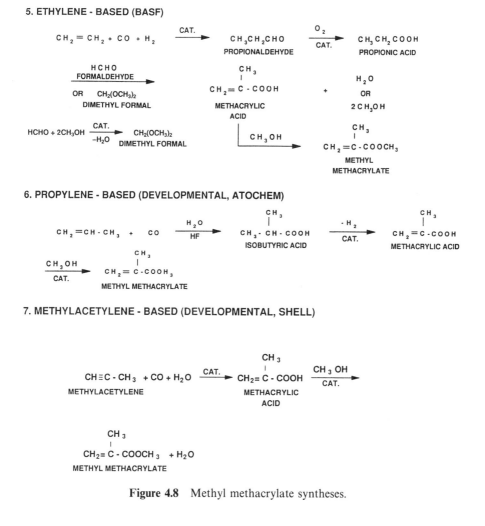

Figure 4.8 Methyl methacrylate syntheses.

satisfactorily solved over a period of 50 years, but ICI brought a process for byproduct treatment on stream in the early 1990s and Mitsubishi Gas Chemical have an entirely new process (see below) under development. Another problem is that about 21% of US hydrogen cyanide production is a byproduct from acrylonitrile manufacture (Section 4.4). New acrylonitrile processes generate decreased amounts of hydrogen cyanide, which means that new, dedicated, and for this process uneconomical, hydrogen cyanide facilities (Section 10.1) may be required. These two problems, particularly acute in Japan where hydrogen cyanide is in short supply, have motivated the development of alternative processes.

The ICI process involves high-temperature pyrolysis of the ammonium bisulfate in the presence of air. The bisulfate decomposes to ammonia and sulfur trioxide which, at the temperature of operation, dissociates partly to sulfur dioxide and oxygen. The ammonia is oxidized by the air to nitrogen and water, and the reaction stream is sent to a sulfuric acid plant where the sulfur dioxide is reconverted to sulfuric acid, which is recycled.

The Mitsubishi Gas Chemical Co. process for methyl methacrylate manufacture demonstrates a high level of ingenuity and was announced in 1989. It solves the two defects that characterize the acetone cyanohydrin process—the need to generate hydrocyanic acid and the production of unwanted ammonium bisulfate. It is noteworthy that the solution to these problems did not come from methyl methacrylate's major suppliers but from a Japanese company.

In the new process, acetone cyanohydrin, produced conventionally from acetone and hydrocyanic acid, is partially hydrolyzed to α-hydroxyisobutyramide (Fig. 4.8 Reaction 2). An ester–amide interchange with methyl formate produces α-hydroxyisobutyrate. Thus, the carboxyl function has been inserted in the molecule without ammonium bisulfate production. Formamide is the coproduct. Dehydration of the butyrate provides the desired methyl methacrylate. Hydrocyanic acid is regenerated for reuse by dehydrating the formamide

$$HCONH_2 \rightarrow HCN + H_2O$$

The dehydration is catalyzed by iron with magnesium, calcium, zinc, or manganese promoters at about $400°C$ at reduced pressure. The need for a dedicated plant for hydrocyanic acid production is thus obviated. The small quantities of hydrocyanic acid required to make up for losses in the plant can be generated from formamide made by reacting methyl formate with ammonia. The methyl formate required for this use, as well as for the exchange reaction, is made by the carbonylation of methanol with a basic catalyst. By early 1990, Mitsubishi Gas Chemical Co. was operating a pilot plant for this process and was scheduled for commercialization by the late 1990s.

A summary of other processes follows. In the Mitsubishi Rayon Process in use in Japan (Fig. 4.8 Reaction 3) *tert*-butanol or isobutene is oxidized to methacrylic acid, via methacrolein which is not isolated. The reaction probably proceeds via the dehydration of the *tert*-butanol to isobutene. Catalysts are similar to those described for the oxidation of propylene to acrolein and acrylic acid (Section 4.3) and are based on molybdenum and bismuth. A typical catalyst for the first stage to methacrolein comprises molybdenum, bismuth, iron, and lithium in a molar ratio of $1:1:0.3:0.1$. The second stage of the oxidation may be carried out with a vanadium phosphomolybdate catalyst with cesium and strontium. A typical composition is: $Mo_{12}P_2VCs_2SrO_{0.05}$. Japan Catalytic in a joint venture with Sumitomo oxidizes isobutene directly.

Asahi Glass in Japan ammoxidizes isobutene to methacrylamide in a converted acrylonitrile plant (Fig. 4.8 Reaction 4). Thereafter the process uses conventional chemistry. This is a multistage process and generates ammonium bisulfate in the hydrolysis of the nitrile to the amide. It is apparently economic because it is carried out in depreciated equipment.

A synthesis announced as a commercial process by BASF (Fig. 4.8 Reaction 5) in the late 1980s is based on the hydroformylation of ethylene to propionaldehyde (Sections 3.11.1 and 4.8), which may be oxidized to propionic acid. An alternate route to propionic acid is the carbonylation of ethylene (Section 3.11.1). Reaction with formaldehyde or with a formaldehyde precursor, dimethyl formal, yields methacrylic acid for esterfication to methyl methacrylate.

Although the process appears cumbersome, it must be recognized that BASF already manufactures propionic acid and processes exist for esterfication of methacrylic acid. Thus, only the intermediate step of inserting the methylene group, which also was used in a proposed acrylic acid process (Section 4.3), is new. Whereas formaldehyde may be used directly, the dimethyl formal, produced from formaldehyde and methanol, will be more compatible and may lead to higher selectivities. On the other hand, the methanol generated in the condensation must be recovered. A variant of this process involves reaction of formaldehyde with propionaldehyde to yield the methylene derivative $CH_2=C(CH_3)CHO$, which can be oxidized to methacrylic acid. This may be the preferred route.

Another approach under development by a number of European companies, particularly Atochem, involves propylene carbonylation in the presence of liquid hydrofluoric acid (Fig. 4.8 Reaction 6), to yield isobutyric acid. Oxidative dehydrogenation to methacrylic acid takes place in the presence of aluminum-modified iron phosphate. Esterification with methanol yields methyl methacrylate.

A seventh synthesis recently proposed for development by Shell starts with propyne (methylacetylene) or a mixture of propyne with allene ($CH_2=C=CH_2$), the latter being in equilibrium with propyne under the reaction conditions.

$$CH_3C\equiv CH \rightleftharpoons CH_2=C=CH_2$$
Propyne   Allene

Propyne and allene are produced during steam cracking (Section 2.2.1) in small quantities and can be recovered by solvent extraction, although they are usually selectively hydrogenated to propylene. In the Shell process, however, the mixture is reacted with carbon monoxide and water (Fig. 4.8 Reaction 7) in the presence of a catalyst comprising palladium acetate, triphenylphosphine, and benzenephosphonic acid. Methacrylic acid results, which is esterified to methyl methacrylate. Alternatively, the carbonylation can be carried out in the presence of methanol rather than water to yield the ester directly. The process is closely related to an obsolete process for acrylic acid production (Section 4.3) from acetylene, which used a conventional nickel carbonyl catalyst.

Another proposed methyl methacrylate process that gives propylene oxide as a coproduct is described in Section 4.7.3.2. Despite all these interesting new processes, most of the world's methyl methacrylate was still produced in the early 1990s by the acetone cyanohydrin route. Because of the problems of ammonium bisulfate disposal, it is likely that the Japanese isobutene oxidation route will dominate in the future.

Most methyl methacrylate is polymerized to poly(methyl methacrylate) from which is made acrylic sheet for use in glazing, lighting, signs, and sanitary items such as bathtubs made by thermoforming. Glazing, in which the transparent poly(methyl methacrylate) replaces glass, is by far the largest market. Molding resins and surface coatings provide the other two large applications. Methyl methacrylate is used with other acrylic monomers to make emulsions for water-based paint (Sections 4.3 and 3.6). It also replaces styrene in certain specialty unsaturated polyester compositions (Section 9.1.1). If fears of styrene toxicity in polyester resins turn out to be well founded, methyl methacrylate consumption will increase. Methyl methacrylate has a high refractive index and critical angle, and hence is capable of conducting light around bends. This is the basis for its growing use in fiber optics and light pipes.

4.6.2   Methyl Isobutyl Ketone and Other Acetone Derivatives

Further products from acetone are shown in Figure 4.9. Practically all the materials are solvents, mostly for the paint industry. The most important, as previously indicated, is methyl isobutyl ketone. The aldol condensation of acetone to diacetone alcohol is base catalyzed and carried out in the liquid phase. Dehydration takes place at about 100°C in the presence of sulfuric and phosphoric acids to yield mesityl oxide. Hydrogenation of mesityl oxide to methyl isobutyl ketone must be accomplished without involving the carbonyl group. Actually, the hydrogenation is carried out with copper and nickel catalysts at 150–200°C and up to 10 bar to give a combination of methyl isobutyl ketone and methyl isobutyl carbinol, which can be separated by distillation.

Isophorone, a high-boiling ketone solvent, results from the base-catalyzed trimerization of acetone (Fig. 4.9). Hydrocyanic acid will add across its double bond to give a nitrile, and reductive amination with ammonia and hydrogen gives "isophorone diamine." Like the toluene diamines, treatment with carbonyl chloride gives a diisocyanate—isophororone diisocyanate or IPDI—which is used in a light stable polyurethane resin.

Hydrogenation of isophorone reduces the double bond and converts the ketone group to an alcohol. Nitric acid oxidation provides 2,2,4-trimethyladipic acid, which may be converted to a dinitrile with ammonia. Further hydrogenation yields 2,2,4-trimethylhexamethylenediamine. Both these diamines, but particularly isophorone diamine, are useful as curing agents for epoxy resins.

4.7   PROPYLENE OXIDE

Propylene oxide (Fig. 4.2e and f) was made traditionally by reaction of propylene with hypochlorous acid, generated in situ from chlorine and water, followed

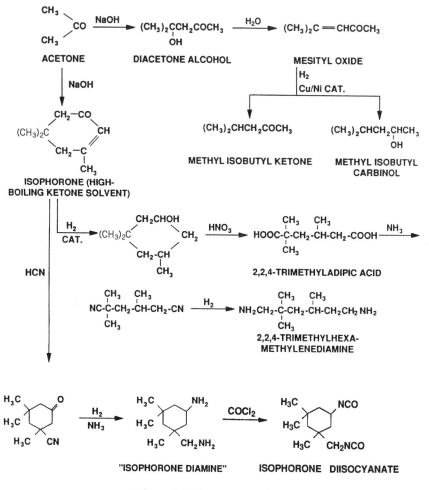

Figure 4.9 Acetone reactions.

by dehydrochlorination with calcium or sodium hydroxide.

$$Cl_2 + H_2O \rightleftharpoons HOCl + HCl$$

$$CH_2{=}CHCH_3 + HOCl \longrightarrow CH_2ClCHOHCH_3 \ (90\%) + CH_2OHCHClCH_3 \ (10\%)$$

$$CH_2ClCHOHCH_3 \quad \text{or} \quad CH_2OHCHClCH_3 + NaOH \longrightarrow H_2C{-}CHCH_3 + NaCl + H_2O$$
$$\diagdown \! \diagup$$
$$O$$

The same process was at one time used for ethylene oxide, and many of these plants were converted to propylene oxide production when the direct oxidation route (Section 3.7) to ethylene oxide appeared. The process with sodium hydroxide is still used by Dow in the United States. Several European and Japanese companies use the calcium hydroxide process. Dow has a highly integrated plant with a chlorine–caustic soda unit to provide chlorine for the chlorohydrin reactor and sodium hydroxide for the dehydrochlorination. In addition, the brine produced by the dehydrochlorination can be reprocessed into caustic and chlorine.

The reaction takes place in dilute solution and 4.0–4.5% propylene chlorohydrins (the α-isomer predominates) are produced in a molar yield of 94% with 4.7% of 1,2-dichloropropane and 1.7% of chloroisopropyl ether. In the dehydrochlorination the yield of propylene oxide is 96%. The process is wasteful of chlorine (or electricity in the integrated plant) and much research was devoted to discovering a direct oxidation route. The first process to be commercialized involved oxidation of isobutane to *t*-butyl hydroperoxide. This then oxidized propylene to propylene oxide.

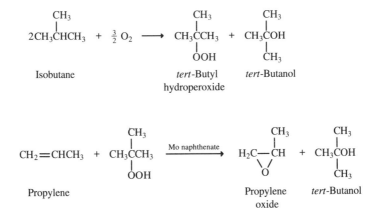

tert-Butanol is a byproduct in both the formation of the peroxide and the epoxidation step. Approximately 2.2 mol of *tert*-butanol per mol of propylene oxide are produced, and the economics hinge markedly on the market for the byproduct. This reaction provides a second route to *tert*-butanol, the first being the in situ hydration of the isobutene in the C_4 fraction from steam or catalytic cracking (Chapter 5). This low melting (mp 25.6°C) white solid can be oxidized to methyl methacrylate as is done in Japan (Section 4.6.1). It aroused interest as an octane improver in nonleaded gasoline, where it may be used in combination with methanol. However, methyl *tert*-butyl ether (Section 5.2.1) is superior. Accordingly, in the late-1980s, ARCO, who pioneered the two-for-one reaction for propylene oxide and *tert*-butanol, started to dehydrate *tert*-butanol to isobutene for conversion to methyl *tert*-butyl ether and quickly became the largest US producer of this important gasoline additive.

Pure oxygen is required for the isobutane hydroperoxidation. The uncatalyzed liquid-phase reaction proceeds at 120–140°C and 34 bar. Acetone forms as a minor byproduct. Water is necessary in the reaction mixture to inhibit oxidation of isobutane to ketones, aldehydes, and acids. The subsequent epoxidation is carried out at 135°C and about 50 bar to keep the reactants in the liquid phase and requires a molybdenum naphthenate catalyst.

A second two-for-one process for propylene oxide, also used by ARCO, provides styrene as a coproduct. Two and one-half pounds of styrene are generated for each pound of propylene oxide. Thus the first plant produced one billion pounds of styrene and 400 million pounds of propylene oxide. Benzene and ethylene give ethylbenzene, which is converted to ethylbenzene hydroperoxide. This epoxidizes propylene to propylene oxide, yielding coproduct phenylmethylcarbinol, whose dehydration provides styrene (Section 3.8).

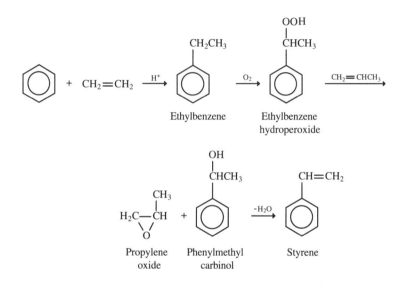

Because of the problems of isolating and recovering styrene, capital investment for this process is higher than that for the propylene oxide–*tert*-butanol process. The process is similar to the oxidation of cumene to cumene hydroperoxide (Section 4.5) and isobutane to *tert*-butyl hydroperoxide. Magnesium carbonate is added to adjust the pH to 7 to reduce the decomposition of the hydroperoxide. Selectivity to about 65% is possible only at a low conversion of 15–17%. Above that concentration, byproducts including phenylmethylcarbinol and acetophenone, increase appreciably.

The epoxidation stage is also similar to the production of *tert*-butyl hydroperoxide, a molybdenum naphthenate catalyst playing an important role. However, this requires milder conditions of 100–130°C in the liquid phase under ambient pressure. Selectivity of propylene to propylene oxide is about 91%. Byproducts include dimers of propylene, whose formation can be inhibited by antioxidants. The vapor-phase dehydration of phenylmethylcarbinol to styrene takes place over a catalyst at 200–280°C. Titania and alumina are typical.

A Shell variation of the ARCO route to propylene oxide and styrene involves a heterogeneous system for the epoxidation, with catalysts such as vanadium, tungsten, molybdenum, or titanium supported on silica and treated with a silylating agent. A plant exists in Europe and another was being built in Singapore in the mid-1990s.

In the United States in the early 1990s, approximately 50% of propylene oxide production was via the chlorohydrin method and the remaining 50% by the coproduct processes. In Europe 66% was produced by the chlorohydrin process and the remainder by the coproduct processes.

4.7.1 Propylene Oxide Applications

Whereas 60% of ethylene oxide is hydrolyzed to ethylene glycol, only about one-third of the propylene oxide produced is converted to propylene glycol. The major use, and the reason for the increased consumption of propylene oxide in the 1970s and early 1980s, is its oligomerization to poly(propylene glycols) for polyurethanes (Sections 7.3.1 and 8.3). The oligomerization is initiated by active hydrogen compounds. To provide the hydroxyl end groups necessary for interaction with isocyanates, propylene glycol is used as the initiator.

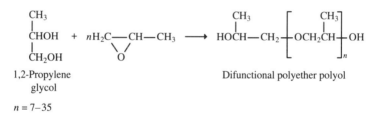

1,2-Propylene glycol

Difunctional polyether polyol

$n = 7-35$

The polymerization proceeds head-to-tail to give mainly secondary hydroxyl end groups. The polyols and polyethers react with diisocyanates, especially MDI or toluene diisocyanate, to give polyurethane resins. If a more highly cross-linked polymer is desired, as is frequently the case, a triol can be prepared by the interaction of glycerol or trimethylolpropane with three mols of propylene oxide. If more active hydroxyl groups are required, reaction with ethylene oxide provides primary hydroxyethyl groups.

The second largest application of propylene oxide is its reaction with water to give 1,2-propylene glycol.

$$H_2C\!\!-\!\!CHCH_3 \;+\; H_2O \;\longrightarrow\; HOCH_2CHOHCH_3$$

1,2-Propylene glycol

In practice, a large excess of water is used to inhibit the formation of oligomers. If the amount of water is decreased, dipropylene and tripropylene glycols form and indeed poly(propylene glycols) with molecular weights of several thousands can be made. Because propylene glycol is not toxic it can be used in cosmetics. The largest use, however, is in the manufacture of unsaturated polyesters (Section 9.1.1). Copolymers of ethylene oxide and propylene oxide are nonionic surfactants.

Useful solvents result when propylene oxide reacts with alcohols to form glycol ethers.

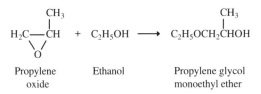

| Propylene oxide | Ethanol | Propylene glycol monoethyl ether |

This is analogous to the reaction of ethylene oxide with alcohols (Section 3.11.6.2). To some extent they are replacing the ethylene oxide-based products, some of which are toxic to humans (see note, Section 3.11.6.2). Because the propylene oxide-based glycol ethers have a branched methyl group, they are probably not as good solvents as the ethylene oxide-based products. Even so, increased demand is expected as the corresponding ethylene oxide-based products are phased out. In 1991, the US production of poly(ether glycols) and glycol ethers based on propylene oxide was 69 and 175 million lb, respectively.

4.7.2 Projected Propylene Oxide–Propylene Glycol Processes

Much work has been done on potential processes for propylene oxide and the corresponding glycol, some of which do not produce coproducts. One of the most interesting involves the direct acetoxylation of propylene to a mixture of propylene glycol mono- and diacetates. The reaction takes place at 180°C and 80 bar over a tellurium oxide–iodine catalyst at 20% propylene conversion. Byproducts include propylene glycol, acetone, carbon dioxide, and water. The diacetoxy compound is hydrolyzed to the monoacetoxy compound, the acetyl group on the secondary hydroxyl being removed preferentially. The liquid-phase hydrolysis is catalyzed by acetic acid at 230°C and 70 bar. The mono-acetoxy compound may then be hydrolyzed further to propylene glycol, or cracked catalytically to propylene oxide and acetic acid.

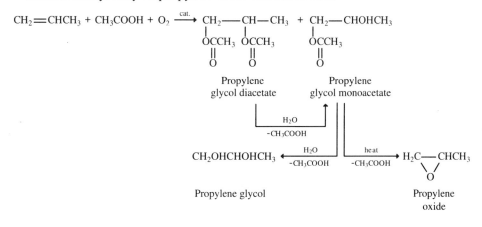

At 380°C and 0.13 bar, selectivity to propylene oxide is about 83 at 42% conversion. Cracking may also be carried out in the liquid phase with a tetrahydrothiophene-1,1-dioxide catalyst to obtain selectivities as high as 93% at 20–30% conversion. Major byproducts are acetone and propionaldehyde, which form by the decomposition of the monoacetates. The reaction system is highly corrosive, and a related process (Section 3.7.2) failed because of corrosion problems.

Another related process involves the hydroformylation of vinyl acetate to α-acetoxypropionaldehyde, which can be reduced to 1,2-propylene glycol monoacetate. On hydrolysis this product gives propylene glycol. This circumvents the need for propylene oxide as a starting material.

$$CH_2{=}CHOCOCH_3 \xrightarrow[\text{cat.}]{H_2/CO} \underset{\underset{CHO}{|}}{CH_3CHOCOCH_3} \xrightarrow[\text{cat.}]{H_2} \underset{\underset{CH_2OH}{|}}{CH_3CHOCOCH_3} \xrightarrow[\text{cat.}]{H_2O}$$

Vinyl acetate   1,2-Propylene
                                                                       glycol monoacetate

$$CH_3CHOHCH_2OH \ + \ CH_3COOH$$

Propylene   Acetic acid
glycol

4.7.3   Other Novel Syntheses of Propylene Oxide

Processes developed to varying degrees for propylene oxide production include:

1. Direct oxidation.
2. Use of peracids or hydrogen peroxide.
3. Electrochemical processes.
4. Biotechnological approaches.

They will be discussed in turn.

4.7.3.1   Direct Oxidation   Direct oxidation of propylene analogous to that of ethylene has not proved feasible because of low selectivity and numerous byproducts. The most promising results have been reported by Russian workers who used a catalyst comprising europium oxide–silver on a silica gel carrier. Selectivity was 90.7% at 32.5 mol% conversion. Typical of US work is that of Union Carbide and Phillips who have developed noncatalytic liquid-phase processes in organic solvents such as acetonitrile and methyl formate at 150–225°C and 5–14 bar. Selectivities of 55–65% were achieved at propylene conversions of less than 25%.

4.7.3.2   Use of Peracids or Hydrogen Peroxide   The use of peracids provides an elegant way of transferring oxygen to the double bond of propylene. The peracid is converted correspondingly to the parent carboxylic acid.

$$\underset{\text{O}}{\overset{\text{O}}{\text{RC}}}\text{—OOH} + \text{CH}_3\text{CH}\!=\!\text{CH}_2 \longrightarrow \underset{\text{O}}{\overset{\text{O}}{\text{RC}}}\text{—OH} + \text{CH}_3\text{CH}\!\!-\!\!\text{CH}_3$$

These processes are related to the previously described hydroperoxide processes. A process involving the coproduction of propylene oxide and methyl methacrylate has been suggested. It starts with isobutyraldehyde, which is converted to perisobutyric acid. This in turn oxidizes propylene to propylene oxide and gives isobutyric acid, which can be dehydrogenated to methyl methacrylate.

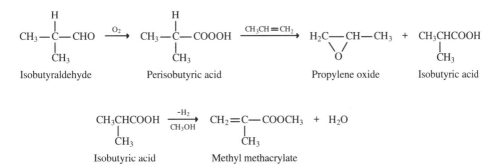

Isobutyraldehyde   Perisobutyric acid   Propylene oxide   Isobutyric acid

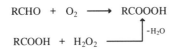

Isobutyric acid   Methyl methacrylate

Isobutyric acid is an intermediate in one of the propylene-based routes to methyl methacrylate (Section 4.6.1). Isobutyraldehyde is an undesired byproduct in the cobalt-catalyzed hydroformylation of propylenes (Section 4.8). Lack of demand, except from Eastman, who converts it to neopentyl glycol, (Section 4.8.1) motivated the development of the elegant rhodium-based catalyst system (Section 4.8) that provides high yields of n-butyraldehyde.

In addition to perisobutyric acid, the peracids of formic, acetic, and propionic acids can be used to make propylene oxide. The peracid can be prepared either by oxidizing the corresponding aldehyde or by reacting the corresponding carboxylic acid with hydrogen peroxide.

$$\text{RCHO} + \text{O}_2 \longrightarrow \text{RCOOOH}$$
$$\text{RCOOH} + \text{H}_2\text{O}_2 \overset{-\text{H}_2\text{O}}{\underset{}{\rule{0pt}{0pt}}}$$

In a one-step process, the percarboxylic acid is formed in situ by the oxidation of the aldehyde in the presence of propylene. The percarboxylic acid that forms epoxidizes the propylene immediately. In a two-step process, the percarboxylic acid is isolated and used to epoxidize propylene in a second step.

$$\text{RCHO} + \text{CH}_2\!=\!\text{CHCH}_3 + \text{O}_2 \longrightarrow \text{RCOOH} + \text{H}_2\text{C}\!\!-\!\!\text{CHCH}_3$$

Although results are excellent, the processes are not economical because of the high cost of hydrogen peroxide. If the peracid is prepared from an aldehyde, economics are better but not on a par with the previously described hydroperoxide processes.

Hydrogen peroxide also forms hydroperoxides with alcohols, which epoxidize propylene with the regeneration of the alcohol. Processes have been devised starting with isopropanol and isoamyl alcohol. The isopropanol process gives acetone as the major byproduct.

$$
\underset{}{H_2O_2} \;+\; \underset{\text{Isopropanol}}{CH_3\overset{\overset{\displaystyle OH}{|}}{C}HCH_3} \;\longrightarrow\; CH_3\overset{\overset{\displaystyle OOH}{|}}{C}HCH_3 \;+\; H_2O
$$

$$
CH_3\overset{\overset{\displaystyle OOH}{|}}{C}HCH_3 \;+\; CH_3CH{=}CH_2 \;\longrightarrow\; CH_3\overset{\displaystyle O}{\overset{\displaystyle\diagup\;\diagdown}{CH{-}CH_2}} \;+\; CH_3\overset{\overset{\displaystyle OH}{|}}{C}HCH_3
$$

$$
\xrightarrow{\text{side reaction}}\; CH_3COCH_3 \;+\; H_2O
$$

4.7.3.3 *Electrochemical Processes*

The electrochemical processes suffer from high energy cost. Typically, brine is electrolyzed in the presence of propylene. The overall reaction is simple, although many reactions are involved. At the anode, chloride ions are discharged to give chlorine, which reacts with water to form hypochlorous acid.

$$2Cl^- \rightarrow Cl_2 + 2e^-$$

$$Cl_2 + H_2O \rightarrow H^+ + Cl^- + HOCl$$

The reaction of the hypochlorous acid with propylene provides a chlorohydrin which is dehydrochlorinated by hydroxyl ions from the caustic produced at the cathode.

$$
CH_3CH{=}CH_2 \;+\; HOCl \;\longrightarrow\; CH_3CHOHCH_2Cl
$$

$$
\xrightarrow{OH^-}\; H_2\overset{\displaystyle O}{\underset{\diagdown\;\diagup}{C{-}CHCH_3}} \;+\; H_2O \;+\; Cl^-
$$

Protons are reduced to hydrogen at the cathode. Their removal increases the basicity of the solution and provides the "caustic" solution to dehydrochlorinate the chlorohydrin:

$$2H^+ + 2e^- \rightarrow H_2$$

$$H_2O \rightleftharpoons H^+ + OH^-$$

The sodium ion is a "spectator" throughout.

Dichloropropane and propylene glycol are byproducts, the first from the direct addition of chlorine to propylene and the second by the hydrolysis of propylene oxide. Propylene chlorohydrin and chloroisopropyl ether are also possible byproducts.

A problem with the original chlorohydrin process is the disposal of the calcium or sodium chloride that results from the dehydrochlorination. As indicated, in a totally integrated process, the sodium chloride can be used as brine for the generation of chlorine. A proposed Lummus process uses a concentrated brine in both the epoxidation and neutralization steps, which can readily be recycled to the electrolysis unit. In the first step chlorine, sodium hydroxide from the caustic cell liquor, and *tert*-butanol are combined to give *tert*-butyl hypochlorite and sodium chloride. The brine, which does not dissolve the *tert*-butyl hypochlorite, can be recycled to the electrolysis unit, and the hypochlorite can be used to convert the propylene to the chlorohydrin and regenerate *tert*-butanol.

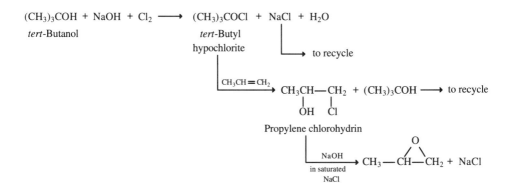

The two products are removed and separated by distillation, and the water is recycled to the hydrochlorination step where it is saturated with the NaCl produced. The dehydrochlorination is also performed in almost saturated brine, which can be recycled to electrolysis. In this way, there is never need to dispose of salt.

4.7.3.4 *Biotechnological approaches*

Biotechnological processes have been proposed for propylene oxide production although they are not practical at the present state of development, and there is no chance of their competing with straightforward chemical processing. A four-step process devised by Cetus starts with the production of hydrogen peroxide in the enzyme-catalyzed conversion of D-glucose to D-arabino-2-hexosulose. The latter can be hydrogenated to D-fructose, which becomes one of the products of the reaction. The hydrogen peroxide converts propylene to the bromohydrin with potassium bromide in an enzyme-catalyzed reaction. The bromohydrin can then be dehydrohalogenated either chemically or enzymatically to propylene oxide.

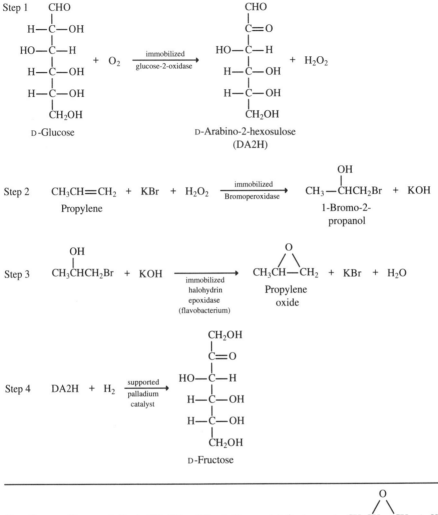

Step 1: D-Glucose + O₂ —(immobilized glucose-2-oxidase)→ D-Arabino-2-hexosulose (DA2H) + H₂O₂

Step 2: CH₃CH=CH₂ (Propylene) + KBr + H₂O₂ —(immobilized Bromoperoxidase)→ CH₃—CHCH₂Br (1-Bromo-2-propanol) with OH + KOH

Step 3: CH₃CHCH₂Br (with OH) + KOH —(immobilized halohydrin epoxidase (flavobacterium))→ CH₃CH—CH₂ O (Propylene oxide) + KBr + H₂O

Step 4: DA2H + H₂ —(supported palladium catalyst)→ D-Fructose

Overall: D-Glucose + O₂ + CH₃CH=CH₂ + H₂ ⟶ D-fructose + CH₃CH—CH₂ (O) + H₂O

The key to this imaginative process was the discovery of a peroxidase that forms the bromohydrin.

An Exxon process is based on a microorganism that consumes methane and can oxidize propylene to propylene oxide. Methylotrophic organisms, however, can only function with a coenzyme that undergoes a redox cycle with a readily oxidizable material such as methanol, which oxidizes to formaldehyde. The fragile enzyme and coenzyme are resident in a monocellular microorganism, and the cell rather than the enzymes can be used. Even so, the scale-up of such a process presents formidable problems.

These ingenious but currently impractical processes are described to emphasize that choice of R & D projects is one of the serious problems facing chemical companies today. Of course, it has always been a serious problem, but viability of projects was much more obvious in the 1950s and 1960s than in the 1990s. The search for the "no-go" factor, and the attempt to resolve it before expensive development is undertaken, is the key to any project evaluation.

4.8   *n*-BUTYRALDEHYDE AND ISOBUTYRALDEHYDE

n-Butyraldehyde was originally prepared by the aldol condensation of acetaldehyde with crotonaldehyde followed by hydrogenation to obtain a mixture of *n*-butyraldehyde and *n*-butanol in a ratio of 70:30 (Section 3.11.3).

The route has been replaced by reaction of propylene (Fig. 4.2 Reaction i) with carbon monoxide and hydrogen (Section 10.4) at 130–175°C and 250 bar over a cobalt carbonyl catalyst to give a mixture of between 3 and 4 mols of *n*-butyraldehyde to 1 mol of isobutyraldehyde. This reaction is called the oxo process or hydroformylation and was invented during World War II by Otto Roelen at Ruhrchemie. The mechanism is shown in Figure 4.10. The cobalt

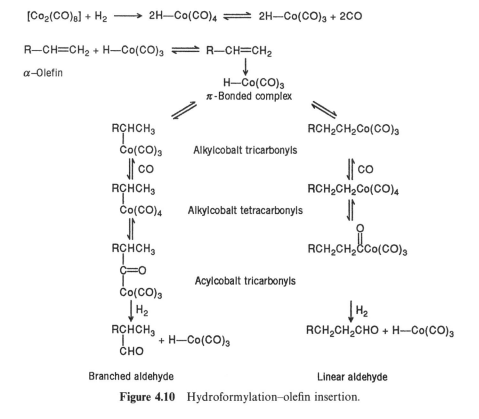

Figure 4.10   Hydroformylation–olefin insertion.

interacts with the olefin to give a π-bonded complex, after which there is a rearrangement in which the propylene inserts itself into the Co–H bond. The manner of the insertion determines whether the final compound will be linear or branched. The two possibilities are shown in the figure.

With a cobalt catalyst, an appreciable amount of branched product forms. If this rearrangement could be influenced so that a branched structure could not form, the yield of linear product would be increased. Collaboration between Union Carbide, Davy McKee, and Johnson Matthey resulted in the adaptation of the Wilkinson catalyst (Section 16.7.2), triphenylphosphine and rhodium chloride, to a homogeneous hydroformylation, in which the ratio of linear-to-branched compounds is 10:1. The mechanism of this linear hydroformylation is shown in Figure 4.11. The reaction takes place at mild temperatures of 100°C and pressures of 10–20 bar. The mild conditions are made possible by the phosphine ligands (represented as L), which increase the activity of the catalyst. It can be seen that the decomposition of the complex (**XIX**) primarily to the linear (anti-Markownikoff) product results because of the steric requirements of the bulky ligands.

Linear hydroformylation is also an important part of the SHOP (Section 3.3.4) process. However, because of the need in this process to hydroformylate internal olefins with the concomitant shifting of a double bond to the terminal position, a cobalt rather than a rhodium catalyst is used with the ligands that promote linearity. This catalyst is somewhat less effective, giving a ratio of normal-to-iso products of 5.7–6.0 to 1 and requiring a temperature of 150–190°C at 40 bar.

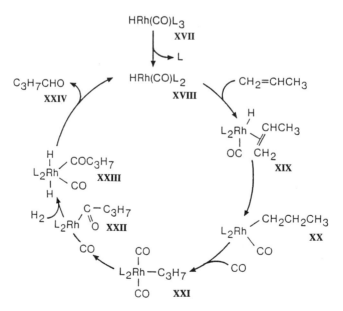

Figure 4.11 Linear hydroformylation.

Newer ligands described by Union Carbide are diphosphine monoxides of the general structure

$$\begin{array}{c} Ar_1 \\ \diagdown \\ \diagup \\ Ar_2 \end{array} P - Y - \overset{\overset{\displaystyle R_1}{|}}{\underset{\underset{\displaystyle R_2}{|}}{P}} = O$$

where the two aryl (Ar) groups may be the same or different, and R_1 and R_2 may be aromatic or aliphatic. Y is $(CH_2)_n$. In one example, $Ar_1 = Ar_2 = R_1 = R_2 = H_3CC_6H_4-$ and $Y = CH_2$. The ligands give a linear-to-branched ratio as high as 33.8:1.

Even newer ligands developed by Carbide are diorganophosphites (see note). These have a marked stabilizing effect on the catalyst and increase the rate of reaction. They make possible the hydroformylation under mild conditions of less active olefins such as 2-butene, isobutene, and vinyl acetate.

Rhodium is expensive, and efficient catalyst recovery is therefore important. Two approaches are noteworthy. In the first, described in a Union Carbide patent, reaction is carried out in a high boiling solvent comprising butyraldehyde aldol condensation products, primarily trimers and tetramers. After reaction, the products can be distilled off, leaving the nonvolatile rhodium complex in the solvent. The second approach involves sulfonated or carboxylated arylphosphine catalysts. After reaction, the organic phase in which the reaction was carried out is treated with alkali. The catalyst is converted to an alkali sulfonate or carboxylate which dissolves in the aqueous layer. Phase separation makes possible catalyst recovery. Another way to cope with the expense of rhodium is to make the catalyst so active that only an insignificant amount is used per pound of product. Today's oxo technology has made important strides in this direction.

4.8.1 Uses for Butyraldehyde and Isobutyraldehyde

The reason high linear-to-branched ratio is desired is because the major use of the aldehydes is for hydrogenation to the corresponding alcohols, which are used as solvents either as such or more often in the form of their acetates. *n*-Butanol, because of its linearity, solvates better than the branched isobutanol and is accordingly preferred, particularly for solvents for surface coatings. It is prepared by hydrogenation of *n*-butyraldehyde in the gas phase with a nickel or copper catalyst or in the liquid phase with a nickel catalyst. The gas phase requires mild pressures and a temperature of 115–120°C with a nickel catalyst or a temperature of 160°C with a copper catalyst. With a nickel catalyst in the liquid phase a temperature of 115°C and a pressure of 80 bar are required.

n-Butyraldehyde will undergo the aldol condensation, in the presence of either sodium hydroxide or a basic ion exchange resin, to provide an intermediate that loses water spontaneously to give an unsaturated aldehyde in quantitative yield. Hydrogenation gives the important plasticizer alcohol, 2-ethylhexanol.

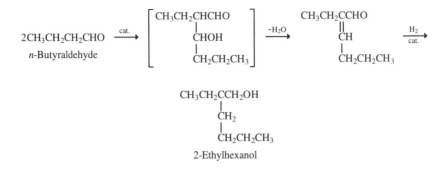

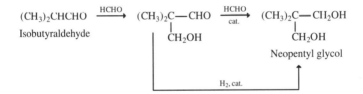

Because isobutyraldehyde was the surplus coproduct in the classical oxo process, uses were devised for it other than hydrogenation to isobutanol. Conversion to neopentyl glycol (2,2-dimethyl-1,3-propanediol) is the most important. Isobutyraldehyde reacts with 2 mol of formaldehyde to yield neopentyl glycol, used largely in unsaturated polyester synthesis to enhance alkali resistance of the cured product. It is also used to prepare polyesters for application in powder coating. Its preparation is analogous to that of pentaerythritol (Section 3.11.3).

4.8.2 Other Oxo Products

Oxo alcohols are used for solvents, surfactants, and PVC plasticizers. The solvent alcohols, n-propanol (Section 3.11.1), butanol and isobutanol (Section 4.8) have already been discussed. A good example of oxo alcohols for surfactants are those produced by the SHOP process (Section 3.3.4) in which hydroformylation is a major step. Plasticizers are usually esters of phthalic anhydride or less often adipic acid with aliphatic alcohols. Added to poly(vinyl chloride) during compounding, they lend softness and flexibility to the polymer. 2-Ethylhexanol is still the most important plasticizer alcohol in the United States, but its use is not growing because of possible toxicity problems. Straight chain alcohols (Sections 2.2.4, 4.2) solvate better so that they have better softening power and compatibility. 2-Ethylhexanol holds its position because of its lower price. Higher molecular weight alcohols are important because their phthalates have better high-temperature properties and lower volatility. This helps the "fogged windshield" problem and decreases the rate of embrittlement of plasticized articles because of volatile loss of plasticizer. Meanwhile, the use of PVC in automobile seating is declining.

The C_6–C_{12} olefins for hydroformylation to the plasticizer alcohols as well as C_{10}–C_{20} olefins for alcohols for nonbiodegradable surfactants come largely from propylene and butene oligomerization (Section 4.2). The most important plasticizer-range alcohols are isodecanol (from propylene trimer) and isononanol (from diisobutylene). Isooctanol (from propylene and butylene) and tridecanol from butylene trimer or propylene tetramer) are less widely used. In the United Kingdom "Alfanol," a mixture of C_7–C_9 alcohols, is important for plasticizers. It is obtained by application of the oxo process to an olefin mixture from the cracking of a petroleum fraction known as slack wax.

A smaller volume use for hydroformylation involves the conversion of 1-octene to pelargonaldehyde. Oxidation of the aldehyde yields pelargonic acid. Reaction of the acid with pentaerythritol (Section 3.11.3), gives the tetra ester useful as a component of synthetic lubricants. The oxo catalyst comprises rhodium chloride with either triphenylphosphine or triphenylphosphine and ferrocene ligand modifiers. The reaction take place at 5 bar and 100°C. Conversion of 95% of the α-olefin is obtained at a selectivity to the normal aldehyde of 87%. The oxidation is catalyzed by cobalt pelargonate at about 8 bar and 100°C. At 75% conversion of the aldehyde, 96% selectivity to the acid is obtained.

$$CH_3(CH_2)_5CH{=}CH_2 \xrightarrow{CO/H_2} CH_3(CH_2)_7CHO \xrightarrow{O_2} CH_3(CH_2)_7COOH$$

$$\text{1-Octene} \qquad\qquad \text{Pelargonic aldehyde} \qquad \text{Pelargonic acid}$$

4.9 MAJOR CHEMICALS FROM PROPYLENE—A SUMMARY

Sections 4.1–4.8 described the major chemicals (US production > 1 billion lb/year) from propylene and their derivatives. These major chemicals and their production volumes in 1977 and 1993, their growth rates and the amount of propylene used in their production are indicated in Table 4.1. As with ethylene, we have included only one polymer, polypropylene since it is the only one made directly from propylene and is the largest consumer of it.

The second largest volume chemical from propylene is cumene the progenitor of phenol (Section 7.1) and acetone. Acetone (position 5) is used as a solvent but its important chemical use is as a starting material for methyl methacrylate (position 11), which in turn is polymerized to poly(methyl methacrylate and related copolymers. Competitive routes based on C_2 and C_4 olefins for methyl methacrylate emerged but are not used in the United States, and the figures here represent total methyl methacrylate production. Acetone and phenol are the starting materials for bisphenol A (position 8) which in turn is combined with another propylene-based chemical, epichlorohydrin, to make epoxy resins.

Propylene oxide enjoys the third largest use. Its hydroxyl-terminated oligomers are the major coreactants with isocyanates for polyurethane polymers. The fourth largest volume chemical from propylene in 1991 was acrylonitrile whose major use is in polymers primarily for fibers. It is also used for

TABLE 4.1 Production of Chemicals from Propylene (1977–1993)—Summary, United States (million lb)

Polymer or Chemical	1977	1993	Average Annual % Increase 1977–1993	Propylene Consumption[a]
Polypropylene	2,747	8,614	7.4	8872
Cumene	2,640	4,490	3.4	1796
(Phenol	2,380	3,720	2.8)[b]	—
Propylene oxide	1,900	3,300	3.5	2607
Acrylonitrile	1,640	2,510	2.7	2761
Acetone	2,140	2,460	0.9	—
Oxo alcohols[c]	1,240	2,150	—	—
Including n-butanol	809	1,328	4.2	1023
2-Ethylhexanol	389[d]	688	10.1	—
Acrylate esters	n.a.[e]	1,322[f]	—	—
Bisphenol A	555[d]	1,286	7.2	—
Isopropanol	1,870	1,236	− 2.6	1051
Acrylic acid	n.a.[e]	1,199[g]	—	769
Methyl methacrylate	592[d]	1,088	5.2	—
Propylene glycol	473[d]	885	5.4	—

[a] This figure is the propylene required to give the 1993 production. Yields are estimated by Chem. Systems Inc.

[b] Phenol is really a benzene derivative. Although coproduced with acetone, it contains none of the propylene moiety.

[c] Includes n-butanol, isobutanol and 2-ethylhexanol.

[d] 1981 figure.

[e] Not available = n.a.

[f] 1992 figure. Includes 318 million lb n-butyl acrylate, 337 million lb ethyl acrylate and 172 million lb 2-ethylhexyl acrylate.

[g] 1992 figure.

elastomers. Oxo alcohols (position 6), some of which are not made from propylene, are used primarily as solvents and for plasticizers. Isopropanol is in position 9. It is the only chemical whose volume has decreased. It is used as a solvent, but some of it is converted to acetone (Section 4.6).

A marked difference can be seen in the growth rates of propylene-based chemicals compared to those of ethylene (Table 3.2). Propylene-based chemicals are less mature and, with the exception of isopropanol, have demonstrated better growth in the past decade. This faster growth was one reason for a brief shortage of propylene in Europe.

The demand these chemicals make on propylene supplies is shown in the final column of Table 4.1. As might be expected, polypropylene is the biggest consumer of propylene, propylene oxide is next and then acrylonitrile. The chemicals where a figure is given in the final column make up only about 74% of propylene consumption and this reflects a substantial number of minor uses, such as epichlorohydrin, ethylene–propylene elastomers, acrolein, allyl chloride, and isopropyl acetate. Some of the smaller volume chemicals from propylene are described in Section 4.10.

4.10 LESSER VOLUME CHEMICALS FROM PROPYLENE

4.10.1 Allyl Chloride and Epichlorohydrin

Allyl chloride is formed by the selective chlorination of the methyl group of propylene at 500°C, at which temperature chlorine addition to the double bond is obviated. Propylene and chlorine conversions are about 24 and 100%, respectively, at an allyl chloride selectivity of 86 mol %. Byproducts include dichloropropenes and minor amounts of isopropyl chloride and other chlorinated compounds.

$$CH_2=CHCH_3 + Cl_2 \xrightarrow{500°C} CH_2=CHCH_2Cl + HCl$$

An as yet uncommercialized oxychlorination route to allyl chloride catalyzed by palladium and cupric chlorides (Section 3.4) has been devised. It proceeds at a much lower temperature of 240°C:

$$CH_2=CHCH_3 + HCl + 0.5O_2 \xrightarrow{PdCl_2, CuCl_2} CH_2=CHCH_2Cl + H_2O$$

The epoxidation of allyl chloride to epichlorohydrin proceeds like the traditional route to propylene oxide (Section 4.7). Treatment of allyl chloride with hypochlorous acid gives 1,3-dichloro-2-hydroxypropane $ClCH_2CHOHCH_2Cl$ and 1,2-dichloro-3-hydroxypropane $ClCH_2CHClCH_2OH$ in a ratio of 9:1. These react below 60°C with a 10–15 wt% aqueous slurry of calcium hydroxide to give almost 100% conversion to epichlorohydrin.

$$CH_2=CHCH_2Cl \xrightarrow{HOCl} \begin{array}{c} ClCH_2CHOHCH_2Cl \\ + \\ ClCH_2CHClCH_2OH \end{array} \xrightarrow{Ca(OH)_2} H_2C\underset{\underset{O}{\diagdown\diagup}}{\quad}CHCH_2Cl + CaCl_2$$

Allyl chloride   Epichlorohydrin

In the overall process for epichlorohydrin from propylene, 75% of the chlorine used is wasted, that is, it does not appear in the final product.

Because epichlorohydrin is made in far smaller quantities than ethylene and propylene oxides, the search for a replacement for the chlorohydrin process was less strongly motivated. Nonetheless, a new process has recently been instituted

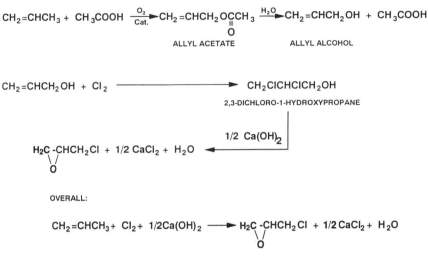

Figure 4.12   Epichlorohydrin–Showa Denko process.

by Showa Denko in Japan. This process wastes only 50% of the chlorine. Propylene is oxidized to allyl acetate by oxygen in acetic acid with a palladium catalyst, as shown in Figure 4.12. Hydrolysis yields allyl alcohol and acetic acid for recycle. The allyl alcohol is treated with chlorine and the resulting 2,3-di-chloro-1-hydroxypropane is dehydrohalogenated to epichlorohydrin.

Epichlorohydrin's major use is its reaction with bisphenol A (Section 7.1.2) to yield epoxy resins. It also plays a role in glycerol synthesis (Section 4.10.2). The major use of allyl alcohol is also in glycerol preparation. It is prepared from allyl chloride by treatment with aqueous caustic alkali. Allyl alcohol also reacts with phthalic anhydride to yield diallyl phthalate, a monomer for a thermoset polymer with good electrical properties, and with cyanuric chloride, to give triallyl cyanurate a polyester for thermoset resins with high heat resistance.

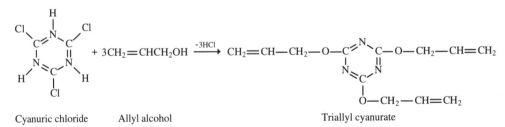

Cyanuric chloride   Allyl alcohol   Triallyl cyanurate

Allyl chloride is converted by ammonia to mono-, di-, and triallylamines.

4.10.2 Glycerol

Natural fats and oils are triglycerides (Section 13) and give glycerol on hydroly-sis. This source accounts for about 75% of the world's glycerol. The remainder comes from chemical synthesis, and the synthetic routes are summarized in Figure 4.13. The most important process (Fig. 4.13, Shell route) involves simply the hydrolysis of epichlorohydrin with 10% aqueous caustic soda at 150°C. Complete conversion is obtained. The reaction goes stepwise, the chlorine atom hydrolyzing first.

A route via acrolein (Fig. 4.13, Daicel route) involves a Meerwein–Pondorff reduction of acrolein to allyl alcohol, which is epoxidized with hydrogen peroxide, hypochlorous acid, or peracetic acid to glycidol, whose hydrolysis yields glycerol. Alternatively, the allyl alcohol may be hydroxylated to glycerol directly. In the early 1990s, Daicel modified its process by oxidizing propylene directly to allyl acetate (Section 4.10.1), which on saponification yields allyl alcohol. This is an efficient route, but one which requires high capital investment. In the FMC route shown in Figure 4.13 but no longer in use, propylene oxide is isomerized with a lithium phosphate catalyst to allyl alcohol. This can be converted to glycerol by the process already described. It is of interest that ARCO has adapted this route to allyl alcohol for a synthesis of 1,4-butanediol (Section 10.3).

Glycerol's uses are reviewed in Section 13.9. The pattern of glycerol manufacture varies around the world depending on indigenous production of fats and oils. In fat- and vegetable oil-rich countries like the United States, most

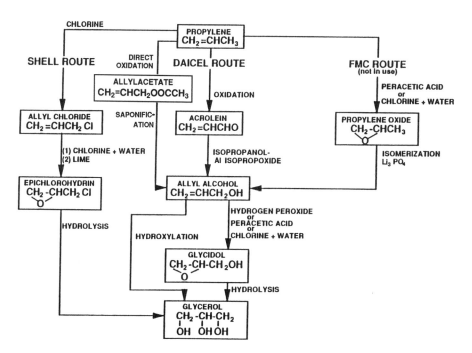

Figure 4.13 Synthetic routes to glycerol.

glycerol comes from hydrolysis of natural fats and oils. In countries poor in agricultural produce, such as Japan, most glycerol is made synthetically.

4.10.3   Acrylamide

Hydrolysis of acrylonitrile with sulfuric acid gives acrylamide:

$$CH_2\!=\!CH\!-\!CN \ + \ H_2SO_4 \ + \ H_2O \ \longrightarrow \ CH_2\!=\!CHCONH_2 \cdot H_2SO_4$$

Acrylonitrile

$$(NH_4)_2SO_4 \ + \ CH_2\!=\!CHCONH_2 \ \xleftarrow{\quad NH_3 \quad}$$

Acrylamide

As in the synthesis of caprolactam (Section 7.2.2) an undesirable low-value byproduct, ammonium sulfate, is generated, whose separation from the acrylamide is complex. This process has largely been replaced by an elegant catalytic hydrolysis using copper chromite. The reaction takes place at 100°C and gives almost 100% selectivity at 60–80% conversion.

$$CH_2\!=\!CHCN \ + \ H_2O \ \xrightarrow{\text{copper chromite}} \ CH_2\!=\!CHCONH_2$$

Nitto in Japan has developed a microbiological conversion of acrylonitrile to acrylamide. It provides the only example of the application of modern microbiology to the production of a reasonably large volume chemical other than fermentation-produced ethanol and high fructose corn syrup (Section 14.2). This conversion is brought about by an immobilized bacterial cell (e.g., species of nocardia, microbacterium and corynebacterium) that contains the hydrolyzing enzyme. A temperature of 10°C and a pH of 7.5 lead to a 15% aqueous solution of acrylamide in quantitative yield.

Acrylamide is polymerized to a water-soluble polymer. Anionic character may be conferred by partial hydrolysis of amide linkages to carboxyl groups. Anionic functionality can also be achieved by copolymerization of acrylamide with monomers such as sodium acrylate. Conversely, monomers such as dimethylaminoethyl acrylate impart cationic character.

A potentially important use for the polymers is in enhanced oil recovery where they increase the viscosity of the water used for the flooding that pushes oil through the rock formation so that it can be pumped. They act as a flocculent in industrial and municipal waste water treatment. They are included in formulations for paper and textile sizing, drilling muds for oil well drilling, and in the coagulation of slimes. With other acrylic monomers, acrylamide provides copolymers useful in the formulation of water-based protective coatings. If the amide group is converted to a methylol group with formaldehyde, a compound results that can cross-link with melamine resins (Section 10.5.1.1).

$$CH_2\!=\!CHCONH_2 \ + \ HCHO \ \rightarrow \ CH_2\!=\!CHCONHCH_2OH$$

4.10.4   Acrolein

The preparation of acrolein as an intermediate in acrylic acid production was described in Section 4.3. Its role in glycerol synthesis was described in Section 4.10.2. The most important use for acrolein, however, is for the manufacture of D,L-methionine, an essential amino acid added to poultry feeds. Methionine is the amino acid produced in third largest volume after L-glutamic acid, whose monosodium salt is a flavor enhancer in foods, and L-lysine, also used in feeds. In the methionine synthesis, acrolein reacts almost quantitatively with methyl-mercaptan in the presence of pyridine to yield β-methylthiopropionaldehyde. The aldehyde is converted to a hydantoin with sodium cyanide and ammonium carbonate.

$$CH_2\!\!=\!\!CHCHO \ + \ CH_3SH \ \xrightarrow{\text{Pyridine}} \ CH_3SCH_2CH_2CHO$$

| Acrolein | Methyl mercaptan | β-Methyl thiopropionaldehyde |

$$CH_3SCH_2CH_2CHO \ + \ NaCN \ + \ 1.5(NH_4)_2CO_3 \ \longrightarrow$$

$$CH_3SCH_2CH_2CH\!\!-\!\!CO \ + \ 0.5Na_2CO_3 \ + \ 2H_2O \ + \ 2NH_3$$

$$\underset{HN \diagdown_{\underset{\parallel}{C}}\diagup NH}{}$$

$$\underset{O}{}$$

Aqueous hydrolysis of the hydantoin with sodium hydroxide and sodium carbonate at 6 bar provides the sodium salt of methionine with the release of ammonia.

$$CH_3SCH_2CH_2CH\!\!-\!\!CO \ + \ NaOH \ + \ H_2O \ \longrightarrow$$

$$\underset{HN \diagdown_{\underset{\parallel}{C}}\diagup NH}{}$$

$$\underset{O}{}$$

$$CH_3SCH_2CH_2CHCOONa \ + \ NH_3 \ + \ CO_2$$

$$\underset{NH_2}{}$$

Sodium salt of
methionine

The sodium salt is then acidified and the methionine precipitates at its isoelectric pH of 5.7. The product of commerce is DL-methionine, the D isomer being converted in the organism to the useful L-structure. Also effective in feeds is the DL-methionine hydroxy analogue prepared by a simpler process, since the amino group is not inserted into the molecule. This process too goes via β-methylthiopropionaldehyde, whose quantitative reaction with hydrocyanic acid at 40°C gives a cyanohydrin that can be hydrolyzed with sulfuric acid at 140–160°C and 3–4 bar to DL-methionine hydroxy analogue. The product is sold in aqueous solution.

$$CH_3SCH_2CH_2CHO \ + \ HCN \xrightarrow{\text{alkali}} CH_3SCH_2CH_2\underset{\underset{OH}{|}}{CH}CN$$

$$CH_3SCH_2CH_2\underset{\underset{OH}{|}}{CH}CN \ + \ 2H_2O \ + \ 0.5H_2SO_4 \xrightarrow{140-160°\,C}$$

$$CH_3SCH_2CH_2\underset{\underset{OH}{|}}{CH}COOH \ + \ 0.5(NH_4)_2SO_4$$

Methionine
hydroxyanalogue

Glutaraldehyde is manufactured in small quantities from acrolein, which undergoes a Diels–Alder reaction with methyl vinyl ether (Section 3.11.11) to yield 2-methoxy-2,3-dihydro-γ-pyran, whose hydrolysis in the presence of acid yields glutaraldehyde.

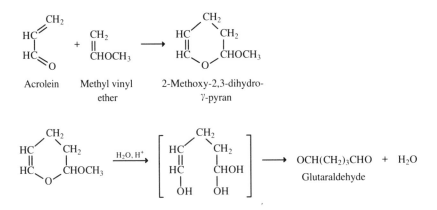

Acrolein   Methyl vinyl   2-Methoxy-2,3-dihydro-
ether   γ-pyran

Glutaraldehyde is one of the two dialdehydes available commercially, the other being glyoxal (Section 3.11.6.6). They are useful for cross-linking hydroxyl-containing materials like leather or cellulose for textile and paper treatment. They are also used in the immobilization of enzymes (Section 16.8).

4.10.5   Acrylonitrile Derivatives

Acrylonitrile production was described in Section 4.4 and its most important uses have been discussed in Sections 4.4.1 and 4.10.3. It also undergoes a reaction called cyanoethylation. A compound with a reactive hydrogen atom undergoes a base-catalyzed Michael addition to its double bond. A quaternary ammonium base such as trimethylbenzylammonium hydroxide is an excellent catalyst. Typical reactions are those with fatty amines and fatty alcohols. Reaction with octadecylamine yields octadecylaminopropionitrile. On hydrogenation this yields octadecylaminopropylamine, a corrosion inhibitor used to protect bits and casings during the drilling of oil wells.

$$C_{18}H_{37}NH_2 \ + \ CH_2{=}CHCN \ \xrightarrow{\text{cat.}} \ C_{18}H_{37}NHCH_2CH_2CN \ \xrightarrow[\text{cat.}]{H_2} \ C_{18}H_{37}NHCH_2CH_2CH_2NH_2$$

Octadecylamine   Octadecylamino-   Octadecylamino-
                              propionitrile   propylamine

The reaction with octadecanol provides cyanopropyloctadecyl ether whose nitrile group can also be hydrogenated to an amine. This compound is a flotation agent for the removal of silica from ion ore.

$$C_{18}H_{37}OH \ + \ CH_2{=}CHCN \ \xrightarrow{\text{cat.}} \ C_{18}H_{37}OCH_2CH_2CN$$

Octadecanol   Octadecyloxypropionitrile

$$\xrightarrow[\text{cat.}]{H_2} \ C_{18}H_{37}OCH_2CH_2CH_2NH_2$$

Octadecyloxypropylamine

The graft polymerization of acrylonitrile onto starch provides a polymer capable of absorbing many times its own weight of water (Section 4.3), once the nitrile groups are partially hydrolyzed to amides and carboxyls.

NOTES AND REFERENCES

Section 4   The small amount of ethylene from catalytic cracking has in the past not been isolated. As the price of ethylene increases, its isolation becomes economically feasible. Thus almost 1.5 billion lb of ethylene was isolated from catalytic cracking in 1991 in the United States. Saudi Arabia opted only to steam crack ethane and exported the propane and butane in their associated gas. This is discussed in Chapter 2.

Section 4.1   The Himont–Mitsui catalyst is described in US Patent 4 226 963 (October 9, 1980) to Montedison S.p.A.

Section 4.3   The reaction of acetic acid with formaldehyde to yield acrylic acid is claimed in US Patent 4 165 438 (August 21, 1979) to Chevron Research.

The catalysts described for the oxidation of propylene to acrylic acid are claimed in British Patent 1 346 626 (February 13, 1974) and in British Patent 1 361 246 (July 24, 1974) both to Nippon Shokubai. Other companies involved in the preparation of catalysts for propylene oxidation include Celanese, Union Carbide, Rohm GmbH, and BASF.

Section 4.4   The catalysts proposed for acrylonitrile formation are described in numerous patents typical of which are US Patent 4 473 506 (September 25, 1984) to Sohio and European Patent Application 0 109 259 (May 23, 1984) to Nitto Chemical Industries.

The technology for preparing acrylonitrile from acetonitrile and methane is found in West German Offen 2 616 275 (October 27, 1977). This provides an interesting example of how unwanted byproducts may be handled. Either the technology can be advanced to the point where their production is eliminated (as has been done for acetonitrile) or else a use can be found for them. The latter approach is demonstrated both by this example with acetonitrile and by the conversion of unwanted isobutyraldehyde to neopentyl glycol (Section 4.8.1).

The mechanism of ammoxidation is discussed in several articles including R. K. Grasselli, *J. Chem. Educ.*, **63**, 216 (1986), and R. K. Grasselli and J. D. Burrington, *Ind. Eng. Chem. Prod. Res. Dev.*, **13**, 393 (1984).

The stability of the allyl radical is worth illustrating. The heat of the reaction of $CH_3CH_2CH_3 \rightarrow C_3H_7 + 0.5H_2$ is 45.8 or 42.4 kcal/mol depending on whether the *n*-propyl or isopropyl radical is formed, whereas the heat of $CH_3CH=CH_2 \rightarrow CH_2 \dot{=}\dot{=} CH \dot{=}\dot{=} CH_2 + 0.5H_2$ is 35.7 kcal/mol, so that the allyl radical is stabilized to the extent of about 10 kcal/mol.

Section 4.4.1   The ICI route to hexamethylenediamine via acrylonitrile coupling has been described in *Chem. Week*, March 7, 1990, p. 9.

Section 4.5   The conversion of cumene, via the hydroperoxide, to acetone and phenol is described in detail by H. Wittcoff, *J. Chem. Educ.*, **56**, 810 (1979).

Section 4.6.1   The Shell process for methyl methacrylate is described in European Patent Spec. 0 186 228 (August 23, 1989) and in European Patent Application 0 279 477 A1 (August 24, 1988).

The Mitsubishi Gas Chemical Process for methyl methacrylate production is described in US Patent 4 613 684 (September 23, 1986).

Catalysts for isobutene oxidation to methyl methacrylate are described in numerous patents. Typical is West German Offen. 2 065 692 (January 2, 1975) to Japanese Geon and US Patent 3 875 220 (April 1, 1975) to Sohio.

The BASF route to methyl methacrylate from ethylene has been described in *Chem. Week*, January 26, 64 (1989). Various steps in the process provide the subject matter for US Patent 4 118 588 (October 3, 1978) to BASF, West German Offen. 3 213 618 A1 (October 27, 1983), British Patent 1 573 272 (August 20, 1986) to BASF, and US Patent 4 118 588 (October 3, 1978) to BASF.

The Asahi process for methyl methacrylate via ammoxidation is described in *Chem. Econ. Eng. Rev.*, **15**, 42 (1983).

Section 4.7.2   The preparation of propylene glycol via the hydroformylation of vinyl acetate is described in US Patent 4 723 036 (February 2, 1988) to Kuraray.

The Russian work on the direct oxidation of propylene to propylene oxide is found in US Patent 3 957 690 (May 18, 1986) issued to Bobulev et al.

The Union Carbide and Phillips work on the direct oxidation of propylene to propylene oxide is described in Canadian Patents 968 364 (May 24, 1975) and 986 127 (March 23, 1976) to Union Carbide and in US Patent 4 380 659 (April 19, 1983) to Phillips Petroleum.

Typical of the peracid processes is one devised by Interox Chemicals using perpropionic acid. It is described in many patents including British Patent 1 591 497 (June 24, 1981) and US Patent 4 177 196 (December 4, 1979).

Typical of the processes for epoxidation of propylene with hydroperoxides is the process described in West German Offen 2 803 757 (August 3, 1978) to Péchiney Ugine Kuhlmann.

The Cetus process is described in an article in *Chem. Eng. News* March 17, 1980, p.15. The Exxon work with the methylotrophic organism is the subject of several patent applications including UK Patent Application 2 081 306 (July 3, 1981).

The acetoxylation of propylene as a means for preparing propylene glycol and propylene oxide is described in several patents including German Patent 2 120 005 (August 31, 1978) to Scientific Design; US Patent 4 045 477 (August 30, 1977) to Chem Systems; US Patent REISSUE 29 597 (March 28, 1987) and 4 012 424 (March 15, 1977) to Chem Systems; US Patent 4 399 295 (August 16, 1983) to Chem Systems; US Patent 4 158 008 (June 12, 1979) to BASF.

Typical of the electrochemical processes for propylene oxide production is one described by Shell, *Chem. Eng. News*, September 28, 1987, p. 29.

The mechanism for linear hydroformylation was advanced by G. W. Parshall in his book cited in the note to Section 16.7.2.

Section 4.8   The Union Carbide development of a high-boiling solvent to prevent loss of rhodium catalyst in hydroformylation is the basis for US Patent 4 148 830 (April 10, 1979).

Carbide's diorganophosphite liquids are described in *Chem. Eng. News*, October 10, 1988, p. 270. The hydrogenation is carried out in two stages with a nickel catalyst. In the first stage, the vapor phase, conversion is about seven percent at 10 bar and 140°C. The second stage is in the liquid phase at 100 bar and 60°C. The product is purified by distillation. Copper catalysts have also been used.

The technology associated with water-soluble rhodium catalysts for hydroformylation is described in European Patent Application 103 810 (March 28, 1984) to Ruhrchemie. Sulfonated triarylphosphines are described by Rhone Poulenc, US Patent 4 483 801 (November 20, 1984)).

Section 4.10.2   The FMC process for hydroxylating allyl alcohol is described in German Offen. 2 439 879 (March 4, 1976).

CHAPTER 5

CHEMICALS AND POLYMERS FROM THE C$_4$ STREAM

We have described the huge volume of chemicals based on ethylene and the somewhat smaller volume based on propylene. Quantitatively the C$_4$ stream provides far less chemicals than ethylene or propylene (Fig. 4.1). Only three C$_4$-based chemicals [butadiene, isobutene, and methyl *tert*-butyl ether (MTBE)] are among the 50 chemicals produced in highest volume in the United States. The C$_4$ stream usage is not in the same league as ethylene's 41 billion lb and propylene's 22 billion lb consumption in 1993. About 3 billion lb of butadiene were consumed. Some 1.1 billion lb of isobutene were required for chemical uses but the surging demand for MTBE in unleaded gasoline meant that of the order of 13 billion lb went into this application, making it the fastest growing chemical ever. Butadiene goes mainly into synthetic rubbers. There is a range of these of both chemical and historical importance. It was the C$_4$ stream that provided the synthetic rubber vital to both sides in World War II. Also the chemistry associated with the C$_4$ stream is interesting and different, so we shall deal with it in comparable detail.

Steam cracking of natural gas gives far less C$_4$ fraction than does naphtha cracking (Section 2.2.3). *n*-Butane, traditionally, has seldom been cracked in the United States, because it occurs in relatively low volume in natural gas and is used largely for fuel for heating and cooking. This application in the United States increased in the mid-1980s and will increase further in the 1990s because butane is being removed from gasoline, particularly in the summer months when it volatilizes and participates in the formation of photochemical smog. A more important source of *n*-butane is LPG (Section 2.2.1), which is finding greater use as cracking feedstock. A C$_4$ unsaturated fraction is also obtained from catalytic cracking (Section 2.2.2). However, the fraction is dilute, and isolation of chemicals from it is more expensive. Most of it is used for alkylation (Section 2.2.5).

Because the United States has traditionally cracked natural gas, there has always been a shortage of butadiene. The shortfall has been accommodated by imports from Japan and Western Europe where naphtha cracking yields an excess of the C$_4$ fraction. It was this shortfall in the United States that motivated the discovery of the metathesis reaction (Section 2.2.8) and the dimerization of ethylene (Section 3.2.1), both aimed at providing C$_4$ chemicals.

The European surplus may disappear in 1995 because five companies have announced plans to convert butadiene to n-butenes, which may be isomerized to isobutene for MTBE (Section 5.2.1). At least one other plant has been hydrogenating butadiene to butanes for cracker feed (see note). This is all an ironic reversal of previous trends, in that butadiene has always been the premium product and isobutene the unwanted byproduct.

The five main components of the C$_4$ stream are n-butane, 1-butene, 2-butene, isobutene, and 1,3-butadiene. Table 5.1 shows a typical composition of the C$_4$ fraction from steam cracking. These can to some extent be interconverted (Fig. 5.1). n-Butane readily isomerizes to isobutane, which can be dehydrogenated to isobutene, a process commercialized in the early 1990s. Commercialized in the mid-1990s was a process for isomerizing n-butenes to isobutene for MTBE. Both n-butane and the butenes can be dehydrogenated to 1,3-butadiene. Dehydrogenation of butenes is practiced in the United States when it is favored by a high butadiene price. It is harder to dehydrogenate n-butane and this is not currently done, although it has been in the past.

Like propylene, the butenes have both fuel and chemical uses. When reacted with isobutane (Section 2.2.5) they provide alkylate for gasoline. When dimerized (Section 2.2.4) they become a component of high-octane polymer gasoline. The dimers can be hydroformylated (Section 4.8) to provide nonanol for plasticizers (Section 9.1.1). Newer methods of dimerization (Section 5.1.3.2) yield more nearly linear products, which give more effective plasticizers. Alkylate has no chemical uses.

The separation of the C$_4$ olefins is complex and is outlined in Figure 5.2. The first step is removal of 1,3-butadiene by extractive distillation with an aprotic

TABLE 5.1 Typical Composition of a C$_4$ Olefin Fraction from Steam Cracking of Naphtha

Component	Percentage
C$_3$	0.5
n-Butane	3
Isobutane	1
Isobutene	23
1-Butene	14
2-Butene	11
Butadiene	47
C$_5$, other	0.5

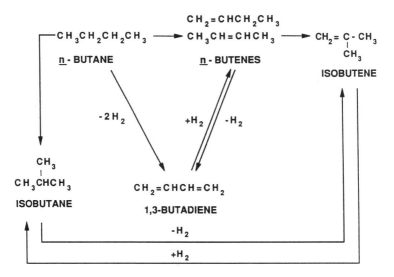

Figure 5.1   Major components of the C$_4$ stream.

solvent, usually acetonitrile, *N*-methylpyrrolidone, or *N,N*-dimethylformamide (DMF). Each of these solvents is effective in depressing the volatility of components other than butadiene. However, a small amount of butadiene is left in the raffinate, which must be removed by after-treatment (selective hydrogenation) if pure 1-butene is desired later in the process.

The butadiene-free raffinate that results is called and traded as Raffinate I. It contains the two *n*-butenes and isobutene. The isobutene is reactive and may be separated by hydration to *tert*-butanol in the presence of sulfuric acid or, in more modern plants, by use of a fixed-bed catalyst. *tert*-Butanol may be used as such or it may be dehydrated to pure isobutene.

In an alternative process, Raffinate I reacts with methanol to give MTBE (Section 5.2.1). This usually is used as such but, like *tert*-butanol, it may be cracked to give pure isobutene. The etherification process to MTBE is highly selective, on a par with the carboxylation of methanol to acetic acid (Section 10.5.2.2). It goes to about 97% conversion and, if all the isobutene must be removed, a second step is necessary in which the Raffinate I is reacted with a large excess of methanol.

Other processes for removing the isobutene involve an acid-catalyzed oligomerization at 100°C to form isomers of 2,2,4-trimethylpentene, useful for the gasoline pool. Another possibility is the polymerization of the isobutene to polyisobutene. The polymerization is catalyzed by aluminum chloride and the polymer is useful in adhesive and sealant formulations and petroleum product additives (Section 5.2.3).

The mixture of 1- and 2-butenes that remains is called and traded as Raffinate II. It may be hydrated to *sec*-butanol (Section 5.3) or oxidized to maleic

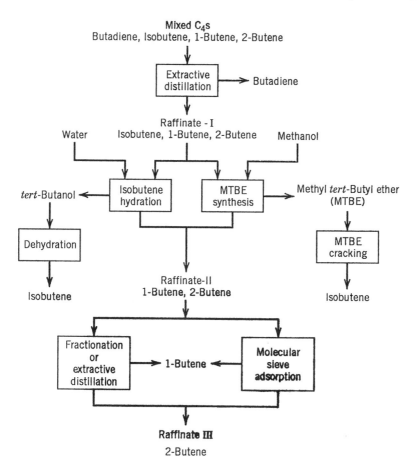

Figure 5.2   The separation of the C$_4$ olefins stream.

anhydride (Section 5.4), although the latter is not done in the United States. In the process, 1-butene isomerizes to the more stable 2-butene.

The two butenes in Raffinate II may also be separated to yield pure 1-butene for linear low-density polyethylene (LLDPE) production (Section 5.3). There are three procedures for separating the isomers. Fractionation requires a high, multiplate, low-temperature column because the boiling points are $-6.47°C$ (1-butene) and $3.73°C$ (cis-2-butene). Extractive distillation and molecular sieve absorption (UOP's Sorbutene process) are also possible. The advantage of the UOP process is that pure 1-butene results even if appreciable isobutene remains in the Raffinate II. The other two processes are useful only if the butadiene and isobutene contents of Raffinate II are low. The cis- and trans 2-butene stream that results is called Raffinate III. The pure isomers do not have many uses, and the material is not ordinarily purified but sent to the alkylation unit.

5.1 CHEMICALS AND POLYMERS FROM BUTADIENE

The United States and Western Europe traditionally have used the C_4 stream differently. In Europe there is an excess of gasoline relative to diesel fuel and fuel oil. Hence, there is only a small market for alkylate gasoline (Section 2.25). Furthermore, naphtha is the major feedstock for steam cracking, and Western Europe's requirements for 1,3-butadiene can be satisfied by extraction from the copious C_4 stream that naphtha cracking produces. Hence, there is little demand for butenes and little difference in value between them and butanes. In the United States, on the other hand, alkylate gasoline is in ever greater demand for unleaded gasoline, and butenes are also required for dehydrogenation to butadiene, which is not provided in sufficient quantities by steam cracking of natural gas liquids. Therefore, the butenes have greater value.

Figure 5.3 shows butadienes's end-use pattern. About 70% goes into synthetic rubbers Table 5.2 lists the major elastomers and their compositions. The most important one is styrene–butadiene rubber (SBR) for tires. This use is not growing. Not only are American automobiles now smaller, requiring smaller tires, but also the number of automobiles produced has decreased in the 1980s because of overseas competition. Adding to butadiene's poor fortune is the longer life of modern tires and the virtual elimination of the spare tire. Modern radial ply tires contain less SBR and more natural rubber than the older cross-ply tires. Thus butadiene consumption in the United States since 1987 has been about 3 billion lb, and growth through the 1990s is unlikely to be greater than 1.5% per year.

Styrene–butadiene copolymers were the original basis for latex paint, but latices with superior properties result from poly(vinyl acetate), polyacrylates, polymethacrylates, and their copolymers, and these have since captured the

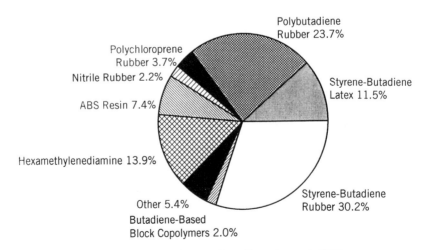

Figure 5.3 United States butadiene demand, 1993.

market. Styrene–butadiene latices are still important as binders for paper and for carpet backing. Styrene–butadiene block copolymers were among the first thermoplastic elastomers and are discussed in Section 15.3.8. A specialty butadiene–styrene block copolymer with high styrene content, one version of which is called "K-Resin," can be blended with styrene homopolymer to increase its impact resistance. It can be used as such in adhesives and molded and extruded items, and is stronger than polystyrene.

Several other synthetic rubbers are based on butadiene. The homopolymer, *cis*-1,4-polybutadiene rubber is blended with natural and other rubbers such as SBR to aid their processing and to impart resilience and less heat buildup to tires. It is also used in tire treads to increase wear resistance, low-temperature flexibility, and resistance to aging. The sophistication of modern polymerization techniques is such that it is possible to polymerize butadiene to give any of the four possible stereoisomeric forms, which vary from elastomers to hard, resinous materials (Section 15.3.10). The elastomeric form is *cis*-1,4-polybutadiene. *trans*-1,4-Polybutadiene is used in the preparation of acrylonitrile–butadiene–styrene (ABS) copolymer (Section 5.1.1). The isotactic and syndiotactic 1,2 structures are not important commercially, although syndiotactic 1,2-polybutadiene can be fabricated into photodegradable film.

The chlorination of butadiene in the vapor phase at 250°C gives a mixture of 3,4-dichloro-1-butene, $CH_2=CH–CHCl–CH_2Cl$, and 1,4-dichloro-2-butene, $ClCH_2–CH=CH–CH_2Cl$. The latter may be isomerized to the former by heat in the presence of cuprous chloride. Dehydrochlorination of 3,4-dichloro-1-butene leads to "chloroprene," the monomer for neoprene or polychloroprene rubbers.

$$CH_2=CH–CHCl–CH_2Cl \xrightarrow{-HCl} CH_2=CH–CCl=CH_2$$

3,4-Dichloro-1-butene   2-Chloro-1,3-butadiene
(chloroprene)

Neoprene was an early synthetic rubber. It lacks the resilience necessary for use in tires, the major consumer of rubber, but it has good resistance to oil and ozone, which suits it for many specialized uses such as roofing and flexible hose. In the early days, it was made by reacting dimerized acetylene (i.e., vinylacetylene) with hydrogen chloride. Here is another example of petrochemicals replacing acetylene, for Du Pont phased out the last acetylene-based polychloroprene plant in the United States in 1974 and in Northern Ireland in 1977.

$$2CH\equiv CH \xrightarrow{CuCl/NH_4Cl} CH_2=CHC\equiv CH \xrightarrow{+HCl} CH_2=CH–CCl=CH_2$$

Acetylene   Vinylacetylene   Chloroprene

Nitrile rubber (Table 5.2) is a copolymer of acrylonitrile and butadiene, a typical ratio being 1:2. It is made with a redox initiator and is characterized by outstanding oil and abrasion resistance, which makes it suitable for oil seals, fuel tank fabrication, oil-resistant hoses, and ink rollers for printing presses.

TABLE 5.2 Major Elastomers

Elastomer	Monomers and/or Structure	Polymerized by	Comments
Natural rubber	cis-1,4-Polyisoprene $\left[\begin{array}{c}CH_2 \quad CH_2 \\ C=C \\ H_3C \quad\quad H\end{array}\right]_n$	Enzymes	Obtained from *Hevea Brasiliensis*
Styrene–butadiene	~ 1 Mol styrene to $CH=CH_2$ (benzene ring) 6 mol butadiene, $CH_2=CH-CH=CH_2$	Emulsion polymerization, free radical initiator often at low temp. (cold rubber)	~ 20% Butadiene units are in 1,2-configuration, ~ 60% are *trans*-1,4 and 20% *cis*-1,4
cis-1,4-Polybutadiene rubber	$\left[\begin{array}{c}CH_2 \quad CH_2 \\ C=C \\ H \quad\quad H\end{array}\right]_n$	Metal complex	High resiliency
Butyl rubber	Isobutene + 1–3% isoprene $$CH_2=C(CH_3)_2$$ $$+$$ $$CH_2=C-CH=CH_2$$ $$CH_3$$	Cationic initiator	High gas impermeability

Synthetic cis-1,4-polyisoprene	Isoprene
	
	$CH_2=C-CH=CH_2$
	     $\mid$
	     CH_3
		Metal complex	Properties are similar to those of natural rubber

Polychloroprene rubber	Chloroprene
	
	$\left[CH_2-CH=C-CH_2 \right]_n$
	                $\mid$
	                Cl
	(cis-1,4-Polychloroprene)
		Free radical initiator	Excellent ozone resistance

Nitrile rubber	Butadiene + 20–40% acrylonitrile $CH_2=CH-CH=CH_2$ + $CH_2=CHCN$	Free radical initiator	Similar to SBR but more oil resistant
Ethylene-propylene rubber (EP and EPDM)	C_2H_4, C_3H_6 in EP. In EPDM ethylidene norbornene or other diene monomer is added	Metal complex	EP vulcanized with peroxide; EPDM with normal vulcanizing agents
Thermoplastic rubber	Block polymer of butadiene and styrene (and many others)	Ionic "living" polymerization	High strength, excellent solvent resistance, processability
Thiokol	$Na_2S_4 + nClCH_2CH_2Cl \rightarrow$ $[CH_2CH_2-S-S-S-S]_n$ $+ 2nNaCl$	Step growth polymerization	Used in sealants

5.1.1   Acrylonitrile–Butadiene–Styrene Resins

Acrylonitrile–butadiene–styrene resins are two-phase systems of styrene–butadiene rubber dispersed in a glassy styrene–acrylonitrile matrix (Section 3.8). The copolymer has the best properties if there is sufficient input of energy in the mixing to cause grafting of the rubbery and glassy phases.

The ABS resins may be used between − 40 and 107°C, not a very wide temperature range but a satisfactory one. They are flammable but fire retardants can be added. Apart from these minor drawbacks, ABS resins have excellent properties. They process reasonably, have high gloss, do not mar, and have excellent dimensional stability and good mechanical properties. In terms of price and tonnage production (2920 million lb in 1993), they fit in a unique position between cheap, high-volume plastics and expensive, low-volume engineering plastics. They can nonetheless be described as the first of the engineering plastics.

Injection molding of ABS resins gives housings for radios, telephones, business machines, pocket calculators, and parts for high quality refrigerators. Extrusion gives high quality pipe and fittings (the biggest use) and also sheet that is thermoformed into vehicle covers and fittings, lawn mower housings, and snowmobile shrouds. Sheet is also used for luggage and the tops of skis. In most of these applications, ABS is replacing metal or, in the case of telephones, a less impact-resistant plastic.

The ABS resins may be painted, metallized, chromium plated, or printed and do not present the problems of a nonadhesive surface associated with the polyolefins. The plated plastic has replaced die cast metal in the radiator grilles of many automobiles.

A polymer related to ABS is methyl acrylate–butadiene–styrene (MBS), produced in considerably lesser quantities than ABS. It is an impact modifier for poly(vinyl chloride) (PVC) for bottles. The PVC bottles are scarcely used in the United States, but in Western Europe they are used to package mineral water, cooking oil and to a lesser extent alcoholic beverages.

5.1.2   Hexamethylenediamine

The remaining major application for butadiene in Figure 5.3 is hexamethylenediamine (HMDA). Only two of the various routes to HMDA start with butadiene, but they include the most economic synthesis, and HMDA is butadiene's most important nonpolymer use. Accordingly, the various routes to HMDA will be discussed here.

The evolution of syntheses for HMDA provides an excellent example of how sophisticated processes develop in the chemical industry. The first route to HMDA involved traditional reactions in which adipic acid (Section 7.2.1) and ammonia form ammonium adipate which, on dehydration first to the amide and then to the nitrile, provides adiponitrile. This in turn may be hydrogenated to hexamethylenediamine.

$$HOOC\!-\!(CH_2)_4\!-\!COOH \xrightarrow{2NH_3} [NH_4OOC\!-\!(CH_2)_4\!-\!COONH_4] \xrightarrow{-2H_2O}$$

Adipic acid   Ammonium adipate

$$\left[H_2N\!-\!\overset{\overset{\displaystyle O}{\|}}{C}\!-\!(CH_2)_4\!-\!\overset{\overset{\displaystyle O}{\|}}{C}\!-\!NH_2 \right] \xrightarrow{-2H_2O} N\!\equiv\!C\!-\!(CH_2)_4\!-\!C\!\equiv\!N \xrightarrow{4H_2} H_2N\!-\!(CH_2)_6\!-\!NH_2$$

Adipamide   Adiponitrile   Hexamethylenediamine

Under appropriate conditions, the reaction of adipic acid and ammonia leads to the nitrile without the isolation of intermediates. The vapor-phase reaction takes place at 275°C in the presence of a mixture of boric and phosphoric acids. It may also be conducted either in the liquid phase or as a melt at a somewhat lower temperature with a phosphoric acid catalyst. Selectivity is higher in the liquid and melt phases. The synthesis is simple but has the disadvantage of starting with expensive adipic acid.

Butadiene entered the picture because of a synthesis in which a mixture of *cis* and *trans* 1,4-dichloro-2-butenes (see above) and 3,4-dichloro-1-butene was treated with sodium cyanide to give 1,4-dicyano-2-butene. The reaction takes place at about 80°C in the presence of cuprous chloride. The 1,4-dichloro compound yields the desired dicyano compound. The 3,4-dicyano-1-butene, which forms from the corresponding dichloro compound, undergoes an allylic rearrangement to give more of the desired 1,4-dicyano-2-butene, in both *cis* and *trans* forms at high selectivity.

$$ClCH_2\!-\!CH\!=\!CH\!-\!CH_2Cl \xrightarrow[CuCl]{NaCN} N\!\equiv\!C\!-\!CH_2\!-\!CH\!=\!CH\!-\!CH_2\!-\!C\!\equiv\!N$$

1,4-Dicyano-2-butene

$$\xrightarrow{H_2} N\!\equiv\!C\!-\!CH_2\!-\!CH_2\!-\!CH_2\!-\!CH_2\!-\!C\!\equiv\!N \xrightarrow{H_2} H_2N\!-\!(CH_2)_6\!-\!NH_2$$

1,4-Dicyanobutane   Hexamethylenediamine

In the final stage—the hydrogenation of adiponitrile to hexamethylene-diamine—excess ammonia is used to depress formation of a triamine, which nonetheless forms in small quantities.

$$2NH_2(CH_2)_6NH_2 \rightarrow NH_2(CH_2)_6NH(CH_2)_6NH_2 + NH_3$$

The electrohydrodimerization of acrylonitrile (Section 4.4.1) was the next synthesis. It was an ingenious departure from classical chemistry. The hydro-dimerization to adiponitrile takes place by a complicated mechanism, which may be summarized as

$$2CH_2\!=\!CHCN + 2H^+ + 2e^- \rightarrow NC(CH_2)_4CN$$

The first step is the reduction of acrylonitrile to its anion radical and this is followed by dimerization and protonation reactions. A necessary condition in the original process was the presence of a tetraalkylammonium salt as the electrolyte; otherwise propionitrile is the main product. The quaternary was thought to adsorb on the electrode to give a hydrophobic layer that prevented immediate protonation of the acrylonitrile anion radical as it was formed. A membrane to divide the cell was also required.

The newest version of the process employs an undivided cell and an emulsion of acrylonitrile with 10–15% disodium hydrogen phosphate solution containing 0.4% of a complex phosphate. The aqueous phase is saturated (7%) with acrylonitrile, and the adiponitrile extracts into the organic phase as it is formed. Conversion of acrylonitrile is 50 mol% and selectivity is about 92%. Byproducts include propionitrile C_2H_5CN, bis(cyanoethyl)ether $(NCC_2H_5)_2O$, and hydroxy-propionitrile $HOCH_2CH_2CN$.

The newer process lowers utility costs. Acrylonitrile, the starting material, is more expensive than butadiene. Nonetheless, adiponitrile is manufactured by this clever Monsanto process in at least two plants, and it is the only example of the use of electrochemistry to manufacture a high-volume organic chemical.

Chemical means for dimerizing acrylonitrile have been studied extensively. Unlike the Monsanto process, the reaction is not a hydrodimerization. Thus it yields a mixture of the two possible 1,4-dicyanobutenes. The reaction proceeds at 1–2 bar and temperatures varying from 30–100°C to give 87% conversion with a selectivity to linear products of 84%.

$$2CH_2=CHCN \rightarrow NCCH_2CH=CHCH_2CN + NCCH=CHCH_2CH_2CN$$

These are hydrogenated to hexamethylenediamine. A branched isomer, methy-
lene–glutaronitrile, $NC-\overset{\overset{\textstyle CH_2}{\|}}{C}-CH_2CH_2CN$, forms as a byproduct and may be isomerized to a linear product. The dimerization catalysts are phosphonates or phosphinates. Isopropyl bis-p-tolylphosphinate is typical.

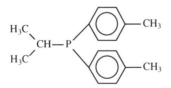

Isopropyl *bis-p*-tolylphosphine

These processes are rivaled by still another process, which turns out to be the most economic. It starts with the 1,4 addition of hydrogen cyanide to butadiene. The catalyst, nickel with triaryl phosphite ligands, $Ni[P(OAr)_3]_4$ is remarkable. It not only effects the 1,4 addition but it also isomerizes the byproduct 2-methyl-3-butenenitrile, which is produced in appreciable quantities, to the desired isomer. In addition, it shifts the double bond to a terminal position and then makes possible the addition of another mole of hydrogen cyanide anti-Markovnikov to the terminal bond. If it were not possible to use the same catalyst for both reactions, the double-bond shift could not be carried out in high yield because of an equilibrium. The resulting adiponitrile may be hydrogenated to hexamethylenediamine.

$$CH_2\!\!=\!\!CH\!-\!CH\!=\!\!CH_2 \xrightarrow{\text{HCN}} CH_2\!\!=\!\!CH\!-\!CH_2\!-\!CH_2\!-\!CN$$

Transition
metal catalyst
HCN | double-bond shift
(anti-Markovnikov)

$$NC\!-\!CH_2\!-\!CH_2\!-\!CH_2\!-\!CH_2\!-\!CN$$
Adiponitrile

This synthesis, as opposed to the one starting with the chlorination of butadiene, eliminates the need for costly chlorine, uses hydrogen cyanide, which is now cheaper than sodium cyanide (but not when the synthesis based on 1,4-dichloro-2-butene was devised), eliminates the handling of carcinogenic 1,4-dichloro-2-butene, and solves the problem of disposing of ecologically unacceptable copper-contaminated sodium chloride.

The mechanism of the HCN addition (Fig. 5.4) may involve the formation of an H–M–CN complex in which M is the metallic catalyst (a). This in turn

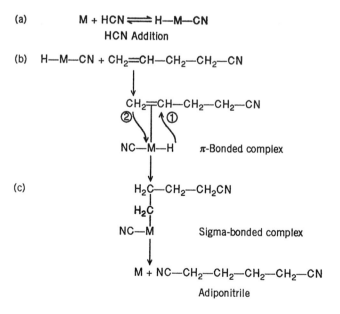

(a)   M + HCN ⇌ H—M—CN
         HCN Addition

(b)   H—M—CN + CH₂=CH—CH₂—CH₂—CN

CH₂=CH—CH₂—CH₂—CN
②   ①
NC—M—H   π-Bonded complex

(c)   H₂C—CH₂—CH₂CN
           H₂C
           NC—M   Sigma-bonded complex

M + NC—CH₂—CH₂—CH₂—CH₂—CN
Adiponitrile

Ligands omitted.

Figure 5.4 Mechanism of HCN addition.

π bonds with the terminal olefin (b) after which a σ-bonded complex forms (c), which subsequently eliminates the metal to provide the adiponitrile.

An obsolete process for hexamethylenediamine production involves the amination of 1,6-hexanediol with ammonia in the presence of Raney nickel. The diol, still manufactured for use in polyurethane formulations, results from the hydrogenolysis of methyl adipate.

$$
\underset{\underset{\substack{|| \\ O}}{}}{H_3COC(CH_2)_4CCH_3} \xrightarrow[\text{cat.}]{H_2} HO\!-\!(CH_2)_6\!-\!OH \xrightarrow[\text{cat.}]{NH_3} NH_2\!-\!(CH_2)_6\!-\!NH_2
$$

Methyl adipate   1,6-Hexanediol

Another route to 1,6-hexanediol is via cyclohexanone, which is oxidized by hydrogen peroxide or peracetic acid to caprolactone (Section 7.2.2). The lactone can then be cleaved with hydrogen in the presence of a reducing catalyst such as copper chromite to 6-hydroxycaproic acid whose ester, on hydrogenolysis, gives 1,6-hexanediol. The amination of the diol takes place in the presence of ammonia and hydrogen at 200°C and 230 bar with a Raney nickel catalyst.

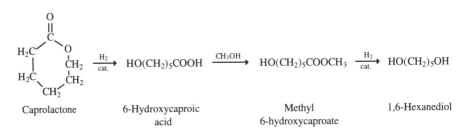

Caprolactone   6-Hydroxycaproic   Methyl   1,6-Hexanediol
                         acid   6-hydroxycaproate

A proposed process for adiponitrile, never commercialized, involves the ammoxidation (Section 4.4) of methylcyclohexane (prepared by toluene hydrogenation) at 43°C over a tin oxide–antimony oxide catalyst. Adiponitrile is produced in relatively modest yields together with byproduct glutaronitrile and succinonitrile.

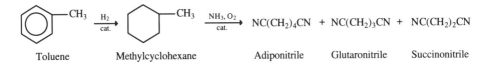

Toluene   Methylcyclohexane   Adiponitrile   Glutaronitrile   Succinonitrile

Hexamethylenediamine's major role is in the production of nylon 6,6 (Section 7.2.1). It can, however, be converted to an isocyanate by reaction with phosgene. This toxic product, used as a trimer to reduce its vapor pressure, is useful for the preparation of resistant, non-yellowing, glossy, low-temperature curing aircraft topcoats.

5.1.3 Lesser Volume Chemicals from Butadiene

Various lesser volume chemicals based on butadiene are generated by cycliz-ation, di- and trimerization, and the Diels–Alder reaction. Carbonylation leads to an adipic acid synthesis.

5.1.3.1 Cyclization Cyclization of two and three mol of butadiene gives a dimer, 1,5-cyclooctadiene, and a trimer, *cis,trans,trans*-1,5,9-cyclododeca-triene, an intermediate for perfumes.

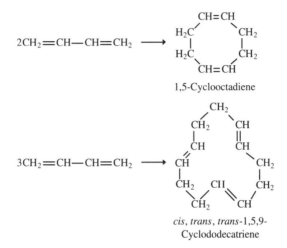

Dimer formation is catalyzed by a nickel salt of 2-ethylhexanoic acid, triethylaluminum, and triphenylphosphine or triphenyl phosphite. A byproduct is another butadiene dimer, 4-vinylcyclohexene, which is also produced in high yield in an uncatalyzed dimerization. It is a precursor of styrene, which forms by dehydrogenation.

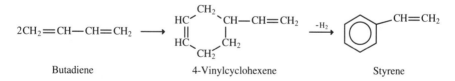

This styrene process has not been commercialized, although Dow described an improved catalyst in the mid-1990s (see note). A catalyst with large ligands that suppresses 4-vinylcyclohexene formation is nickel in combination with *tris-O-*phenylphenyl phosphite. At 80°C and one bar, selectivity to 1,5-cyclooctadiene is 96%.

Butadiene trimerizes over a titanium tetrachloride–dialkylaluminum chloride catalyst at 70°C and one bar to provides the *cis,trans,trans* isomer. The co-products are minor amounts of the all *trans* isomer, polybutadiene, 1,5-cyclo-octadiene, and 4-vinylcyclohexene. The *trans,trans,trans* isomer results if a nickel catalyst is used, but this is not done commercially. The catalyst is similar to the Ziegler polyethylene catalyst.

5.1.3.2 *Dimerization and Trimerization*   Linear dimers of butadiene are formed in several ways. One is 1,3,7-octatriene, which forms with a nickel catalyst and appropriate ligands.

$$2CH_2{=}CH{-}CH{=}CH_2 \rightarrow CH_2{=}CH{-}CH{=}CH{-}CH_2{-}CH_2{-}CH{=}CH_2$$
$$\text{1,3,7-Octatriene}$$

More important are the products intended for hydorformylation to yield plasticizer alcohols (Section 4.2).

Dimers with terminal functionality, such as 1,7-octadiene, result from 2 mol of butadiene, with formic acid as the source of hydrogen.

$$2CH_2{=}CH{-}CH{=}CH_2 + HCOOH \rightarrow CH_2{=}CH(CH_2)_4CH{=}CH_2 + CO_2$$
$$\text{Formic} \qquad\qquad \text{1,7-Octadiene}$$
$$\text{acid}$$

The catalyst is palladium acetate with triethylamine. The reaction is run in an aprotic solvent, DMF. Ligands such as triethylphosphine must be present. Conversion and selectivity are high, 95 and 93%, respectively.

In the late 1980s Shell developed routes to two terminal olefins, 1,5-hexadiene and 1,9-decadiene. The first compound results from the metathesis (Section 2.2.9) of 1,5-cyclooctadiene with ethylene.

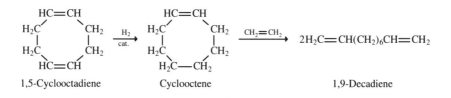

1,9-decadiene is made by the reaction of cyclooctene with ethylene, the starting material resulting again from the selective hydrogenation of 1,5-cyclooctadiene. The heterogeneous reaction makes use of a rhenium catalyst at 5–20°C and 1–2 bar. Yields are over 90%.

1,5-hexadiene can also be prepared by the metathesis of 1,5,9-cyclododecatriene with 3 moles of ethylene.

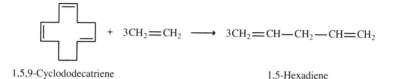

1,5,9-Cyclododecatriene   1,5-Hexadiene

1,5,9-Cyclododecatriene can be hydrogenated to cyclododecane. Treatment with nitrosyl chloride gives laurolactam, the monomer for nylon 12.

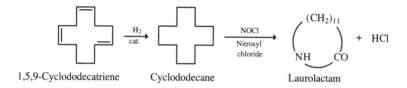

1,5,9-Cyclododecatriene   Cyclododecane   Laurolactam

The same process is one of several used to obtain caprolactam from cyclohexane, the monomer for nylon 6 (Section 7.2.2).

The partly hydrogenated triene, cyclododecene, can be oxidized with nitric acid to dodecanedioic acid. This was part of a now defunct Du Pont process to make the nylon "Qiana," a condensate of dodecanedioic acid with *p,p′*-di-aminodicyclohexylmethane, which in turn was made by hydrogenation of the adduct from the condensation of aniline and formaldehyde.

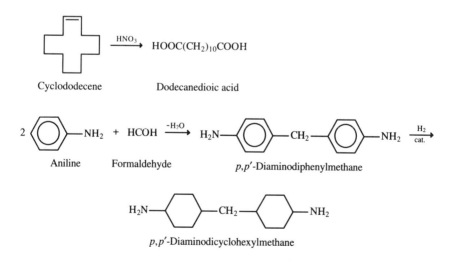

Qiana was said to have the drape and feel of silk, but still did not succeed in the marketplace.

5.1.3.3 Diels–Alder Reactions   In contrast to the laboratory, the Diels–Alder condensation is used industrially only on a small scale. None the less, butadiene is the classic diene for the reaction. It reacts with cyclopentadiene (Section 6.3) to give ethylidene norbornene, a monomer useful in EPDM rubber (Section 3.2.5). It also reacts with maleic anhydride to give tetrahydrophthalic anhydride, used in polyester and alkyd resins and as an ingredient in the fungicide, captan.

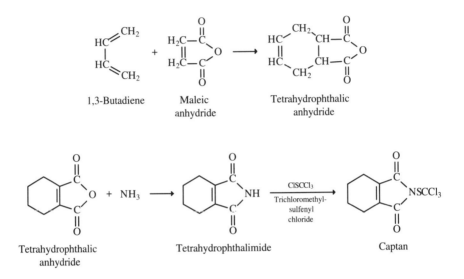

Diels–Alder addition of sulfur dioxide to butadiene at room temperature followed by hydrogenation yields tetramethylene sulfone (sulfolane), which is used to extract aromatic compounds (Section 2.2.3) from catalytic reformate in petroleum refineries.

$$CH_2{=}CH{-}CH{=}CH_2 \ + \ SO_2 \ \longrightarrow \ CH_2{-}CH{=}CH{-}CH_2 \ \xrightarrow{\ H_2\ }$$

with SO_2 bridging the terminal carbons

Sulfolene

$$CH_2{-}CH_2{-}CH_2{-}CH_2$$

with SO_2 bridging the terminal carbons

Sulfolane

It is also mixed with diisopropanolamine and water in the so-called Sulfinol process for the removal of hydrogen sulfide, carbon dioxide, carbonyl sulfide, and other acid components from so-called "sour" gas. The acid components are absorbed chemically by the amine and physically by the sulfolane. Being aprotic, sulfolane is also useful in extractive distillations, for example, for the separation of butadiene from other C$_4$ isomers (Section 5).

The Diels–Alder reaction between maleic anhydride and hexachloro cyclo-pentadiene is discussed later (Section 6.3). The Diels–Alder reaction between two molecules of butadiene yields vinylcyclohexene (Section 5.1.3.1).

5.1.3.4 *Adipic Acid* Adipic acid is produced in large quantities by the classical synthesis based on cyclohexane (Section 7.2.1). It is being made by BASF in the 1990s on a pilot plant scale by dicarbonylation of butadiene. This technology appears competititve with the classical route. The reaction proceeds in two stages since the dicarbonylation is difficult. Carbon monoxide and methanol react with butadiene in the presence of dicobalt octacarbonyl in quinoline or a related base at 600 bar and 120°C to give methyl-3-pentenoate at 98% selectivity. The second step takes place at a higher temperature of 185°C but at a lower pressure of 30 bar in the presence of additional CO and methanol.

$$CH_2=CH-CH=CH_2 + CO + CH_3OH \xrightarrow{cat.} CH_3CH=CHCH_2COOCH_3$$

Methyl 3-pentenoate

$$CH_3CH=CHCH_2COOCH_3 + CO + CH_3OH \xrightarrow{cat.} CH_3OOC(CH_2)_4COOCH_3$$

Dimethyl adipate

$$CH_3OOC(CH_2)_4COOCH_3 + 2H_2O \xrightarrow{H^+} HOOC(CH_2)_4COOH + 2CH_3OH$$

Adipic acid

Selectivity at this point is about 85% to the desired dimethyl adipate. The migration of the double bond in the second step (cf. a similar shift in the hexamethylenediamine synthesis from butadiene, Section 5.1.2) makes possible linearity. Other catalyst systems include platinum and palladium halides, phosphine-promoted cobalt complexes and rhodium complexes. Byproducts include methyl 3-pentenoate, methyl glutarate, methyl ethyl succinate, dimethyl and diethyl succinate and methyl pentenoates. The first two materials are obtained in largest amounts.

5.1.3.5 *1,4-Butanediol* Butadiene is the basis for one of the important routes to 1,4-butanediol by way of a double acetoxylation. This is described in Section 10.3.1.

5.1.3.6 *trans*-1,4-Hexadiene *trans*-1,4-Hexadiene (Section 16.7.1) results from the condensation of ethylene and butadiene in the presence of a rhodium catalyst.

$$CH_2=CH-CH=CH_2 + CH_2=CH_2 \xrightarrow{cat.} CH_2=CH-CH=CH-CH_2-CH_3$$

The compound is a useful comonomer for the production of ethylene propylene diene monomer (EPDM) rubber since the terminal double bond enters into the polymerization and the less reactive *trans* double bond is pendant and provides a site for cross-linking or vulcanization with sulfur compounds.

Although consumed on a scale almost an order of magnitude less than ethylene, butadiene has no fewer nor less interesting reactions. They are carried out, however, on a smaller scale.

5.2   CHEMICALS AND POLYMERS FROM ISOBUTENE

Isobutene was a lackluster chemical until the 1980s. Between 1982 and 1985, however, United States production more than doubled to over 1 billion lb/year and soared to 6 billion lb by the early 1990s. Its growth results from its role as a precursor for methyl-*tert*-butyl ether (MTBE), the preferred oxygen-containing additive for increasing the octane number of unleaded gasoline. Other uses for isobutene include formation of butyl rubber, polyisobutenes, alkylates for gasoline, *tert*-butanol, and methyl methacrylate.

The production of isobutene from the mixed C$_4$ stream was described in Section 5. A second major source of isobutene is the dehydration of the *tert*-butanol produced as a coproduct in a propylene oxide process (Section 4.7).

Because there was insufficient isobutene in the late 1980s to satisfy the need for MTBE, processes were devised for dehydrogenating isobutane to isobutene in the same way that the butenes may be dehydrogenated to butadiene. The dehydrogenation goes more readily because isobutane has a tertiary carbon atom. The first step, the isomerization of *n*-butane to isobutane, is a well-defined refinery reaction. By 1990 many plants around the world to dehydrogenate isobutane to isobutene were in the planning stage. The major sources of isobutene for MTBE and the associated capacities for both MTBE and isobutene are shown in Table 5.3. A process for isomerizing butenes to isobutene was being commercialized in the mid-1990s (Section 5).

A related route to octane-improving tertiary ethers is BPs Etherol Process. It is intended to produce the higher analogues of MTBE by making use of the isoalkenes from catalytic cracking. These include 3-methyl-1-butene and 3-methyl-1-pentene. The catalyst, a Pd-impregnated ion exchange resin, isomerizes these to 2-methyl-2-butene and 3-methyl-2-pentene and catalyzes the addition of methanol to the newly created, reactive double bond. Also, in the presence of hydrogen, it selectively catalyzes the hydrogenation of diolefins to monoolefins. Thus, it has three functions. An example is the formation of *tert*-amyl methyl ether (TAME) from 3-methyl-1-butene.

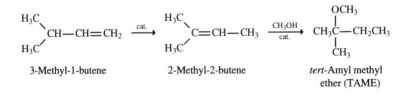

3-Methyl-1-butene   2-Methyl-2-butene   *tert*-Amyl methyl
                                                 ether (TAME)

TABLE 5.3 United States MTBE Demand and Capacity

(Million lb)	1981	1982	1984	1986	1987	1989	1990	1995
MTBE Capacity	1,213	1,543	2,425	5,181	7,275	8,201	9,038	28,440
MTBE Demand	1,213	1,543	2,425	5,181	7,275	8,201	10,140	36,820[a]
MTBE Provided by isobutene from								
Steam cracking	972	1,169	1,433	1,455	1,554	1,808	1,808	1,808
Catalytic cracking	242	374	992	1,653	1,830	2,138	2,399	2,399
tert-Butanol dehydration				1,301	3,099	3,483	3,549	4,728[b]
Isobutane dehydrogenation				772	772	772	1,262	19,508
Imports							1,121	8,377

[a] Equates to 344,565 bbl/d.
[b] Includes Texaco's announced plant.

5.2.1 Methyl *tert*-Butyl Ether

The role of methyl *tert*-butyl ether as an octane improver in gasoline has been mentioned repeatedly. Its production and isolation directly from Raffinate I, a portion of the C$_4$ fraction from either steam or catalytic cracking, was discussed in Section 5. It can be made similarly from isolated isobutene by reaction with methanol in a highly selective reaction in the presence of an acid catalyst, usually an acidic ion exchange resin.

5.2.2 Butyl Rubber

The second largest outlet for isobutene is in butyl rubber, a copolymer of isobutene with 2–5% of isoprene. Butyl rubber was used for inner tubes before the advent of the tubeless tire because of its impermeability to air. It is still used as an inner liner for tubeless tires, for truck inner tubes, for tire sidewall components, and for air cushions and bellows. Without the isoprene, the polymer would contain no double bonds and would not vulcanize with sulfur. The isoprene comonomer leaves only a small number of double bonds to be cross-linked and the resulting polymer contains little if any unsaturation. For this reason, it resists aging and is therefore useful in constructions such as convertible tops.

5.2.3 Polyisobutenes and Isobutene Oligomers

Isobutene can be polymerized to polyisobutene under the influence of Friedel–Crafts catalysts such as boron trifluoride or aluminum chloride. The lower molecular weight liquid products are adhesives and tackifiers. The higher molecular weight products are used in caulking compounds and as a chewing gum base.

An oligomer produced in much higher volume, about 750 million lb/year in the United States, is so-called polybutenes, made by polymerizing Raffinate I (Section 5) with an acid catalyst such as aluminum chloride and with hydrochloric acid as an activator. The *n*-butenes act largely as chain stoppers since the isobutene polymerizes much more rapidly. The resulting low molecular weight (1000–1500) polymers are widely used as additives for lubricating oils and gasoline, usually functionalized by reaction with maleic anhydride.

Like propylene and the butenes, isobutene is used for alkylates for gasoline (Section 2.2.5) and it may be dimerized or oligomerized (Section 2.2.4), to give diisobutenes, unsaturated precursors of 2,2,4-trimethylpentane. Both the unsaturated and subsequently hydrogenated compounds increase the octane number of gasoline, the latter of course being more stable.

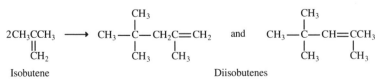

Isobutene   Diisobutenes

The diisobutenes may be hydroformylated (Section 4.8) to branched-chain aldehydes that can be hydrogenated to branched nonanol. They are also used to alkylate phenol to provide branched-chain octylphenol for the preparation of nonionic surfactants. The usage of these are decreasing because of possible lack of biodegradability.

5.2.4 *tert*-Butanol

Isobutene may be hydrated to *tert*-butanol with an acid catalyst such as 60% sulfuric acid at a low temperature of 10–30°C. This is a major route for isolating isobutene from Raffinate I (Section 5). *tert*-Butanol is also produced concurrently with propylene oxide as described in Section 4.7. A third route is acidification of *tert*-butylhydroperoxide, which in turn results from the reaction of isobutane with oxygen. This is not done commercially since the major interest in *tert*-butanol is for dehydration to isobutene, which is more extensively obtained by dehydrogenating isobutane.

   tert-Butanol is a solid that dissolves readily in gasoline and has been used as an octane improver. It may be mixed with an equal quantity of methanol, which by itself is not useful in gasoline because it separates out as a second layer in the presence of moisture. However, neither the *tert*-butanol nor the methanol-*tert*-butanol combination is as effective as MTBE. The most important chemical use for *tert*-butanol is its dehydration to isobutene.

5.2.5 Methacrylic Acid

Both isobutene and *tert*-butanol can be oxidized to methacrylic acid (Section 4.6.1). Methacrylaldehyde is an intermediate. Both undergo ammoxidation to methacrylonitrile, an intermediate for the preparation of methacrylic acid in one Japanese process (Section 4.6.1).

5.2.6 Lesser Volume Chemicals from Isobutene

Two small volume but widely used products based on isobutene are butylated hydroxytoluene (BHT) and butylated hydroxyanisole (BHA). The BHT results from the interaction of isobutene and *p*-cresol with a silica catalyst.

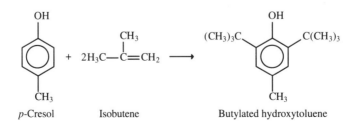

| *p*-Cresol | Isobutene | Butylated hydroxytoluene |

BHA results from the reaction of 1 mol of isobutene with the monomethyl ether of hydroquinone.

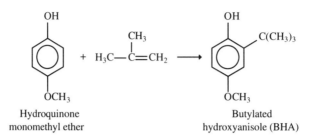

Hydroquinone   Butylated
monomethyl ether   hydroxyanisole (BHA)

BHT and BHA are oil-soluble antioxidants for rubbers and plastics and for fatty foods such as shortenings. Small quantities protect huge tonnages of products as diverse as elastomers and shortening from deterioration and thus have great economic significance. The food usage has been investigated repeatedly by the Food and Drug Administration (FDA) but thus far no evidence against them has been found. Even so, their use in food products is decreasing.

Liquid-phase chlorination of isobutene at 400–500°C leads to methallyl chloride. Treatment with aluminum and hydrogen gives triisobutylaluminum and diisobutylaluminum hydride, both important as Ziegler–Natta polymerization catalysts. The hydride is also a reducing agent.

$$CH_2=\underset{\underset{CH_3}{|}}{C}-CH_3 \xrightarrow{Cl_2} CH_2=\underset{\underset{CH_3}{|}}{C}-CH_2Cl \;+\; HCl$$

Isobutene   Methallyl chloride

$$CH_2=\underset{\underset{CH_3}{|}}{C}-CH_3 \xrightarrow[H_2]{Al} [(CH_3)_2CH_2CH_2]_3Al \;+\; [(CH_3)_2CH_2CH_2]_2AlH$$

Isobutene   Triisobutylaluminum   Diisobutylaluminum
                                                          hydride

Isobutene figures prominently in a synthesis of isoprene as discussed in Section 6.1.

Isobutene and ammonia yield *tert*-butylamine with a catalyst of ammonium iodide promoted with chromium chloride.

$$CH_3\underset{\underset{CH_3}{|}}{C}=CH_2 \;+\; NH_3 \xrightarrow{cat.} (CH_3)_3CNH_2$$

Isobutene   *tert*-Butylamine

Conversion is 19% at 100% selectivity. The reaction is probably not yet used industrially. This is an interesting demonstration of the high activity of the double bond of isobutene, since ammonia addition to a double bond normally takes place only with difficulty, and with specialized homogeneous catalysts. *tert*-Butylamine is normally made by the Ritter reaction of isobutene with hydrogen cyanide, which involves the addition of hydrogen cyanide and water to a double bond, followed by hydrolysis of the resultant formamide.

$$\underset{\substack{\text{Isobutene}}}{\overset{\overset{\displaystyle CH_3}{|}}{CH_3C}=CH_2} + \underset{\substack{\text{Hydrogen}\\ \text{cyanide}}}{HCN} + H_2O \xrightarrow{\text{cat.}} \underset{\substack{\textit{N-tert}\text{-Butyl-}\\ \text{formamide}}}{(CH_3)_3CNHCH} \xrightarrow{H_2O} \underset{\substack{\textit{tert}\text{-Butyl-}\\ \text{amine}}}{(CH_3)_3CNH_2} + \underset{\substack{\text{Formic}\\ \text{acid}}}{HCOOH}$$

tert-Butylamine is used in the formulation of lubricating oil additives. The labile nature of the *tert*-butyl group makes isobutene valuable in chemical synthesis. An example is the use of isobutene as a shielding group for the synthesis of *p,p′*-biphenol, a monomer for liquid crystal polymers. Phenol is alkylated with isobutene to give 2,6-di-*tert*-butylphenol (**I**). Then, 2 mol can be coupled in the para position to give a diquinone, which can be hydrogenated to the substituted biphenol. The protective *tert*-butyl groups can be removed by acid catalysis and the recovered isobutene can be recycled.

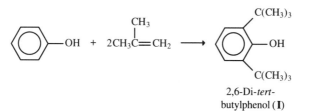

2,6-Di-*tert*-
butylphenol (**I**)

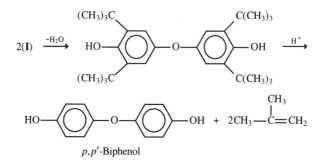

p,p′-Biphenol

5.3 CHEMICALS AND POLYMERS FROM 1- AND 2-BUTENES

Most uses for the butenes are relatively low volume. Pure 1-butene plays an important role in the production of linear low-density polyethylene and to a lesser extent of high-density polyethylene (Section 3.1.1).

In combination with isobutene, the mixture of 1- and 2-butenes is polymerized to so-called polybutenes (Section 5.2.3). 1-Butene can be polymerized to poly(1-butene) with a Ziegler–Natta catalyst. Like the polypropylene of commerce (Section 4.1) it has an isotactic structure. Its advantages over polypropylene are that it is less prone to stress crack and has less cold flow. Its major use, which is small, is in construction of pipes and vessels for handling and storing corrosive liquids at high temperatures and pressures. It is not useful in molding because of postcrystallization, which leads to shrinking.

Like propylene and isobutene, 1-butene and 2-butene mixtures can be hydrated with concentrated sulfuric acid to give 2-butanol.

$$CH_2{=}CH{-}CH_2{-}CH_3 \xrightarrow[H_2SO_4]{H_2O} CH_3{-}CH_2{-}\underset{\underset{\displaystyle OH}{|}}{CH}{-}CH_3$$

or

$$CH_3{-}CH{=}CH{-}CH_3$$

n-Butenes

2-Butanol

$\downarrow$ -H$_2$

$CH_3{-}CH_2{-}\underset{\underset{\displaystyle O}{\|}}{C}{-}CH_3$

Methyl ethyl ketone
(2-Butanone)

Direct hydration, as with ethylene and propylene, is also possible (Section 3.9). 2-Butanol has a few minor uses as a solvent, but most if it is dehydrogenated to methyl ethyl ketone (MEK), an important solvent for vinyl and nitrocellulose lacquers and acrylic resins. In the United States a vapor-phase process with a zinc oxide or zinc–copper catalyst is used, while in Europe a liquid-phase reaction with finely divided nickel or copper chromite is preferred.

The mixture of 1- and 2-butenes can also undergo the oxo reaction to give a mixture of pentanols, which are useful as solvents as such or as their esters. As described previously (Section 4.8) catalysts are available that lead to a high percentage of *n*-pentanol from 1-butene.

Like propylene, 1- and 2-butenes have uses in gasoline, for they may be alkylated with isobutane to provide branched-chain structures (Section 2.2.5). Indeed, they are preferred to propylene for alkylate because they provide more highly branched molecules. They also may be components of polygas (Section 2.2.4). If the oligomerization takes place with propylene, heptenes are produced, which can be used as the basis for C$_8$ oxo alcohols (Section 4.8).

2-Butene is the starting material for a new process, as yet uncommercialized, for isoprene (Section 6.1).

A small amount of 1-butene is converted to butene oxide presumably by epoxidation with a peracid. Like isobutene (Section 5.2.6) it forms aluminum derivatives such as tributylaluminum, a useful catalyst.

1- and 2-Butenes may be dehydrogenated to butadiene (Section 5.1) at a temperature of 650°C in the presence of steam, which not only minimizes butadiene polymerization but also inhibits carbon deposition on the catalyst. The dehydrogenation is endothermic and thus a pressure of 0–1 bar is used at a contact time of 0.2 s. The catalyst comprises calcium nickel phosphate. Oxidative dehydrogenation is also possible and is in fact preferred since it makes possible operation at lower temperatures of around 550°C with a fixed-catalyst bed. The hydrogen released reacts with oxygen to form water, an exothermic reaction that compensates for the heat lost by the endothermic dehydrogenation.

5.4 MALEIC ANHYDRIDE

1- and 2-Butenes may be oxidized to maleic anhydride, the 1-butene isomerizing to the more thermodynamically stable 2-butene. The original catalyst was vanadium pentoxide. Maleic anhydride traditionally has been prepared by the oxidation of benzene over the same catalyst (Section 7.5). A benzene shortage in the early 1970s motivated the development of an alternate route based on butenes, and the new route also had the advantage that two carbon atoms were not lost. n-Butenes are not as important for refinery reactions in Europe as they are in the United States and therefore are not much more expensive than n-butane.

n-Butane, cheaper in the United States, was investigated as a feedstock for maleic anhydride and unexpectedly proved adequate.

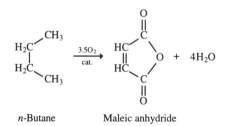

n-Butane   Maleic anhydride

The pathway for the oxidation of 2-butene to maleic anhydride is analogous to the oxidation of propylene to acrylic acid (Section 4.3), but there is no analogous reaction involving the oxidation of a saturated hydrocarbon in reasonably high selectivity with only carbon dioxide and water as byproducts. The oxidation of n-butane or naphtha to acetic acid (Section 10.5.2.2) is not analogous because numerous byproducts are produced.

It is impressive that a catalyst systems can be found that will oxidize methyl groups to carboxyls and insert a double bond while refraining from promoting C–C bond scission. By the early 1980s n-butane oxidation had become the preferred method for maleic anhydride manufacture. The reaction may be carried out either in a fixed or fluid-bed reactor, the latter requiring an attrition–resistant catalyst. The secondary products, as in ethylene oxidation to ethylene oxide (Section 2.7), are carbon dioxide and water.

The fluid-bed reaction, instituted in Japan in the early 1990s, is carried out with a butane concentration in air of 4.5% at a temperature of 440°C with a contact time of about 11 s. The catalyst is based on vanadium pentoxide and various phosphoric acids. Other catalysts described in the patent literature contain boron, lithium, and zinc as activators.

A dramatic improvement in maleic anhydride/maleic acid technology is described in a Du Pont patent. The catalyst comprises mixed oxides of vanadium and phosphorus with promoters, primarily silicon and one other element such as indium, antimony, tantalum. It is formulated to be highly attrition resistant. The unique aspect, however, which increases the yield from the conventional 50–52 to 72%, is that no oxygen is used in the oxidation other than that which is part of the catalyst. The catalyst is thus a reactant and its double role is recognized by the term cataloreactant. In this way the side reactions in which the *n*-butane is oxidized to water and carbon dioxide are limited.

The conversion is, of course, determined by the amount of oxygen available from the catalyst and is about 8.5%. The catalyst is inactivated in each run and must be removed for reactivation and replaced with fresh catalyst. This is accomplished by a so-called transport bed process, which requires the catalyst to be attrition resistant.

The process was scheduled for commercialization in Spain in 1995. Its economics are comparable to those of the fluid-bed process, when solvent is used in both processes for the extraction of the products. Water may be used for the extraction, but its evaporation increases operating costs. However, Du Pont intends to use water, since the aqueous solution without isolation of the maleic acid can be subjected to hydrogenation/hydrogenolysis to yield tetrahydrofuran (THF) (Section 10.3.1) with a newly developed catalyst that makes possible the use of milder temperatures and pressures than are commonly required for hydrogenolyses.

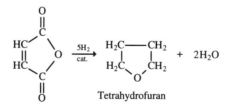

Du Pont requires tetrahydrofuran for oligomerization to a polybutene ether for use in their spandex polymer. Spandex is a block copolymer comprising hard urethane blocks and flexible polyether blocks, useful for swimsuits, sports clothes, and elastic in undergarments. Although it is an old polymer, its use increased rapidly in the late-1980s and early-1990s when tights became fashionable.

Maleic anhydride, which is produced at a level of about 400 million lbs a year in the United States, finds its greatest market in unsaturated polyester resins (Section 9.1.1). Its other important uses are conversion to fumaric acid for use as a food acidulant and in the formulation of agricultural chemicals and lube oil additives. Its newest use is for conversion to 1,4-butanediol (Section 10.3.1).

NOTES AND REFERENCES

Butenes are known trivially as butylenes and this terminology is consistent with the terms ethylene and propylene. However, the systematic name butene has made deep inroads into industrial chemical literature. Therefore we use it here, along with the systematic names for the other higher olefins, that is pentenes, hexenes and so on.

The most recent specialist book in this area is J. Schulze and M. Homann, *C₄-Hydrocarbons and Derivatives—Resources, Production and Marketing*, Springer-Verlag, Berlin, 1989. It provides an authoritative overview of the technology, chemical engineering, marketing and economics of the C_4 stream.

One estimate indicates that 1.3 billion lb/year of butadiene will be hydrogenated in Western Europe by the year 2000. This will help reduce the European surplus. The technology of butadiene hydrogenation has been discussed by B. Torck, *Chem. Ind.* October 4, 1993, p. 742. This article also presents a valuable overview of the state of the C_4 market.

Section 5.1   Syndiotactic 1,2-polybutadiene is discussed by Y. Takeuchi et al., in Chapter 3 of *New Industrial Polymers*, R. D. Deanin, Ed., American Chemical Society, ACS Symposium Series 4, Washington, DC, 1972.

Section 5.1.1   A detailed discussion of the development of processes for HMDA manufacture has been provided by H. A. Wittcoff, *J. Chem. Educ.*, **56**, 654 (1979).

The ICI dimerization of acrylonitrile is described in *Chem. Week*, March 7, 1990, p. 9, in European Patent Publication 035 200 7A2 (January 24, 1990), and in US Patent 4 958 042 (September 18, 1990). It is also discussed in three West German Offen. 2 694 904 (May 17, 1977); 2 720 279 (November 24, 1977); and 2 721 808 (December 8, 1977). Related work by Halcon is described in West German Offen. 2 559 185-7 (July 1, 1976).

The Du Pont process for preparing hexamethylene diamine by the hydrocyanation of butadiene is described in US Patents 4 339 395 (July 13, 1982) and 4 080 374 (March 21, 1978). It is also described by V. D. Luedeke in *Encyclopedia of Chemical Processing and Design*, Vol. 2, J. J. McKetta and W. A. Cunningham, Eds., Dekker, New York, 1977, p. 146.

Section 5.1.2   The ammoxidation of methylcyclohexane is described in US Patent 3 624 125 (November 30, 1971) to Imperial Chemical Industries. Also described is the ammoxidation of cyclohexane [West German Offen. 1 807 354 (July 3, 1969)]. In fact it is the cyclohexane that undergoes ammoxidation, the methylcyclohexane being dehydroalkylated in the process.

The mechanism for the hydrodimerization of acrylonitrile is explained well by D. Pletcher, *Industrial Electrochemistry*, 2nd ed., Chapman & Hall, London, 1990 (paperback, Blackie, 1993).

Section 5.1.3.1 The preparation of 1,5-hexadiene and 1,9-decadiene have been described in three patents; Great Britain Patent 1 482 745 (August 10, 1977) to Shell Research; US Patent 3 792 102 (February 12, 1984) and US Patent 3 878 262 (April 15, 1975) both to Phillips Petroleum.

The Dow process for conversion of butadiene to styrene involves dimerization of butadiene to vinylcyclohexene followed by an oxidative dehydrogenation of the vinylcyclohexene to styrene. The dimerization is catalyzed by copper loaded Y zeolite. The oxidative hydrogenation uses proprietary mixed oxide catalysts. A DSM process has three steps, the first being dimerization and the second dehydrogenation to ethylbenzene followed by further dehydrogenation to styrene. The first dehydrogenation uses a Pd/MgO catalyst. The ethylbenzene–styrene conversion is conventional. Typical patents are US Patents 5 196 621 (March 23, 1993) and 5 329 057 (July 12, 1994) and World Patent 94/01385 (January 20, 1994), all to Dow Chemical Co.

Section 5.1.3.2 The dimerization of butadiene to 1,5-cyclooctadiene is described in US Patent 3 250 817 (May 10, 1966) to Chevron Research. The reaction is also discussed by P. W. Jolly and G. Wilke, *The Organic Chemistry of Nickel*, Vol. 2, Academic, New York, 1975.

Section 5.1.3.6 The mechanism of the formation of *trans*-1,4-Hexadiene is described by G. W. Parshall, *Homogeneous Catalysis*, Wiley, New York, 1980, pp. 63–65.

Section 5.2.1 The isomerization of butenes to isobutene is described in two patents to ARCO, US Patent 4 778 943 (October 18, 1988) and 4 654 463 (March 31, 1987).

Section 5.2.6 BASF's process for adipic acid by the dicarbonylation of butadiene is described in numerous patents including US Patent 4 310 686 January 12, 1982; US Patent 4 360 695, November 23, 1982; US Patent 4 350 572 (September 21, 1982); and West German Offen. 2 630 086 (January 12, 1978); and 2 646 955 (April 20, 1978).

Section 5.3 The direct hydration of *n*-butenes to *sec*-butanol is described in West German Offen. 3 512 518 A1 (October 9, 1986), to Deutsche Texaco AG.

Section 5.4 The Du Pont process for achieving high yields of maleic anhydride is described in US Patents 4 371 702 (February 1, 1983) and 4 442 226 (April 10, 1984) and in European patent application 0 189 261 A1 (July 30, 1986).

CHAPTER 6

CHEMICALS AND POLYMERS FROM THE C₅ STREAM

The C_5 olefins are not a major source of chemicals, and US consumption in the mid-1990s was less than 2 billion lb/year. Consumption in Western Europe and Japan is considerably less. Usage, however, is scheduled to increase to as much as 8 billion lb by the year 2000 because of the need for 3-methyl-1-butene (isopentene or *tert*-amylene, Section 5.2) for conversion to the octane improver *tert*-amyl methyl ether (TAME).

Like propylene and the C_4 olefins, the C_5 olefins are produced by both catalytic and steam cracking (Sections 2.2.1 and 2.2.2). Just as catalytic cracking produces butenes but no butadiene, it produces pentenes but practically no diolefins such as isoprene and cyclopentadiene. The steam cracking of naphtha or gas oil, on the other hand, produces isoprene equivalent to 2.5% of the ethylene formed. The C_5 fraction from gas oil cracking as well as from high severity naphtha cracking contains more cyclopentadiene (15–25%) than isoprene, which is present to the extent of 10–15%. From naphtha cracked at low severity, there is more isoprene than cyclopentadiene.

Steam cracking of liquids produces a fraction called pyrolysis gasoline (Section 2.2.1), which contains both the C_5 olefins and an aromatic fraction. It may all be sent to the gasoline pool. Alternatively, the C_5 fraction is recovered by distillation, and the remainder hydrotreated to remove any remaining olefins. After this it is processed like catalytic reformate (Section 2.2.3) to remove benzene and toluene. The residue is then sent to the gasoline pool.

Since the United States has cracked ethane and propane predominantly, pyrolysis gasoline has not been a major source of C_5 olefins, which accounts for the fact that in 1993 only 484 million lb of isoprene were produced.

The number of hydrocarbon isomers increases with increasing chain length. One can write three pentanes, six pentenes plus cyclopentane and methylcyclobutane, six pentadienes plus cyclopentene, methylcyclobutanes, cyclopentadienes, and so on. In the C$_5$ fraction there are appreciable quantities of 11 components, as shown in Figure 6.1. Of these, isoprene and cyclopentadiene are isolated as pure compounds.

The isopentenes (methylbutenes) are isolated as mixed isomers. One use is dehydrogenation to isoprene. Isopentane can also be dehydrogenated to iso-prene, and this is believed to be done in the CIS. For dehydrogenation, the olefin–alkane mixture can be used, the olefin dehydrogenating first. In the United States, the 3-methyl-1-butene is isomerized to 2-methyl-2-butene and 2-methyl-1-butene. These may be extracted as mixed isomers as described below, just as isobutene is isolated from the C$_4$ olefin fraction.

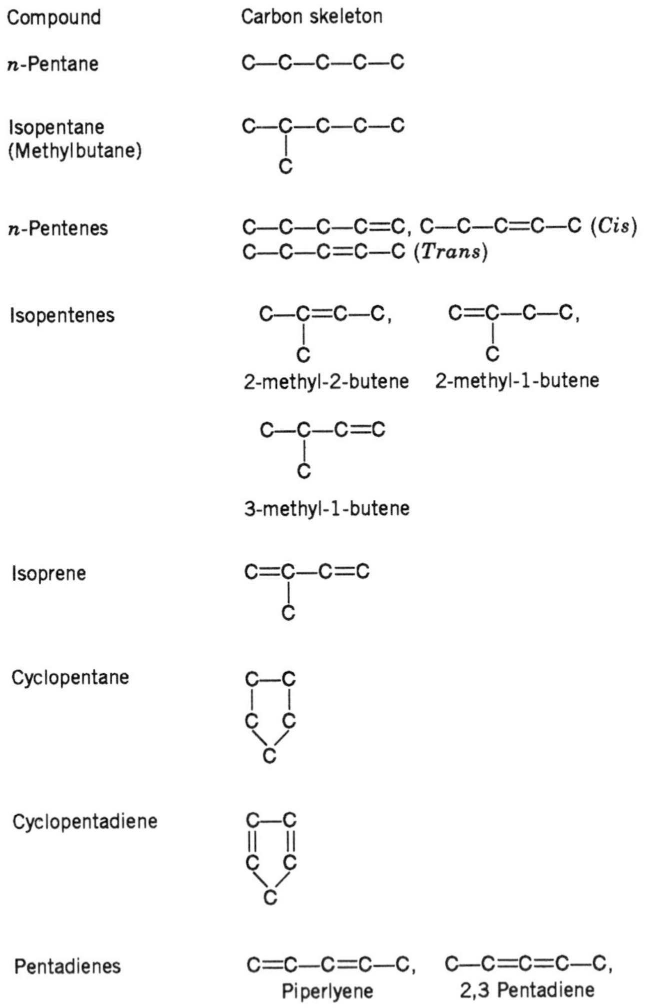

Compound	Carbon skeleton
n-Pentane	C—C—C—C—C
Isopentane (Methylbutane)	C—C—C—C—C with C branch
n-Pentenes	C—C—C—C=C, C—C—C=C—C (*Cis*) C—C—C=C—C (*Trans*)
Isopentenes	C—C=C—C with C branch (2-methyl-2-butene), C=C—C—C with C branch (2-methyl-1-butene) C—C—C=C with C branch (3-methyl-1-butene)
Isoprene	C=C—C=C with C branch
Cyclopentane	cyclic C—C / C C / C
Cyclopentadiene	cyclic C=C / C C / C
Pentadienes	C=C—C=C—C (Piperlyene), C—C=C=C—C (2,3 Pentadiene)

Figure 6.1   Components of the C$_5$ fraction.

Isopentane is not isolated from the C_5 cracking fraction because separation from C_5 olefins is too difficult. Instead it is obtained from the C_5 fraction of straight run gasoline (Section 2.1). The separation of isopentane from C_5 olefins can be brought about with molecular sieves, but this process is not yet practiced industrially. The separation from n-pentane is brought about either by distillation or with molecular sieves. Isopentane is useful as a solvent.

6.1 SEPARATION OF THE C_5 STREAM

The separation of the many C_5 components is difficult and is done differently around the world. Figure 6.2 provides a conceptualized scheme that combines processes used in several refineries. It comprises four major steps: cyclopentadiene removal, extractive distillation, solvent regeneration and product distillation.

Cyclopentadiene must be removed first because it polymerizes and can foul the reboilers of the downstream distillation columns. Its isolation depends on a facile and reversible dimerization that takes place when the C_5 fraction is heated at 100°C for 2–3 h or at 150°C under pressure for a shorter period. The dimer, dicyclopentadiene, forms by a Diels–Alder reaction and normally has the endo form. Since it boils 130°C higher than the monomer, it is readily separated from the other components by vacuum distillation. At 350°C in a tubular reactor, the monomer is rapidly regenerated.

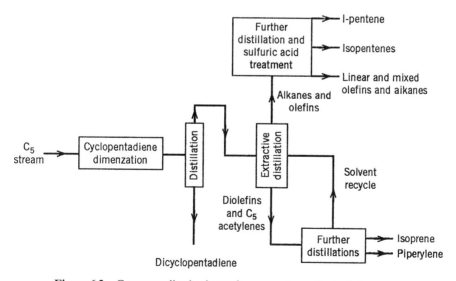

Figure 6.2 Conceptualized scheme for separation of mixed C_5 stream.

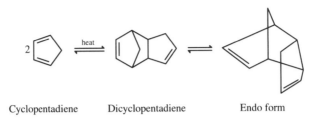

Cyclopentadiene   Dicyclopentadiene   Endo form

Isoprene is removed next by extractive distillation with the aprotic solvents acetonitrile or N-methylpyrrolidone. Almost all the alkanes and most of the olefins distil overhead, while the solvent, the diolefins and the C$_5$ acetylenes appear at the bottom of the tower. Successive distillations of this bottom stream regenerate the solvent for recycle, split the hydrocarbon stream into acetylenes and diolefins and split the diolefins into isoprene and piperylene (1,3-penta-diene). Polymerization grade isoprene requires a further distillation for adequate purification.

The overhead stream from the extractive distillation can be further distilled to give 1-pentene, isopentenes (methylbutenes), and a mixture of the remaining C$_5$ alkanes and olefins. The isopentenes, mainly 3-methyl-1-butene, are isomerized by 65% sulfuric acid primarily to 2-methyl-2-butene $(CH_3)_2C{=}CHCH_3$ and to a lesser extent to 2-methyl-1-butene $(CH_2){=}C(CH_3)CHCH_3$. These react readily with water in the presence of sulfuric acid at 0–10°C to yield a tertiary alcohol, just as isobutene does in the process for separating the C$_4$ olefins (Section 5). The 2-methyl-2-butanol is separated and heated to 350°C to regenerate the alkene. The methylbutanol can be dehydrogenated to isoprene with a catalyst based on iron, chromium oxides and potassium carbonate at 60°C. Selectivity is 85% at 35% conversion. Just as with isobutene, methanol may be used instead of water (Section 5.2.1). With 2-methyl-2-butene, TAME results.

$$(CH_3)_2C{=}CHCH_3 \;+\; CH_3OH \;\rightarrow\; (CH_3)_2C(OCH_3)CH_2CH_3$$

The residual C$_5$ stream contains linear and mixed olefins. They can be separated without difficulty but there is little demand for them. The mixture can be polymerized to give an inexpensive hydrocarbon resin (Section 6.3) useful in coatings and adhesives.

6.2 ISOPRENE

There are three possible sources of isoprene starting with the C$_5$ fraction, all mentioned in Section 6.1. The first is isolation from the C$_5$ fraction from steam cracking, the second is dehydrogenation of isopentenes and the third is dehydrogenation of isopentane.

There are several synthetic routes to isoprene. In an Italian process devised by SNAM Progetti and still in use, acetylene reacts with acetone to yield 2-methyl-

3-butyn-2-ol. Hydrogenation gives 2-methyl-3-butenol, which on dehydration provides isoprene.

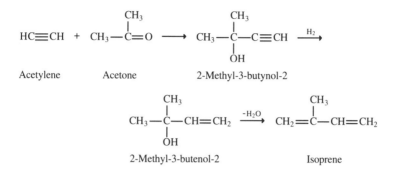

2-Methyl-3-butenol-2 → Isoprene

The condensation of acetylene and acetone is carried out in liquid ammonia at low temperatures with potassium hydroxide as the catalyst. The hydrogenation must be carried out selectively so that only a double bond forms. This is reminiscent of the hydrogenation of 1,4-butynediol to 1,4-butenediol (Section 10.3.1). The dehydration goes well at 250–300°C with alumina.

A second process, devised by the Institut Français du Pétrole and said to be used in the CIS, involves the reaction of isobutene with formaldehyde, via the Prins reaction, to give dimethyldioxane, which can be cracked to isoprene.

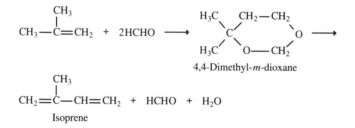

The Prins reaction is catalyzed by strong acid or acid ion exchange resin. Raffinate I (Section 5.1) may be used instead of pure isobutene. The second step of the reaction requires a high temperature with a phosphoric acid–charcoal or calcium phosphate catalyst with water.

Another synthetic process, used only briefly by Goodyear because it proved uneconomical, involves the dimerization of propylene to 2-methyl-1-pentene (Section 4.2). The reaction goes in 90% yield at about 300°C and elevated pressures with a tri-n-propylaluminum catalyst. The dimer is isomerized at about 100°C with a silica–alumina catalyst to the more stable 2-methyl-2-pentene, which in turn is demethanated to isoprene and methane at 660°C in the presence of superheated steam and catalytic amounts of hydrogen bromide.

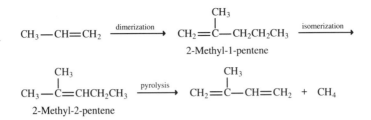

An attractive process for isoprene formation, which has never been commercialized, involves a metathesis reaction between isobutene and 2-butene to yield propylene and 2-methyl-2-butene which, as indicated above, can be dehydrogenated to isoprene.

CH$_3$—C=CH$_2$ + CH$_3$CH=CHCH$_3$ $\xrightarrow{\text{cat.}}$ CH$_3$—C=CHCH$_3$ + CH$_3$CH=CH$_2$

Isobutene   2-Butene   2-Methyl-2-Butene

$\downarrow$ -H$_2$

CH$_2$=C—CH=CH$_2$
         |
        CH$_3$

Isoprene

A newer synthesis of isoprene, also not commercialized, involves the hydroformylation of 2-butene with a rhodium catalyst such as rhodium (cyclooctadiene) acetylacetonate with triphenylphosphine at 115°C and 8 bar. A branched aldehyde, 2-methylbutanal, results.

CH$_3$CH=CHCH$_3$ + CO + H$_2$ $\xrightarrow{\text{cat.}}$ CH$_3$CHCH$_2$CH$_3$ $\xrightarrow[\text{cat.}]{-\text{H}_2\text{O}}$ CH$_2$=C—CH=CH$_2$
                                              | |
                                             CHO   CH$_3$

2-Butene   2-Methylbutanal   Isoprene

With a cobalt catalyst, a linear compound would form because cobalt facilitates migration of the double bond to the terminal position, a property important in the hydroformylation step of the Shell SHOP process (Section 3.3.4). The 2-methylbutanal is dehydrated with crystalline boron phosphate catalyst. Conversion to isoprene is 80%, with 85% selectivity. A proposed mechanism for the dehydration is shown in Figure 6.3.

The major use for isoprene is its stereospecific polymerization to *cis*-1,4-polyisoprene, a polymer closely related to natural rubber. Its preparation was not possible prior to the advent of Ziegler–Natta catalysis. The catalyst may comprise a titanium tetrachloride–trialkylaluminum combination or an alkyl lithium. Modern radial ply tires require increased amounts of natural rubber because it has greater resilience and lower hysteresis losses than SBR. The

Enolization:

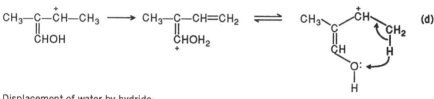

(a)
$$CH_3\text{—}CH\text{—}CH_2\text{—}CH_3 \ (CHO) + B^{3+}PO_4^{3-} \longrightarrow CH_3\text{—}\overset{-}{C}\text{—}CH_2\text{—}CH_3 \ (CHO) + B^{3+}HPO_4^{2-}$$
Enolate

(b)
$$CH_3\text{—}\overset{-}{C}\text{—}CH_2\text{—}CH_3 \ (CHO) + B^{3+}HPO_4^{2-} \longrightarrow CH_3\text{—}C\text{—}CH_2\text{—}CH_3 \ (CHOH) + B^{3+}PO_4^{3-}$$
Enol

Hydride abstraction:

(c)
$$CH_3\text{—}C\text{—}CH_2\text{—}CH_3 \ (CHOH) + B^{3+} \longrightarrow CH_3\text{—}C\text{—}\overset{+}{C}H\text{—}CH_3 \ (CHOH) + BH^{2+}$$

Hydride shift:

(d)
$$CH_3\text{—}C\text{—}\overset{+}{C}H\text{—}CH_3 \ (CHOH) \longrightarrow CH_3\text{—}C\text{—}CH{=}CH_2 \ (\overset{+}{C}HOH_2) \rightleftharpoons$$

Displacement of water by hydride:

(e)
$$CH_3\text{—}C\text{—}CH{=}CH_2 \ (\overset{+}{C}HOH_2) + BH^{2+} \xrightarrow{\text{(from Eq. c)}} CH_3\text{—}C\text{—}CH{=}CH_2 \ (CH_2) + B^{3+} + H_2O$$

Figure 6.3 Proposed mechanism for the production of isoprene by dehydration of 2-methylbutanal.

demand can equally be satisfied with synthetic *cis*-1,4-polyisoprene, provided that it is competitive in price with natural rubber.

The second important use for isoprene is in styrene–isoprene–styrene block copolymers similar to S–B–S copolymers (Section 5). These contain 85% isoprene and 15% styrene and are used largely in hot melt and pressure-sensitive adhesives. Another use for isoprene is in the manufacture of butyl rubber, a copolymer of isobutene and 3–5% isoprene (Section 5.2.2), which shows very low gas permeability.

A small volume isoprene derivative is *trans*-1,4-polyisoprene. The trans double bond makes the molecule crystalline, removing all the elastomeric properties of the *cis* isomer. The product resembles two natural materials, balata and gutta percha, and is useful for golfball covers. This use has declined because of the superior properties of ionomers such as "Surlyn" (Section 3.2.3). Gutta percha is still used by dentists for temporary fillings for teeth, particularly in "root canal" work. Balata and gutta percha are both polyisoprenes and are believed to be stereoisomers.

6.3 CYCLOPENTADIENE

Cyclopentadiene accounts for 15–25% of the C$_5$ olefin fraction from high severity steam cracking of naphtha. If gas oil is cracked, it is the predominant olefin. It also occurs in distillate from coke ovens. Like butadiene, it is not formed in catalytic cracking. United States production in 1992 was 195 million lb.

There are two main uses for cyclopentadiene. One is in the formulation of inexpensive low molecular weight (under 2000) oligomers, which fall into the category of hydrocarbon resins. There are two types, linear aliphatic and cycloaliphatic. The former are made from unsaturated C$_5$ cuts containing primarily pipcrylcnc (1,3-pentadiene) and pentenes. The cycloaliphatics contain these and cyclopentadiene. The resins are used in thermoplastic and contact adhesives and in printing inks. In rubber formulations, they increase hardness and flex life. The second important use is in unsaturated polyester resins (Section 9.1.1) instead of styrene. The product is said to be more brittle but to have a smoother surface.

Because dicyclopentadiene has two double bonds of varying reactivity, it can be used as a monomer in ethylene–propylene diene monomer elastomers (Section 3.2.5). In this application it competes with *trans*-1,4-hexadiene (Section 5.1.3.6) and 5-ethylidenenorbornene. The latter results from a Diels–Alder condensation of cyclopentadiene and butadiene. The adduct is isomerized with an alkaline earth metal catalyst to 5-ethylidenenorbornene.

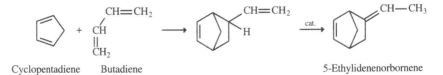

Cyclopentadiene   Butadiene   5-Ethylidenenorbornene

In all these compounds, the active double bond enters into the copolymerization leaving the less active double-bond pendant for cross-linking or vulcanization with sulfur-containing compounds.

The liquid-phase chlorination of cyclopentadiene provides hexachlorocyclopentadiene, which undergoes a Diels–Alder reaction with maleic anhydride to form so-called chlorendic anhydride, used as a flame retardant in unsaturated polyester resins (Section 9.1.1).

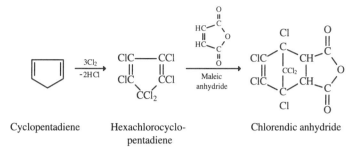

Cyclopentadiene   Hexachlorocyclo-   Chlorendic anhydride
                              pentadiene

Hexachlorocyclopentadiene undergoes a similar Diels–Alder reaction with 1,4-dihydroxy-2-butene (Section 10.3.1) to give a diol adduct, which reacts with thionyl chloride to give the insecticide, endosulfan.

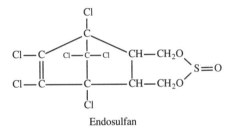

Endosulfan

Like most organochlorine compounds, endosulfan is suspect because its persistence leads to its accumulation in the food chain.

In a new application, which was withdrawn from the market in the early 1990s but is nonetheless interesting, dicyclopentadiene is a monomer for the production of a thermoset polymer useful in reaction injection molding (RIM). In conventional injection molding, the size of moldings made from viscous polymers like polyethylene is limited by the amount of pressure available to push the molding compound into the mold. In RIM, low molecular weight components react in a mold almost instantaneously. The advantage is that large moldings can be made. Two molecules of dicyclopentadiene undergo a metathesis reaction (Section 2.2.9) in the mold in the presence of a catalyst comprising tungsten hexachloride in *p-tert*-butylphenol to give a cross-linked polymer.

Dicyclopentadiene can be oxidized to maleic anhydride but this reaction has never been commercialized.

NOTES AND REFERENCES

Section 6.2   The process for isoprene production via 2-methylbutanal is covered in European Patent Application 0 080 449 (January 6, 1983) to Monsanto.

Section 6.3   The use of polycyclopentadiene in reaction injection molding is described in US Patent 4 400 340 (August 23, 1983) to Hercules.

CHAPTER 7

CHEMICALS AND POLYMERS FROM BENZENE

In the United States, most benzene comes from the catalytic reforming of naphtha (Section 2.2.3), which yields a mixture of benzene, toluene, and the xylenes (BTX). Benzene is also a volatile byproduct of the conversion of coal to coke. Coal distillate benzene is available only to the extent that coke is required by the steel industry, and this has not expanded to accommodate the expanding benzene market. Furthermore, coal distillate benzene contains impurities such as thiophene, which can be removed only with difficulty and render it unacceptable for many purposes. In 1949 all US benzene was produced from coal tar; by 1959 the proportion had dropped to 50%. In 1972 it was down to only 6.4%, and in the early 1990s it was less than 2%. Somewhat more benzene comes from coke oven distillate in Europe than in the United States.

That is not the whole story, however. In Europe and Japan pyrolysis gasoline (Section 2.2.1) is the major source of benzene and toluene. It results from the steam cracking of liquid hydrocarbons, particularly naphtha and gas oil. Xylenes also occur in pyrolysis gasoline but are not easily isolated because of a high concentration of ethylbenzene. In the United States only about 20% of benzene comes from this source. Figure 7.1 shows the sources of benzene in the United States and Western Europe.

However, this is also not the whole story. Catalytic reforming typically leads to a BTX mixture containing about 50% toluene, 35–45% xylenes, but only about 10–15% benzene, the end product most in demand for chemical production. Toluene and the xylenes, on the other hand, have been favored for raising the octane number of unleaded gasoline, because of their low toxicity. Since there have always been large surpluses of chemical toluene, a process was devised to convert toluene to benzene by hydrodealkylation (Section 8.1). In the early 1990s this accounted for about 24% of US and 28% of West European toluene production, although the amount of toluene dehydroalkylated fluctu-

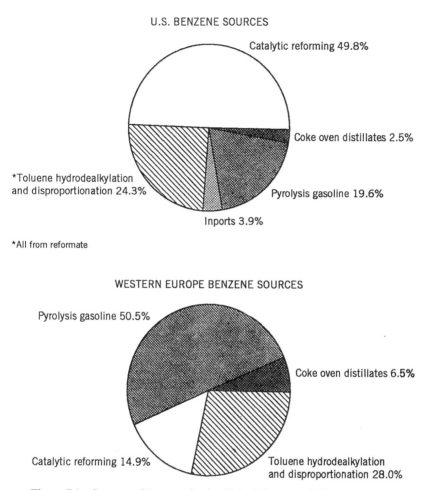

U.S. BENZENE SOURCES

Catalytic reforming 49.8%

Coke oven distillates 2.5%

*Toluene hydrodealkylation
and disproportionation 24.3%

Pyrolysis gasoline 19.6%

Inports 3.9%

*All from reformate

WESTERN EUROPE BENZENE SOURCES

Pyrolysis gasoline 50.5%

Coke oven distillates 6.5%

Catalytic reforming 14.9%

Toluene hydrodealkylation
and disproportionation 28.0%

Figure 7.1   Sources of benzene in the United States and Western Europe.

ates because it is a function of the price differential between benzene and toluene. Toluene must be at least 15% cheaper to warrant dehydroalkylation.

The Clean Air Act, which came into effect in 1995, changed the aromatics supply picture since it stipulates that benzene in gasoline must be decreased to 1% and that the remaining aromatics content must be no greater than 25%. In 1990, gasoline contained as much as 3% benzene and 36% total aromatics. This means that catalytic reforming (Section 2.2.2) will have to be carried out as usual for benzene, toluene, and xylenes (BTX) for the chemical industry. For gasoline, naphtha, stripped of the C_5 and C_6 fraction, will have to be lightly reformed to produce as little benzene as possible (see note).

A minor source ($\sim 1\%$) of benzene is toluene disproportionation although this reaction is becoming more popular as need for p-xylene for terephthalic acid

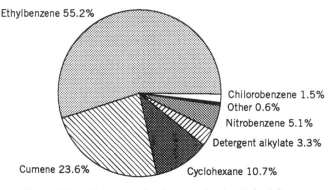

Figure 7.2   Major uses for benzene in the United States.

expands because of burgeoning demand in the late-1980s and early-1990s in the Far East (Section 8.1).

In 1993, about 15 billion lb of benzene was used in the United States as a chemical feedstock as compared to 22.4 billion lb of propylene (Section 1.5). The corresponding figures for Western Europe were 12.2 and 22.3 billion lb. This makes benzene the third most important basic chemical.

The major uses for benzene are shown in Figure 7.2. In addition to these, we shall discuss a number of small volume chemicals from benzene. The biggest use for benzene is its reaction with ethylene to give ethylbenzene for dehydrogenation to styrene (Section 4.2) and for peroxidation to ethylbenzene hydroperoxide for propylene oxide and styrene (Section 4.2). These processes have already been described. Next is its reaction with propylene to give cumene for conversion to phenol and acetone (Section 4.5). In 1993 the production in the United States of 11.8 billion lb of ethylbenzene and 4.5 billion lb of cumene accounted for 80% of the total benzene consumption.

7.1   PHENOL

Phenol (Section 4.5) is yet another chemical for which a variety of processes is available. The cumene hydroperoxide process, by far the dominant process, was described in Section 4.5, but there are various obsolete routes. The earliest process was the sulfonation of benzene to benzenesulfonic acid followed by fusion of the sodium salt of the acid with alkali (Fig. 7.3a). During the Boer War (1899–1902) the British used picric acid (trinitrophenol), an uncertain and unreliable explosive, in their shells. The phenol in coal tar did not provide enough picric acid for the war effort, and the benzenesulfonate process thus became the first tonnage organic chemical process to be operated. Large amounts of byproduct Na_2SO_3 and $NaHSO_3$ made this process cumbersome. To be economical, a benzenesulfonate plant requires cheap sulfuric acid and caustic soda and a nearby paper mill that can use the sodium sulfite byproduct

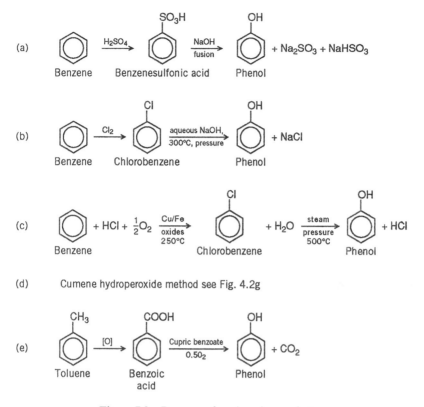

Figure 7.3 Processes for phenol manufacture.

for pulping. Labor costs are high but capital costs are low. No synthetic phenol is made in this way now in either the United States or Western Europe.

The second route appeared in 1924 (Figure 7.3b) and involved the direct chlorination of benzene to chlorobenzene, which was then hydrolyzed to the sodium salt of phenol by means of sodium carbonate or sodium hydroxide. The chlorobenzene process is expensive because of chlorine and alkali usage and must be operated on a very large scale to be economical. The closing of Dow–Midland's massive 100,000 tons/year plant means that no US synthetic phenol is made in this way.

In a later process, the Raschig–Hooker regenerative process (Fig. 7.3c), chlorobenzene was prepared from hydrogen chloride, air, and benzene [a reaction comparable to the oxychlorination of ethylene (Section 3.4) at 250°C with a $CuCl_2/FeCl_3$ catalyst supported on alumina]. Hydrolysis by steam in the vapor phase with a $Ca_3(PO_4)_2/SiO_2$ catalyst at 450°C yields phenol, and hydrogen chloride for recycle. The oxychlorination is run to a conversion of only 10–15% to diminish dichlorobenzene formation, which takes place to the extent of 6–10%. In the hydrolysis of the chlorobenzene, diphenyl ether and *o*- and *p*-hydroxydiphenyl are formed as byproducts. The dichlorobenzenes are recycled to the chlorination reaction, their presence preventing the formation of additional more highly chlorinated products. The Raschig–Hooker process requires highly corrosion-resistant equipment because of the presence of hydrogen chloride at high temperatures. It is not as wasteful of chlorine and alkali as the chlorobenzene route, but conversions per pass are low, and therefore capital costs are high. Also the high temperatures and acid conditions cause corrosion problems, and the high pressures increase operating costs. The process has been obsolete since 1971.

Another reaction was motivated by surplus toluene (Fig. 7.2e), which is first oxidized in the liquid phase to benzoic acid. Molten benzoic acid then reacts with air and steam in the presence of cupric benzoate promoted with magnesium benzoate as catalyst. The volatile phenol is removed by distillation. This process, which gives a selectivity approaching 90%, has the advantage of starting with a lower cost raw material. However, the process was only a limited success, in part because the economics of the cumene process benefited from the acetone coproduct credit (see note). The economics of toluene oxidation depend to a degree on whether toluene provides greater value in this reaction or in hydrodealkylation (Section 8.1). In the early-1990s, only about 70 million lb of phenol were manufactured by this route in the United States, and 110 million in Western Europe. The process is only economically feasible because valuable benzaldehyde and benzoic acid byproducts are formed, but if acetone consumption drops because of the general reduction in solvent use, it could become more attractive.

Although the mechanism for the conversion of benzoic acid to phenol is uncertain, it may proceed through phenyl benzoate, as shown in Figure 7.4. The phenyl benzoate decomposes to phenol and benzoic acid, and the cuprous benzoate is reoxidized to cupric benzoate. An alternative mechanism postulates the formation of salicylic acid or *p*-hydroxybenzoic acid, either of which may decompose to phenol and carbon dioxide.

The dominant cumene process thus has much in its favor. No expensive chlorine, sodium hydroxide, or sulfuric acid is wasted; conditions are mild; and utility costs are low. Its sole drawback is that a demand must exist for both chemicals in the ratio in which they are produced. Market disruptions during the mid-1980s motivated research for a new process that produces phenol only. Thus, a plant was built in Australia in which a cyclohexanol–cyclohexanone mixture (Section 7.2.1) was dehydrogenated to phenol, but it was soon closed as uneconomical. A Japanese process (Section 4.5) operates with cumene made from benzene and isopropanol rather than propylene. The isopropanol can be made by hydrogenation of the surplus acetone produced in the process.

There has been research on direct air oxidation of benzene to phenol, the most attractive involving the reaction of benzene with acetic acid in the presence of palladium and oxygen to give phenyl acetate. This reaction is analogous to the formation of acetaldehyde (Section 3.5) by the Wacker reaction or vinyl acetate (Section 3.6) by a related reaction.

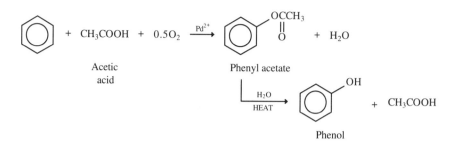

As in the Wacker reaction, the palladium is rendered catalytic by the incorporation of an oxidant. As in the vinyl acetate process, the reaction may be carried out in the vapor phase over a supported palladium metal catalyst. The phenyl acetate may then be hydrolyzed to phenol and acetic acid, which is recycled.

Ethylbenzene hydroperoxide (Section 4.7) may be decomposed with a nickel catalyst to phenol and acetaldehyde in very high yield. However, this too is a coproduct process, and the demand for acetaldehyde had decreased by the mid-1980s (Section 3.5).

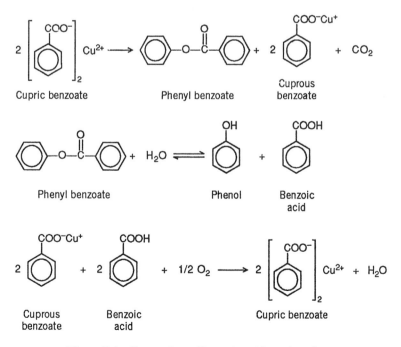

Figure 7.4 Conversion of benzoic acid to phenol.

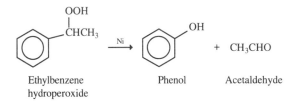

Ethylbenzene   Phenol   Acetaldehyde
hydroperoxide

7.1.1 Phenolic Resins

The major applications for phenol are shown in Figure 7.5. The biggest use, the manufacture of phenol–formaldehyde resins (Section 15.4.1), consumes 35% of the United States and 32% of Western Europe's phenol production. Phenolic resin production in the United States was 3.1 billion lb in 1993.

Phenolic resins have good chemical, heat, and water resistance, good dielectric properties, and high-surface hardness and dimensional stability. They are cheap and can be formulated to meet the needs of the electrical, automotive, appliance, and adhesives industries. Adhesives comprise the largest market for phenolic resins, with bonding of plywood, particle board, and hardboard the largest single application. Phenolic resins are also used in the fabrication of laminates such as "Formica" in which a phenolic bound laminate, which has an unattractive brown color, is faced or coated with a layer of a high-quality, colorless, nonyellowing melamine–formaldehyde resin. Under the layer is placed the pattern—often an actual photograph of wood grain or, in cheaper formulations, a printed picture of wood grain. Phenolic resins are frequently compounded with other adhesives to provide water resistance and tack or stickiness.

Phenolic moldings are inevitably a poor color and moldings can be produced only in dark colors. They are used for the characteristic electrical plugs and ashtrays. Other applications include insulation, abrasives, and foundry and shell

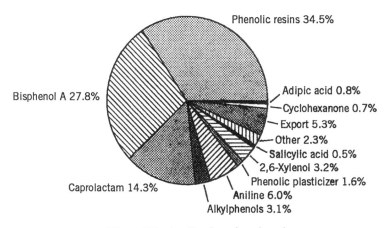

Figure 7.5 Applications for phenol.

moldings. They are used in brake linings, clutch facings, and other friction parts where a high-temperature-resistant binder is required. Phenolics are also cross-linking agents for epoxy resins in the formulation of structural adhesives and resistant coatings, particularly for can linings. Although usage is large, growth has been inhibited, as it has been for all thermoset resins, because of the difficulty of fabrication and the low-molding speeds. The water and heat resistance of the resins is a positive point but the poor color a negative one.

7.1.2 Bisphenol A

Phenol condenses with acetone to give bisphenol A, which in turn reacts with epichlorohydrin to give epoxy resins and with phosgene to give polycarbonates. For epoxy resins a mixture of the *ortho* and *para* isomers is acceptable, but the other polymers require the *para*, *para'* isomer.

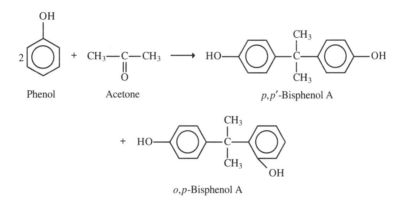

Polycarbonate grade bisphenol A is 99.5% pure, whereas the epoxy grade is 95% pure. The condensation may be catalyzed with either mineral acid or with a sulfonated cation exchange resin. A hydrogen chloride catalyzed reaction requires a temperature of about 50°C, whereas the ion exchange resin is used at 70–90°C. A variety of byproducts is produced and these can be rearranged to BPA at 75°C in the presence of a catalyst comprising a cation exchange resin partially esterified with a mercaptoalcohol.

The separation of the *ortho*, *para* isomer from the *para*, *para'* isomer is accomplished by a combination of distillation and crystallization. In the United States 52% of bisphenol A is required for production of polycarbonate resins and 37% for epoxy resins. In Western Europe epoxy resins still provide the major application.

7.1.2.1 *Epoxy Resins* Epoxy resins are oligomers resulting from the condensation of bisphenol A and epichlorohydrin.

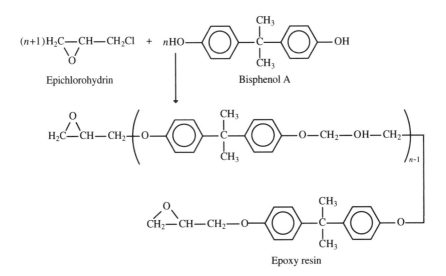

Epoxy resin

When cross-linked they are hard, chemically resistant, and dimensionally stable, and have superb electrical properties. Their largest use is for protective coatings for metal containers, appliances, and ships, as well as for general maintenance where resistance to severe corrosion is required. They are used in the computer industry for "potting" electrical components. The second largest use is in fiber-reinforced composites for circuit boards, aerospace components, and sporting equipment. Other uses include adhesives, sealants, patching and flooring compounds, and castings.

7.1.2.2 Polycarbonate Resins   Polycarbonate resins result from the condensation of bisphenol A and phosgene. They are thermoplastics with exceptional clarity, impact strength, flame resistance, and low warpage. The problem of reacting a high-melting solid, the sodium salt of bisphenol A, with a gas, phosgene, has been solved by the use of phase-transfer catalysis (Section 16.10). The salt is slurried in an organic phase such as dichloromethane while the phosgene is dissolved in water (25% of it hydrolyzes) containing a catalytic amount of a base such as pyridine. The phosgene–pyridine complex is sufficiently lipophilic so that it migrates to the organic layer where the condensation takes place. The pyridine returns to the aqueous layer to repeat the operation. This was probably the first industrial use for phase-transfer catalysis.

The largest application for polycarbonates is in the electrical and electronics sector for compact disks, business machine enclosures, connectors and plugs, telephones and electrical distribution devices. The second largest application is for window glazing and related applications such as binoculars, where a virtually unbreakable molding with exceptional optical properties is required. Polycarbonates are also blended with poly(ethylene terephthalate) (Section 9.3.3) or poly(butylene terephthalate) (Section 9.3.4) to make impact resistant polymer alloys useful, for example, for automobile bumpers and even for side panels.

Phosgene is highly toxic; indeed it was used as a poison gas in World War I. In the early 1990s, General Electric built a plant in Japan using a non-phosgene route to polycarbonate resins. A possible route, shown in Figure 7.6, starts with methanol, carbon monoxide and oxygen. These combine to give dimethyl carbonate, in the presence of a copper-based catalyst or in a two–for–one reaction involving the aqueous methanolysis of ethylene carbonate (Section 3.7.1). An exchange reaction between phenol and dimethyl carbonate provides diphenyl carbonate, which reacts with bisphenol A acetate in an ester interchange in the presence of a catalyst such as tetraphenyl titanate to give a polycarbonate oligomer. This is heat treated to effect further condensation to a higher molecular weight polycarbonate. The phenol is released for recycle.

7.1.2.3 *Lesser Volume Uses for Bisphenol A*   Bisphenol A is used in the synthesis of several engineering polymers because its stiffness contributes strength. The low-volume polysulfone and polyether sulfone polymers require less than 2% of the bisphenol A produced. An example is the condensate of bisphenol A and *p, p'*-dichlorodiphenylsulfone. The hydrogen chloride produced in the condensation is neutralized with base as the reaction progresses.

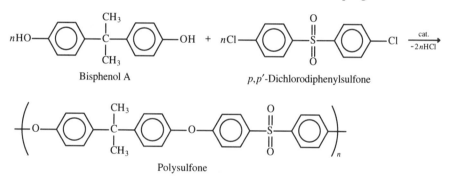

Bisphenol A   *p,p'*-Dichlorodiphenylsulfone

Polysulfone

These polymers are used in electrical and electronics applications because of their thermal stability and excellent resistance properties. Their superior corrosion resistance and stability to hydrolysis make them useful for medical instruments that require repeated sterilization. They are also used for ultrafiltration membranes in biotechnology and for the "Prism" separators by which hydrogen is separated from argon and nitrogen in the most recent modification to the Haber process. The Haber process gives low yields per pass and much hydrogen and nitrogen must be recycled, but there is argon left from the air, which accumulates. To prevent an unacceptable level, there had to be a purge in which nitrogen, argon, and valuable hydrogen were taken off and used as fuel. The modern Monsanto process puts the recycled gases over a microporous polysulfone membrane. The hydrogen diffuses rapidly through the pores and is recycled without waste.

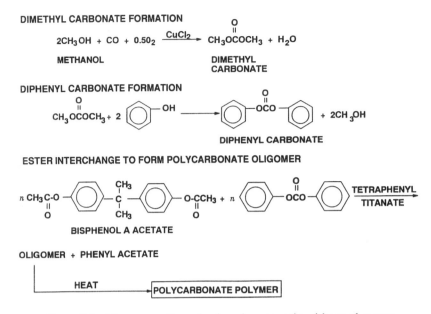

Figure 7.6   The preparation of polycarbonate resin without phosgene.

Produced in even smaller quantities than polysulfones are the polyarylates and the poly(ether imides). Polyarylates are condensates of bisphenol A with isophthalic and terephthalic acids. They have excellent resistance to ultraviolet radiation and good impact resistance. They are used in fog lamp lenses and to a larger degree in the rear tail lights of automobiles that are at eye level.

Reaction 1 (Formation of bis(ether substituted phthalimide))

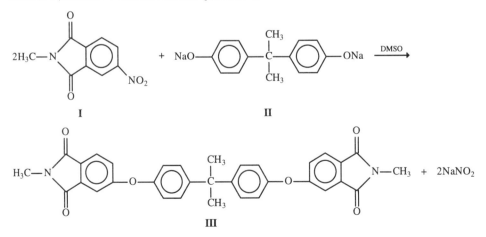

Reaction 2 (Formation of bis(ether phthalic acid))

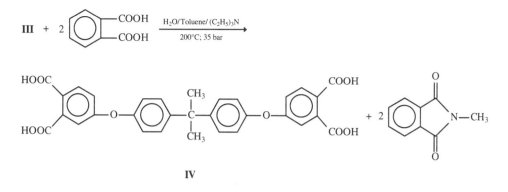

IV

Poly(ether imides) (General Electric's "Ultem") have been designed to have some of the excellent properties of a polyimide but with greater tractability. They are condensates of *m*-phenylene diamine with a bisphenol-based dianhydride, made by a nucleophilic displacement between the sodium salt of bisphenol A and 4-nitrophthalic anhydride.

Reaction 3 (formation of dianhydride)

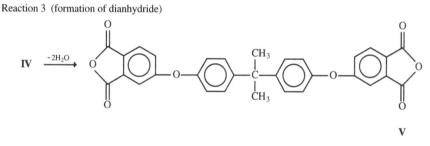

V

Reaction 4 (formation of poly(ether imide))

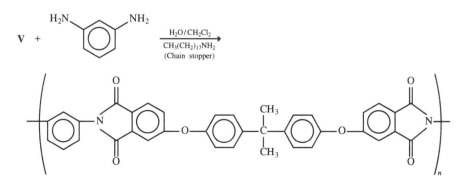

Bisphenol A is also the basis for a specialty corrosion-resistant, unsaturated polyester based on its reaction with phthalic and maleic anhydrides. Bromination of bisphenol A provides tetrabromobisphenol A, a flame retardant useful in its own right and as a monomer for flame retardant epoxy resins.

7.1.3 Cyclohexanone

The third largest use for phenol in both the United States and Western Europe is for caprolactam. Approximately 25% of caprolactam comes from this source and the remainder from cyclohexane oxidation (Section 7.2.2). The starting material for caprolactam is cyclohexanone, and this could come from phenol by a straightforward two-stage process in which the phenol is first hydrogenated to cyclohexanol and then dehydrogenated to cyclohexanone.

By a unique hydrogenation, however, phenol may be converted directly to cyclohexanone in high yield in a single step with a supported palladium catalyst at 12–13 bar and 200°C. Selectivity is about 97%. Presumably the cyclohexanol converts to a keto form, which on hydrogenation provides cyclohexanone:

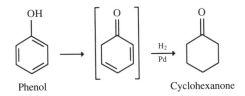

Conversion to caprolactam is described in Section 7.2.2.

7.1.4 Alkylphenols

Phenol may be alkylated with olefins by a Friedel–Crafts reaction (Section 5.2.3) to provide alkylphenols. When alkylated with propylene trimer (Sections 2.2.4 and 4.2) nonylphenol results which, when reacted with ethylene oxide, provides an ethoxylate. This has been the most important surfactant for industrial detergents and has also been used in liquid dishwashing detergents. It has low-foam properties and good detergency. Because of poor biodegradability, however, it is being replaced by other surfactants including the more expensive ethoxylated fatty alcohols, alcohol sulfates and alcohol ethoxysulfates. It is also converted to antioxidants for rubbers and plastics and for lube oil additives.

Octylphenol is made (Section 5.2.3) by alkylating phenol with diisobutene. Ethoxylation gives nonionic surface cleaners. Reaction with formaldehyde provides an oil-soluble phenolic resin. Dodecylphenol results from the condensation of propylene tetramer (Section 4.2) with phenol. Most of the product is used for lube oil additive formulations.

7.1.5 Chlorinated Phenols

The most important chlorinated phenol is 2,4-dichlorophenol, the raw material for the herbicide, 2,4-dichlorophenoxyacetic acid or 2,4-D.

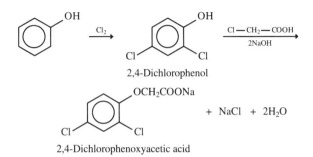

2,4-Dichlorophenol

2,4-Dichlorophenoxyacetic acid

Pentachlorophenol, made by chlorinating phenol exhaustively, is a wood preservative and slimicide. Like other chlorinated compounds it has been outlawed in the United States because of its adverse environmental affects. The chlorine atoms make it highly lipophilic and when it enters the body via the food chain it accumulates in fatty tissue. The same is not true of 2,4-D, whose carboxyl group may facilitate its excretion.

7.1.6 2,6-Xylenol/Cresols

The synthesis of 2,6-xylenol (2,6-dimethylphenol) provides a small application for phenol. It is made by methylation of phenol at about 500°C at almost atmospheric pressure with a methanol/phenol molar ratio of 6:1 and a catalyst comprising magnesium oxide on an inert carrier pretreated with methanol vapor. Ten per cent water provides continuous catalyst regeneration. o- and p-Cresol are byproducts.

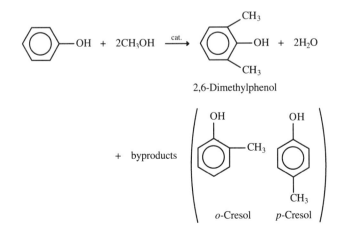

2,6-Dimethylphenol

o-Cresol   p-Cresol

The xylenol is converted by oxidative coupling to poly(phenylene ether) whose recurring unit is

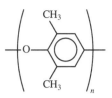

The *o*-cresol finds use in the manufacture of pesticides.

7.1.7 Aniline from Phenol

Aniline is normally prepared from nitrobenzene (Section 7.3). An alternative route involves the ammonolysis of phenol. Both of these processes are described in Section 7.5.

7.2 CYCLOHEXANE

Small quantities of cyclohexane are extracted from naphtha (Section 2.1) but most is produced by hydrogenation of benzene. The process is operated in refinery complexes using hydrogen from catalytic reforming (Section 2.2.3).

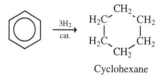

Cyclohexane

Either nickel or platinum catalysts may be used, but both require benzene feed with a sulfur content of less than 1 ppm. Hydrogenation is carried out at an upper temperature of 230°C at 1.5–2.5 bar. Although liquid-phase hydrogenations are more common, gas-phase processes are available. These operate at temperatures as high as 600°C with moderate pressures of 25–50 bar.

Although some cyclohexane is used as a solvent, its major market is for the production of adipic acid and caprolactam.

7.2.1 Adipic Acid

Benzene is the source of the two most important nylons, nylon 6,6 and nylon 6. Nylon 6,6 is the polymer formed when adipic acid condenses with hexamethylene diamine. Nylon 6 is the self-condensation product of caprolactam, which is the dehydration product of 6-aminocaproic (6-aminohexanoic) acid. The numbers used to designate nylons refer to the number of carbon atoms in the diamine and dibasic acid in that order. A single number indicates that the amino and carboxyl functions are in one molecule. Thus $\{NH(CH_2)_6NHCO(CH_2)_4CO\}_n$ is nylon 6,6, $\{NH(CH_2)_5CO\}_n$ is nylon 6, and $\{NH(CH_2)_6NHCO(CH_2)_8CO\}_n$ is nylon 6,10, a specialty nylon made with sebacic rather than adipic acid.

In the early days of nylon, Du Pont created a mystique about their product by pointing out that it was made from coal, air, and water. Coal at that time was the source of the benzene or phenol feedstocks. Now both are derived from petroleum, and "made from petroleum, air and water" does not sound nearly so good.

Nylon 6,6 is the preferred polymer in both Western Europe and United States. In Europe in 1991 it enjoyed 53% of the market. In the United States, there is twice as much capacity for manufacturing nylon 6,6 as nylon 6. In Japan, however, nylon 6 is more popular. Many of the producers of caprolactam are fertilizer manufacturers who can use the byproduct ammonium sulfate of which Japan is a particularly large consumer.

The first step in the manufacture of the adipic acid needed for nylon 6,6 is the hydrogenation of benzene to cyclohexane. Thereafter the cyclohexane may be oxidized directly to adipic acid with nitric acid or with air over cobalt acetate, but yields are low, production of valueless byproducts is high, and large amounts of nitric acid are consumed. Instead, a two-stage process is used. The initial oxidation at about 150°C and 10–15 bar over cobalt or manganese naphthenate or octanoate gives a cyclohexanol–cyclohexanone "mixed oil." In the first stage of the reaction, cyclohexyl hydroperoxide forms and this is converted catalytically to the ol/one.

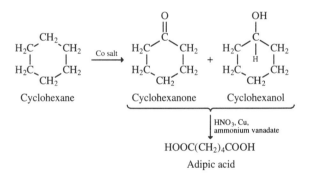

Conversions of the order of 10–12% are accepted in order to minimize the formation of adipic acid, which at this stage presents processing problems and, more important, because of the degradation reactions that provide glutaric (C_5) and succinic (C_4) acids. The selectivity to the ol/one does not surpass 85%, but unreacted cyclohexane can be recycled. The desired ol/one mixture is purified by distillation.

The ratio of cyclohexanol–cyclohexanone may be increased from between 1:1 and 2:1 to about 9:1 by the addition of boric acid, which esterifies the "ol" as it is formed and prevents its further oxidation to cyclohexanone. The boric acid also increases the selectivity of the reaction. The cyclohexanol–cyclohexanone "mixed oil" is then oxidized to adipic acid with nitric acid at high selectivity over a copper catalyst with ammonium vanadate. Conversions per pass are of the order of 12%. No way of circumventing the use of nitric acid has yet been found, and this is the only example in industrial chemistry of the use of nitric acid as an oxidant for the production of a large volume chemical.

The above oxidation process has been used for adipic acid since the early 1940s. Recently, however, BASF has commissioned a process that depends on the carbonylation of butadiene (Section 5.1.2).

A Gulf patent, not commercialized, describes a one-step oxidation of cyclohexane to adipic acid in the presence of Co^{3+} ions in an aqueous solution of acetic acid. Benzoyl peroxide is included to convert Co^{2+} to Co^{3+} ions. The patent claims that the addition of water is critical not only in facilitating high yield but in preventing degradation of the adipic acid to glutaric and succinic acids.

A route to adipic acid not currently used starts with tetrahydrofuran (Sections 10.3.1 and 14.1):

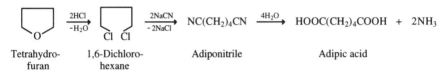

| Tetrahydrofuran | 1,6-Dichlorohexane | Adiponitrile | Adipic acid |

The waste of chlorine, which does not appear in the end product, and the use of poisonous sodium cyanide makes the process unattractive. Tetrahydrofuran was originally an agricultural product derived from furfural (Section 14.1.1) obtained in turn by hydrolysis of the pentosan content of cereal hulls. Today it is largely made from 1,4-butanediol (Section 10.3.1).

A microbiological process for the production of adipic acid proposed by Celanese is not likely to be commercialized in the near future but is mentioned here because, like the microbiological process for propylene oxide (Section 4.7.2), it demonstrates the potential of modern biotechnology. Nylon 6,6 salt (an equimolar compound of adipic acid and hexamethylene diamine) is produced from toluene via muconic acid. Toluene is first converted to muconic acid by a mutant strain of *Pseudomonas putida*.

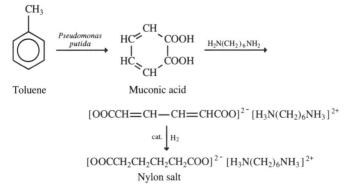

Toluene   Muconic acid

$$[OOCCH{=}CH{-}CH{=}CHCOO]^{2-}\,[H_3N(CH_2)_6NH_3]^{2+}$$

$$\text{cat.} \downarrow H_2$$

$$[OOCCH_2CH_2CH_2CH_2COO]^{2-}\,[H_3N(CH_2)_6NH_3]^{2+}$$
Nylon salt

Muconic acid is much less soluble than adipic acid and is neutralized with hexamethylene diamine as it is formed in the fermenter to yield a solution of hexamethylene diammonium muconate in concentrations as high as 3.9%, which is separated from the cells by ultrafiltration. The salt may be precipitated by addition of isopropanol. Thereafter it is hydrogenated to give nylon salt for nylon preparation. The key to the process is the ability to isolate the product from dilute aqueous solution without expensive removal of water by distillation or other means.

7.2.1.1 *Nylons from Adipic Acid*   Once adipic acid has been made, it must be reacted with hexamethylene diamine, whose manufacture has already been described (Section 5.1.3) to obtain nylon 6,6.

$$HOOC(CH_2)_4COOH \ + \ H_2N(CH_2)_6NH_2 \ \longrightarrow \ \overline{\left[OC(CH_2)_4CONH(CH_2)_6NH\right]}_n$$

| Adipic acid | Hexamethylene diamine | Nylon 6,6 |

Nylon 4,6 is another adipic acid containing polymer, whose amine is 1,4-diaminobutane. It is used for engineering plastic parts rather than as a fiber. It has a higher melting point than either nylon 6,6 or nylon 6, with better toughness and temperature stability, but with the high water absorption that bedevils nylons. 1,4-Diaminobutane results from the addition of hydrocyanic acid to acrylonitrile to provide succinonitrile, which on hydrogenation gives the diamine.

$$NCCH{=}CH_2 + HCN \rightarrow NCCH_2CH_2CN \xrightarrow{\ H_2/cat.\ } H_2N(CH_2)_4NH_2$$

| Acrylonitrile | Succinonitrile | 1,4-Diaminobutane |

The addition of HCN to the highly activated double bond of acrylonitrile takes place at 80°C in the presence of an alkaline catalyst such as triethylamine. The hydrogenation to 1,4-diaminobutane may take place with a cobalt oxide catalyst in a tetrahydrofuran solvent at 100°C and 190 bar.

Nylon raw materials cannot be made sufficiently pure to form a satisfactory polymer. Therefore, the diamine and dicarboxylic acid are reacted at mild temperature to form a salt, which can be purified further by crystallization. Thereafter, the salt is heated further to polymerize it.

7.2.2 Caprolactam

There are competing routes to caprolactam. In the initial and most widely used one, cyclohexanone is the starting material. The cyclohexanone component of "mixed oil" (Section 7.2.1) may be separated by distillation, the cyclohexanol may be dehydrogenated to cyclohexanone at 425°C and atmospheric pressure with metallic catalysts such as zinc or copper, and cyclohexanone may also be obtained from phenol (Section 7.1.3).

The cyclohexanone (Fig. 7.7a) reacts with hydroxylamine sulfate to give cyclohexanone oxime and sulfuric acid, which is converted to ammonium sulfate by the ammonia injected into the reaction mixture to drive the reaction to the right. Oximes undergo the Beckmann rearrangement, and this one is no exception; treatment with sulfuric acid gives caprolactam and ammonium sulfate.

The production of hydroxylamine also gives an ammonium sulfate byproduct. It is made from ammonium bicarbonate and ammonia which react to give ammonium carbonate. Oxidation of ammonia gives an NO/NO_2 mixture that reacts with the carbonate to give ammonium nitrite for conversion with sulfur dioxide and ammonium hydroxide solution to hydroxylamine sulfate and ammonium sulfate.

$$NH_4HCO_3 + NH_3 \rightarrow (NH_4)_2CO_3 \xrightarrow{NO/NO_2} 2NH_4NO_2$$

Ammonium   Ammonium   Ammonium
bicarbonate   carbonate   nitrite

$$NH_4NO_2 + SO_2/NH_4OH \rightarrow [HONH_3]^+HSO_4^- + NH_4HSO_4 \xrightarrow{NH_3}$$

Hydroxylamine   Ammonium
sulfate   bisulfate

$$NH_2OH + (NH_4)_2SO_4$$

Hydroxyl-   Ammonium
amine   sulfate

Each pound of caprolactam produced generates 4.4 lb of ammonium sulfate byproduct. The yield is 1.6 lb from the production of hydroxylamine sulfate, 1.1 pounds from the production of the oxime, and 1.7 pounds from the Beckmann rearrangement. This large amount of ammonium sulfate finds its way to the fertilizer market, particularly if a fertilizer plant is close to the source of the ammonium sulfate. In a new ICI process (Section 4.6.1) the ammonium sulfate is pyrolyzed and converted to sulfuric acid.

The production of cumbersome amounts of unwanted ammonium sulfate, which, with ammonium bisulphate, is also a byproduct of methyl methacrylate manufacture, stimulated the search for alternative methods for caprolactam production. One of them involves the use of phosphoric rather than sulfuric acid to effect the Beckmann rearrangement, because ammonium phosphate has greater value as a fertilizer.

A Japanese (Toray) process (Fig. 7.7b) involves treatment of cyclohexane with nitrosyl chloride and hydrogen chloride under actinic light (500 μm) to give the oxime hydrochloride.

$$2H_2SO_4 + N_2O_3 \rightarrow 2HNOSO_4 + H_2O$$
$$HNOSO_4 + HCl \rightarrow NOCl + H_2SO_4$$

The nitrosyl chloride is made in three steps. Ammonia is burned in air to

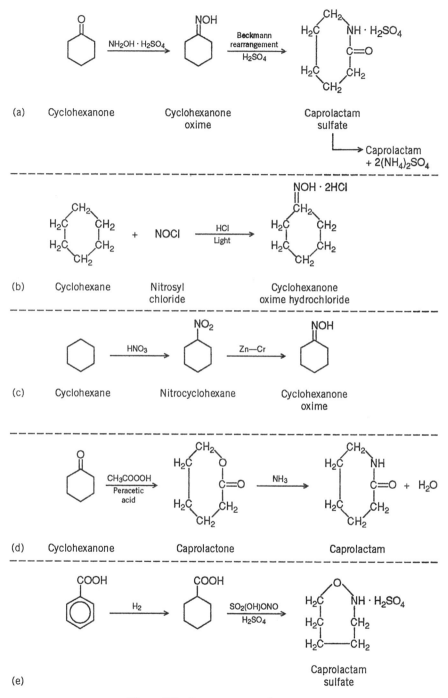

Figure 7.7 Routes to caprolactam.

nitrogen trioxide; the trioxide is absorbed in sulfuric acid to give nitrosylsulfuric acid; hydrogen chloride is then added to the nitrosylsulfuric acid:

$$2NH_3 + 3O_2 \rightarrow N_2O_3 + 3H_2O$$

$$2H_2SO_4 + N_2O_3 \rightarrow 2HNOSO_4 + H_2O$$

$$HNOSO_4 + HCl \rightarrow NOCl + H_2SO_4$$

This process is wasteful of chlorine since the cyclohexanone oxime hydrochloride must be converted to the free oxime. On the other hand, the sulfuric acid that results in the NOCl reaction is recycled so that ammonium sulfate production is virtually eliminated from that step. Nonetheless, it still results from the Beckmann rearrangement.

The major problem with photonitrosation is the engineering of an appropriate mercury light source. This same reaction is useful on cyclododecane (Section 5.1.2.2) to provide the monomer for nylon 12.

Another process (Fig. 7.7c), now largely of historical interest but used at one time by Du Pont, involves nitration of cyclohexane to nitrocyclohexane. This can be reduced over a zinc–chromium catalyst to cyclohexanone oxime, the precursor of caprolactam. The manufacture of hydroxylamine is avoided, and about 66% of the ammonium sulfate is eliminated. The nitration step is akin to the nitration of propane to a mixture of nitroparaffins (Section 11.2) and is difficult to carry out in high selectivity.

A fourth route to caprolactam (Fig. 7.7d) uses peracetic acid (from acetaldehyde and air) to convert cyclohexanone, at 50°C and atmospheric pressure, to caprolactone, which on reaction with ammonia provides caprolactam. Only the first step is now in use for the production of caprolactone.

A toluene-based route to caprolactam has been pioneered by Snia Viscosa and was used in Italy until the early 1990s. Benzoic acid (Fig. 7.7e), prepared by the cobalt-catalyzed oxidation of toluene, is hydrogenated to hexahydrobenzoic acid, over a palladium catalyst at 170°C and 15 bar. This is treated with nitrosylsulfuric acid to obtain caprolactam sulfate. This process also eliminates 66% of the ammonium sulfate. In a variation of it, the caprolactam sulfate is diluted with water to dissociate the salt, after which the caprolactam is extracted from the 50% sulfuric acid solution with toluene or an alkylphenol. Thus the byproduct is sulfuric acid rather than the salt. The acid can be pyrolyzed to sulfur dioxide for conversion via sulfur trioxide to concentrated sulfuric acid. This may be the most economical of all the processes.

A clever approach to the problem of eliminating ammonium sulfate formation is found in a DSM process in which a buffered solution of hydroxylamine reacts with cyclohexanone in solution to produce the oxime. The buffered hydroxylamine solution is produced by hydrogenation of nitrate ions to hydroxylamine in the presence of a phosphate buffer with a palladium catalyst. This solution reacts with hydroxylamine, and the oxime can be extracted with toluene. The aqueous solution can be recycled.

Of these processes, only three are in use, the cyclohexane-based process first described, the DSM variant, and the photonitrosation. While this book was in press, two new processes were announced and are described on p. 270. All these processes exemplify the imagination that chemists bring to bear on troublesome problems, in this case the formation of unwanted byproduct ammonium sulfate.

7.3 ANILINE

Next in line of benzene based chemicals is aniline. Production in 1992 in the United States was 1.0 billion lb, 75% of which went into isocyanates (Section 7.3.1). In the traditional process, benzene is first nitrated with mixed acids (H_2SO_4/HNO_3). These form a nitronium ion (NO_2^+), which attacks the benzene ring. The reaction is exothermic and the mixture must be cooled to maintain a temperature of about 50°C. An adiabatic process has been described, in which 65 rather than 98% sulfuric acid is used. The water in the acid absorbs the heat eliminating the need for external cooling. In the nitration of benzene, a small amount of *meta* isomer is obtained.

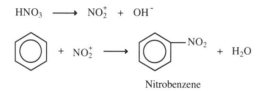

Nitrobenzene

Nitrobenzene is reduced to aniline in almost quantitative yield by vapor-phase hydrogenation at 270°C and 1.25 bar in a fluidized bed of a copper-on-silica catalyst. Also feasible is a vapor-phase hydrogenation over a fixed bed of nickel sulfide on alumina.

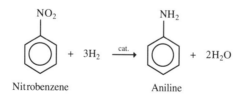

An older process employed iron turnings and hydrochloric acid as a source of hydrogen. This liquid-phase process, in which the iron was converted to Fe_3O_4, useful as a pigment, is little used today. Another process involving the ammonolysis of chlorobenzene was wasteful of chlorine and has not been used since 1967.

The newest method for aniline preparation involves the ammonolysis of phenol.

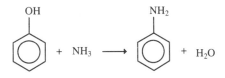

This reversible reaction is driven to the right by a high ammonia/phenol ratio, which also retards formation of diphenylamine. The reaction takes place at 200 bar and 425°C. The catalyst is proprietary, although initially a mixture of alumina and silica was proposed. Oxides of magnesium, aluminum, and tungsten are also effective in the presence of cocatalysts such as cerium and vanadium. Aniline selectivity is lower (~ 90%) than in the nitrobenzene process, and diphenylamine and carbazole form as byproducts by dehydrogenation. The lower capital cost of this process as compared to the nitrobenzene process is partly offset by higher raw material costs. On the other hand, the process eliminates ecological problems associated with the use of sulfuric and nitric acids.

The direct amination of benzene via intermolecular dehydrogenation has been studied by Du Pont but has never been commercialized. Obviously, such a process could provide the lowest possible raw material costs. The "cataloreactant," a term apparently coined for this reaction, comprises nickel with small amounts of rare earth metal oxides together with a stoichiometric amount of nickel oxide, which reacts with the released hydrogen. The reaction takes place at 350°C and 290 bar with a high selectivity of 97% and a low conversion of 10% benzene per pass.

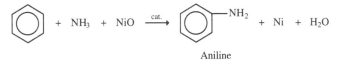

Aniline

Aniline's major use is for conversion to 4,4'-diphenylmethane diisocyanate (MDI) discussed below. The second most important use is for rubber chemicals (see note). A third use is as an intermediate for dyes, drugs such as antihistamines (bamipine and thenaldine), and intermediate chemicals. It is also used in the synthesis of riboflavin.

A small use is for the preparation of hydroquinone (Section 7.7) the classic developer for black-and-white photography, since it is capable of reducing light-exposed crystals of silver halide in emulsion to silver without affecting unexposed crystals:

HO—⟨benzene ring⟩—OH + 2AgBr → O=⟨ring⟩=O + 2HBr + 2Ag

Color photography depends on chromogenic (dye-forming) developers, which are also aniline derivatives, such as *p*-diethylaminoaniline, 2-methyl-4-diethyl-aminoaniline and *N-p*-anilinopyrrolidine.

A new use for hydroquinone is as a component of a high-performance plastic, poly(ether ether ketone) (PEEK) a condensate of hydroquinone and *p,p'*-difluorobenzophenone.

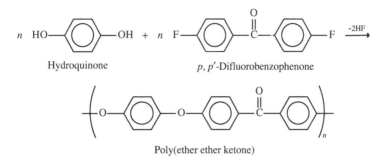

Poly(ether ether ketone)

7.3.1 4,4′-Diphenylmethane Diisocyanate

4,4′-Diphenylmethane diisocyanate (MDI, standing for methylene diphenylene diisocyanate) is produced by reaction of aniline hydrochloride with formaldehyde to form the 4,4′, 2,4′, and 2,2′ isomers of diaminodiphenylmethane. The equation shows only the 4,4′ isomer:

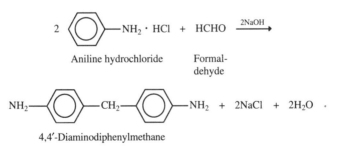

4,4′-Diaminodiphenylmethane

The diamine reacts with additional formaldehyde to give trimers, tetramers, and higher oligomers. The diisocyanate from the diamine is known as MDI, whereas the isocyanate from the oligomers is known as poly MDI or PMDI. When MDI is required, the diamine is removed from the mixture by distillation and phosgenated. Conversely, the entire mixture may be phosgenated and the MDI separated from the mixed isocyanates by distillation.

For the condensation of aniline and formaldehyde, aniline is treated with a stoichiometric amount of hydrochloric acid, and the hydrochloride is reacted with 37% formaldehyde for a few minutes at 70°C. The condensation is completed at 100–160°C for 1 h. The mixture of di- and higher amines is recovered and phosgenated by reaction with phosgene in chlorobenzene solution. The carbamoyl chloride forms at 50–70°C and this is decomposed to the isocyanate at 90–130°C.

Non-phosgene routes have been proposed for the preparation of MDI. An ARCO process developed further by Japanese companies involves the direct carbonylation of nitrobenzene in the presence of a lower alcohol with a sulfur or selenium catalyst. A carbamate forms that can be reacted with formaldehyde to give dimers and oligomers, after which the carbamate groups are converted to isocyanates by heating with or without a catalyst. The recovered alcohol is recycled.

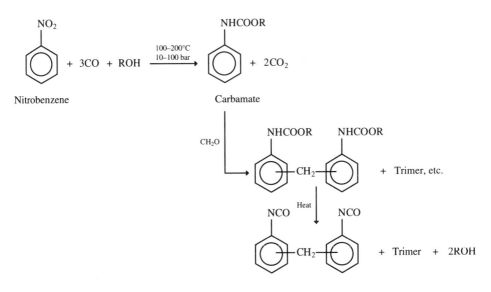

ARCO never commercialized the process, presumably because residual selenium could not be completely removed from the product. The catalysts proposed by the Japanese companies are said to be superior and include iodide-promoted palladium metal and a ruthenium carbonyl complex.

An Asahi process starts with aniline, which makes necessary an oxidative carbonylation. The catalyst is palladium metal with an alkali iodide promoter. Ethylene phenylcarbamate forms, which reacts with formaldehyde to give a dicarbamate, which in turn decomposes to the diisocyanate. An advantage of this process is that it is said to give a minimum of polymeric products in the formaldehyde condensation.

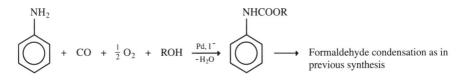

In an even newer process developed by Catalytica Associates, Nippon Kokan, and Haldor Topsoe, a mixture of aniline and nitrobenzene is carbonylated in the presence of an alcohol to give methyl N-phenyl carbamate. Since oxygen is not present, the nascent hydrogen that forms reacts immediately with nitrobenzene to give aniline, which is further carbonylated.

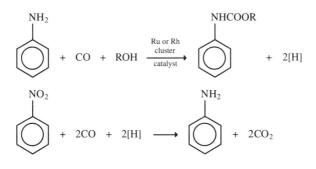

The overall reaction is:

No net aniline need be supplied since the nitrobenzene is reduced to aniline in the process. A redox catalyst, which often leads to corrosion, is not required. Instead the process uses a cluster catalyst (an agglomeration of metal atoms) based on rhodium or ruthenium carbonyl complexes incorporating biphosphino or poly tertiary-amino ligands.

Urea may also provide the CO for carbonylating the amine. Thus an amine, urea, and an alcohol will form a carbamate, which can be converted to an isocyanate.

$$RNH_2 + H_2NCONH_2 + R'OH \rightarrow RNHCOOR' + 2NH_3$$

$$RNHCOOR' \rightarrow RNCO + R'OH$$

This process was being commercialized in 1995 by Hüls for the manufacture of isophorone diisocyanate (Fig. 7.8).

Polyurethanes are made by reaction of diisocyanates with hydroxyl-containing compounds, primarily polyether polyols (Section 4.7.1). If the hydroxyl compound is bifunctional, a linear polymer results, but usually a polyfunctional alcohol is used to give a cross-linked, thermosetting resin. The two major diisocyanates are MDI/PMDI and toluene diisocyanate (TDI) (Section 8.3) and their application is for foams. MDI/PMDI provides most of the rigid foams, which are largely used for insulation, whereas TDI provides the flexible ones, which are used in upholstery, bedding, and automobiles. Of particular interest is RIM, an acronym for reaction injection molding. In this process the isocyanate and a polyol are pumped into a mold together with a catalyst so that they may react in the mold. This makes large moldings possible.

There are also noncellular applications for isocyanates, mostly in corrosion-resistant maintenance coatings, which use primarily MDI/PMDI if yellowing is not a problem. Non-yellowing urethane coatings are based on aliphatic iso-cyanates derived from hexamethylene diamine, bis-aminocyclohexylmethane, "isophorone diamine," and xylylene diamine. A recent addition is tetramethyl-xylylene diisocyanate. These are all termed aliphatic, even though two of them contain aromatic rings. The isocyanate groups, however, are attached to aliphatic carbon atoms. The structures are shown in Figure 7.8.

Non-yellowing isocyanates are useful for aircraft topcoats over epoxy primers. A more recent application, which could develop into large volume usage, is clear automotive topcoating.

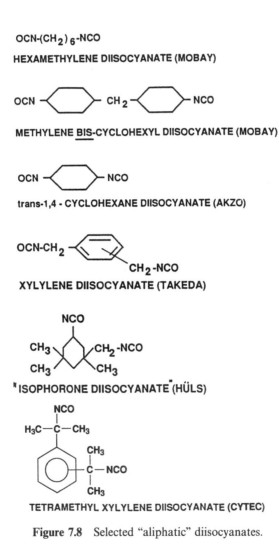

Figure 7.8 Selected "aliphatic" diisocyanates.

7.4 ALKYLBENZENES

The most important alkylbenzenes are those with C_{10}–C_{14} side chains. They are sulfonated to provide the alkylbenzenesulfonate surfactants, useful in detergent formulations. The equation is written for an internal olefin:

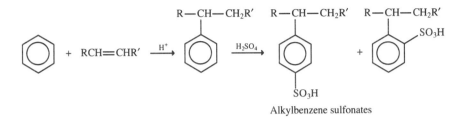

Alkylbenzene sulfonates

There are several sources of the side chains for the Friedel–Crafts alkylation of benzene for surfactants. Initially propylene tetramer, a highly branched dodecene, was used and this is still important in developing countries (Section 4.2). In the United States and Western Europe, it has been outlawed because branched-chain surfactants biodegrade very slowly with resultant foaming in rivers and sewage plants.

The move to biodegradable detergents made necessary the manufacture of linear olefins. High molecular weight paraffins (C_{20}–C_{30}), which occur in the wax separated from lubricating oil, can be separated into linear and branched molecules by use of molecular sieves, particularly in UOP's Molex process. The linear fraction can then be steam-cracked, just as lower hydrocarbons can, to provide olefins with both even and odd numbers of carbon atoms with chain lengths varying from C_6 to C_{18}. These are impure but suitable for surfactants. Alternatively, paraffins from C_6 to C_{19} may be dehydrogenated catalytically to internal olefins.

The Friedel–Crafts alkylation of benzene can also be carried out with alkyl chlorides (monochloroparaffins) made by chlorinating *n*-alkanes with 10–15 carbon atoms. The dehydrochlorination of the monochloroparaffins also provides useful olefins. The supply of wax, however, is limited, which makes necessary the use of olefins from ethylene oligomerization (Section 3.2). Either linear or internal olefins with chain lengths of C_{10}–C_{13} can be used, the major source of the latter being the Shell SHOP process (Section 3.3.3). From internal olefins, as in the equation above, a molecule with two side chains results, but since both of these are linear they are biodegradable.

7.5 MALEIC ANHYDRIDE

Most maleic anhydride is made by the oxidation of butane or, less frequently, 1- and 2-butenes (Section 5.4). An older method still used in Europe is based on the oxidation of benzene and is analogous to the process for phthalic anhydride from naphthalene (Sections 9.1.1 and 12.1). The vapor-phase oxidation is carried out with a supported vanadium pentoxide catalyst at 400°C. A mixture of maleic acid and maleic anhydride results, and the acid may be dehydrated to the anhydride directly without separation. The product is purified by batch vacuum distillation. The uses for maleic anhydride were described in Section 5.4.

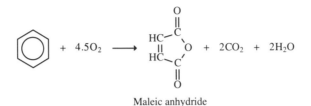

Maleic anhydride

7.6 CHLORINATED BENZENES

Chlorobenzene results from the liquid-phase chlorination of benzene at slightly above room temperature with a ferric chloride catalyst. Small amounts of di- and trichlorobenzene form. Oxychlorination achieves the same result (Section 3.4) with hydrogen chloride and air instead of chlorine, and an alumina-supported cupric chloride-ferric chloride catalyst. This was in fact the first use for oxychlorination, which today makes a key contribution to vinyl chloride manufacture (Section 3.3). Conversion is kept at 10–15% to suppress the formation of di- and polychlorobenzenes and to control the reaction's exothermicity.

Chlorobenzene is used primarily as a solvent (cf. phosgenation of methylenedianiline, Section 7.2.4). It may be converted to aniline (Section 7.2.3) and has been used in the past as a raw material for phenol (Section 7.1). Chlorobenzene condenses with trichloroacetaldehyde (chloral) to yield the insecticide 1,1,1-trichloro-2,2-bis(p-chlorophenyl)ethane (DDT).

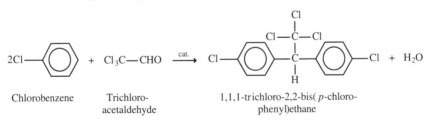

| Chlorobenzene | Trichloro-acetaldehyde | 1,1,1-trichloro-2,2-bis(p-chloro-phenyl)ethane |

The DDT was remarkably effective in mosquito control but was banned because its stability and lipophilicity caused it to accumulate in body fat. Many strains of mosquitos have anyhow developed enzymes for DDT hydrolysis, making it useless in most places.

Dichlorobenzenes are small-volume chemicals which, like many chlorine derivatives, are being phased out. p-Dichlorobenzene is used as a moth repellent and deodorant. On reaction with sodium sulfide, the specialty polymer, poly(phenylene sulfide) results. Its recurring unit is

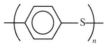

o-Dichlorobenzene is a solvent for toluene diisocyanate.

7.7 DIHYDROXYBENZENES

Of the three dihydroxybenzenes—hydroquinone, resorcinol, and catechol—hydroquinone is the most important. Numerous processes have been proposed for its preparation. One is analogous to the cumene-based phenol process (Section 4.5). *p*-Diisopropylbenzene is prepared by alkylating cumene with isopropyl chloride in the presence of aluminum chloride at 90°C to achieve a 98% molar yield. Peroxidation takes place at 90–100°C with air in the presence of dilute sodium carbonate. A mixture of mono- and dihydroperoxides results. Extraction with aqueous alkali separates the two. The oxidation gives 10% conversion and the ultimate molar yield of dihydroperoxide is 65%. It is decomposed in acetone solution with dilute sulfuric acid at 90°C and 2 bar. Hydroquinone is formed in an overall yield of 60 mol%. Acetone is a coproduct.

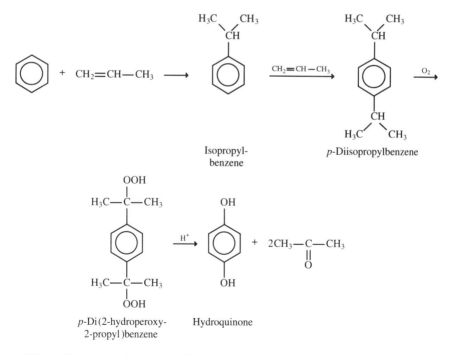

The earliest procedure for hydroquinone manufacture but still in use involves aniline sulfate oxidation (Section 7.3) to benzoquinone in the presence of manganese dioxide. Reduction of the quinone with iron and sulfuric or hydrochloric acids provides hydroquinone and iron salts, which can be converted to ferric oxide, useful as a pigment but whose consumption is declining.

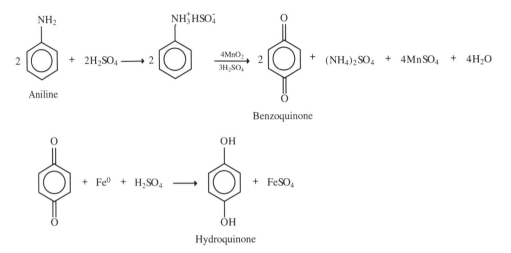

Aniline   Benzoquinone

Hydroquinone

The initial oxidation is carried out below 10°C. Chromic acid may be used instead of manganese dioxide. In either instance, ammonium sulfate residues provide a disposal problem.

The electrolytic oxidation of benzene to hydroquinone has never been commercialized although it appears attractive economically. The benzene is oxidized continuously at the anode of an electrolytic cell to produce benzoquinone, which is concurrently reduced to hydroquinone at the cathode.

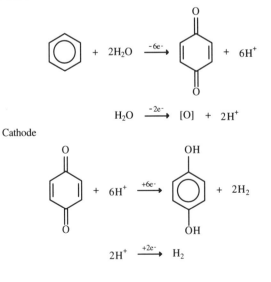

Overall

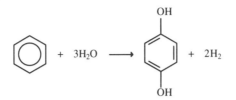

The direct hydroxylation of phenol with hydrogen peroxide provides a mixture of catechol and hydroquinone. The process is believed to be used in Europe and Japan.

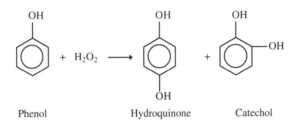

A process based on bisphenol A involves its conversion to *p*-isopropenylphenol either by a base-catalyzed or thermal decomposition. Reaction of *p*-isopropenylphenol with hydrogen peroxide yields hydroquinone and acetone.

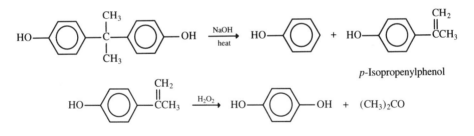

A Reppe process involves simultaneous cyclization and carbonylation of acetylene with iron or cobalt catalysts. A more modern version makes use of ruthenium or rhenium catalysts at 600–900 bar with hydrogen instead of water. If carbonyl cluster catalysts are used, lower pressures are possible at temperatures of 100–300°C. These processes have not been commercialized.

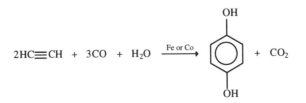

A biotechnological route to hydroquinone and benzoquinone is described in the literature (see note to Section 16.8).

The major uses for hydroquinone are as an antioxidant and antiozonant in rubber manufacture and in chemical formulations for the development of photographic films. It is also a polymerization inhibitor in the manufacture and storage of monomers such as acrylonitrile, vinyl acetate, and acrylic and methacrylic esters. It is a precursor of butylated hydroxyanisole (BHA) a food grade antioxidant prepared by alkylating the monomethyl ether of hydroquinone with isobutene. A related food grade antioxidant is butylated hydroxytoluene (BHT) (Section 5.2.6). Growth in hydroquinone usage is not expected because tires are becoming smaller and new photographic films require less hydroquinone for developing.

Resorcinol is made by benzene disulfonation analogous to an obsolete process for phenol manufacture (Section 7.1). Isomers are avoided because of the meta directiveness of the sulfonic acid group. Monosulfonation is carried out at 100°C with 100% sulfuric acid made by adding oleum to 96% acid (Fig. 7.9). Oleum is used at 80°C for the second sulfonation because the water produced dilutes the acid and, if the strength drops below 80%, sulfonation ceases. The disulfonic acid is neutralized with sodium sulfite or sodium carbonate to produce SO_2 or CO_2. Sodium sulfite is preferred because it is a byproduct of the overall reaction and can be recycled.

Sodium sulfate is also formed as a byproduct because excess sulfuric acid is used, which must be neutralized. A variation of the neutralization step, which eliminates the excess sulfuric acid, involves the use of calcium carbonate to complete the 90% neutralization achieved with sodium sulfite. A 9:1 mixture of sodium and calcium salts of the benzenedisulfonic acid is obtained, which is further treated with sodium carbonate. The sodium displaces the calcium and produces calcium carbonate together with the completely neutralized sodium salt of benzenedisulfonic acid.

This sodium salt is fused with sodium hydroxide in a batch process at 300°C at a 4:1 molar ratio of caustic to sodium salt. The resulting mixture is neutralized either with sulfuric acid or with the sulfur dioxide evolved in the initial neutralization of the benzenedisulfonic acid. A mixture of resorcinol, sodium sulfate, and sodium sulfite results. Resorcinol may be extracted from the aqueous solution by several solvents including diisopropyl ether.

Resorcinol is used in the formulation of high-performance adhesives; primarily resorcinol–formaldehyde condensates, for the rubber and wood product industries and for bonding tire cord to the rubber tire matrix. These adhesives have better properties than phenol–formaldehyde resins (Section 7.1.1) for the production of laminated beams where high strength is required and the application of heat is impractical. Resorcinol is also a raw material for the production of pharmaceuticals, dyes, and ultraviolet absorbers. Typical is its conversion to m-aminophenol by reaction with ammonia, a reaction analogous to the amination of phenol to aniline (Section 7.1.5).

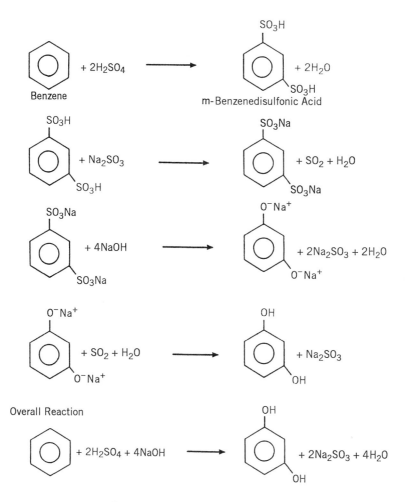

Figure 7.9 Resorcinol from benzene.

o-Dihydroxybenzene (catechol or pyrocatechol) results from the caustic hydrolysis of *o*-chlorophenol in the presence of copper powder at 190–230°C and 3–6 bar. In the process, a chlorine atom is replaced by hydroxyl. Conversion is 69%, of which 89% is catechol, and 11% resorcinol and higher phenols.

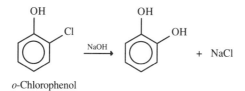

o-Chlorophenol

A second route to catechol involves the direct hydroxylation of phenol with hydrogen peroxide in the presence of formic and phosphoric acids. A mixture of catechol and hydroquinone results in a ratio of 3:2. Formic acid alone favors increased formation of hydroquinone. The two products are separated by distillation.

A third synthesis involves the chlorination of cyclohexanone to 2-chloro-cyclohexanone. Caustic hydrolysis of the chlorine provides 2-hydroxycyclo-hexanone, which can then be dehydrogenated to catechol.

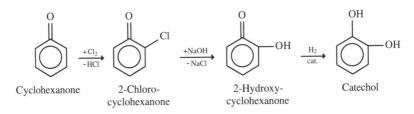

Cyclohexanone   2-Chloro-   2-Hydroxy-   Catechol
                     cyclohexanone   cyclohexanone

The major use for catechol is for the preparation of a carbamate insecticide, propoxur, used to kill household pests.

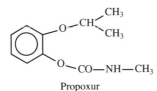

Propoxur

7.8 ANTHRAQUINONE

The Friedel–Crafts condensation of benzene and phthalic anhydride (Section 9.1) in the presence of aluminum chloride at 25–60°C produces benzoylbenzoic acid. The product is cyclized with oleum to anthraquinone, which is purified by sublimation at 380°C. Yields are of the order of 95%.

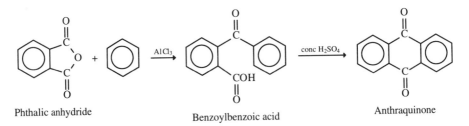

Phthalic anhydride   Anthraquinone
                     Benzoylbenzoic acid

A second route to anthraquinone involves the direct oxidation of anthracene, a process still practiced in Europe, where high purity anthracene has tradition-ally been available from coal tar fractionation.

In a third process, styrene is dimerized to methylphenylindane, which can be oxidized in the vapor phase to anthraquinone. The dimerization is catalyzed by sulfuric acid at reflux temperature. The vapor-phase conversion to anthraquinone takes place over a promoted vanadium pentoxide catalyst.

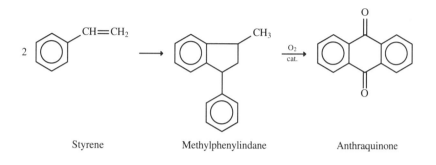

Styrene   Methylphenylindane   Anthraquinone

Anthraquinone has been used primarily for the production of dyes and pigments. Tetrahydroanthraquinone is used as a catalyst to facilitate the chemical pulping of wood for paper manufacture. This technology has been pioneered in Japan where it is used primarily with Kraft pulping, which employs sodium hydroxide and sodium sulfite to disengage lignin from the wood and leave cellulose fibers. In conventional pulping, the 1,4-glycosidic linkages of cellulose are hydrolyzed and reduce the yield of pulp. Tetrahydroanthraquinone apparently accelerates the delignification process while inhibiting attack on the cellulose. Higher yields of pulp result.

Ethylanthraquinone is the catalyst in the auto-oxidation route for the manufacture of hydrogen peroxide. It is reduced by hydrogen on palladium to the hydroquinone, and then reoxidized with air to give hydrogen peroxide and the original anthraquinone.

NOTES AND REFERENCES

The standard work on benzene is still *Benzene and its Industrial Derivatives*, E. G. Hancock, Ed., Benn, London, 1975. More recent is H-G. Franck and J. W. Stadelhofer, *Industrial Aromatic Chemistry*, Springer, Berlin, 1988. There is a slim paperback *Benzene*, VCH Publishers, 1993, but no author is listed.

Catalytic reforming (Section 2.2.3) works best with C_7 and C_8 molecules but less well with C_6 alkanes, which tend to form cracked products, thus reducing the yield of benzene. Also higher temperatures are required to reform C_6 alkanes. Thus, by using lower temperatures, the C_7 and C_8 molecules reform preferentially. If the naphtha is stripped before reforming most of the C_6 alkanes are removed, so that benzene cannot form. Even so, benzene results in the reforming process from hydrodealkylation of the toluene and xylenes.

In the early 1990s, Chevron instituted a catalytic reforming process called Aromax, based on a metal-doped zeolite catalyst, which reforms the C_6 fraction more effectively to give a higher yield of benzene.

Section 7.1   The Japanese process for producing phenol only from cumene has been described by Mitsui Petrochemical in Japanese Patents 300 300 (November 28, 1988) and 328 722 (December 26, 1988).

Benzene acetoxylation has been described by J. Davidson and C. Triggs, *Chem. Ind.* 1361 (1976)' and in West German Offen. 1 643 355 (1967).

Section 7.1.2   The rearrangement of bisphenol A byproducts to bisphenol A is described in US Patent 3 221 061 (November 30, 1965) to Union Carbide.

Section 7.1.2.2   The preparation of dimethyl carbonate is described in US Patent 4 360 477 (November 23, 1982) to General Electric. The use of this reagent in the polycarbonate synthesis is described in US Patent 4 452 968 (June 5, 1984) to General Electric. See also *Chem. Brit.*, **30**, 970 (1994)

Section 7.1.2.3   Membrane processes for hydrogen removal are described by J. Haggin, *Chem. Eng. News* June 6 (1988) p. 7.

Section 7.2.1   The microbiological route from toluene to adipic acid is described in Celanese patents including US Patent 4 355 107 (October 19, 1982) and European Patent Applications 0 074 168 (March 16, 1983) and 0 117 048 (August 29, 1984).

The Gulf process for adipic acid formation in one step is described in US Patent 3 231 608 (January 25, 1966) and US Patent 4 263 453 (April 21, 1981).

Section 7.2.1.1   Nylon 4,6 is described in US Patent 4 408 036 (October 4, 1983) to Stamicarbon BV.

Section 7.2.2   An ENI process, announced while this book was in press, involves the direct conversion of cyclohexanone to an oxime with ammonia and hydrogen peroxide.

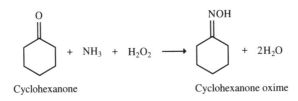

Cyclohexanone   Cyclohexanone oxime

DSM-Du Pont also announced a new process based on butadiene. The first step, a carbonylation, is the same as the first step in BASF's adipic acid synthesis (Section 5.1.3.4) and gives methyl 3-pentenoate (**I**). This is hydroformylated with a cobalt catalyst rather than the more efficient rhodium catalyst, because it shifts the double bond to the terminal position to provide the aldehyde (**II**). Reductive amination introduces an amine group to give an aminoester (**III**), and this cyclizes to caprolactam (**IV**) with the elimination of methanol. No ammonium sulphate is produced.

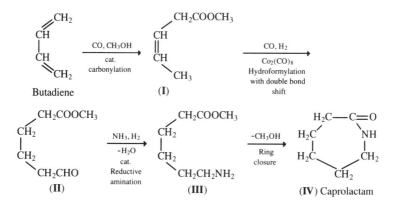

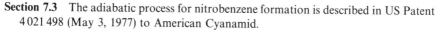

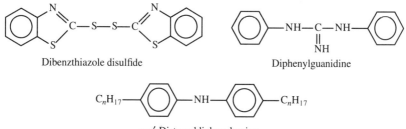

Section 7.3   The adiabatic process for nitrobenzene formation is described in US Patent 4 021 498 (May 3, 1977) to American Cyanamid.

Aniline is the basis for several rubber compounding agents. Thus dibenzthiazole disulfide is a primary accelerator. Diphenylguanidine is a secondary accelerator, and *p, p'*-distearyldiphenylamine is an antioxidant and antiozonant.

Dibenzthiazole disulfide   Diphenylguanidine

p,p'-Distearyldiphenylamine

Section 7.3.1   The ARCO process for carbonylation of nitrobenzene is described in a number of patents including US patents 3 595 054 (July 15, 1975); 3 962 302 (June 8, 1976); 4 041 139 (May 8, 1977); and 4 038 377 (August 1, 1977).

The Asahi process is described in numerous Japanese Patents including 58-159751 (1983); 57-158746-8 (1982); 58-67660 (1983); and 58-96054 (1983).

Section 7.7   The electrolytic oxidation of benzene to hydroquinone is a URBK-Wesseling process described in M. Fremery, H. Höver and G. Schwarzlose, *Chem Ing. Tech.*, **46**, 635 (1974). Related processes have been patented by Du Pont [US Patent 3 884 776 (June 20, 1975)]. The electrolytic oxidation of phenol is described in patents to Union Carbide [French Patent 1 5444 350 (October 31, 1968)] and to Eastman Kodak [US Patent 4 061 548 (December 6, 1977)]. Phenol oxidation with hydrogen peroxide is also described by Goodyear in US Patents 3 870 731 (March 11, 1975) and 3 859 317 (January 7, 1975). The process involving *para*-isopropenylphenol is patented by Upjohn in West German Offen. 2 214 971 (November 16, 1972).

Section 7.8   The use of anthraquinone for pulping of wood has been described by J. M. MacLoed, *TAPPI Procs*, 1983 Pulping Conference.

CHAPTER 8

CHEMICALS AND POLYMERS FROM TOLUENE

Toluene is the major product from catalytic reforming (Section 2.2.3). The chemical industry requires considerably more benzene than toluene, but toluene is preferred for unleaded gasoline because of its low toxicity and high octane number. Catalytic reforming is the major source of toluene in the United States but in Western Europe pyrolysis gasoline (Section 2.2.1) is the major source.

Just as propylene has traditionally been cheaper than ethylene, toluene has been cheaper than benzene. Propylene's ready availability led to the development of a series of brilliant chemical reactions. Attempts to find broad usage for toluene were less successful, and its main chemical use is hydrodealkylation to benzene. This is a "swing" operation and occurs when the price differential is sufficient. It consumes 40–55% of the toluene isolated from catalytic reforming in the United States.

With a US consumption in 1993 of about 6.4 billion lb for chemical uses, toluene is the smallest of the seven basic chemicals. The fact that so much of it is used to augment benzene and to a lesser extent xylenes output (by disproportionation, Section 8.1) indicates that even this figure of 6 billion lb exaggerates its chemical importance.

8.1 HYDRODEALKYLATION AND DISPROPORTIONATION

In situ hydrodealkylation during catalytic reforming was discussed in Section 2.2.3 as an additional source of benzene in that reaction. Hydrodealkylation in a dedicated plant may be purely thermal or it may be catalyzed by metals or supported metal oxides. Hydrogen is always present. Typical reaction conditions are 600°C and 40–60 bar over oxides of chromium, platinum, molybdenum, or cobalt, supported on alumina. The uncatalyzed reaction requires a higher temperature of up to 800°C and a pressure up to 100 bar.

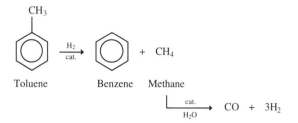

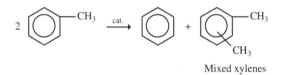

Methane forms and in one version of the process is subjected to the synthesis gas reaction (Section 10.4) to provide hydrogen for the hydrodealkylation. Despite the cost of catalyst, the catalytic method is preferred because it allows for higher conversion and gives greater selectivity so that 92+ % benzene results.

A variation of hydrodealkylation is disproportionation. Two molecules of toluene react to give one of benzene and one of mixed xylenes.

Disproportionation is important as a source of mixed xylenes from which the *para* isomer, the one most in demand, can be isolated. The reaction is carried out in the vapor phase in the presence of a non-noble metal catalyst. Even a casual estimate of the energetics of the reaction indicates that the equilibrium constant is close to unity, so large volumes of reactants must be recycled from the product-recovery section of the plant to the reactor. The advantage of the process, compared with catalytic reforming is that the reactor effluent is free from ethylbenzene. This makes the separation of the xylene isomers easier.

A dramatic improvement is Mobil's development of a liquid-phase process based on an antimony-doped zeolite ZSM-5 catalyst at 300°C and 45 bar. A xylene fraction results with about 80% of the desired *p*-xylene. The disproportionation takes place inside a zeolite cage whose "mouth" is shaped such that benzene, toluene and *p*-xylene have ready access but *o*- and *m*-xylenes cannot escape (Section 16.9). Accordingly, these isomers remain in the cage to undergo further isomerization and to supply the equilibrium quantities required each time 2 mols of toluene disproportionate.

Disproportionation has assumed importance in the United States in the mid-1990s and 3 billion lb of benzene and xylenes were produced in 1995 by this process. The figure is expected to rise to 4.2 billion by the year 2000.

Another disproportionation, called transalkylation, involves the interaction of toluene with trimethylbenzene or mesitylene, a compound produced during catalytic reforming.

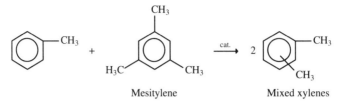

Mesitylene   Mixed xylenes

A methyl group migrates from a trimethylbenzene molecule to a toluene molecule giving two molecules of mixed xylenes. In practice, benzene is also produced, but its volume can be kept low by use of a high ratio of trimethylbenzene to toluene.

8.2 SOLVENTS

Toluene's second largest use is as a solvent mainly for coatings. The finding in the United States that benzene is carcinogenic has increased toluene's solvent usage at the expense of benzene, although the latter is in any case small. Toluene is a primary solvent for coatings based on medium and short oil alkyd resins but is a so-called latent solvent for nitrocellulose lacquers, which require as primary solvents polar compounds such as esters, ketones and glycol ethers.

8.3 DINITROTOLUENE AND TOLUENE DIISOCYANATE

The largest outlet for toluene in which its chemical properties are of value in their own right is as a raw material for a mixture of 2,4- and 2,6-toluene diisocyanate (tolylene diisocyanate, diisocyanatotoluene, TDI) used for polyurethane resins. TDI is made by chemistry similar to that used for 4,4'-diphenylmethane diisocyanate (MDI) (Section 7.3.1). Some of the non-phosgene routes described there are applicable to TDI, although less development work has been done.

Commercial TDI contains about 80% of 2,4- (*o, p*) isomer and 20% of the 2,6- (*o, o*) isomer. The three-step process comprises the dinitration of toluene; the hydrogenation of the nitro compounds to diaminotoluenes; and the reaction of these with phosgene to commercial grade TDI. Toluene is nitrated with a mixture of nitric and sulfuric acids in two stages. The first produces the three isomers of mononitrotoluene.

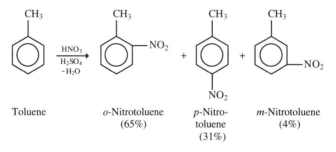

| Toluene | *o*-Nitrotoluene (65%) | *p*-Nitrotoluene (31%) | *m*-Nitrotoluene (4%) |

These are subsequently nitrated further to obtain the six possible dinitro-toluene isomers. The 2,4 isomer comprises 74–76% of the mixture and the 2,6 isomer about 19–21%. The concentrations of the other four isomers are minimal. The 3,4-compound forms at a level of 2.4–2.6%. The 2,3, 2,5, and 3,5 isomers are present at a level of no more than 1.7%. The concentration of the acids is carefully controlled so that very little of the trinitro isomers forms.

The dinitrotoluenes are dissolved in methanol and hydrogenated continuously to diaminotoluenes by reaction with hydrogen in the presence of Raney nickel at 150–180°C and 65–130 bar. Numerous other catalysts such as supported platinum or palladium may be used.

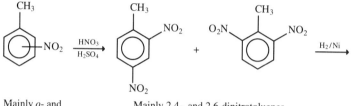

Mainly *o*- and   Mainly 2,4- and 2,6-dinitrotoluenes
p-nitrotoluenes

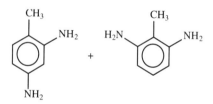

Mainly 2,4- and 2,6-diaminotoluenes

The possibility of eliminating the nitration step by an intermolecular amination of toluene with ammonia is of some interest. Similar chemistry has been discussed for the possible preparation of aniline (Section 7.3) and *m*-aminophenol (Section 7.7).

The carbonylation with phosgene occurs in two stages, the first yielding a carbamoyl chloride and the second the isocyanate. In both reactions, hydrogen chloride is liberated. The reaction is carried out with a 10–20% solution of the diamine mixture in chlorobenzene (Section 7.6). This is combined with a chlorobenzene solution of phosgene.

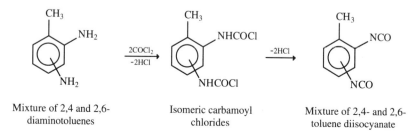

Mixture of 2,4 and 2,6-   Isomeric carbamoyl   Mixture of 2,4- and 2,6-
diaminotoluenes   chlorides   toluene diisocyanate

Carbamoyl chloride formation occurs at 0–30°C and isocyanate formation at 160–180°C. Conversion is about 80%. Treatment of the residue with alkali provides unreacted diaminotoluene for recycle.

Dinitrotoluene is also used as a gelatinizing and waterproofing agent in explosive compositions. Additional nitration gives trinitrotoluene (TNT), an explosive formerly used in military and civilian applications but now of importance only to the military. It is safer than picric acid (Section 7.1) because it does not form detonation-sensitive salts with metals and has a lower melting point (80°C) so that it can be conveniently loaded into shells in the molten state. For civilian explosives, such as those used in mining, ammonium nitrate is the preferred material.

8.4 LESSER VOLUME USES OF TOLUENE

Toluene may be converted to phenol (Section 7.1). It also provides the starting material for one route to caprolactam (Section 7.2.2). In a reaction analogous to the formation of allyl chloride from propylene (Section 4.10.1) the methyl group of toluene can be chlorinated to yield benzyl chloride, which may be used to quaternize tertiary amines such as lauryldimethylamine to give germicidal compounds. Its main use is as a raw material for the minor poly(vinyl chloride) (PVC) plasticizer, butyl benzyl phthalate. The presence of the aromatic ring appears to confer stain resistance to PVC floor coverings. Dichlorination of toluene leads to benzal chloride, which on hydrolysis provides benzaldehyde, an ingredient of flavors and perfumes. This is the most important route to benzaldehyde, which is also a byproduct of the toluene-to-phenol process (Section 7.1).

Benzal chloride   Benzaldehyde

It may also be made by direct oxidation of toluene. This is usually accomplished in the liquid phase with air at about 100°C and 3 bar with cobalt salt catalysts. In another process the oxidation is carried out in acetic acid with cobalt and manganese acetates as catalysts. This is analogous to the process used for the preparation of terephthalic acid (Section 9.3).

The oxidation of toluene gives not only benzaldehyde but also benzyl alcohol, just as the oxidation of propylene provides acrolein and allyl alcohol (Sections 4.3 and 4.10.2), although the mechanisms are different. These normally occur as recyclable byproducts during benzoic acid formation. If benzyl alcohol is required by direct oxidation, the reaction is carried out in the presence of acetic acid, which captures the alcohol as benzyl acetate before it is oxidized further to benzoic acid.

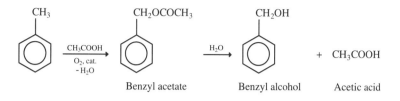

The most important route to benzyl alcohol, however, is by the hydrolysis of benzyl chloride.

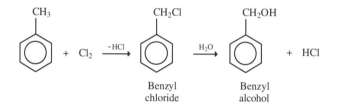

Benzoic acid is a halfway stage in a minor process for phenol manufacture (Section 7.1), but benzoic acid itself has a few small uses. Diethylene glycol dibenzoate is a useful nonstaining plasticizer for PVC flooring where it competes with butyl benzyl phthalate. Butyl benzoate is a perfume ingredient. Benzoic acid is an intermediate in a caprolactam synthesis (Section 7.2.2) and in a process for the production of terephthalic acid known as the Henkel II reaction. This involves the disproportionation of potassium benzoate to benzene and the potassium salt of terephthalic acid. It is noteworthy because of the *para* directiveness of the potassium ion. The process has been used in Japan.

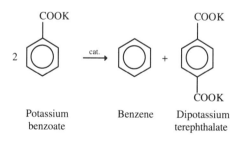

The conversion of toluene to *p*-methylstyrene has already been described (Section 3.8). There is, however, an uncommercialized but chemically interesting process for styrene from toluene. Toluene is dehydrocoupled to stilbene followed by a metathesis (Section 2.2.9) reaction with ethylene. The first step takes place at 600°C in the presence of a lead magnesium aluminate catalyst and the second at 500°C with a calcium oxide–tungsten oxide catalyst on silica.

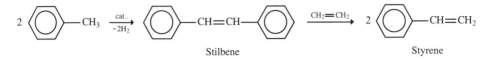

The carbonylation of toluene leads to *p*-tolualdehyde, which can be oxidized to terephthalic acid (Section 9.3.2). The reaction is catalyzed by a combination of HF and BF$_3$ to give selectivities of about 97 at 85% conversion. This process has not been commercialized, in part because in the 1970s there was excess capacity for terephthalic acid production, which continued until 1989 when a large market for poly(ethylene terephthalate) evolved in the Far East.

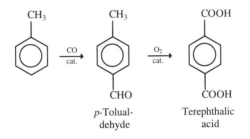

p-Tolual- dehyde	Terephthalic acid

NOTES AND REFERENCES

Toluene is discussed in Franck and Stadelhofer, cited in the bibliography to Chapter 7. There is yet another standard work edited by E. G. Hancock, *Toluene, the Xylenes and their Industrial Derivatives*, Elsevier, Amsterdam, 1982.

Section 8.1   Typical of the many patents describing the disproportionation of toluene to *p*-xylene is West German Offen. 2 633 881 (June 2, 1977) to Mobil Oil.

Section 8.4   The conversion of toluene to styrene is described in numerous patents to Monsanto, including the following: West German Offen. 2 500 023 (July 10, 1975), US Patent 3 965 206 (June 22, 1976), US Patent 4 091 044 (May 23, 1978), West German Offen. 2 748 018 (May 11, 1978), British Patent 1 578 994 (October 26, 1977), US Patent 4 243 825 (January 6, 1981), US Patent 4 247 727 (January 27, 1981), US Patent 4 254 293 (March 3, 1981), US Patent 4 255 602 (March 10, 1981), US Patent 4 255 603 (March 10, 1981), US Patent 4 255 604 (March 10, 1981), US Patent 4 268 703 (May 19, 1981), US Patent 4 268 704 (May 19, 1981).

The oxidative coupling of toluene to stilbene and stilbene's subsequent metathesis with ethylene to styrene is described in US Patent 3 980 580 and *CHEMTECH*, **7**, 140 (1977).

The production of terephthalic acid by toluene carbonylation has been claimed in three patents issued to Mitsubishi Gas Chemical Co. as follows: West German Offen. 2 422 197 (November 28, 1974), 2 460 673 (July 3, 1975), and 2 425 571 (December 12, 1974). The oxidation of the tolualdehyde to terephthalic acid is claimed in West German Offen. 1 943 510 (June 18, 1970), also assigned to Mitsubishi Gas Chemical Co.

CHAPTER 9

CHEMICALS AND POLYMERS FROM XYLENES

Xylenes are produced mainly by catalytic reforming (Section 2.2.3). This is so even in Europe where most of the benzene and about one-half of the toluene comes from pyrolysis gasoline (Section 2.2.1). The xylenes are isolated only with difficulty from pyrolysis gasoline because it contains about 50% ethylbenzene with a similar vapor pressure. Accordingly, the C_8 fraction of pyrolysis gasoline is usually returned to the gasoline pool.

A comparison of the C_8 fractions from catalytic reforming and pyrolysis gasoline was shown in Table 2.10 together with the end use requirement. The chemical industry needs to separate the isomers, isomerize the unwanted ones to an equilibrium mixture, and repeat the process to extinction.

The separation of the three xylene isomers and ethylbenzene from each other is an awesome task. Their physical constants are shown in Table 9.1. The boiling points of all four compounds are within 9°C. o-Xylene boils more than 5°C above the others, however, and may be separated by fractional distillation on a huge column with 150–200 plates and a high-reflux ratio. The mixture at the top of the column contains about 40% m-xylene, 20% p-xylene, and 40% ethylbenzene. If required, the low-boiling ethylbenzene can be removed by an

TABLE 9.1 Physical Constants of the C_8-Stream

	Melting Point (°C)	Boiling Point (°C)
o-Xylene	− 25.2	144.4
m-Xylene	− 47.9	139.1
p-Xylene	13.2	138.3
Ethylbenzene	− 95.0	136.2

involved extractive distillation. Energy costs are high, and the ethylbenzene is sometime allowed to remain for further processing.

The isomers in the intermediate fraction differ markedly in melting point and, in the oldest process, are separated by low-temperature cyrstallization. The mixture is carefully dried to avoid icing of the equipment and then it is cooled. Crystallization of p-xylene starts at $-4°C$ and continues until $-68°C$, at which point the p-xylene/m-xylene eutectic starts to separate so the procedure is halted. The first stage only raises the p-xylene concentration in the crystalline mass to 70% but a series of melting, washing, and recrystallization steps eventually raises this to 99.5%.

If m-xylene is required, it may be extracted from the C_8 stream by complex formation. Treatment of the stream with $HF-BF_3$ gives two layers. The m-xylene selectively dissolves in the $HF-BF_3$ layer as the complex $C_6H_4(CH_3)_2 \cdot HBF_4$. The phases are separated and the m-xylene regenerated by heating.

The drawbacks of the low-temperature crystallization are the high energy requirements for cooling and the problems of handling the solid p-xylene which, for example, deposits on the walls of the cooling vessel and reduces the rate of heat transfer. Nonetheless, it is still in use.

The more recent processes for separating the m- and p-xylenes make use of molecular sieves for which the feed components show small differences in affinity. The UOP Parex process is the most widely used. It is analogous to the one used to separate linear and branched-chain hydrocarbons (Section 7.4). It is based on a continuous countercurrent flow of liquid and solid adsorbent. This is achieved in a novel way. The solid bed of adsorbent cannot easily be moved countercurrent to the liquid flow. Instead, countercurrent flow is simulated with a stationary bed of adsorbent by periodically displacing the positions at which the process streams enter and leave the bed. That is, the positions of the liquid feed and withdrawal points are shifted in the same direction as the fluid flow down the bed.

The exit streams from both the low-temperature crystallization and the adsorption processes contain unwanted products, primarily m-xylene, ethylbenzene (if it was not removed earlier) and the portion of o-xylene that the market does not require. They are catalytically isomerized in the presence of acid catalysts to provide another equilibrium mixture, which is somewhat more favorable in that it contains about 48% m-xylene, 22% each of o- and p-xylenes, and 8% ethylbenzene. Acidic catalysts include silica–alumina, and silica with $HF-BF_3$, the same material that complexes with m-xylene. The drawback of silica–alumina is that it promotes disproportionation and transalkylation. If platinized alumina is added to the silica–alumina, as in a dual function catalyst, the system will isomerize ethylbenzene as well as the xylenes.

The most important isomerization catalyst today is the zeolite, ZSM-5. This is the same zeolite catalyst useful for toluene disproportionation to benzene and p-xylene (Section 8.1). The silica–alumina/platinized alumina catalyst operates at 23–33 bar in a hydrogen atmosphere with substantial recycle. The ZSM-5 process has a major economic advantage because it operates at low pressures either in the vapor or liquid phase and requires less or even no hydrogen and much less recycle than does the noble metal catalyst. Its one disadvantage is that it does not isomerize ethylbenzene but rather dealkylates it to benzene. Because the ZSM-5 process requires less capital investment and provides lower operating costs, it has been widely accepted. A recently announced but not commercialized process makes use of non-zeolite molecular sieves, primarily silicoaluminophosphates. This process is claimed not only to isomerize ethylbenzene to xylenes but also to provide a higher level of the desired *p*-xylene.

Mixed xylenes are used as solvents, particularly in the paint industry, and are valued components of the gasoline pool because of their high octane number. The major chemical use for the individual xylenes is oxidation to terephthalic acid, isophthalic acid, and phthalic anhydride.

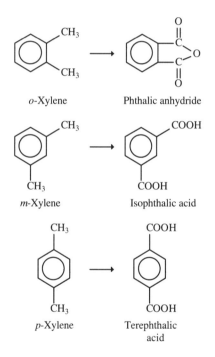

o-Xylene   Phthalic anhydride

m-Xylene   Isophthalic acid

p-Xylene   Terephthalic acid

9.1 *o*-XYLENE AND PHTHALIC ANHYDRIDE

o-Xylene may be oxidized to phthalic anhydride in gas- or liquid-phase processes and in fixed or fluidized beds. The liquid-phase oxidation makes use of metal salt catalysts that are soluble in the reaction medium. The gas-phase fluidized bed process takes place at 375–410°C over a vanadium pentoxide catalyst. The process is highly exothermic, and the fluidized bed offers better temperature control and less risk of explosion. It also enables the handling of phthalic anhydride as a liquid (mp 130.8°C). Yields are rather low, probably less than 80%, and there are side reactions leading to *o*-toluic acid, phthalide, benzoic acid, and maleic anhydride as well as the complete oxidation to carbon dioxide and water.

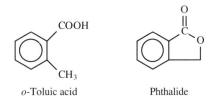

o-Toluic acid   Phthalide

As a result of the low yields, an older route to phthalic anhydride, based on naphthalene (Section 12.1) has remained viable. The oxidation is similar, again with a vanadium pentoxide catalyst. Indeed some plants can operate with either feedstock. The reaction is analogous to the oxidation of benzene to maleic anhydride (Section 5.4).

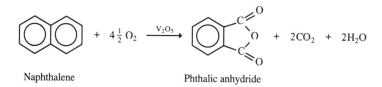

Naphthalene   Phthalic anhydride

The *o*-xylene process is somewhat less exothermic (265 kcal mol^{-1} as against 429 kcal mol^{-1}) because there are fewer carbon atoms to be oxidized. The loss of two carbon atoms from naphthalene as carbon dioxide is an economic drawback of the process, which is balanced by the higher yields.

Naphthalene is available from the distillate from coke ovens and is one of the few coal tar chemicals still of importance. It is also made by the hydrodealkyation of the methylnaphthalenes produced by the catalytic reforming (Section 2.2.3) of gas oil fractions. About 8.0% of US phthalic anhydride comes from naphthalene. The proportion is higher in Europe.

9.1.1 Uses of Phthalic Anhydride

About one-half of phthalic anhydride production is used in plasticizers (Section 4.8.2). One-quarter goes into alkyd resins and an equal amount into unsaturated polyesters. All three of these applications involve esterification of phthalic anhydride. As with all anhydrides, the esterification proceeds in two stages. The first, shown below with 2-ethylhexanol, the most important plasticizer alcohol, goes readily to form the half ester, whereas the second step proceeds only with the aid of a catalyst such as *p*-toluenesulfonic acid and a temperature of 160°C.

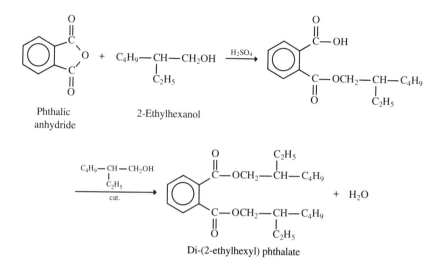

Di-(2-ethylhexyl) phthalate

At a temperature between 185 and 205°C, a noncatalytic esterification takes place, obviating the catalyst removal steps. With both routes color development must be avoided. The plasticizer has a high boiling point and, once formed, is difficult and expensive to distil or otherwise purify.

Alkyd resins are oligomers in which polyester functions have been inserted into natural "drying oils." Before 1929, gloss paints were based on drying oils themselves. These are triglycerides containing unsaturated fatty acids (Section 13.1). Linseed oil, for example, is mainly linolenic acid triglyceride. Linolenic acid is

$$CH_3-CH_2-CH{=}CH-CH_2-CH{=}CH-CH_2-CH{=}CH-(CH_2)_7-COOH$$

while linoleic acid, found in soybean oil, has one fewer double bond:

$$CH_3-CH_2-CH_2-CH_2-CH_2-CH{=}CH-CH_2-CH{=}CH-(CH_2)_7-COOH$$

The drying oil was dissolved in a solvent and mixed with pigment. An oil-soluble metal salt such as lead or cobalt naphthenate was added as an initiator. It induced peroxide formation when the paint was spread as a thin film. The peroxide in turn catalyzed the polymerization of the double bonds in the linseed oil to give a paint film. The drying oils used were primarily linseed oil and tung oil. Soybean oil could not be used because it does not contain enough unsaturation. It is valuable in alkyds (see below) because, whereas the oil has only 3 fatty acid residues, the alkyd has 6–10.

These paints based on natural drying oils had little resistance to solvents, chemicals, or ultraviolet light. They were replaced by oil-modified alkyd resins, which are made by heating or interesterifying a drying oil with a polyol, such as glycerol, and esterifying the product with a dibasic acid or anhydride, such as phthalic anhydride. If linoleic or linolenic acid is written L–COOH then the triglyceride is

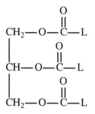

Interesterification with glycerol gives a mixture of esters

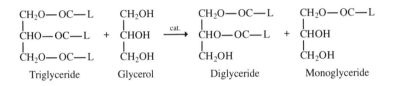

These react with phthalic anhydride to give a more or less linear polymer, the chain length of which is determined by the ratio of mono- to diglycerides in the mixture. (Monoglyceride units are shown in bold; diglyceride units in italic). This polymer is called an alkyd.

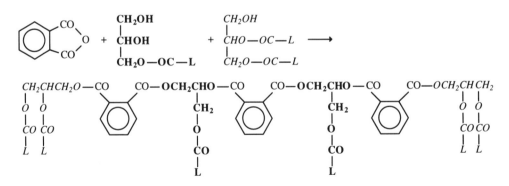

The incorporation of the polyester function gives the paint film greater solvent and ultraviolet resistance and also imparts somewhat greater corrosion resistance.

Most alkyd resins are made either from soybean oil, which is rich in a mixture of oleic and linoleic acids, or from tall oil acids (Section 13.1), which are also mixtures of these acids. Over 400 different alkyds are available commercially, varying in the type and amount of fatty acid present. Also, instead of the glycerol nature provides in fats and oils, the more highly functional pentaerythritol (Section 3.11.3) may be used. Maleic anhydride or occasionally other dibasic acids can be included. In another variation, part or all of the phthalic anhydride can be replaced by toluene diisocyanate (Section 8.3) to impart solvent and chemical-resistant urethane linkages.

Although alkyd resins are the most important vehicles for oil-based paints, their market is decreasing because of competition from water-based paints, corrosion-resistant paints based on epoxy, polyurethane and vinyl polymers, and solventless coatings such as radiation-cured and powder coatings.

Unsaturated polyester resins are oligomers that result from the condensation of phthalic anhydride, maleic anhydride, and a glycol such as diethylene glycol. The unsaturated oligomers are copolymerized with styrene or less frequently with methyl methacrylate in the presence of a peroxide initiator.

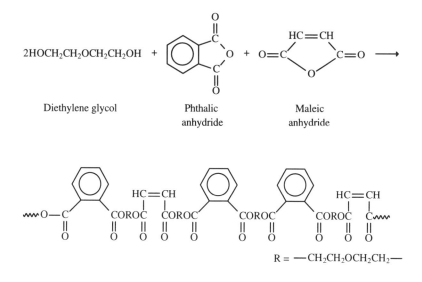

Diethylene glycol	Phthalic anhydride	Maleic anhydride

$R = -CH_2CH_2OCH_2CH_2-$

Unsaturated polyesters reinforced with glass cloth or fiber were used for the classic automobile, the Corvette. They are used as a metal replacement for the fabrication of pipe and storage tanks, and in the manufacture of small boats, and even large minesweepers, where the absence of metal is essential. The fabrication, however, must be done largely by hand (the lay-up and spray-up technique), which introduces an expensive labor component. Some of the hand labor can be eliminated by the use of a process called sheet molding, which makes use of impregnated glass cloth and an initiator, which does not become effective until the assembly is heated beyond a threshold temperature. Nonetheless, the technique is primarily of value for short runs of products.

Among the lesser volume uses for phthalic anhydride is its reaction with benzene to form an intermediate which, on dehydration, leads to anthraquinone (Section 7.8). Phthalic anhydride reacts with phenol in the presence of sulfuric acid to form the pH indicator, phenolphthalein.

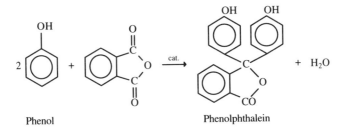

Phenol   Phenolphthalein

On reaction with ammonia, phthalic anhydride gives phthalimide which, on further heating with ammonia, dehydrates to give phthalonitrile, the starting material for phthalocyanine dyes.

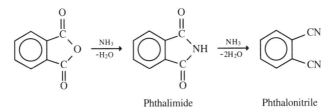

Phthalimide   Phthalonitrile

Esterification with allyl alcohol gives diallyl phthalate, a monomer for high-performance thermoset polymers, useful for glass-reinforced plastics. The esterification takes place in two steps as with di-(2-ethylhexyl)phthalate (Section 9.1.1).

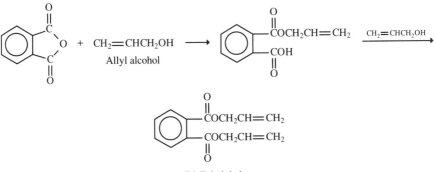

Diallyl phthalate

A high-performance poly(ether imide) "Ultem," developed by General Electric, involves the condensation of *m*-phenylene diamine with an *o*-phthalic-based dianhydride related to bisphenol A (Section 7.1.2.3).

9.2 *m*-XYLENE AND ISOPHTHALIC ACID

m-Xylene undergoes ammoxidation (Section 4.4) to yield isophthalonitrile which, on chlorination of the ring positions, gives the fungicide Daconil.

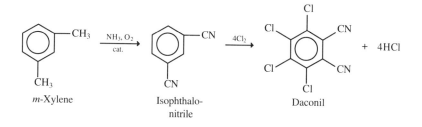

m-Xylene	Isophthalo-nitrile	Daconil

Hydrogenation of the nitrile provides *m*-xylylene diamine, which is a curing agent for epoxy resins.

Isophthalic acid is made by the oxidation of *m*-xylene by the Amoco–Mid-Century process (Section 9.3.1). The highly exothermic reaction takes place at about 200°C and 12 bar with a yield of 95 mol%. Excess oxygen is used.

9.2.1   Uses of Isophthalic Acid

Isophthalic acid has a number of low-volume uses, the major one in unsaturated polyesters (Section 9.1.1) where the greater stability to alkali of the ester linkages enhances the corrosion resistance of the final product. To a lesser extent, it is used in alkyd resins. It goes into specialty plasticizers and its acid chloride, isophthaloyl chloride, is condensed with *m*-phenylene diamine to give Nomex, a high temperature-resistant aramid polymer useful for fire resistant cloth for fire fighters and race car drivers' uniforms and more recently for plant workers' clothing.

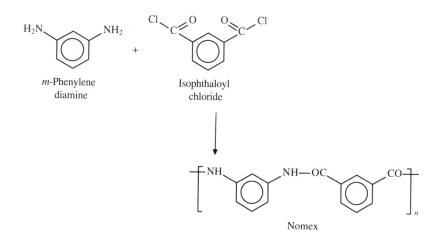

Polybenzimidazoles are also temperature and fire resistant and made from the diphenyl ester of isophthalic acid and a tetramine such as 3,3'-diaminobenzidine (3,3',4,4'-tetraaminobiphenyl).

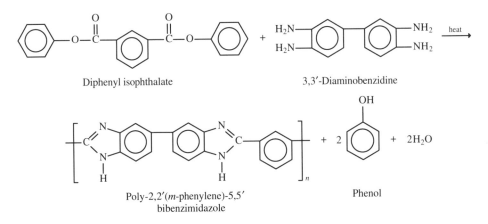

Diphenyl isophthalate   3,3'-Diaminobenzidine

Poly-2,2'(*m*-phenylene)-5,5'
bibenzimidazole

Phenol

9.3 *p*-XYLENE AND TEREPHTHALIC ACID/DIMETHYL TEREPHTHALATE

The major use for *p*-xylene is oxidation to terephthalic acid which, as such or as its methyl ester, is reacted with ethylene glycol to give polyester resins for fibers, films, molding resins, and most recently for biaxially oriented bottle resins. Terephthalic acid/dimethyl terephthalate is the most widely used xylene-based chemical by a wide margin, 7.8 billion lbs having been produced in the United States in 1993 compared with 854 million lbs of phthalic anhydride, the next largest xylenes derivative.

9.3.1 Oxidation of *p*-Xylene

Although *o*-xylene oxidizes readily, *m*- and *p*-xylenes present a problem. The *m*-and *p*-toluic acids formed in the first stage of oxidation contain a methyl group

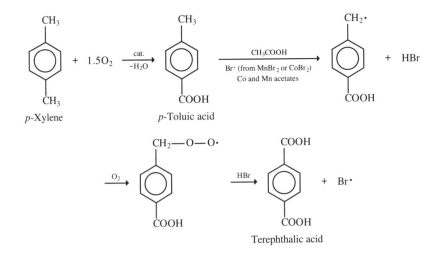

p-Xylene   *p*-Toluic acid

Terephthalic acid

that defies further oxidation, because the carboxyl group is electron withdrawing. There are several ways to overcome this, the most important being the
Amoco Mid-Century process. The oxidation is carried out in acetic acid solution with a catalyst comprising a manganese or cobalt salt with a bromine
promoter, which may actually be bromine itself but is usually manganous or
cobaltous bromide. The bromide converts the recalcitrant methyl group to a free
radical, which is then much more susceptible to oxidation.

Acetic acid is used as a solvent because terephthalic acid (TPA) is much less
soluble in it than are the intermediate products, and this allows for the separation of relatively pure TPA. Much of the acetic acid is oxidized and must be
replaced. This is now an important use for acetic acid (Section 10.5.2.2). Even in
its presence, impurities are formed in the reaction, the most important of which
is *p*-carboxybenzaldehyde. Crude terephthalic acid containing this impurity in
aqueous solution is subjected to hydrogenolysis with a palladium catalyst at
about 250°C and 36 bar. The impurity is converted to *p*-toluic acid, which, in
turn, is separated from the terephthalic acid by fractional crystallization.

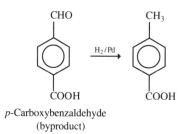

p-Carboxybenzaldehyde
(byproduct)

This hydrogenolysis provides the most important industrial use for palladium
catalysts. It is this step that made possible the production of sufficiently pure
terephthalic acid to be used directly in the esterification reaction to produce
polyester fibers and resins. The purified acid is called PTA (purified terephthalic
acid).

Although the mechanism that causes the cobalt bromide to function as
a cocatalyst is not completely understood, it has been suggested that the
aromatic molecule becomes a free radical because of electron extraction by the
cobalt cation. The free radical expels a proton and interacts with oxygen to form
a peroxy radical, which then maintains the chain reaction by hydrogen abstraction. Reaction chains are probably very short and the bromide is present to
produce oxygen-containing radicals at an optimum rate.

Prior to the development of this process for pure terephthalic acid, it was
necessary to use dimethyl terephthalate, which can be purified by distillation, for
polyester formation. In early processes, nitric acid oxidized *p*-xylene to crude
terephthalic acid, which was then converted for purification to the methyl ester.
In another process, *p*-xylene was oxidized to toluic acid, which was esterified to
give methyl toluate. The oxidation of the second methyl group is then possible,
and the second carboxyl group may then be esterified to give dimethyl terephthalate.

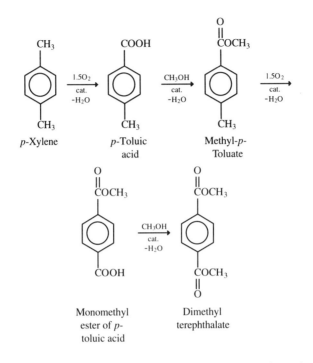

Currently, all new facilities produce purified terephthalic acid rather than the ester, and the latter is being phased out. In 1991, 55% of capacity in both the United States and Western Europe produced terephthalic acid. There has been burgeoning growth of polyester fiber production in the Far East in the late 1980s that will continue into the year 2000.

Emphasis on recycling of plastics has made discarded polyester bottles a source of dimethyl terephthalate or related polyester oligomers. Two main processes are available. In the first, poly(ethylene terephthalate) scrap is treated with methanol to generate dimethyl terephthalate and ethylene glycol, which may be purified by distillation. The second process achieves only a partial depolymerization. A small amount of ethylene glycol is added to the scrap to give a low-viscosity oligomer, which is purified by adsorption and filtration. It is then repolymerized under partial vacuum, which removes volatile contaminants and the ethylene glycol formed during the reaction. Costs of collecting, sorting, and cleaning scrap are a major part of the recycling cost, but it appears that the overall process is similar in cost to the conventional process starting with virgin material. Plastics recycling has been most successful with polyesters.

9.3.2   Alternate Sources for Terephthalic Acid

There are various other sources for terephthalic acid. The disproportionation of toluene to p-xylene, which in turn may be oxidized to terephthalic acid, has been described in Section 8.1 as was the carbonylation of toluene in Section 8.4, and the *para*-directed disproportionation of potassium benzoate in Section 8.4. The last of these is known as the Henkel II process.

The Henkel I process is related, and involves the isomerization of dipotassium *o*-phthalate to the *p*-isomer, in the presence of a zinc–cadmium catalyst at temperatures somewhat above 400°C and a modest pressure below 20 bar. Potassium is an expensive material and must be recovered. This can be done by reacting the potassium terephthalate with phthalic anhydride to yield more of the dipotassium *o*-phthalate.

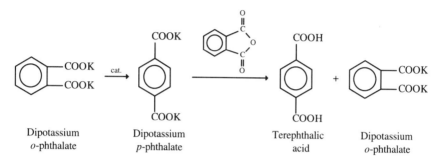

| Dipotassium *o*-phthalate | Dipotassium *p*-phthalate | Terephthalic acid | Dipotassium *o*-phthalate |

Like *m*-xylene, *p*-xylene can be ammoxidized. Terephthalonitrile results, which can be hydrolyzed to terephthalic acid. This process is not in use. The ammoxidation is carried out without added oxygen, the vanadium pentoxide–alumina catalyst giving up its oxygen in the process. The concept of an oxide as both a catalyst and a coreactant for oxidations has been developed recently by Du Pont in a new process for the oxidation of *n*-butane to maleic anhydride (Section 5.4) and in an unused process for aniline (Section 7.3).

9.3.3 Poly(ethylene terephthalate)

The formation of poly(ethylene terephthalate) was described in Section 3.7.1. In 1994, about two thirds of the production was used for polyester fibers. Another 17% was used for bottle resin, largely for containers for carbonated beverages. This application grew at almost 15%/year during the 1980s and is scheduled to grow at 8–10%/year in the 1990s. The polyester is biaxially oriented, that is it is drawn in two dimensions by blowing rather than in one dimension by being drawn into a fiber. It provides both resistance to transmission of carbon dioxide and sufficient mechanical strength to withstand the pressure, but it does not have sufficient resilience to withstand the sharp bend required at the bottom of the bottle. For this reason, the first bottles made from polyester had a hemispherical end, and the bottle was set into a polyethylene cup. A later version distributed the stresses created in making the sharp bend by molding a bottom comprising four convolutions, all of which have rounded edges. The elimination of the polyethylene makes recycling easier.

About 9% of poly(ethylene terephthalate) goes into a strong, biaxially oriented film known as Mylar. It is used in photographic and X-ray film, magnetic tapes for sound and video recordings, and containers for microwave and boil-in-the-bag cooking.

A small but growing amount of polyester is used as a molding compound. Because the polymer has a glass transition temperature of about 80°C, the molding is dimensionally unstable after cooling. Crystallization can be induced by nucleating the polymer, which makes molding possible in specialized applications. Dimensional stability can also be achieved by addition of glass fiber.

9.3.4 Lower Volume Polymers from Terephthalic Acid

Poly(butylene terephthalate), made by condensing terephthalic acid or its dimethyl ester with 1,4-butanediol (Section 10.3.1), is a relatively inexpensive molding resin useful for pipe, automotive parts, and toothbrush bristles. It is also alloyed with poly(phenylene oxide) to give a plastic useful for automobile bumpers and even for side panels.

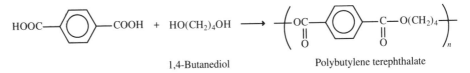

1,4-Butanediol   Polybutylene terephthalate

The condensation of terephthaloyl chloride with p-phenylene diamine leads to an aramid known as Kevlar. That it must be spun into fibers from concentrated sulfuric acid solution is an indication of its properties. The resulting fibers are stronger than steel. Kevlar was probably the first of the liquid crystal polymers, a category of polymers whose rigid chains allow them to maintain a crystalline structure either in the melt form or in solution. It finds its major application as an asbestos replacement. Other uses include cord for tires for large vehicles and reinforcement for bulletproof vests and military helmets. It is used combined with epoxy resin for canoe manufacture, and as cloth for boat sails. Nomex (Section 9.2.1) is a related polymer.

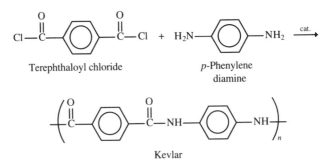

Kevlar

Dimethyl terephthalate may be subjected to hydrogenolysis to convert the ester groups to alcohols. At the same time the ring is hydrogenated. 1,4-Bis(hydroxymethyl)cyclohexane, known as cyclohexane dimethanol, results.

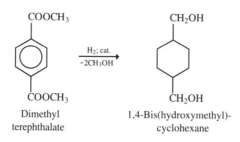

Dimethyl
terephthalate

1,4-Bis(hydroxymethyl)-
cyclohexane

This diol may exist in the cis or trans form, and the ratio must be carefully controlled if the product is to be used for the formation of polyester fibers by condensation with terephthalic acid. Eastman's "Kodel" contains this diol. It is also useful for the preparation of specialty polyurethane resins.

NOTES AND REFERENCES

Xylenes are discussed in H. G. Franck and J. W. Stadelhofer, cited in the bibliography to Chapter 7. The standard work is *Toluene, the Xylenes and their Industrial Derivatives*, E. G. Hancock, Ed., Elsevier, Amsterdam, 1982.

Section 9   The use of silicoaluminophosphates for xylene isomerization is described in European Patent Application 0 249 914A1 (December 23, 1987), to Union Carbide.

The UOP process for separation of *m*- and *p*-xylenes is described in detail by P. Spitz, p. 192, cited in the general bibliography (Section 0.4.2).

Section 9.1   For a more detailed discussion of both alkyd and polyester resins and for a chapter on plasticizers, see H. A. Wittcoff and B. G. Reuben, Part 2, cited in the general bibliography.

Section 9.3.1   Recycling of poly(ethylene terephthalate) bottles is described in *Plast Eur.*, December 1992, p. 750.

CHAPTER 10

CHEMICALS FROM METHANE

So far, we have described six natural gas and petroleum-derived products as raw materials for chemicals: ethylene, propylene, the C_4 stream, benzene, toluene, and the xylenes. To complete the picture we must add methane, whose major use is conversion to synthesis gas, $CO + H_2$. Lesser uses include chlorination, conversion to acetylene, and ammoxidation to hydrocyanic acid.

The major source of methane is natural or associated gases (Section 2.4). It is also available from refinery gases but is usually used for fuel and not recovered for chemicals. Although there has been an "explosion" in the discovery of natural gas in many places the world, notably the CIS, the reserves in the United States in the early 1990s appeared to be diminishing faster than new sources were being discovered. If necessary, methane may be synthesized from oil (Section 10.4.2) or coal (Section 12.5).

10.1 HYDROCYANIC ACID

Hydrocyanic acid (hydrogen cyanide or HCN) is produced in the United States at a level of about 1.3 billion lb/year. It was formerly made by the reaction of sodium, carbon (charcoal), and ammonia. In the first stage of the process, ammonia and sodium reacted for form sodamide, $NaNH_2$. This reacted with carbon to form sodium cyanamide, $NaHCN_2$, which in turn reacted with more carbon to provide sodium cyanide, NaCN, at temperatures as high as 850°C. Yields based on sodium and ammonia were high. Acidification with sulfuric acid provided hydrocyanic acid.

$$2Na + 2C + 2NH_3 + \xrightarrow{-3H_2O} 2NaCN \xrightarrow{H_2SO_4} HCN$$

The process has been superseded by the more economical Andrussow reaction between methane, air, and ammonia at 1000°C and slightly above atmospheric pressure over a platinum catalyst with 10–20% rhodium to prevent volatilization.

$$2CH_4 + 3O_2 + 2NH_3 \xrightarrow{\text{Pt–Rh 1000°C}} 2HCN + 6H_2O$$

In a modified process used by DeGussa, air is omitted, and methane (or any other hydrocarbon feedstock including naphtha) reacts directly with ammonia at 1400°C.

$$CH_4 + NH_3 \xrightarrow{\text{Pt–Rh 1400°C}} HCN + 3H_2$$

This provides one of the few examples in industrial chemistry of an intermolecular dehydrogenation and is possible because of the high temperature.

Hydrocyanic acid is also a byproduct of the ammoxidation of propylene to acrylonitrile (Section 4.4). It was this lower cost source of the chemical that motivated the more economical methane-based route. The byproduct hydrocyanic acid enjoys about one-half of the total market.

Another route to hydrocyanic acid is the dehydration of formamide (Section 10.8), a process that has been revived as a basic step in a proposed synthesis of methyl methacrylate (Section 4.6.1). Hydrocyanic acid also results from the interaction of carbon monoxide and ammonia.

$$CO + NH_3 \rightarrow HCN + H_2O$$

Hydrocyanic acid and its sodium salt undergo a number of industrially important reactions. Its largest application, hexamethylene diamine production, is described in Section 5.1.1. Its second largest use is for methyl methacrylate (Section 4.6.1). It has been used in two obsolete routes to acrylic acid (Section 4.3) and one to acrylonitrile (Section 4.4).

The powerful chelator, sodium nitrilotriacetic acid, is used to replace the sodium tripolyphosphate "builder" in some detergents. It is made from ammonia, formaldehyde, and hydrocyanic acid. In the first step 37% formaldehyde is reacted with HCN in the presence of sulfuric acid to give glycolonitrile. This reacts at 60°C with ammonia to give tris(cyanomethyl)amine. Hydrolysis with aqueous sodium hydroxide at 140°C and 3 bar at a pH of 14 gives sodium nitrilotriacetate. Acidification with sulfuric acid provides nitrilotriacetic acid.

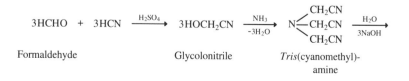

| 3HCHO + 3HCN | Formaldehyde | | 3HOCH₂CN Glycolonitrile | | Tris(cyanomethyl)-amine | |

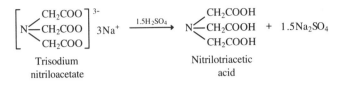

Its use as a builder in the United States has been inhibited because of the suggested toxicity of the heavy metal chelates, such as iron chelate, which could form in a washing machine. These are now believed to be nontoxic, and the product is one of the several phosphate replacements used more in Europe than in the United States. The others include zeolites, which act as ion exchangers, and polyacrylic acids (Section 4.3).

Other uses for hydrocyanic acid include conversion to sodium cyanide for gold recovery and preparation of ethylenediaminetetraacetic acid (EDTA). The latter is made via a modified Mannich reaction in which ethylene diamine is condensed with formaldehyde and hydrocyanic acid. Hydrolysis of the nitrile gives EDTA.

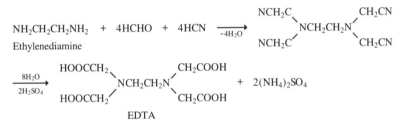

Single-step processes are also used, either with sodium hydroxide or with an aqueous solution of sodium cyanide. In each instance the sodium salt results.

$$H_2NCH_2CH_2NH_2 + 4HCHO + 4NaCN + 4H_2O \xrightarrow{\text{NaOH 80°C}}$$

$$(NaOOCCH_2)_2NCH_2CH_2N(CH_2COONa)_2 + 4NH_3$$

Another use for hydrogen cyanide is in the production of cyanuric chloride. Cyanogen chloride is first produced from chlorine and aqueous hydrocyanic acid at 40°C. The gas is trimerized to cyanuric chloride in the gas phase at over 300°C with activated charcoal and metal salts as catalysts. Cyanuric chloride is a starting material for the important triazine herbicides and fiber-reactive dyes. It is also used for the manufacture of triallyl cyanurate, useful for specialty polyesters and for certain pharmaceutical syntheses.

$$HCN + Cl_2 \longrightarrow \underset{\substack{\text{Cyanogen} \\ \text{chloride}}}{CNCl} + HCl \qquad\qquad 3CNCl \xrightarrow{\text{cat.}}$$

Cyanuric
chloride

An important use for hydrocyanic acid, or more frequently sodium cyanide, is in the synthesis of amino acids by the Strecker reaction. The most important application of this reaction is for the synthesis of DL-methionine, a poultry feed additive (Section 4.10.4).

Oxamide is a specialty fertilizer releasing ammonia slowly into the soil. It melts at 419°C without decomposition, making it one of the highest melting of all organic compounds. It results from the hydrolysis of cyanogen, which in turn is made by the cupric nitrate catalyzed oxidation of hydrocyanic acid with nitrogen dioxide at 0–5°C.

$$2HCN + NO_2 \xrightarrow{Cu(NO_3)_2} \underset{\text{Cyanogen}}{NCCN} + NO + H_2O$$

$$\xrightarrow[\text{cat.}]{2 H_2O} \underset{\text{Oxamide}}{NH_2\overset{\overset{O}{\|}}{C}-\overset{\overset{O}{\|}}{C}NH_2}$$

The cyanogen hydrolyzes quantitatively with hydrochloric acid at room temperature. It has been proposed that the process be used with the byproduct hydrocyanic acid from ammoxidation (Section 4.4). However, this use is precluded by the imbalance between fertilizer usage, which is very large, and hydrocyanic acid production, which is relatively small. Because fertilizers must be cheap, large capital investment in expensive hydrocyanic acid production is not warranted.

10.2 HALOGENATED METHANES

Chlorination of methane yields a mixture of mono-, di-, tri-, and tetra-chloromethanes (see note). These can be separated by distillation and this provides a route to their preparation, but it is not the preferred one. Chloromethane (methyl chloride) chlorinates more readily than methane, the formation of dichloromethane becoming significant after 18% of the methane has been converted to chloromethane. Indeed a high conversion to chloromethane can be obtained only if the methane to chlorine ratio is greater than 10:1. Consequently, the preferred route to chloromethane is the reaction of methanol with hydrogen chloride.

$$CH_3OH + HCl \rightarrow CH_3Cl + H_2O$$

In contrast, chloroethane (ethyl chloride) can be made by the chlorination of ethane since dichlorination does not start until about 75% of chloroethane has been produced.

10.2.1   Chloromethane

In 1993, 849 million lb of chloromethane was produced. Its major use in the past has been for conversion to dichloromethane (Section 10.2.2) whose use has declined markedly in the 1990s because of possible toxicity. About 80% of chloromethane production is used to make silicone resins, (or more appropriately polysiloxanes), which result from the hydrolysis of dimethyldichlorosilane. This monomer is made by reaction of chloromethane with silicon, which is usually in the form of a copper alloy.

$$2CH_3Cl + Si(Cu) \rightarrow (CH_3)_2SiCl_2 + Cu$$

Silicon/copper   Dimethyl
alloy   dichlorosilane

The silane hydrolyzes with water to provide dimethyldihydroxysilane, which converts to a polysiloxane. Trimethylhydroxysilane from trimethylchlorosilane acts as a chain stopper.

$$(CH_3)_2SiCl_2 + 2H_2O \longrightarrow (CH_3)_2Si(OH)_2 + 2HCl$$

Dimethyl-
dihydroxysilane

$$n(CH_3)_2Si(OH)_2 \longrightarrow \left(\begin{array}{c} CH_3 \quad CH_3 \\ | \qquad | \\ -Si-O-Si-O- \\ | \qquad | \\ CH_3 \quad CH_3 \end{array} \right)_{n/2} + \; nH_2O$$

Polysiloxane

The type of organic groups attached to the siloxane backbone and the extent of cross-linking between the chains determines whether the silicone will be an oil, an elastomer, or a resin. Silicones are used in water-repellent coatings, mold release agents, and slip agents. They are formulated as elastomers, caulking compounds, and sealants and as resins for lamination with glass. Silicone oils are used as antifoam agents in detergents, and in aerobic fermentation processes such as penicillin production.

A smaller volume use for chloromethane is in the preparation of methylcellulose (Section 14.3). Its use in the manufacture of tetramethyllead for gasoline had practically been phased out in the United States in the early 1990s. It is used in the preparation of quaternary ammonium compounds such as difatty dimethylammonium chloride (Section 13.1) and has a small use as a catalyst and solvent in the production of butyl rubber.

10.2.2 Dichloromethane

Dichloromethane (methylene chloride) is made by the chlorination of chloromethane. Further chlorination of the mixture of chlorinated methanes provides chloroform and carbon tetrachloride. Dichloromethane production in 1993 was 354 million lb, down from 607 million lb in 1984.

$$CH_3Cl + Cl_2 \rightarrow CH_2Cl_2 + 2HCl$$

Chloro-   Dichloro-
methane   methane

Dichloromethane has been used as a paint stripper, especially for jet aircraft, which must be stripped and examined for cracks at intervals. It has the advantage over alkaline paint removers of not attacking aluminum. On the other hand, many chlorinated hydrocarbons are health hazards. Dichloromethane is described as a possible carcinogen–mutagen but it is not nearly as dangerous as tri- and tetrachloromethane because its appreciable water solubility means it does not accumulate in body fat. Nonetheless, dichloromethane's use as a paint stripper has declined precipitously. A possible substitute is *N*-methylpyrrolidone (Section 10.3.1). It has also been important in solvent degreasing, an application that is similarly being eliminated. Thirty-five percent of dichloromethane production was for synthesis of chlorofluorocarbons, which are being phased out as is the application for solvent extraction of caffeine, cocoa and edible oils.

10.2.3 Trichloromethane

Trichloromethane (chloroform), whose production in 1994 was about 400 million lb, is made by chlorination either of methane or chloromethane. The chlorination of methane can be carried out either thermally with chlorine without catalysts or by oxychlorination (Section 3.4) with hydrogen chloride and a potassium chloride–cupric chloride catalyst. Both processes are highly exothermic. The oxychlorination is carried out at 400–450°C at slightly elevated pressure. The chlorination may be initiated by photons or by the homolysis of chlorine molecules that occurs on heating. It was once an important anesthetic.

At one time, 97% was used for production of chlorofluorocarbon refrigerants and aerosol propellants, applications scheduled to be phased out by 1995 because the products destroy the ozone layer. Fluorocarbons are made from both chloroform and carbon tetrachloride by stepwise displacement of the chlorine atoms with hydrofluoric acid. The higher the temperature and pressure, the more substitution results. Thus tetrachloromethane yields $CFCl_3$, CF_2Cl_2, and CF_3Cl. Trichloromethane yields $CHFCl_2$, CHF_2Cl, and CHF_3. Catalysts for the reaction are fluorides or oxyfluorides of aluminum or chromium.

Substitutes for phased-out fluorocarbon refrigerants and propellants generally contain fluorine and hydrogen, but no chlorine. In some replacements, a reduced number of chlorine atoms is considered acceptable. These too are believed to destroy the ozone layer, but at a lower rate. Thus their use is considered a temporary measure, and they are scheduled to be phased out in the early part of the twenty-first century. The most important fluorocarbons and their applications and proposed replacements are shown in Table 10.1.

Trichloromethane reacts with hydrofluoric acid as noted above to give chlorodifluoromethane which, on pyrolysis at 700°C, gives tetrafluoroethylene and hexafluoropropylene. The former is the monomer for Teflon.

$$2CHCl_3 + 2HF \xrightarrow[-2HCl]{} 2CHClF_2 \xrightarrow{700\,°C} F_2C=CF_2 + 2HCl$$

| Trichloro-methane | Chlorodifluoro-methane | Tetrafluoro-ethylene |

$$3CHClF_2 \xrightarrow{700\,°C} F_3CCF=CF_2 + 3HCl$$

Hexafluoropropylene

10.2.4 Tetrachloromethane and Carbon Disulfide

Tetrachloromethane (carbon tetrachloride) is the raw material for trichlorofluoromethane and dichlorofluoromethane, discussed above. It may be made by exhaustive chlorination of methane or chloromethane, but there are two more important routes. The most widely used is a "two-for-one" reaction that produces both tetrachloromethane and perchloroethylene by the chlorolysis of a propane–propylene mixture at about 500°C:

$$CH_2=CHCH_3 + 7Cl_2 \rightarrow CCl_4 + Cl_2C=CCl_2 + 6HCl$$

| Propylene | Carbon tetrachloride | Perchloroethylene |

$$CH_3CH_2CH_3 + 8Cl_2 \rightarrow CCl_4 + Cl_2C=CCl_2 + 8HCl$$

Propane

The quantity produced is dependent on the demand for perchloroethylene and the shortfall is made up by a route starting with methane but proceeding via carbon disulfide, another important methane-based chemical. An iron catalyst at 30°C brings about the chlorination:

$$CS_2 + 2Cl_2 \rightarrow CCl_4 + 2S$$

Excess chlorine gives sulfur monochloride, also of interest industrially:

$$CS_2 + 3Cl_2 \rightarrow CCl_4 + S_2Cl_2$$

TABLE 10.1 Fluorocarbons and their Replacements

Application	Fluorocarbon Currently Used	Boiling Point [°C]	Code Number[a]	Proposed Replacement[b]	Comments on Replacement
Aerosol propellant	$CF_2Cl_2 + CFCl_3$	−29.8; 23.8	F12 + F11	Propane + butane	Highly flammable
Refrigerant	CF_2Cl_2 or CF_2HCl	−29.8; −40.8	F12 or F22	CH_3CH_2F NH_3 Propane, butane	Not suitable for existing equipment Pungent odor/toxicity Highly flammable
Foaming agent	$CFCl_3$	23.8	F11	CH_3CFCl_2 CH_3CH_2F	Suspected toxicity Suspected carcinogenicity
Cleaning electronic parts etc.	$CFCl_2CF_2Cl$	47.6	F113		No single direct replacement. Ultra-pure water and polyester microfiber cloth used in some applications

[a] Fluorocarbons are referred to as Freon $ABCD$ or $FABCD$ where $ABCD$ is a number with up to four digits. D is the number of fluorine atoms in the molecule; C is the number of hydrogen atoms plus one; B is the number of carbon atoms minus one; and A is the number of double bonds. If $A = 0$ or $B = 0$, they are omitted.

[b] All the compounds proposed except ammonia have significant global warming potential, although CH_3CH_2F decomposes readily in the atmosphere.

Carbon disulfide is made by reaction of methane and sulfur vapor in the presence of a catalyst:

$$CH_4 + aS_x \rightarrow CS_2 + 2H_2S$$

where $ax = 4$ and x is between 2 and 8 since it is an equilibrium mixture of S_2, S_6, and S_8.

An important use for carbon disulfide is in the manufacture of regenerated cellulose for rayon, cellophane, and synthetic sponges (Section 14.3), all mature products.

10.2.5   Bromomethane

Bromomethane (methyl bromide) is made from methanol and hydrobromic acid:

$$CH_3OH + HBr \rightarrow CH_3Br + H_2O$$

World production in 1993 was about 66,000 tonnes. It is widely used as a soil and grain fumigant and less widely to fumigate private residences infested with termites. It was a latecomer to the list of chemicals believed to affect the ozone layer, and its future is still in doubt although the US Environmental Protection Agency is to phase out production and imports by 2001. Not only are people uncertain about how seriously it may damage the ozone layer, but also they are worried about the effect its withdrawal would have on the proportion of crops eaten by pests. Living without hairspray is one thing; living without food is another.

10.3   ACETYLENE

Acetylene rose in importance as a chemical feedstock after World War II, reached a peak in the mid-1960s, and has since declined. Coal was not only a source of aromatics, by way of coke oven distillate (Chapter 7), but was also the raw material for acetylene and hence the source of many early plastics and aliphatic organic chemicals.

In the oldest process, coke and lime are heated in an electric furnace to 2000°C to give calcium carbide. This is hydrolyzed by water to give acetylene.

$$CaO + 3C \rightarrow CaC_2 + CO$$

$$CaC_2 + 2H_2O \rightarrow Ca(OH)_2 + CH{\equiv}CH$$
$$\text{Acetylene}$$

The process has many of the characteristics of nineteenth century industry. It is a batch process. It is labor intensive because of the handling of solids, and energy intensive because of the electric furnace. It is environmentally unattractive because every ton of acetylene produced is accompanied by 2.8 tons of calcium hydroxide, usually in a slurry with 10 times as much water. The capital cost of the furnace is high and it has a short life because of the "heroic" conditions used.

Even in the days of cheap energy, the problems with the carbide process provided an impetus for its replacement. In the early 1960s, many attempts were made to derive acetylene from petroleum. Thermodynamically, it is the stable hydrocarbon above 1300°C, but the pyrolysis of other hydrocarbons is not so simple. It is difficult to quench the reaction mixture rapidly enough to prevent reverse reactions and to keep residence times short enough to prevent coking. Separation of acetylene from byproduct hydrogen provides an additional problem.

These difficulties were addressed in the Wulff and Sachsse processes, which in theory can start with naphtha or natural gas. The naphtha-based plants, mainly European, gave continuous trouble and were closed after short lives.

Today there are four sources of acetylene. The calcium carbide route is one, but much of US acetylene produced in this way is for the nonchemical application of arc welding. Calcium carbide has played an honorable role in the history of the US chemical industry and even gave its name to the Union Carbide Corporation. Its derivative, calcium cyanamide, gave its name to the American Cyanamid Corporation. Its energy requirements, however, mean that today it can only be made economically in countries such as Norway, which have cheap hydroelectric power and unproductive land where the waste calcium hydroxide can be dumped.

For most chemical processes in the United States, acetylene is obtained from three further sources. The first is as a byproduct from thermal or steam cracking (Section 2.2.1). The cracking of propane can yield up to 2% of acetylene at high-severity cracking, that is at cracking under conditions that produce a maximum amount of ethylene. Steam cracking of ethane–propane accounts for about half of US acetylene production. Before the C_2 fractionation, the acetylene in the C_2 stream is selectively absorbed in dipolar aprotic solvents such as dimethylformamide or N-methylpyrrolidone. If the acetylene is not wanted, it must be selectively hydrogenated to ethylene, and most crackers operate in this way.

A second process is thermal, the above-mentioned Wulff process, operated not with naphtha but with natural gas, which is primarily methane. The reactor consists of a furnace containing a hot brick lattice. The bricks are heated for a minute to about 1300°C by burning a gaseous fuel. The feed methane is then pyrolyzed for a minute. The same sequence of operations is then carried out in the opposite direction through the furnace, giving a 4-min cycle with a final temperature below 400°C. The product is quenched and comprises a mixture of acetylene and ethylene in a ratio of 1:2.

The third process requires oxygen and is known as the partial oxidation process. There are several variations, but the Sachsse process is the most important of them. Methane and oxygen are preheated to 600°C and reacted in a special burner with flame formation. A temperature of about 1500°C is achieved. After a residence time of a few milliseconds, the mixture is quenched to 40°C with water or quench oil to inhibit soot and hydrogen formation. This yields a gaseous product containing 8–10% acetylene, corresponding to 30% of the carbon in the feedstock. Synthesis gas (Section 10.4) is the major product and about 1.5% of the initial carbon emerges as soot.

The Wulff and Sachsse processes, although not as hostile as the carbide process, are energy intensive. Thus acetylene has become steadily more expensive in comparison with ethylene, and almost everything that used to be done with acetylene can now be done more cheaply with olefins.

The list of obsolete or nearly obsolete acetylene-based processes is impressive—vinyl chloride from acetylene and hydrogen chloride; vinyl acetate from acetylene and acetic acid; acrylonitrile from acetylene and hydrocyanic acid; acetaldehyde from acetylene and water; chloroprene from acetylene dimer (vinylacetylene); acrylate esters by way of acetylene, an alcohol, and carbon monoxide; and perchloroethylene (Section 10.2) by a multistage process. Between 1967 and 1974, 25 acetylene plants in the United States closed, and consumption for chemical use dropped from 1.23 to 0.40 billion lb between 1962 and 1977.

The major area for the remaining acetylene use is in Reppe chemistry. Reppe was a German chemist who studied the interaction of acetylene with aldehydes, ketones, alcohols, and carbon dioxide, during World War II. The acrylate process mentioned above (Section 4.3) is a Reppe reaction as is the 1,4-butanediol process below.

10.3.1 1,4-Butanediol

Acetylene reacts with formaldehyde to give 1,4-butynediol, which may be hydrogenated stepwise to butenediol and butanediol.

$$HC\equiv CH + 2HCHO \rightarrow HOCH_2-C\equiv C-CH_2OH \xrightarrow{H_2}$$

Acetylene   Formalde-   1,4-Butynediol
            hyde

$$HOCH_2-CH\equiv CH-CH_2OH \xrightarrow{H_2} HOCH_2-CH_2-CH_2-CH_2OH$$

1,4-Butenediol   1,4-Butanediol

The condensation of 37% aqueous formaldehyde and acetylene takes place at about 100°C and 5 bar in the presence of a cuprous acetylide catalyst deposited on magnesium silicate. In this process, the formation of byproduct propargyl alcohol $CH\equiv CCH_2OH$ is minimized. The hydrogenation is carried out in two stages because a purer product results. The first state is operated at 50–60°C at 7–15 bar with a Raney nickel catalyst to yield 1,4-butenediol, which finds application in the manufacture of the pesticide endosulfan (Section 6.3). The second stage requires a temperature of 120–140°C with hydrogen at 150–200 bar with a nickel–copper–manganese catalyst on a silica gel carrier and yields 1,4-butanediol.

About 50% of 1,4-butanediol production is converted to tetrahydrofuran. This is converted to poly(tetramethylene ether glycol) in the same way that ethylene oxide can be converted to poly(ethylene oxide). The product is used in the formulation of polyurethanes and elastomeric fibres such as spandex (Section 5.4) and a thermoplastic elastomer, Hytrel (Section 15.3.8).

In 1990 Du Pont announced a process involving a proprietary catalyst for preparation of tetrahydrofuran by the direct hydrogenation–hydrogenolysis of maleic anhydride (see below).

1,4-Butanediol is used as such in specialized polyurethane compositions for the manufacture of wheels for skates. It is also the basis for the so-called acetylenic chemicals γ-butyrolactone, pyrrolidone, and N-vinylpyrrolidone. Butyrolactone and pyrrolidone are both aprotic solvents. The former can be converted to N-methylpyrrolidone, a replacement for chlorinated solvents.

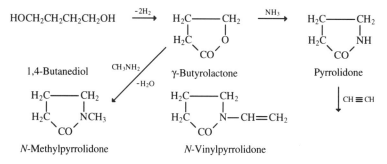

Pyrrolidine's major use is for conversion to N-vinylpyrrolidone. This can be polymerized to poly(N-vinylpyrrolidone) or copolymerized with vinyl acetate. These polymers are formulated into hair sprays. During World War II poly-(N-vinylpyrrolidone) was of interest in Germany as a replacement for blood plasma. The other important use for 1,4-butanediol is in the formulation of poly(butylene terephthalate) (Section 9.3.4).

Although 1,4-butanediol has a relatively modest market (US production in 1992 was about 400 million lb) numerous alternative routes to it have been developed. In Japan it is prepared from butadiene by an acetoxylation reminiscent of an unsuccessful process for the manufacture of ethylene glycol (Section 3.7.2). The acetoxylation proceeds more readily with the conjugated butadiene than with ethylene because the double-bond system is more reactive (Section 3.7.1).

Thus it is possible to use a palladium–tellurium catalyst without the iodine promoter required for ethylene, alleviating much of the corrosion, which was the Achilles' heel of the earlier process. After the initial acetoxylation, the 1,4-diacetoxy-2-butene is hydrogenated to 1,4-diacetoxybutane, which is saponified to the desired product and acetic acid for recycle.

A Mitsubishi process involved 1,4-chlorination of butadiene. Hydrolysis of the chlorine atoms then gives 1,4-dihydroxy-2-butene, which on hydrogenation provides 1,4-butanediol.

$$CH_2{=}CHCH{=}CH_2 \ + \ Cl_2 \ \longrightarrow \ ClCH_2CH{=}CHCH_2Cl \xrightarrow[-2NaCl]{2NaOH}$$

Butadiene   1,4-Dichloro-2-butene

$$HOCH_2CH{=}CHCH_2OH \xrightarrow{H_2} HOCH_2CH_2CH_2CH_2OH$$

1,4-Dihydroxy-2-butene   1,4-Butanediol

It is related to an early Du Pont process for hexamethylene diamine (Section 5.1.1) and, like it, was uneconomical because of the waste of chlorine.

A process, commercialized in the United States by ARCO in 1990, starts with propylene oxide, which may be isomerized to allyl alcohol with a lithium phosphate catalyst. This was a step in an obsolete gylcerol process (Section 4.10.2). Hydroformylation of the allyl alcohol yields an aldehyde, which on hydrogenation gives 1,4-butanediol. The hydroformylation is difficult and does not provide maximum selectivity to a linear product. Thus the branched-aldehyde $CH_3CH(CHO)CH_2OH$ forms, which on hydrogenation provides a coproduct, 2-methyl-1,3-propanediol, $CH_3CH(CH_2OH)_2$.

$$CH_3{-}CH{-}CH_2 \xrightarrow{Li_3PO_4} CH_2{=}CHCH_2OH \xrightarrow[cat.]{H_2,\ CO}$$

Propylene oxide   Allyl alcohol

$$OHCCH_2CH_2CH_2OH \xrightarrow{H_2}_{cat.} HOCH_2CH_2CH_2\overset{\smile}{C}H_2OH$$

A process developed by Davy McKee and instituted in Japan in 1993 starts with maleic anhydride (Section 5.4), which is converted to ethyl maleate in two steps. This unsaturated ester is hydrogenated and subjected to hydrogenolysis in one step to give the desired product and ethanol for recycle.

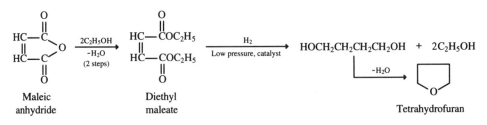

Maleic anhydride   Diethyl maleate   Tetrahydrofuran

A new Du Pont process for which a plant was completed in Spain in 1995 uses a proprietary catalyst that makes possible the conversion of maleic acid to tetrahydrofuran in one step. The process apparently works better with the acid than with the anhydride and Du Pont has therefore integrated backwards to a maleic acid process using a transport bed (Section 5.4). The Du Pont process accomplishes in one step what the Davy McKee process does in several.

10.3.2 Lesser Uses for Acetylene

Trichloroethylene (for degreasing) and perchloroethylene (for dry cleaning) may be made from acetylene. Most perchloroethylene, however, is produced from simultaneous chlorination and pyrolysis of hydrocarbons or by chlorination or oxychlorination of ethylene dichloride (Sections 3.11.7, and 10.2.4)

Vinyl fluoride results from the addition of hydrogen fluoride to acetylene (Section 3.11.8). It can be polymerized to poly(vinyl fluoride), a specialty polymer, with outstanding weathering properties.

$$n\text{CH}{=}\text{CH} + n\text{HF} \rightarrow n\text{CH}_2{=}\text{CHF} \rightarrow \{\!\!\{\,\text{CH}_2{=}\text{CHF}\,\}\!\!\}_n$$

Partial carbonylation of acetylene with formaldehyde gives propargyl alcohol, used in the petroleum and metallurgical industries and as an intermediate in the manufacture of the miticide, propargite; the antibacterial, sulfadiazine, and various propargyl carbonate fungicides.

$$\text{HCHO} + \text{HC}{\equiv}\text{CH} \xrightarrow[\text{100°C, 5 bar}]{\text{Cu acetylide}} \text{HC}{\equiv}\text{CCH}_2\text{OH}$$
Propargyl alcohol

Higher carbonyl compounds also give acetylenic alcohols, and these form part of synthetic routes to terpenes including vitamins A and E and various perfumes and steroids.

Other small uses for acetylene include vinyl ester formation by the zinc or mercury salt-catalyzed reaction between an acid and acetylene. Thus the so-called neo or Koch acids may be converted to vinyl esters as may fatty acids such as stearic or oleic.

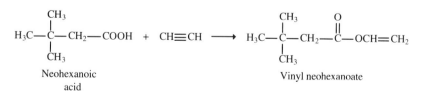

Neohexanoic   Vinyl neohexanoate
acid

Vinyl esters may also be made by an exchange reaction or transvinylation between an acid and a vinyl compound such as vinyl acetate or vinyl chloride.

The major use of these materials is as internal plasticizers or modifiers for polymers based on vinyl chloride and vinyl acetate.

$$RCOOH \; + \; CH_2{=}CHOCOCH_3 \; \xrightarrow{\text{cat.}} \; RCOOCH{=}CH_2 \; + \; CH_3COOH$$

Vinyl acetate   Vinyl ester   Acetic acid

In a related reaction, vinyl ethers result from the interaction of alcohols and acetylene. The reaction is catalyzed by alkali. These too are used as comonomers for the modification of vinyl polymers.

$$n\text{-}C_{18}H_{37}OH \; + \; CH{\equiv}CH \; \longrightarrow \; n\text{-}C_{18}H_{37}OCH{=}CH_2$$

Stearyl   Vinyl stearyl ether
alcohol

10.4 SYNTHESIS GAS

Synthesis gas is the name given to a variety of mixtures of carbon monoxide and hydrogen, or nitrogen and hydrogen. It is made from methane from natural gas if the latter is available, as it is in the United States and most of Europe. Countries lacking natural gas (e.g., Japan and Israel) make synthesis gas from naphtha. It may, however, be made from virtually any hydrocarbon (e.g., so-called resid, the residue from petroleum distillation). Indeed, it may be made from almost any carbonaceous material including coal, peat, wood, biomass, agricultural residues, and municipal solid waste.

Coal was an important feedstock prior to 1960. During World War II, it was used in Germany to provide synthesis gas for the manufacture of fuel and chemicals by the Fischer–Tropsch process (Section 12.2). The same technology is applied today in South Africa and makes coal the basis for at least half that country's energy needs. Coal is also gasified in China.

Different applications for synthesis gas require different mixtures. The Fischer–Tropsch reaction and methanol manufacture require $CO/H_2 = 1:2$; hydroformylation [the oxo process (Section 4.8)] requires $CO/H_2 = 1:1$ and the Haber process for ammonia requires $N_2/H_2 = 1:3$ without any carbon monoxide at all. Other organic chemical syntheses require pure hydrogen. Properly adjusted, nonetheless, the basic synthesis gas processes can give all these mixtures.

10.4.1 Steam Reforming of Methane

The most widely used synthesis gas process is the steam reforming of hydrocarbons, with partial oxidation of hydrocarbons as another possibility. Coal-based routes are also of significance and will be described in Section 12.2. The dominant synthesis gas process is the steam reforming of methane.

Before steam reforming can take place, the methane feedstock is desulfurized by passage over a zinc oxide catalyst at 360–400°C. Sulfur levels less than 2 ppm are required if the nickel steam-reforming catalyst is to have an adequate life. It is then mixed with steam in a molar ratio of between 2.5 and 3.5 of steam to methane and passed over an alkali-promoted nickel catalyst, supported on potassium oxide or alumina. The reaction is carried out at about 800°C and 35 bar; it is endothermic and the heat is supplied externally. The principal reaction is

$$CH_4 + H_2O \rightarrow CO + 3H_2 \qquad \Delta H_{800°C} = 227 \text{ kJ/mol}$$

The exit gases contain about 7% unchanged methane plus some carbon dioxide and traces of nitrogen present in the natural gas. The excess steam minimizes coking of the catalyst. The higher the steam/carbon ratio, the higher the CO_2/CO ratio in the products.

$$C + H_2O \rightarrow CO + H_2$$
$$C + 2H_2O \rightarrow CO_2 + 2H_2$$

If the synthesis gas is to be used to make ammonia, nitrogen must be added and carbon monoxide removed. This is done by addition of an amount of air calculated to provide a N_2/H_2 ratio of 1:3. The oxygen in the air reacts with some of the carbon monoxide to give carbon dioxide.

$$2CO + O_2 \rightarrow 2CO_2$$

The gases then pass to a second reforming unit at 370°C, possibly with additional steam. In the presence of an iron oxide catalyst, the water–gas conversion or shift reaction takes place.

$$CO + H_2O \rightarrow CO_2 + H_2 \qquad \Delta H_{370°C} = -38 \text{ kJ/mol}$$

These reactions are summarized in Figure 10.1.

The product gases contain hydrogen, nitrogen, carbon dioxide, and traces of methane, carbon monoxide, and argon. They are compressed and scrubbed with aqueous monoethanolamine and diethanolamine (a variety of other processes is available) to remove carbon dioxide. Some of the carbon dioxide dissolves in the water — at high pressure, the solubility is high — and some reacts with the amine to give an unstable salt. The ethanolamine may then be recovered by steam stripping, which decomposes the salt. A further shift reaction may improve yields, and other processes are employed to reduce carbon monoxide to a very low level. The product is then delivered to an ammonia plant.

COAL:

$2C + O_2 \rightleftharpoons 2CO$

$C + H_2O \rightleftharpoons CO + H_2$

$CO + H_2O \rightleftharpoons H_2 + CO_2$ (SHIFT REACTION)

$C + CO_2 \rightleftharpoons 2CO$ (BOUDOUARD REACTION)

$C + 2H_2 \rightleftharpoons CH_4$

$CO + 3H_2 \rightleftharpoons CH_4 + H_2O$

METHANE:

$CH_4 + 1/2O_2 \rightleftharpoons CO + 2H_2$

$CH_4 + H_2O \rightleftharpoons CO + 3H_2$

NAPHTHA:

$-CH_2- + 1/2O_2 \rightleftharpoons CO + H_2$

$-CH_2- + H_2O \rightleftharpoons CO + 2H_2$

CARBON
FORMATION:

$2CO \rightleftharpoons C + CO_2$

$CO + H_2 \rightleftharpoons C + H_2O$

$CH_4 \rightleftharpoons C + 2H_2$

Figure 10.1   Important reactions in synthesis gas formation.

10.4.2   Variants of Steam Reforming

If synthesis gas is required for other purposes, the addition of nitrogen is not necessary. The shift reaction will then give CO/H_2 mixtures rich in hydrogen or it will even give pure hydrogen. Alternatively, if carbon monoxide-rich materials are needed, then carbon dioxide may be added at the shift conversion stage so that the equilibrium in the shift reaction is pushed to the left and, instead of carbon monoxide converting water to hydrogen, the reverse occurs.

$$CO_2 + H_2 \rightarrow CO + H_2O$$

There are many variants to this process. Propane or naphtha may be the feedstock. With propane as the example, the following occurs.

$$C_3H_8 + 3H_2O \rightarrow 3CO + 7H_2 \qquad \Delta H_{800°C} = 552 \text{ kJ/mol}$$

$$C_3H_8 + 6H_2O \rightarrow 3CO_2 + 10H_2 \qquad \Delta H_{800°C} = 435 \text{ kJ/mol}$$

If a higher steam/hydrocarbon ratio is used, the product has a higher CO/H_2 ratio than is possible with methane. It is thus less suitable for ammonia. Flexibility is achieved, however, by adjustment with the shift reaction.

In a further application of the steam reforming reaction, methane rather than synthesis gas is produced. It may be obtained fairly readily from any hydrocarbon feedstock that can be vaporized, for example, naphtha. The process is operated at a lower temperature than if synthesis gas is required. Typically, with nonane as an example,

$$C_9H_{20} + 4H_2O \rightarrow 7CH_4 + 2CO_2$$

The product is called Substitute Natural Gas (SNG). The process was used in the United Kingdom before North Sea discoveries and is still used in Japan where natural gas is lacking. Various processes are available to give methane from heavier hydrocarbon feedstocks, but the huge discoveries of natural gas in the former Soviet Union in the early 1980s mean that oil is now thought likely to be depleted before natural gas. Hence, interest in SNG from oil has diminished.

10.4.3 Partial Oxidation of Hydrocarbons

A second route to synthesis gas is partial oxidation of carbonaceous materials reacted with steam. If the feedstock can be vaporized, the reactions are carried out simultaneously, but if it is a solid they must be carried out sequentially.

A vaporizable hydrocarbon feedstock (e.g., methane, propane, or naphtha) is burned in a flame in the presence of about 35% of the stoichiometric amount of oxygen.

$$CH_4 + O_2 \rightarrow CO_2 + 2H_2 \qquad \text{Rapid} \qquad \Delta H = 318 \text{ kJ/mol}$$

A little water is also formed by conventional combustion reactions. Excess hydrocarbon can now react.

$$CH_4 + \tfrac{1}{2}O_2 \rightarrow CO + 2H_2, \qquad \text{Rapid} \qquad \Delta H = -36 \text{ kJ/mol}$$

$$CH_4 + CO_2 \rightarrow 2CO + 2H_2, \qquad \text{Slow} \qquad \Delta H = 247 \text{ kJ/mol}$$

$$CH_4 + H_2O \rightarrow CO + 3H_2, \qquad \text{Slow} \qquad \Delta H = 227 \text{ kJ/mol}$$

The flame temperature is 1300–1400°C at a pressure of 60–80 bar with a residence time of 2–5 s. The rapid initial reactions provide the heat required to drive the subsequent endothermic reactions. These reactions are slower than the initial reactions, however, and as incomplete combustion is always accompanied by carbon formation, a finely divided carbon is always present in the products and must be removed by washing.

Desulfurization is unnecessary. Sulfur in the feedstock is converted primarily to hydrogen sulfide but also to carbonyl sulfide (COS). Nitrogen compounds end up as elemental nitrogen or ammonia. A plant must be built, however, to produce the pure oxygen for the process. This is generally an economic drawback but confers a slight advantage in ammonia manufacture in that the nitrogen coproduced with the oxygen can be added to the synthesis gas to provide the correct composition for ammonia production. Some of the shift reaction and carbon dioxide removal stages can be avoided.

10.4.4 Solid Feedstocks

The two-stage, partial oxidation of coal has been widely used as a route to synthesis gas, and the method could be extended to other carbonaceous feedstocks. A bed of coke or coal is burned in a stream of air until it reaches about 1000°C. The air is then replaced by steam and two coal gasification reactions occur.

$$C + H_2O \rightarrow CO + H_2 \qquad \Delta H_{1000°C} = 130 \text{ kJ/mol}$$

$$C + 2H_2O \rightarrow CO_2 + 2H_2 \qquad \Delta H_{1000°C} = 88 \text{ kJ/mol}$$

The product is called water gas because it is made from water plus coke, or blue gas, because it burns with a characteristic blue flame caused by chemiluminescence. It can be converted to synthesis gas by techniques already described.

If air and steam are passed over the coke simultaneously, the product gases are diluted with nitrogen and are known as producer gas or low Btu gas. Producer gas is obsolete in the United States but is burned in situ as a cheap fuel gas in some parts of the world.

Much effort has been expended on the development of gasification processes for coal to produce synthesis gas for chemicals. Most important is the work done by Eastman, which led to a coal-based acetic anhydride process (Section 10.5.2.3) based on the Texaco gasifier.

10.4.5 Hydrogen

Hydrogen is manufactured by steam reforming and partial oxidation, but large amounts are also obtained in the refinery as byproducts of cracking and catalytic reforming reactions (Sections 2.2.1 and 2.2.3). Minor sources include coke oven gases, as well as the electrolysis of water, brine (which gives principally sodium hydroxide and chlorine), and hydrogen chloride and hydrogen fluoride (to give chlorine and fluorine).

When hydrogen is obtained from synthesis gas or refinery processes, it is purified by washing successively at 180°C and about 20 bar with liquid methane, to remove nitrogen and carbon monoxide, and with liquid propane to remove methane.

About 60% of all hydrogen is used for the production of ammonia. The second largest use is in refinery processes in hydrotreating, hydrocracking (Section 2.2.6), hydrodesulfurization and toluene hydrodealkylation (Section 8.1). Most of this hydrogen is produced internally in other refinery processes. Among organic chemicals made outside the refinery, methanol is the largest consumer of hydrogen. Other applications include the conversion of benzene to cyclohexane (Section 7.2), nitrobenzene to aniline (Section 7.3), and unsaturated fats to saturated or hard fats (Chapter 13).

10.5 CHEMICALS FROM SYNTHESIS GAS

Ammonia is by far the most important chemical made from synthesis gas and consumes about 5% of the world's natural and associated gas production. Even though it is not organic, it is produced from and used to make organic chemicals. Methanol is the second important chemical from synthesis gas and is the basis for methyl *tert*-butyl ether (Section 5.2.1), formaldehyde, and acetic acid. It consumes about 1% of natural and associated gas production. All except methyl *tert*-butyl ether are described here.

10.5.1 Ammonia and Its Derivatives

Ammonia is prepared by the Haber–Bosch process, which requires a synthesis gas of composition $N_2/3H_2$. The mixture is passed over a promoted iron oxide catalyst at about 450°C and 250 bar. Conversions are low, about 10% per pass, requiring a large recycle of unreacted synthesis gas.

Although the process is formally the same as that developed by Haber, the technology has been modified. More active catalysts have reduced the required operating temperature. The main change, however, has been the replacement of reciprocating pumps by centrifugal pumps. Reciprocating pumps operate like bicycle pumps with an in-and-out action and can achieve very high pressures; the early Haber plants operated at 1000 bar. Centrifugal pumps are like giant electric fans and involve only rotary motion. They are cheaper than reciprocating pumps, require less frequent maintenance, do not require lubrication so that oil contamination of the catalyst is avoided, and can be driven by turbines operated on the steam generated from waste heat or from the steam-reforming process. Their single drawback is that they are limited to pressures of about 250 bar. Their advantages, however, are so great that it is more economic to operate at this relatively low pressure and tolerate the low conversions.

This new technology has made possible the scaling-up of ammonia plants from the 200 tons/day common in 1960 to 2000–3000 tons/day now. Steam crackers for ethylene also depend on centrifugal pumps, which have made possible the scale-up to 1500 tons/day. Another development is the use of microporous polysulfone membranes to separate unreacted hydrogen from the product stream so that it can be recycled without inert contaminants such as argon and without the waste involved in a purge stream (Section 7.1.2.3).

It is not likely that further basic improvements in ammonia technology will be made. The hope of the future is biological nitrogen fixation, whereby non-leguminous plants such as corn, wheat, and oats can be made nitrogen-fixing by bacteria created by gene-splicing techniques. Should this objective be achieved, it might provide the basis for a second "Green Revolution" for food production.

About 75% of ammonia production is used for fertilizers, mostly as ammonia, some in ammonium salts but also, in substantial quantities, in the form of urea.

Ammonia reacts with carbon dioxide to give ammonium carbamate, which in turn dehydrates to urea.

$$NH_3 + CO_2 \rightarrow \underset{\substack{\text{Ammonium} \\ \text{carbamate}}}{NH_2COONH_4} \xrightarrow{-H_2O} \underset{\text{Urea}}{NH_2CONH_2}$$

Urea contains 46% fixed nitrogen and 75% of US production is used as fertilizer. Urea plants are always built next to ammonia plants, which supply the raw materials and also have excess heat available.

Urea was first synthesized by Wöhler in 1828, who heated an aqueous solution of ammonium cyanate at about 100°C.

$$\underset{\substack{\text{Ammonium} \\ \text{cyanate}}}{NH_4OCN} \rightarrow \underset{\text{Urea}}{NH_2CONH_2}$$

This seminal reaction, in which an inorganic compound was converted to an organic material, proved that organic chemicals were chemicals like any other. The discovery gave rise to the discipline of organic chemistry. It had previously been thought that organic chemicals could only be made by living systems. Indeed, Wöhler defined an organic compound as one associated with life processes. Today the definition is much broader, for organic compounds are simply those that contain carbon.

Wöhler's reaction may have commercial application in that nitric oxide, carbon monoxide and hydrogen have been found to combine at 60°C to give ammonium cyanate.

$$2NO + CO + 4H_2 \rightarrow NH_4OCN + 2H_2O$$

The reaction is catalyzed by a variety of metals including platinum and copper–nickel, although a platinum–rhodium gauze appears to be best. The ammonium cyanate in turn is converted quantitatively to urea, at 100°C. This reaction is not economically feasible as a route to urea because of the cost of NO but it has been considered for removing emissions from automobile exhaust.

Ammonia may be oxidized to nitric acid, which is used, sometimes mixed with sulfuric acid, to produce a variety of nitro compounds and their derivatives. Most explosives [e.g., cellulose nitrate, TNT (trinitrotoluene), Tetryl, (2,4,6-trinitrophenylmethylnitramide), picric acid, pentaerythritol tetranitrate, Cyclonite (cyclotrimethylene trinitramine), nitroglycerol, and ammonium nitrate] are nitro compounds, as are nitromethane, nitroethane, and the nitropropanes (Section 11.1), which are used as propellants, chemical intermediates, and solvents.

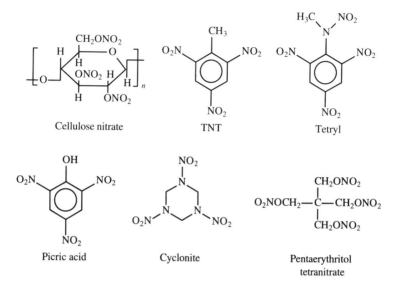

Nitrobenzene is the precursor of aniline and MDI (Sections 7.3 and 7.3.1). Dinitrotoluene is the precursor of toluene diisocyanate (Section 8.3). Nitrocyclohexane was an intermediate in an obsolete caprolactam process (Section 7.2.2) and the most important caprolactam process uses ammonia-based hydroxylamine. Ammonia is important in ammoxidation (Section 4.4) and was involved in the first synthesis of hexamethylenediamine (Section 5.1.1). It is used similarly to convert fatty acids to amides (Section 13.2). It also reacts with alkyl halides, alcohols, and phenols to give amines (Section 7.3). Chemicals produced from ammonia are shown in Table 10.2.

10.5.1.1 Urea and Melamine–Formaldehyde Resins

Urea reacts with formaldehyde to give thermoset urea–formaldehyde (U/F) resins (Section 15.4.1). The reactions are complex, involving first of all the formation of methylolurea and N,N'-dimethylolurea.

$$H_2NCONHCH_2OH \qquad OC(NHCH_2OH)_2$$

Methylolurea $\qquad N,N'$-dimethylolurea

These, by a series of condensations that include ring formation, provide thermoset polymers.

The major use of urea-formaldehyde resins is as a binder in particle board. They are also used to creaseproof fabrics, to impart wet strength to paper, and, in the form of a foam, as insulation for buildings. Residual formaldehyde in such foams is a problem that can be avoided by inclusion of melamine as a formaldehyde scavenger.

TABLE 10.2   Major Chemicals from Ammonia—United States, 1993

Chemical	Production (billion lb)	Ammonia content[a] (billion lb)
Ammonia	34.50	
Diammonium phosphate	27.9	7.19
Nitric acid	17.07	4.61
Ammonium nitrate	16.69	3.57[b]
Urea	15.66	8.88
Monoammonium phosphate	5.94	0.88
Ammonium sulfate[c]	4.80	1.24
Acrylonitrile	2.51	0.81
Caprolactam[d]	1.36	1.58
Aniline	1.01	0.18
Hexamethylenediamine	0.75	0.22
Ethanolamines[e]	0.71	0.16

[a] Assuming 100% yields.
[b] Excluding the ammonia going into the nitric acid.
[c] Does not include byproduct ammonium sulfate from caprolactam manufacture.
[d] Assuming exclusive use of the oxime route and ignoring the ammonia required for the hydroxylamine. About 5.4 billion lb of byproduct ammonium sulfate is produced.
[e] Mono-, di- and triethanolamines.

Compression molding of U/F resins gives items such as electrical fittings and toilet seats. Molding powders are invariably formulated with fillers. They are less heat and water resistant than P/F resins but can be fabricated in a range of cheerful colors.

Urea is also the source of melamine, which is made in the United States by a one-step process involving trimerization of molten urea at 400°C with release of ammonia and CO_2 in the presence of an aluminosilicate catalyst. Both low- and high-pressure processes are used and, in the former, cyanic acid HNCO is an intermediate.

An obsolescent process started with calcium carbide from coal and went by way of dicyandiamide. Some plants in Europe still use this process.

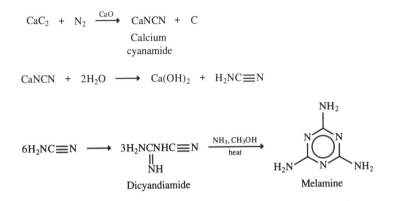

Like urea, melamine reacts with formaldehyde to give complex thermosetting resins. Melamine–formaldehyde (M/F) resins are high quality premium products and are used for dinnerware, the top layer of laminates such as "Formica," and in industrial and decorative coatings. They make urethane foams fire retardant. Like the U/F resins, M/F resins are useful for textile and paper treatment, adhesives, and molding powders.

10.5.2 Methanol

Ninety percent of world methanol is based on synthesis gas from natural gas or methane. Seventy percent of methanol produced in 1990 was used for chemicals. The remaining 30% was used for fuels, particularly for the production of methyl *tert*-butyl ether (Section 5.2.1). Fuel usage will increase appreciably in the mid-1990s, as MTBE replaces a large proportion of the aromatics and lead compounds in gasoline.

Methanol is made from synthesis gas.

$$CO + 2H_2 \rightleftharpoons CH_3OH$$

Excess hydrogen is added for kinetic reasons although the CO/H_2 ratio may be as low as 2.02:1. The steam reforming of naphtha, which as indicated earlier, is not widely used in the United States or Western Europe, produces synthesis gas close to the desired composition. The steam reforming of methane or natural gas yields a synthesis gas with a H_2/CO ratio close to 3:1. There are two ways of adjusting the ratio. The first is to purge the excess hydrogen, and the purged gas is normally burned as reformer fuel. The second is to add carbon dioxide from an external source such as an ammonia plant to take up the excess hydrogen.

$$CO_2 + 3H_2 \rightarrow CH_3OH + H_2O$$

There is recent evidence that the mechanism of the reaction is one in which the carbon monoxide is converted by the shift reaction to carbon dioxide, after which the carbon dioxide reacts with hydrogen as shown above, to yield methanol. A recently announced but uncommercialized process for methanol is based solely on carbon dioxide (see note).

Initially methanol was prepared by BASF by a high-pressure process (320–380°C and 340 bar) with a $ZnO-Cr_2O_3$ catalyst with Zn/Cr ratio of 70:30. In 1972 ICI commercialized a low-pressure process (240–260°C and 50–100 bar), which made use of a copper–zinc catalyst on an alumina support. The energy savings are large but the process requires synthesis gas almost completely free of chlorine and sulfur compounds. There have since been a number of medium-pressure processes (100–250 bar) with copper oxide added to the Zn–Cr catalyst system. Techniques for purifying synthesis gas are sufficiently good to have made the ICI process dominant. Although the catalyst is more expensive, the reaction is more selective, giving higher methanol yields and purer product. The major side reaction is the formation of dimethyl ether from methanol. Aldehydes, ketones, esters, and higher alcohols are formed in minute quantities. The methanol is refined by distillation.

At least four new processes are under development for methanol production, none of which has been commercialized. In one process, a homogeneous catalyst, typically ruthenium carbonyl, operates at temperatures as low as 120°C where the $H_2 + CO/CH_3OH$ equilibrium strongly favors methanol. Intermetallic catalysts have been explored extensively in a second process but thus far have not found commercial application. A barium–copper intermetallic compound makes possible methanol formation at 280°C and 60 bar. A zinc promoted Raney–copper catalyst also has been shown to have higher activity than the Cu–Zn–alumina catalyst used in the standard low-pressure process.

An experimental liquid-phase methanol process makes use of a copper–zinc catalyst supported on alumina. The reaction can take place at 250°C and 50 bar. The reaction is carried out in an inert liquid such as a hydrocarbon oil. Once the methanol forms it is vaporized and purified. The possibility of producing methanol by direct oxidation of methane will be discussed in Section 10.9.

Methyl *tert*-butyl ether (Section 5.2.1) consumes a rapidly rising proportion of methanol output. Excluding this end use, about 35% of methanol is converted to formaldehyde. Acetic acid and acetic anhydride (Sections 10.4.2.2 and 10.4.2.3) syntheses comprise the second largest use followed by methyl methacrylate (Section 4.6.1) terephthalic acid/dimethyl terephthalate (Section 9.3), methylamines, solvents, and a variety of lower volume uses including dimethyl sulfate, and dimethyl carbonate (Section 7.1.2.2). Methanol is also used in the preparation of single cell protein. Applications in chlorinated methanes (Section 10.2) and glycol ethers (Section 3.11.6 and 4.7) are either banned or declining rapidly.

10.5.2.1 Formaldehyde
The major route to formaldehyde is the dehydrogenation–oxidation of methanol. Some producers use pure oxygen over a

ferric oxide–molybdenum oxide catalyst. A small amount of formaldehyde is produced by the partial oxidation of lower petroleum hydrocarbons. Proposed methods include the hydrogenation of carbon monoxide, the pyrolysis of formates, and the direct oxidation of methane (Section 11.1).

The dehydrogenation–oxidation process uses a stationary bed silver catalyst and a mixture of methanol vapor and air at approximately atmospheric pressure and 700°C. Some water is added to facilitate methanol conversion and to prevent catalyst deactivation. Less than the stoichiometric amount of oxygen is used. The resulting gases are absorbed in water. It is believed that two gas-phase reactions take place, one involving dehydrogenation and the other oxidation.

$$CH_3OH \rightleftharpoons HCHO + H_2 \qquad \Delta H = 20 \, kcal/mol$$

$$CH_3OH + 0.5H_2 \rightarrow HCHO + H_2O \qquad \Delta H = -38 \, kcal/mol$$

The first reaction, like all dehydrogenations, is endothermic and the second is exothermic. Properly carried out, a favorable heat balance results. There is some evidence that a second reaction does not in fact take place as written above but rather that the exothermicity results from the oxidation of some of the methanol to carbon dioxide and water. Conversions as high as 75% have been reported with net molar yields of 83–92%. The byproduct hydrogen plus traces of formaldehyde can be burned as fuel.

The oxidation process uses a ferric oxide–molybdenum oxide catalyst in a ratio of about 1:4. Reaction takes place at 300–400°C with almost quantitative methanol conversion at selectivities above 90%. Compared with the silver-catalyzed process, the lower temperature reduces corrosion problems, and higher formaldehyde concentrations are obtained without distillation. The drawbacks are that the great excess of air means higher capital and energy costs, and the waste gas containing traces of formaldehyde is incombustible and must be specially purified. Consequently, most producers currently prefer silver catalysts.

The gaseous formaldehyde is dissolved in water and, when prepared in this way, always contains 1–2% of methanol, which serves as a stabilizer. Commercial formaldehyde is sold in several forms. Aqueous solutions contain up to 60% formaldehyde (37% is common) as a hydrate or as low molecular weight oxymethylene glycols $H[OCH_2]_nOH$. A second form is as a cyclic trimer called trioxane, and a third is "paraformaldehyde," which forms when water is evaporated from an aqueous solution of formaldehyde. This polymeric form and the oxymethylene glycols are both readily "unzipped" by heat or acid.

Of the 2.5 billion lb of formaldehyde produced in the early 1990s in the United States, one-quarter was used for the preparation of phenol–formaldehyde resins (Section 7.1.1). An equal amount was used for urea–formaldehyde resins (Section 10.4.1.1). A smaller amount was used for melamine–formaldehyde resins (Section 10.4.1.1).

Polyacetal resins are polymers of formaldehyde with the recurring unit $-OCH_2-$. They are strong stiff polymers classified as "engineering" plastics, useful for replacing metal in, for example, valves, hose, and tube connectors, machine housings, and many structural parts (Section 15.2).

A smaller use for formaldehyde is in the preparation of MDI/PMDI (Section 7.3.1). Formaldehyde combines with acetylene to give 1,4-butynediol, which can be hydrogenated to 1,4-butanediol, and a series of related chemicals (Section 10.3.1). With ammonia it yields hexamethylene tetramine (Section 15.4.1). "Hexa" is an intermediate in the manufacture of RDX, the explosive that replaced TNT in "blockbuster" bombs in both World War II and the Korean War. Its principal use is as a convenient source of formaldehyde under alkaline conditions for the curing of B-stage or partially cured phenolic resins (Sections 7.1.1 and 14.5.1).

Formaldehyde with acetaldehyde yields pentaerythritol (Section 3.11.3). A related material, trimethylolpropane, is made by the condensation of formaldehyde and butyraldehyde (Section 4.8). Similarly, neopentyl glycol results from the condensation of formaldehyde, and isobutyraldehyde (Section 4.8).

Nitrilotriacetic acid (Section 10.1) requires formaldehyde. One synthesis of isoprene depends on the condensation of formaldehyde and isobutene (Section 6.2). These miscellaneous uses account for about 15% of formaldehyde consumption. A proposed process for ethylene glycol involves the dimerization of formaldehyde (Section 3.7.2).

10.5.2.2 *Acetic Acid*

Acetic (ethanoic) acid can be made by a number of routes, the most important of which is Monsanto's methanol carbonylation.

$$CH_3OH + CO \rightarrow CH_3COOH$$
Acetic acid

The reaction is catalyzed by iodine-promoted rhodium at 200°C and 1–3 bar and gives 99 + % selectivity to acetic acid, based on methanol. This high selectivity is exceeded in industrial chemistry only by the reactions in which isobutene is reacted with water or methanol to give *tert*-butanol or methyl *tert*-butyl ether (Sections 5.2.1 and 5.2.4).

The Monsanto process followed on the heels of an older BASF process using cobalt iodide CoI_2 as a catalyst. This synthesis, however, required a temperature of 250°C and a pressure of 60 bar to give much lower selectivities than the Monsanto process. Numerous byproducts complicated the purification of the acetic acid.

In the United States about 70% of all acetic acid is made by the Monsanto process. In Western Europe in 1992 about 43% was made by methanol carbonylation, 39% by acetaldehyde oxidation, and 18% from light naphtha (see below).

A second method for making acetic acid, which is obsolete in the United States and obsolescent in Western Europe because of methanol carbonylation, is the oxidation of acetaldehyde (Section 3.5). It involves treating a 5–15%

acetaldehyde solution in acetic acid with air in the presence of dissolved cobalt or manganese acetates at 50–70°C. As indicated earlier (Section 3.4) methanol carbonylation has been responsible for a marked decline in acetaldehyde production (Section 3.5).

The third process is based on liquid oxidation of hydrocarbons, namely a mixture of propane and n-butane in the United States and a light naphtha fraction called primary flash distillate in Europe, especially in the United Kingdom. The process requires large amounts of water and yields a dilute acetic acid solution whose concentration is energy intensive. The oxidation of n-butanes takes place at 175°C and 54 bar with a cobalt acetate catalyst. Many byproducts are produced, the major one being methyl ethyl ketone. The oxidation of naphtha takes place at 70–90°C at 40 bar. As might be expected, even more byproducts form than in the oxidation of butane. Important ones are propionic and formic acids. Succinic acid forms in substantial quantity but lacks a market. The hydrocarbon oxidation process makes use of the cheapest raw materials but the separation of the acetic acid is complicated and expensive. Even so, the byproducts have value in their own right and contribute to the economic success of the process. Thus the plants will continue to operate through the 1990s.

Not yet commercialized are processes for making acetic acid by the direct combination of CO and hydrogen. One such process pioneered by Texaco makes use of a bimetallic ruthenium–cobalt system promoted by iodide. Selectivity, however, is far less than with the rhodium-catalyzed process.

Of the 3 billion lb of acetic acid consumed in the United States in 1993, over one-half is used for the preparation of vinyl acetate (Section 3.5). The 1993 end-use pattern is shown in Figure 10.2. Approximately 10% is used to make cellulose acetate (Section 14.3). A miscellany of commercial acetate esters includes butyl, ethyl, n-propyl, and isopropyl. However, the largest volume esters are acetates of glycol ethers (Section 3.11.6.2).

About 8% of acetic acid is also used as a solvent in the Amoco process for terephthalic acid (Section 9.3.1). Because much of it is oxidized, continual replacement is necessary. About 5% is converted to acetic anhydride and the remainder finds application in a number of small volume uses including conversion to chloroacetic acid used for the preparation of carboxymethylcellulose (Section 14.3) and 2,4-dichlorophenoxyacetic acid, a common herbicide.

Glycine and thioglycolic acid are also made from chloroacetic acid. Acetic acid is used in several textile operations including textile dying, and in the synthesis of photographic chemicals, rubber chemicals, pharmaceuticals, and herbicides. It is also used as a grain fumigant.

10.5.2.3 Acetic Anhydride Acetic anhydride is made by three processes. In one, acetic acid (or acetone, but that is uneconomical) is pyrolyzed to ketene, which in turn reacts with acetic acid. The pyrolysis takes place at a 700–800°C in the presence of triethyl phosphate at a very low residence time of 0.25–0.5 s. Molar yields are 85–89%.

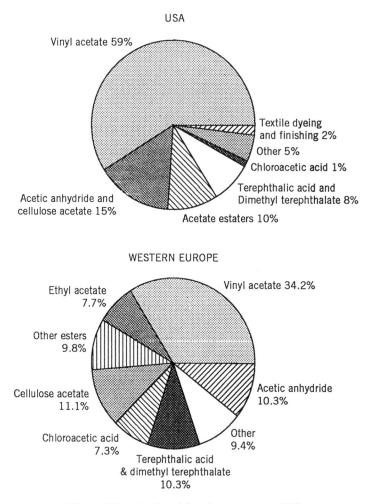

Figure 10.2 Acetic acid end-use pattern, 1993.

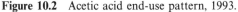

The second procedure involves the in situ production of peracetic acid from acetaldehyde, which in turn reacts with more acetaldehyde to yield the anhydride. This process is probably not in use. Ethyl acetate may be used as a solvent, and the reaction is catalyzed by a mixture of cobalt and copper acetates. The molar yield of acetic anhydride is about 75%.

$$CH_3CHO + O_2 \longrightarrow CH_3COOOH$$

Acetaldehyde   Peracetic
                                                  acid

$\xrightarrow{\quad CH_3CHO \quad}$ $(CH_3CO)_2O + H_2O$

Eastman Chemical working with Halcon have developed a novel process involving the carbonylation of methyl acetate with a catalyst comprising rhodium chloride and chromium hexacarbonyl, $Cr(CO)_6$, in an acetic acid solvent. In one reaction described in a patent (see note), β-picoline is a catalyst modifier and methyl iodide is a promoter. The hydrogen/carbon monoxide ratio is important, since increase in the hydrogen content provides a corresponding increase in the production of byproduct ethylidene diacetate. This compound is the basis for a proposed process for vinyl acetate (Section 3.5) but is undesirable if high selectivity to the anhydride is desired. Since this reaction is new, the possible mechanism is of interest. Only the rhodium is shown as the catalyst. In the first step acetyl iodide is formed. Reaction of this product with methyl acetate yields acetic anhydride.

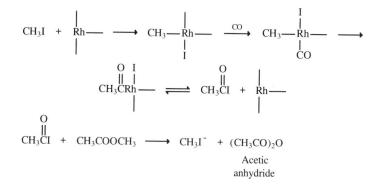

The Eastman–Halcon process is significant not only because of the imaginative chemistry involved but because the synthesis gas (Section 10.4) on which the process depends comes from coal. The Eastman plant in Tennessee is located close to the coal mines, which eliminates expensive transportation costs. The coal is gasified in an Eastman-modified Texaco gasifier operating at a high temperature to provide the synthesis gas with very little of the methane, which is undesirable for chemical operations. The acetic anhydride production is completely synthesis gas based since the carbon monoxide and hydrogen react to form methanol. This in turn esterifies recovered acetic acid (see below) to provide the methyl acetate that is further carbonylated to acetic anhydride.

The acetic anhydride is used by Eastman for the esterification of cellulose to cellulose acetate (Section 14.3). In the process, a mole of acetic acid is released and is recycled, obviating the need for a dedicated acetic acid plant.

$$Coal + O_2 + H_2O \longrightarrow CO + H_2$$

$$CO + 2H_2 \longrightarrow CH_3OH$$

$$CH_3COOH + CH_3OH \longrightarrow CH_3COOCH_3 + H_2O$$

$$CH_3COOCH_3 + CO \longrightarrow (CH_3CO)_2O$$

Acetic
anhydride

$$(CH_3CO)_2O + HO-Cellulose \longrightarrow CH_3COO-Cellulose + CH_3COOH$$

Cellulose acetate

This is the only modern process in which coal has replaced petroleum. It provides confidence in the belief that coal can be used to manufacture all of the chemicals the world needs, should supplies of petroleum and natural gas be depleted. Full plant cost of the acetic anhydride produced by Eastman is 30% cheaper than the same product produced by conventional techniques. However, the investment for the coal-based process, because of the high cost of the gasifier, is roughly three times that of the investment for the conventional ketene-based process. Thus the use of coal requires a trade-off between high investment and low cost raw material. Also the cost of coal-based chemicals will be increased because of the high cost of transporting coal.

Acetic anhydride's main use is in the production of cellulose acetate. Other uses are small and include the formation of various esters such as acetylsalicylic acid (aspirin) and amides of which acetaminophen (paracetamol) is a prime example.

Ketene, the intermediate in one acetic anhydride process, is a powerfully lachrymatory gas with a choking smell. It is invariably used in situ. Apart from its role in acetic anhydride production, it can be dimerized over trimethyl phosphate to diketene. Ammonia is added to inhibit the back reaction.

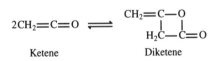

Ketene   Diketene

Diketene reacts with methanol and ethanol to give methyl and ethyl acetoacetates.

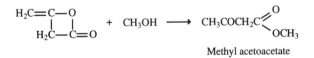

Methyl acetoacetate

It also reacts with aromatic amines to give acetoacetarylamides. With aniline, acetoacetanilide is produced. This and its homologs are used in production of azo dyes.

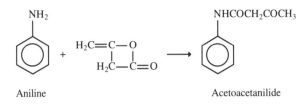

Aniline   Acetoacetanilide

10.5.2.4 *Methanol to Gasoline* In addition to the acetic anhydride process described above, there are two additional uses for methanol—conversion to gasoline, described below, and as a substrate for single-cell protein, which is discussed in Section 14.5.

The process for converting methanol to gasoline, known as the Mobil MTG process, is operated in New Zealand. It is a simple although high-investment process. Like the Fischer–Tropsch process, it starts with the generation of synthesis gas. Methanol is generated by the ICI low-pressure process (Section 10.5.2). The methanol is then brought into contact with a fixed bed of the acid form of the zeolite catalyst ZSM-5 (known as HZSM-5, see Section 16.9) at temperatures of the order of 380°C. There is a separation of carbon from oxygen, the former being polymerized into hydrocarbon chains and the latter emerging as water. The same reaction takes place over many acidic catalysts, but there is rapid coking and loss of catalytic activity. The geometrical selectivity of ZSM-5, however, does not allow for the formation of the linked aromatic rings that are the precursors of coke, hence the catalytic activity is maintained.

Water and CO_2 are the only oxygenated products of this highly exothermic reaction. To control the exotherm, the reaction is carried out in two stages. In the first, a dehydration catalyst such as CuO/γ-Al_2O_3 promotes methanol dehydration to dimethyl ether.

$$2CH_3OH \rightarrow CH_3OCH_3 + H_2O$$
$$\text{Dimethyl}$$
$$\text{ether}$$

The resulting equilibrium mixture of methanol, water, and dimethyl ether is fed to the second reactor. The dimethyl ether appears to lose water, which diffuses easily out of the zeolite structure, to leave CH_2 radicals. The CH_2 radicals polymerize within the zeolite pore system to give alkanes and aromatic hydrocarbons up to the geometrical selectivity limit (C_{10}) imposed by the pore structure. A mixture of aliphatic and methylated aromatics results with a high octane number. Typical compositions at different temperatures are shown in Table 10.3.

This MTG process without modification would not be suitable for the production of aromatics for the chemical industry unless catalyst changes increased benzene content appreciably. Petroleum's low price during the late 1980s made this process uneconomical until the 1990 Persian Gulf crisis. New

TABLE 10.3   Composition (mol%) of Gasoline from Methanol

Reaction temperature (°C)	370	538
C_1–C_4 Aliphatics	28.84	60.50
C_5 + Aliphatics	33.83	3.50
Benzene	0.96	
Toluene	4.69	
Xylenes	12.33	35.90
C_9 Aromatics	12.25	
C_{10} Aromatics	7.10	
C_{11} + Aromatics		0.10
Total aromatics	37.33	36.00

Zealand had announced in 1989 that it would divert one-half of the methanol used for the process to the merchant market. The Gulf crisis and increased petroleum prices caused this decision to be reversed, at least temporarily.

10.5.2.5 Lower Volume and Proposed Uses for Methanol Lower volume uses of methanol include the production of methylamines, glycol ethers, (Sections 3.11.6.2, and 4.7.1) dimethyl sulfate, dimethyl carbonate, and methylated phenols. Methylamine, dimethylamine, and trimethylamine are made by the vapor phase ammonolysis of methanol at 450°C with an alumina gel catalyst. Conversion based on ammonia is 13.5% to primary, 7.5% to secondary, and 10.5% to tertiary amine. The separation of the three products is difficult. Trimethylamine can be separated from the mixture as an ammonia azeotrope. The mono- and diamines may be separated by fractionation. Another method of separation involves removing the dimethylamine, which is the most basic of the three, as the hydrochloride. Extractive distillation is also possible.

Dimethylamine is the product used in largest amount—about 100 million lb/year—but produced in smallest amount in the amination reaction. Thus, the methylamine and trimethylamine are recycled without separation. Dimethylamine's largest uses are for conversion to N,N-dimethylformamide by carbonylation and to dimethylacetamide by reaction with acetic acid, acetic anhydride, or acetyl chloride. Both are solvents for the spinning of acrylic fibers and serve also as aprotic solvents for synthesis and extractive distillation.

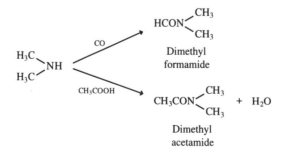

Other applications include the synthesis of dodecyldimethylamine oxide, a detergent foam stabilizer, as shown in the following equations. It is also used for rubber chemicals such as thiuram accelerators, and for conversion to the dimethylamine salt of 2,4-dichlorophenoxyacetic acid, an important herbicide (Section 7.1.5).

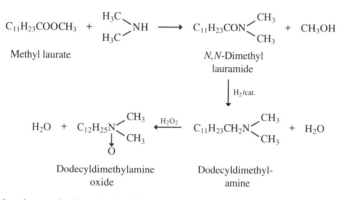

Methyl laurate   N,N-Dimethyl lauramide

Dodecyldimethylamine oxide   Dodecyldimethyl-amine

Methylamine and trimethylamine enjoy approximately equal use of about 40 million lb/year. A major application for methylamine is in the synthesis of the insecticide carbaryl, the compound that was manufactured at Bhopal. Methylamine reacts with phosgene to give methyl isocyanate, the chemical that gave rise to the worst disaster in the history of the chemical industry. The methyl isocyanate, in normal circumstances, reacts with 1-naphthol to yield carbaryl.

$$CH_3NH_2 + COCl_2 \longrightarrow CH_3NCO + 2HCl$$

Methyl
isocyanate

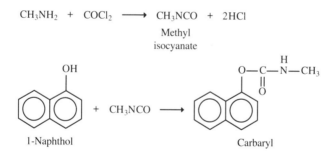

1-Naphthol   Carbaryl

Trimethylamine's major use is in the synthesis of choline chloride, an important animal feed supplement, by reaction with ethylene oxide.

$$(CH_3)_3N + H_2C-CH_2 \xrightarrow[\text{(ii) HCl}]{\text{(i) 100°C}} [(CH_3)_3NCH_2CH_2OH]^+Cl^-$$

Trimethyl-amine   Ethylene oxide   Choline chloride or "choline"

Dimethyl sulfoxide, an important aprotic solvent, is manufactured by the oxidation of dimethyl sulfide, which can be obtained either as a byproduct from the Kraft paper pulping process or by the reaction of methanol with hydrogen sulfide, a reaction that also gives methyl mercaptan. The latter vapor phase reaction takes place at above 300°C and requires a dehydration catalyst. The oxidation to dimethyl sulfoxide can be carried out with nitrogen dioxide or, better, with oxygen containing a small amount of nitrogen dioxide. With the former, the reaction takes place at 40–50°C and the product is purified by distillation.

$$CH_3OH \ + \ H_2S \ \xrightarrow{\text{cat.}} \ (CH_3)_2S \ + \ CH_3SH \ + \ H_2O$$

Dimethyl   Methyl
sulfide   mercaptan

$$(CH_3)_2S \ + \ NO_2 \ \longrightarrow \ (CH_3)_2SO \ + \ NO$$

Dimethyl
sulfoxide

$$2NO \ + \ O_2 \ \rightleftharpoons \ 2NO_2$$

Dimethyl carbonate is made by the oxidative carbonylation of methanol in the liquid phase with a cuprous chloride catalyst. The reaction can also be carried out continuously at a pressure of about 25 bar with synthesis gas rather than pure CO, which is the only material consumed.

$$2CH_3OH \ + \ \tfrac{1}{2}O_2 \ + \ CO \ \xrightarrow{\text{cat.}} \ \begin{matrix} CH_3O \\ \\ CH_3O \end{matrix}\!\!\!C{=}O \ + \ H_2O$$

Dimethyl
carbonate

At conversions of 30–35% the yield of dimethyl carbonate is 100% based on methanol and 90–95% based on carbon monoxide. A newer process makes use of a solid copper methoxide with a nitrogen-containing cocatalyst capable of coordination, such as pyridine. Dimethyl carbonate is of current interest because it may be used for alkylations in place of the more corrosive dimethyl sulfate, and because it can replace phosgene in the preparation of isocyanates and carbamates (see note). A General Electric patent describes a synthesis for polycarbonate resins that replaces phosgene with dimethyl carbonate (Section 7.1.2.2). A plant was under construction in Japan in the mid-1990s. Dimethyl carbonate has also been suggested as an octane improver for gasoline.

Methanol is an alkylating agent. At 50 bar and a carefully controlled temperature of 300°C, it will alkylate phenol to o-cresol or 2,6-xylenol at about 50 bar over an alumina catalyst. Selectivity is high and only small amounts of ethers or *meta* or *para* alkylated products result. The ratio of o-cresol to 2,6-xylenol can be varied by controlling pressure and temperature, but these variations may promote the formation of byproducts such as 2,4-xylenol. The initial and insufficient source of o-cresol is coal tar distillate (Section 12). 2,6-Xylenol is the monomer for an engineering polymer, poly(phenylene oxide).

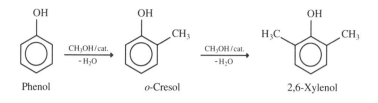

10.5.2.6 *C₁-Based Development Processes* The elaborate chemistry of the

10.5.2.6 *C₁-Based Development Processes*   The elaborate chemistry of the olefin and aromatic feedstocks already described provides facile methods for obtaining and using organic compounds from C_2 upwards. The great divide is between C_1 compounds and those containing several carbon atoms, and much research was focused on so-called C_1 chemistry in the 1970s and 1980s when methanol was cheap. In the late 1980s its price increased because of the demand for its use in methyl *tert*-butyl ether.

Meanwhile, C_1 chemistry gave rise to a number of processes that merit discussion. The chain reaction between methanol and formaldehyde to yield ethylene glycol was described in Section 3.7.2. Other reactions based on methanol include cracking to olefins, conversion to aromatics, synthesis of vinyl acetate via methyl acetate, and homologation to ethanol and higher alcohols.

The cracking of methanol to olefins takes place in the presence of a zeolite catalyst such as ZSM-5 (Section 10.5.2.4). Modification with metals is possible. Manganese doping provides 50% selectivity to ethylene and 30% to propylene. Selectivity of 60% to ethylene is possible with a shape-selective catalyst known as Linde AW-500 zeolite. The ratio of olefins obtained depends on the catalyst. Thus the Mobil process with ZSM-12 provides a selectivity to propylene of almost 60% at a methanol conversion of 91%. In 1991 Montedison announced a new process for propylene production. Although the chemistry was not disclosed, it could involve zeolite-catalyzed cracking of methanol.

A comparison of the steam cracking of naphtha with that of methanol catalyzed by a zeolite shows a marked advantage to methanol cracking, as shown in Table 10.4. The economics, however, do not favor methanol because its price is much higher than that of naphtha or ethane/propane. It nonetheless

TABLE 10.4 Comparison of Steam Cracking of Naphtha and Methanol

	Methanol	Naphtha
Catalyst	Zeolite type	None
Reaction temperature (°C)	325–425	850
Steam feed (tonnes per tonne)	0–1.0	0.5–0.8
Ethylene yield (based on carbon selectivity)%	30–60	26–32
Ethylene + propylene yield (based on carbon selectivity)%	70–80	44–50

provides another route to chemicals from coal via synthesis gas should oil and natural gas supplies be depleted.

Just as methanol may be cracked to olefins over one set of zeolite catalysts, it may be aromatized over others. One example has already been given for the MTG process (Section 10.5.2.4). The highly methylated aromatics are useful for gasoline but would have to be hydrodealkylated to give the benzene that the organic chemical industry needs. Further catalyst development is required.

Vinyl acetate (Section 3.6) can be prepared from methyl acetate by a process devised by Halcon reminiscent of the Eastman–Halcon acetic anhydride process (Section 10.5.2.3). Methyl acetate reacts with carbon monoxide and hydrogen to give ethylidene diacetate, which on pyrolysis provides vinyl acetate and acetic acid.

$$
\underset{\text{Methanol}}{CH_3OH} \;+\; \underset{\text{Acetic acid}}{CH_3\overset{\overset{\displaystyle O}{\|}}{C}-OH} \;\longrightarrow\; \underset{\text{Methyl acetate}}{CH_3\overset{\overset{\displaystyle O}{\|}}{C}-OCH_3} \;+\; H_2O
$$

$$
\underset{\text{Methyl acetate}}{2CH_3\overset{\overset{\displaystyle O}{\|}}{C}-OCH_3} + 2CO + H_2 \;\longrightarrow\; \underset{\text{Ethylidene diacetate}}{CH_3CH(O-\overset{\overset{\displaystyle O}{\|}}{C}CH_3)_2} + \underset{\text{Acetic acid}}{CH_3\overset{\overset{\displaystyle O}{\|}}{C}-OH}
$$

$$
\underset{}{CH_3CH(O-\overset{\overset{\displaystyle O}{\|}}{C}-CH_3)_2} \;\longrightarrow\; \underset{\text{Vinyl acetate}}{H_2C\!=\!\overset{\overset{\displaystyle H}{|}}{C}-O-\overset{\overset{\displaystyle O}{\|}}{C}-CH_3} + \underset{\text{Acetic acid}}{CH_3\overset{\overset{\displaystyle O}{\|}}{C}-OH}
$$

Since the acetic acid is recycled, the net reaction is

$$
2CH_3OH + 2CO + H_2 \;\longrightarrow\; H_2C\!=\!\overset{\overset{\displaystyle H}{|}}{C}-O-\overset{\overset{\displaystyle O}{\|}}{C}-CH_3 + 2H_2O
$$

The carbonylation in the presence of hydrogen is accomplished with a rhodium chloride catalyst modified with β-picoline and promoted with methyl iodide. Acetic anhydride and acetaldehyde are obtained as byproducts. The economics of the Halcon process seem reasonable although capital investment is considerably higher than for the conventional process.

The homologation of methanol to higher alcohols is of interest first of all as a route to ethanol whose price, conventionally, is several times that of methanol (e.g., in 1993, $1.80/gal of ethanol vs. $0.46/gal of methanol). It also provides a route to higher alcohols conceivably useful as oxygenates to augment octane numbers in unleaded gasoline, although methyl *tert*-butyl ether appears superior to any of them.

The homologation of methanol to ethanol proceeds with a dicobalt octacarbonyl catalyst.

$$
CH_3OH + 2H_2 + CO \xrightarrow{\text{cat.}} CH_3CH_2OH + H_2O
$$

At 200 bar and 365°C, 76% conversion and a 40% selectivity to ethanol are obtained. A variety of other catalysts has been tested, including cobalt catalysts with phosphine ligands. A variant of the process involves the reaction of methanol with synthesis gas to give acetaldehyde, which can then be hydrogenated to ethanol. Typically, higher selectivities result when a reaction is carried out in two steps with the isolation of an intermediate. Many catalysts have been proposed for the preparation of acetaldehyde by this route based on cobalt, nickel, palladium, ruthenium, and tungsten. In most instances, ligands are required.

The economics of these processes are attractive. Less favorable are the economics associated with either the homologation of methanol to alcohols higher than ethanol or to the production of higher alcohols from synthesis gas directly. Nonetheless, the high selectivity possible for obtaining specific alcohols from synthesis gas is demonstrated by work in which CO and hydrogen are combined to provide 44% selectivity to isobutanol, which may be dehydrated to isobutene for conversion to methyl *tert*-butyl ether. The technology is not practiced because the reaction sequence *n*-butane → isobutane → isobutene, is more economical (Section 5.2.1). There are also several methanol-based routes to ethylene glycol (Section 3.7.2).

10.6 CARBON MONOXIDE CHEMISTRY

Carbon monoxide is the basic C_1 molecule. Much of its chemistry has already been discussed. Current C_1 industrial processes are based largely on methane (Section 10.1–10.5) from natural gas. The other C_1 molecules comprise methanol, carbon monoxide, and formaldehyde. Interesting chemistry is in the wings based on these molecules, which are obtainable if necessary from coal.

The largest application of C_1 chemistry is the production of methanol (Section 10.4.2). Striking examples of the replacement of classical chemistry by C_1 chemistry are the Monsanto acetic acid process (Section 10.5.2.2) and the Eastman acetic anhydride process (Section 10.5.2.3). The use of CO in hydroformylation has been discussed under propylene (Section 4.8). The use of CO for the preparation of dimethylformamide and dimethylacetamide was described above (Section 10.5.5).

With chlorine, carbon monoxide yields phosgene. The reaction is carried out over activated charcoal at 250°C.

$$CO + Cl_2 \xrightarrow{\text{cat.}} COCl_2$$
Phosgene

About 85% of US phosgene is consumed in the production of diisocyanates (Sections 7.3.1 and 8.3) and the bulk of the remainder for polycarbonate preparation (Section 7.1.2).

10.6.1   Proposed Processes Based on Carbon Monoxide

Proposed uses of carbon monoxide in industrial processes have been mentioned throughout this book, especially the possibility of making ethylene glycol directly from CO and hydrogen (Section 10.6.1) and non-phosgene routes to isocyanates (Sections 7.2.3 and 8.4). Much novel carbon monoxide chemistry is associated with glycols, glycol ethers, and glycol carbonates.

A proposed route to glycol ethers involves the combination of an alcohol with formaldehyde, carbon monoxide, and hydrogen. The reaction takes place at 180 bar and 160°C with a homogeneous catalyst comprising dicobalt octacarbonyl and donor ligands such as diphenyl sulfide and diphenyl oxide.

$$CH_3OH + HCHO + CO + 2H_2 \rightarrow CH_3OCH_2CH_2OH + H_2O$$

Diethylene glycol bis(allyl carbonate), also termed allyl diglycol carbonate, is a specialty polycarbonate polymer, whose major use is for molding eyeglass lenses by in situ polymerization. A classical process involves the reaction of diethylene glycol with phosgene to give a bis chloroformate which, on further reaction with allyl alcohol, gives the desired product.

$$HOCH_2CH_2OCH_2CH_2OH + COCl_2 \longrightarrow Cl\overset{O}{\overset{\|}{C}}OCH_2CH_2OCH_2CH_2O\overset{O}{\overset{\|}{C}}Cl$$

Diethylene glycol

$$\xrightarrow[\text{Allyl alcohol}]{CH_2=CH-CH_2OH} CH_2=CH-CH_2-O-\overset{O}{\overset{\|}{C}}OCH_2CH_2OCH_2CH_2O\overset{O}{\overset{\|}{C}}-O-CH_2-CH=CH_2$$

Diethylene glycol bis(allyl carbonate)

An alternative process operated in Japan reacts diethylene glycol and allyl chloride with carbon dioxide in the presence of sodium carbonate. The reaction probably proceeds in two steps.

$$HOCH_2CH_2OCH_2CH_2OH + 2Na_2CO_3 + 2CO_2 \longrightarrow$$

$$NaO\overset{O}{\overset{\|}{C}}OCH_2CH_2OCH_2CH_2O\overset{O}{\overset{\|}{C}}ONa + 2NaHCO_3 \xrightarrow[\text{Allyl chloride}]{CH_2=CH-CH_2Cl}$$

Sodium diethylene glycol
carbonate

$$CH_2=CH-CH_2-O-\overset{O}{\overset{\|}{C}}OCH_2CH_2OCH_2CH_2O\overset{O}{\overset{\|}{C}}-CH_2-CH=CH_2$$

The idea of using carbon dioxide instead of phosgene could conceivably be expanded to the preparation of other carbonate polymers.

Carbon monoxide based routes to pelargonic, malonic, phenylacetic, phenyl-pyruvic, and oxalic acids have been developed but mostly not commercialized. Currently there are two processes for pelargonic acid, one from natural sources (Chapter 13) and one based on 1-octene (Section 4.8.2) but it may also be made by dimerization of butadiene in the presence of carbon monoxide. The carbonylation is carried out homogeneously with carbon monoxide in alcohol in the presence of a palladium–phosphine complex. Hydrogenation and saponification provide pelargonic acid.

$$2CH_2\text{==}CH\text{---}CH\text{==}CH_2 \ + \ CO \ + \ ROH \ \longrightarrow \ CH_2\text{==}CHCH_2CH_2CH_2CH\text{==}CHCH_2COOR$$

Butadiene

$$\xrightarrow[\text{(2) Saponification}]{\text{(1) Hydrogenation}} \ CH_3(CH_2)_7COOH \ + \ ROH$$

Pelargonic acid

Palladium acetate or palladium acetylacetonate may be used as the source of palladium, with acetonitrile as solvent and methanol as the esterifying agent. The catalyst must be free of halide in order to effect dimerization of the butadiene. If halide is present, the butadiene is also carbonylated to give a C_5 unsaturated acid.

A malonic ester synthesis is based on the reaction of carbon monoxide, ethanol, and ethyl chloroacetate in the presence of cobalt tetracarbonyl at 55°C and 8 bar. The molar yield is 94%. The process is said to be in use in Japan.

$$ClCH_2COOC_2H_5 \ + \ CO \ + \ C_2H_5OH \ \longrightarrow \ H_2C\underset{COOC_2H_5}{\overset{COOC_2H_5}{\diagdown}} \ + \ HCl$$

Ethyl
chloroacetate

Diethyl malonate
(malonic ester)

Another carbon monoxide-based process for malonic ester involves its inter-action with ketene and a nitrous acid ester. The reaction, which makes use of a platinum or platinum salt catalyst, proceeds at 115°C at atmospheric pressure but gives lower yields than the ethyl chloroacetate process.

$$CH_2\text{==}C\text{==}O \ + \ CO \ + \ 2C_2H_5ONO \ \xrightarrow{\text{cat.}} \ H_2C\underset{COOC_2H_5}{\overset{COOC_2H_5}{\diagdown}} \ + \ 2NO$$

Ethyl nitrite

Malonic ester

Phenylacetic acid results from the phase-transfer carbonylation of benzyl chloride.

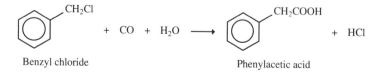

Benzyl chloride

Phenylacetic acid

The aqueous phase comprises 40% aqueous sodium hydroxide solution. The organic phase contains a quaternary ammonium compound and a dicobalt octacarbonyl catalyst. The quaternary ammonium salt transfers the cobalt carbonyl salt $[Co(CO)_4]^-$ from the aqueous phase to the organic phase. The benzyl chloride, which is added continuously, is carbonylated under pressure.

Bis carbonylation of benzyl chloride gives phenylpyruvic acid. The reaction again requires a dicobalt octacarbonyl catalyst and proceeds in the presence of calcium hydroxide at 85°C and 60 bar in a *tert*-butanol–water mixture. Phenylpyruvic acid is formed in a molar yield greater than 90%. Enzymatic amination of this compound yields L-phenylalanine, a key component of the noncaloric sweetener, aspartame.

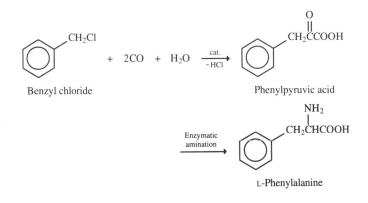

A Japanese process possibly in use for oxalic acid involves the oxidative coupling of carbon monoxide.

$$2CO + 2RONO \xrightarrow{Pd} ROOCCOOR + 2NO$$
Oxalic acid ester

$$2NO + 2ROH + 0.5O_2 \longrightarrow 2RONO + H_2O$$

The reaction is carried out homogeneously at 110°C and 60 bar. In a corresponding heterogeneous reaction, palladium on carbon may be used at 120°C and atmospheric pressure. The oxalic acid ester on hydrogenolysis gives ethylene glycol. This process was mentioned earlier as an experimental one explored jointly by Union Carbide and UBE in Japan as a step in the proposed conversion of synthesis gas to ethylene glycol (Section 3.7.2).

α-Olefins (Section 3.3) result when synthesis gas is passed over zeolite ZSM-5 catalyst impregnated with Fischer–Tropsch catalysts such as iron nitrate. The products have a chain-length distribution of 2–27 carbon atoms but it is heavily skewed to the C_2–C_5 compounds. It is noteworthy that both odd and even number carbon compounds result. Small concentrations of non-linear olefins, alcohols, aldehydes, and ketones are also formed (Section 12.2). The process will require much more development before commercialization. Sasol, however, has commercialized an α-olefin process for C_5–C_8 products based on the Fischer–Tropsch process (Section 12.2).

NOTES AND REFERENCES

C_1 Chemistry attracted much attention in the early 1980s and C_1 *Molecular Chemistry: An International Journal* was published to focus attention on it. The American Chemical Society published the results of a Symposium, *Industrial Chemicals via C_1 Processes*, D. A. Fahey, Ed., American Chemical Society, Washington, DC, 1987. The Japanese Minstry of International Trade and Industry sponsored a 7-year program for C_1 chemistry, some of which was reported in *Progress in C_1 Chemistry in Japan*, Ed., Research Association for C_1 chemistry, Elsevier, Amsterdam, 1990. General books about methane include L. H. Clever, Methane, Pergamon, Oxford, UK, 1987 and J. C. Murrell, *Methane and Methanol Utilizers*, Plenum, NY 1992.

Various of the topics in this chapter are covered in *Petrochemical Processes: Vol. 1, Synthesis Gas Derivatives and Major Hydrocarbons; Vol. 2 Major Oxygenated, chlorinated and Nitrated Derivatives*, Inst. Français du Pétrole Publications, Editions Technip, Paris, 1989.

Section 10.2 In this section we generally use the systematic rather than the trivial names for the chloromethanes. Thus we use chloromethane for methylene chloride, dichloromethane for methylene chloride, trichloromethane for chloroform, and tetrachloromethane for carbon tetrachloride. The technical literature is less consistent and cheerfully mixes terms such as dichloromethane with chloroform.

Among the many books on the problems of fluorocarbons, we note World Health Organization, *Partially Halogenated Chlorofluorocarbons*, WHO, 1992.

Section 10.2.1 For an excellent brief review of silicone properties and uses, the reader is referred to R. B. Seymour and C. E. Carraher, Jr., *Polymer Chemistry*, 2nd ed. Dekker, New York, 1988.

Section 10.2.5 The methyl bromide issue is presented by B. Chakrabarti and C. H. Bell, *Chem. Ind.* (1993) 984 and 992.

Section 10.5.1 The process for converting NO to urea is described in US Patent 3 986 849 (1975) to Val Laboratories.

Section 10.5.2. A number of books have been published on methanol synthesis and possible fuel use including Wu-Hsun Cheng, *Methanol Production and Use*, Dekker, New York, 1994; S. Lee, *Methanol Synthesis Technology*, CRC Press, Boca Raton, FL, 1990; J.H. Perry, *Methanol: Bridge to a Renewable Energy Future*, University Press of America, 1990; and J. C. Fahy, *New Prospects for Methanol: Fuel or Chemical?*, Financial Times, London, 1990.

The use of homogeneous catalysis for methanol formation has been described by R. J. Klinger et al., *Exploring Catalytic Methanol Synthesis Using Soluble Metal Oxide Complexes*, Symposium on Chemicals from Syn Gas, Methanol Division of Fuel Chemistry, American Chemical Society, New York, April 1986.

Homogeneous catalysts for methanol formation have been described by E. G. Baglin et al., *Ind. Eng. Chem. Prod. Res.* **20**, 87–90 (1981).

C. D. Chang's article in *Catalysis Rev*, **25** (1) (1983) was reprinted as *Hydrocarbons from Methanol*, Dekker, New York, 1983.

Promoted Raney–copper catalysts have been described in European Patent Application 109-702 (October 31, 1983) to Shell.

For methanol formation from carbon dioxide see J. Haggin, *Chem. Eng. News*, March 28, 1994, p. 29 and D. Rotman, *Chem. Week* March 23, 1994, p. 14.

Section 10.5.2.2 The formation of acetic acid from CO and hydrogen has been described by J. F. Knifton, *Hydrocarbon Processing*, 113–117, December (1981).

Section 10.5.2.3 The Eastman/Halcon process for acetic anhydride is described in West German Offen 2 610 035 (October 3, 1976) and 2 610 036 (September 23, 1976) both to Halcon.

Section 10.5.2.5 The dimethyl carbonate process using a cuprous chloride catalyst is employed by ENI in Italy. The solid copper catalyst has been developed by Dow and is described in US Patent 4 604 242 (August 5, 1986).

Conceivably carbaryl, the pesticide whose manufacture was responsible for the Bhopal disaster, could be made with dimethyl carbonate instead of phosgene by the following reactions

$$2ROH \ + \ (CH_3O)_2CO \ \longrightarrow \ (RO)_2CO \ + \ 2CH_3OH$$

$$2(RO)_2CO \ + \ CH_3NH_2 \ \longrightarrow \ ROCONHCH_3 \ + \ ROH$$

where R = alpha naphthyl =

Makhteshim, in Israel, avoids the use of methyl isocyanate by the reaction sequence

$$CH_3NH_2 \ + \ COCl_2 \ \longrightarrow \ CH_3NHCOCl \ + \ HCl$$

$$ROH \ + \ CH_3NHCOCl \ \xrightarrow{-HCl} \ ROCONHCH_3$$

General Electric's process for preparing polycarbonate resins based on dimethyl carbonate is described in US Patent 4 452 968 (June 5, 1984).

The cracking of methanol to olefins has been described in numerous patents and papers, typical of which are US Patent 4 049 573 (September 20, 1977) and US Patent 4 025 571, 4 025 572 (May 24, 1977) to Mobil Oil; B. B. Singh et al. *Chem. Eng. Commun.* **4**, 749–758 (1980).

The process claiming high selectivity to propylene by methanol cracking is described in European Patent Application 010 5591 (April 8, 1984) to Mobil Oil.

Section 10.5.2.6 The conversion of CO and hydrogen to isobutanol is described in European Patent Disclosure 0 208 102 82 (May 23, 1986) to Union Rheinische Braunkohlen Kraftstoff, A.G.

The homologation of methanol is described by B. Juran and R. V. Porcelli, *Hydrocarbon Processing*, October 1985, p. 85.

Section 10.6.1 The formation of glycol ethers from carbon monoxide is claimed in US Patent 4 308 403 (December 29, 1981) to Texaco.

The process for preparing diethylene glycol bis(allyl carbonate), in use in Japan, is described in US Patent 4 217 298 (August 12, 1980) to Tokuyama Soda.

The preparation of pelargonic acid from butadiene and CO is described by J. Tsuji et al., *Tetrahedron*, **28**, 3721 (1972); W. E. Billoups et al., *Chem. Commun*, 1067 (1971); and US Patent 4 246 183 (January 20, 1981) to Texaco Development Corp.

The conversion of CO and hydrogen to α-olefins is described in European Patent Application 0 037 213 (October 7, 1981) to Mobil.

The carbonylation of benzyl chloride is described in CHEMTECH **18** 317 1988.

The bis carbonylation reaction to yield phenylpyruvic acid is claimed in US Patent 4 492 798 (January, 1985) to Ethyl.

The formation of oxalic acid esters by the coupling of carbon monoxide is the subject of US Patent 4 229 589 (October 21, 1980) to Ube Industries.

The Val Laboratories process for converting NO to urea is described in US Patent 3 986 849.

CHAPTER 11

CHEMICALS FROM ALKANES

Alkanes occur as such in natural gas and petroleum, and accordingly are the cheapest raw materials for chemicals. They are the feedstocks for cracking (Sections 2.2.1 and 2.2.2) and catalytic reforming (Section 2.2.3). Methane is the main source for synthesis gas (Section 10.4) via steam reforming. The higher alkanes can be subjected to the same process if desired, or the steam reforming process can be redirected to give methane. An important process is pyrolysis of hydrocarbons to carbon black, which is discussed at the end of this chapter.

Apart from pyrolysis, these reactions are endothermic. They are all non-selective and take place at high temperatures. There are few examples of alkane functionalization, that is, of the use of alkanes directly for downstream chemicals. The most important are the conversion of *n*-butane to maleic anhydride (Section 5.4), the oxidation of *n*-butane or naphtha to acetic acid (Section 10.5.2.2), the oxidation of isobutane to *tert*-butylhydroperoxide (Section 4.7) and the chlorination of methane (Section 10.2). Lesser volume uses involve ammoxidation of methane to hydrocyanic acid (Section 10.1), conversion of methane to acetylene (Section 10.3), and nitration of propane.

Any alkane may be nitrated. In practice only propane is used as feed and, from its nitration result nitromethane, nitroethane and 1- and 2-nitropropane. The nitration takes place at 420°C, and the products are separated by distillation. They are used as additives for gasoline for racing cars, as solvents especially for poly(cyanoacrylates), and as stabilizers of chlorinated solvents. Du Pont developed a process for the nitration of cyclohexane to nitrocyclohexane as a step in a caprolactam synthesis (Section 7.2.2) but it is not used today.

In the early 1990s a propylene shortage, primarily in Europe, motivated development of processes for the dehydrogenation of propane (Section 2.2.7). *n*-Butane also may be dehydrogenated to butadiene (Chapter 5) but it is more energy efficient to use *n*-butenes. The dehydrogenation of ethane to ethylene

338

(Section 11.2.2) has not been commercialized, primarily because of a high selectivity to a coproduct, acetic acid. Very important is the dehydrogenation of isobutane to isobutene for methyl *tert*-butyl ether (MTBE).

The functionalization of alkanes is a research goal not only because of the economic advantage of circumventing the cracking process but also because methane—and to a lesser extent ethane, propane, and *n*-butane from natural or associated gas—are now believed to be a longer lasting source of chemicals than petroleum. The current route to chemicals from methane is via synthesis gas (Section 10.4) but the reaction is energy intensive and the aim is to functionalize methane by a direct process.

The strong and equivalent C–H bonds in methane make "bond activation" difficult. Hence, the functionalization of methane provides a challenge to the chemist. If it can be successfully accomplished, the reward is great, for it will provide the chemical industry with the lowest cost raw material possible. Also it will not be necessary in the shorter run to make the shift to coal with its inherent ecological and economic problems.

11.1 FUNCTIONALIZATION OF METHANE

Early research on the functionalization of methane yielded only marginal results. On the basis of these efforts, it was easy to predict that chemistry would never be discovered to make methane the chemical industry's basic building block. The 1980s, however, saw major advances in catalysis. Methane functionalization attracted intense research in the 1980s, which accelerated in the 1990s. Data are accumulating so rapidly that we can do little more here than provide some insights into the approach (see notes).

Three reactions provide the goals. These are the direct oxidation of methane to methanol and/or formaldehyde; the dimerization of methane to ethane, ethylene, or higher hydrocarbons; and the aromatization of methane.

11.1.1 Methane to Methanol/Formaldehyde

The oxidation of methane to methanol and formaldehyde is burdened by the fact that formaldehyde is 21 times more susceptible to oxidation than methane at 670°C. Methanol is even more sensitive to oxidation. Of the scores of patents issued, one to Hüls is typical. Methane and oxygen are mixed at 300–600°C at a pressure of 400 bar. Residence time is a critical 10^{-3} s. Conversions per pass, however, are not greater than 3%. The process is not currently economical because of the high capital investment, which the short residence time and the low-conversion rates necessitate. These negative factors provide the incentives for further research.

Nitrous oxide appears to be a particularly good oxidant for methane and two Japanese patents (see notes) claim its use with catalysts such as Mo_3/SiO_2 and V_2O_5 at 450–500°C. A 93% selectivity to formaldehyde was obtained but at a conversion of only 0.5%. At 11% conversion, a 98% selectivity was obtained to a mixture of methanol and formaldehyde. At this latter conversion, the process shows some promise, but its practicality is questionable because the molar ratio of nitrous oxide to methane must be 2:1. Nitrous oxide is an expensive oxidant because it cannot be made directly from nitrogen and oxygen and requires a roundabout route via ammonium nitrate.

Typical of more recent work is a Catalytica process in which methane reacts with sulfuric acid in the presence of a mercury catalyst to yield methyl hydrogen sulfate CH_3OSO_3H, which can be easily hydrolyzed to methanol and recoverable sulfuric acid.

11.1.2 Dimerization of Methane

The dimerization of methane to ethane and ethylene has been extensively studied. In the Benson process, methane is burned in chlorine in a highly exothermic reaction at an adiabatic flame temperature of 700–1700°C. As might be expected, huge amounts of hydrogen chloride are obtained, which must be reconverted to chlorine. This seriously inhibits commercialization of the process.

Catalytic oxidative coupling of methane has also been explored. Thus it has been shown that lithium-doped magnesium oxide in the presence of oxygen will extract hydrogen from methane to form methyl radicals, which in turn combine to produce ethane and ethylene at 720°C. Conversion is about 38% with a selectivity of about 50%. This is promising. ARCO has been able to obtain conversions up to 15% with selectivities of 78% to C_2–C_7 compounds, mostly ethane and ethylene. One catalyst described is manganese acetate with sodium promoters. In other work nitrous oxide was used.

Recent, more dramatic results from the University of Minnesota suggest the feasibility of oxidative coupling of methane to give ethylene at 830°C in the presence of oxygen and samarium oxide Sm_2O_3. Yields of 60% are claimed as opposed to previously achieved values of 25%. The key appears to be the shifting of equilibrium by the rapid removal of oxygen, methane and ethylene.

The direct conversion of methane to acetic acid with oxygen, rhodium chloride, and water at 100°C has been accomplished at Pennsylvania State University, although reaction rates are very low.

11.1.3 Aromatization of Methane

The conversion of methane to aromatics has been studied by British Petroleum. The aromatization is accomplished in the presence of an oxidant, nitrous oxide, with an acid catalyst, a gallium-doped H-form zeolite. Methane conversion of 39% per pass has been reported with selectivities to aromatics of 19%. Although such results are promising, the economics of the reaction are doubtful, because 1 mol of nitrous oxide is consumed for every two C–H bonds that are broken. Also, additional oxidant is consumed in the undesirable formation of carbon oxides. Obviously, it is necessary to find a cheaper oxidant or ways of using the oxidant catalytically rather than stoichiometrically.

As mentioned above, methane can be halogenated under mild conditions. In a proposed process for converting methane to gasoline, chloromethane is dehydrohalogenated under conditions such that coupling takes place simultaneously to give C_2-C_5 olefins which, under the conditions of the reaction, recombine to provide gasoline-range paraffins and aromatics. Some olefins are also produced. In the process hydrogen chloride is evolved and is reused for oxychlorination of the methane:

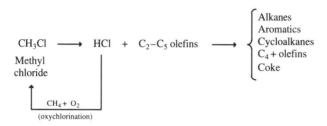

Research from the University of Minnesota indicates that platinum or rhodium catalysts effect the conversion of methane to synthesis gas (Section 10.4) at ambient temperatures. Selectivity and conversions are reported to be high.

11.2 Functionalization of C_2–C_4 Alkanes

The commercialized reactions in which alkanes have been oxidized were listed at the beginning of Chapter 11. Several interesting processes, not yet commercialized, have been described in the literature.

11.2.1 Oxidation of C_2–C_4 Alkanes

The oxychlorination of ethane to vinyl chloride may be carried out with a metallic silver–manganese catalyst in combination with other compounds such as lanthanum salts, either in the particulate form or impregnated on a zeolite, to provide vinyl chloride at 400°C and atmospheric pressure.

$$CH_3CH_3 + 0.5O_2 + Cl_2 \rightarrow CH_2{=}CHCl + H_2O + HCl$$

Contact time is 1–2 s. Complete conversions can be obtained with selectivity to vinyl chloride as high as 50%. This process, patented by ICI, shows promise. An earlier process devised by Lummus using conventional oxychlorination catalysts yielded selectivities of 37% at 28% conversion.

Monsanto has studied the oxychlorination of ethane to vinyl chloride in a vapor phase fluidized-bed reactor with a catalyst comprising alumina-supported copper halide and potassium phosphate at 550°C. Ethyl chloride is a byproduct, which can subsequently be oxidatively dehydrogenated to vinyl chloride. Ethylene dichloride byproduct can be cracked by conventional means to vinyl chloride. Conversions of the order of 85–90% based on hydrogen chloride can be achieved with selectivities as high as 87% based on ethylene.

The ammoxidation of propane to acrylonitrile has been targeted for commercialization in the late 1990s by BP America. The reaction is postulated to proceed by way of a propylene intermediate. One proposed catalyst in an early patent issued to another developer, Monsanto, comprises a mixture of antimony and uranium oxides with a halogen promoter such as methyl bromide. Uranium was a second generation catalyst for the ammoxidation of propylene (Section 4.4). Reaction takes place at 500°C to give 71% selectivity at 85% conversion for a per pass yield of 60%. Raw material savings are somewhat eroded by the higher capital costs of the process. Newer patents describe catalysts comprising a mixture of vanadium, antimony, phosphorus, and cobalt as well as bismuth, vanadium, molybdenum, chromium, and zinc. These provide lower yields per pass but may have other advantages.

In a reaction analogous to the oxidation of butane to acetic acid, propane or a propane-butane mixture can be oxidized at about 450°C and 20 bar to acetaldehyde and a large number of other oxygenated compounds. The reaction can be conducted either in the liquid or gaseous phase and has been used commercially in the United States. In a related development ICI has oxidized ethane in the presence of hydrogen chloride to acetaldehyde. A silver manganate catalyst $AgMnO_4$ is used at 360°C. Conversion is 14% and selectivity to acetaldehyde is 71%. Chlorinated byproducts such as methyl and ethyl chloride can be recycled to inhibit their additional production.

11.2.2 Dehydrogenation of C_2–C_4 Alkanes

Dehydrogenation of ethane, propane, or butane to the corresponding olefins is an alternative to steam cracking, which requires higher temperatures and greater capital investment.

The dehydrogenation of butenes to butadiene (Chapter 5) is carried out commercially in the United States. n-Butane, as mentioned earlier, is seldom if ever dehydrogenated because of the greater energy input required. The dehydrogenation of propane was commercialized in 1991 in Thailand. A European plant came on stream in 1992 to provide propylene for a Statoil–Himont joint venture polypropylene plant. Propane dehydrogenation is also of interest in Saudi Arabia, where ethane cracking provides so little coproduct propylene that its isolation is not economic.

There are at least three processes for propane dehydrogenation, differing largely in operating conditions and catalysts. Generally, however, the reaction takes place at 500–670°C at 0.1–0.3 bar. The Houdry process, used in Europe, is cyclic and employs a fixed-bed catalyst comprising alumina impregnated with 18–20% chromia. Conversion is 50% and selectivity to propylene is over 80%.

The UOP process, used in Thailand, is a continuous fixed-bed operation with a catalyst comprising platinum on an inert support. Conversions are believed to be of the order of 30–40% with selectivities as high as 95%. The temperature is 600–700°C at atmospheric or slight positive pressure. The Phillips process makes use of a supported platinum and tin catalyst with promoters. The process is semicontinuous with relatively high conversions of about 50% per pass.

An important dehydrogenation reaction currently in use and growing rapidly involves conversion of isobutane to isobutene (Section 5.2.1).

Petroleum wax fractions from lubricating oil dewaxing can be dehydrogenated to α-olefins (Section 3.3). These processes could be improved if oxidative dehydrogenation were possible. Thus far successful processes have not evolved because the products oxidize more readily than the starting material.

The dehydrogenation of ethane to ethylene is considerably more difficult than that of the higher hydrocarbons. Phillips Petroleum has worked in this field. Union Carbide has devised an oxidative dehydrogenation in the vapor phase to produce a mixture of ethylene and acetic acid. The ratio of the two products can be varied from 1:1 to 5:1. The catalyst comprises molybdenum and vanadium doped with niobium, antimony, and one other metal, which can be calcium, magnesium, or bismuth. The reaction takes place at 330–435°C and 20 bar with conversions of approximately 30% and selectivities as high as 90%. Higher hydrocarbons are transformed to carbon dioxide and water. A disadvantage is the acetic acid coproduct. Although this might provide a convenient route to acetic acid without requiring the investment for dedicated plants, the inbalance between ethylene and acetic acid demand (41 versus 3.7 billion lb in 1993) means that it could never be the sole source of ethylene, although one or two plants could be accommodated.

11.2.3 Aromatization of C$_2$–C$_4$ Alkanes

Since ease of aromatization increases with molecular weight, ethane aromatizes more readily than methane, and propane and butane more readily than ethane. The dehydrocyclization of alkanes, primarily propane and butane or liquefied petroleum gas (LPG) to aromatics, provides the basis for BPs Cyclar process for which a demonstration plant was operated in the early 1990s. The process was subsequently licensed by Saudi Arabia.

The aromatization of ethane takes place with a gallium or platinum-doped ZSM-5 zeolite catalyst. The shape selective property of the catalyst (cf. Section 8.1, toluene disproportionation) promotes the selective formation of cyclohexane and methylcyclohexane. Dehydrogenation provides benzene and toluene. The inlet temperature is about 700°C. Because the reaction is endothermic, a high input of heat per pound of ethane converted is required. The reaction is carried out at 2 bar and the ethane conversion per pass is about 33%. Of this, 30 mol% is methane and 60% is benzene and toluene. Less than 2% is higher hydrocarbons.

TABLE 11.1 Aromatics Yield from Aromatization of Propane and Butane (%)

Aromatic Compounds	Propane	Butane
Benzene	32.0	27.9
Toluene	41.1	42.9
Xylenes	18.9	21.8
C_9 and C_{10} Aromatics	8.1	7.4

The Cyclar process uses a propane–butane (LPG, Section 2.1) mixture in a reactor maintained at 535°C and 6 bar with a contact time of 14 s. Twenty-nine percent conversion with 95% selectivity to aromatics results. The process will provide a source of chemical benzene in areas like Saudi Arabia where pyrolysis gasoline (Section 2.2.1) is not available and where very little catalytic reforming (Section 2.2.3) is done. The BTX distribution with propane and butane as feeds is shown in Table 11.1.

11.3 CARBON BLACK

Many petrochemical processes are seriously hindered by the formation of carbon, which poisons catalysts, blocks furnace tubes, and so on. Thermodynamically, alkanes are inclined towards carbon and hydrogen as the most stable products. The production of carbon from alkanes, including methane, is thus relatively easy.

Carbon black is an amorphous graphite or soot consisting of highly aromatic carbon structures of colloidal size. It is made by partial combustion or combustion plus thermal cracking of hydrocarbons at 1300–1400°C. The feedstock may be an alkane or olefin or almost any hydrocarbon. Methane was once used widely but is no longer economic. Gas oil and residual oils are now popular especially from sources high in aromatics.

Many grades of carbon black are produced, varying primarily in particle size but also in other properties important in rubber compounding. About 60% is used in tires to provide abrasion resistance and mechanical strength to the rubber. The remaining 40% goes into other elastomers, printing inks, paints, and plastics.

Automobile tires now last longer. More automobiles are imported into the United States and US automobiles are smaller than in the past. Hence, the carbon black market has declined from about 3.5 billion lb/year in the early 1970s to about 2.5 billion lb/year now. It nonetheless remains an important chemical.

NOTES AND REFERENCES

Nomenclature again: The term alkane is generally preferred to paraffin in the industrial chemical literature. The unsaturated counterpart of an alkane is an alkene, and to be consistent this term should also be used. However, the industry overwhelmingly prefers olefin which, from the point of view of nomenclature, is the unsaturated counterpart of a paraffin. We had little choice in this chapter but to follow the industry practice of using the word olefin but not the corresponding word paraffin, which is virtually unknown.

Section 11.1   In addition to the references on methane conversion either to methanol or ethylene, the following, which are but a small sample of the literature available, provide examples of the interest the chemical trade press has taken in this topic. *Chem. Eng. News*, April 10, 1989, p. 5; *ibid.*, July 4, 1988, p. 22; *ibid.*, January 18, 1988, p. 27; *ibid.*, May 9, 1988, p. 45; *ibid.*, September 28, 1987, p. 23; *ibid.*, September 14, 1987, p. 19; *ibid.*, June 1, 1987, p. 22; *Chem. Week*, October 21, 1987 p. 49, *CHEMTECH*, **17** p. 501 (1987); N. D. Parkyns, *Chem. Br.* **26**, 841 (1990). G. J. Stiegal and R. D. Srivastava, *Chem. Ind.* (1994) 854.

Section 11.1.1   The Catalytica and University of Minnesota work have been reported in *Chem. Eng. News*, January 18, 1993, p. 6, *ibid.* October 11, 1993, p. 4, and *Science* **262**, 221 (1993).

The Hüls oxidation of methane to methanol is described in German Patent 2 743 113 (1979).

Japanese patents 189, 249–250 (November 27, 1981) to I. Masakazu describe the oxidation of methane with nitrous oxide to methanol and formaldehyde.

Section 11.1.2   For the University of Minnesota work, see reference in Section 11.1.1. The conversion of methane to acetic acid is described in *Chem. Week*, April 20, 1994, p. 8.

The process in which methane is burned in chlorine is described in US Patent 4 199 533 (April 22, 1980) to the University of Southern California.

The lithium-doped magnesium oxide catalyst for methane coupling has been explored by T. Ito and J. Lunsford, *Nature (London)* **314**, 25 (1985).

The conversion of methane to gasoline, described in a patent to Atlantic Richfield [US Patent 4 849 751 (July 18, 1989)], is accomplished by oxidatively coupling methane to convert it to a mixture of ethylene, carbon monoxide and hydrogen. The mixture is then treated with a dual function catalyst comprising ZSM-5 and an oxide of cobalt, ruthenium, copper, zinc, chromium or aluminum. The zeolite oligomerizes the ethylene to gasoline-sized molecules whereas the oxide, as in the Fischer–Tropsch reaction (Section 12.2), converts CO and H_2 to gasoline range molecules. This is a clever approach, typical of what will be needed if methane functionalization is to become practicable.

Typical of the ARCO patents that claim methane conversion to higher hydrocarbons, are US Patents 4 443 644–649 (April 17, 1984); US Patents 4 544 784–787 (October 1, 1985); US Patents 4 547 607–64 (October 15, 1985); US Patent 4 499 322 (February 12, 1985); US Patent 4 517 398 (May 14, 1984); US Patent 4 523 049 (June 11, 1985); US Patent 4 523 050 (June 11, 1985); US Patent 4 556 749 (December 3, 1985); US Patent 4 560 821 (December 24, 1985); US Patent 4 568 785 (February 4, 1986); and US Patent 4 554 395 (November 19, 1985).

Section 11.1.3   The conversion of methane to aromatics is described in European Patent Application 0 093 543 (November 9, 1983).

The catalyst for the conversion of chloromethane to gasoline-range hydrocarbons comprises zeolites doped with cations such as zinc, gallium, or silver. Reaction takes place at about 325°C and 3 bar. The conversion is described in two International Patent Applications WO85/02608 (June 20, 1985) and WO85/04863 (November 7, 1985) to British Petroleum.

Section 11.2.1   The ICI process for oxychlorination of ethane is described in two British Patent Applications (2 095 242A and 2 095 245A) (September 29, 1982). The older Transcat process is claimed in US Patent 3 775 229 (January 19, 1971) to Lummus.

The Monsanto process for vinyl chloride is claimed in US Patent 4 300 005 (November 11, 1981).

The ICI process for acetaldehyde formation with a sliver catalyst is claimed in US Patent 4 415 757 (November 15, 1983).

Standard Oil of Ohio's (BP America) patents describing propane ammoxidation include US Patent 4 873 215 (October 10, 1959) and Europen Patent Application 0 282 314 (March 10, 1988). An early Monsanto patent is West Germany Patent 2 056 326.

Section 11.2.2   Carbide's ethane dehydrogenation process is the basis for US Patent 4 524 236 (June 18, 1985) to Union Carbide.

The Cyclar process is described in a number of patents including UK Patent Application GB 2 082 157A (March 3, 1982); European Patent Application 0 202 000 A1 (November 20, 1986); US Patent 4 613 716 (September 23, 1986), and US Patent 4 642 402 (February 10, 1987). All of these patents are held by British Petroleum except the last one, issued to UOP, which cooperated with BP on the development.

Section 11.2.3   The conversion of ethane to aromatics is described in a Mobil patent (US Patent 4 350 835, September 21, 1982).

CHAPTER 12

CHEMICALS FROM COAL

We have described the derivation of chemicals from petroleum and natural gas. But about 10% of organic chemicals come from other sources: coal, fats, oils, and carbohydrates. Historically these sources are important because it was from them that the modern chemical industry evolved. Their present applications are also significant, particularly for specialty chemicals. Furthermore, because fats, oils, and carbohydrates are renewable resources, they represent an insurance policy for the future.

Coal, although a nonrenewable source of chemicals and energy, occurs on earth in much larger quantities than petroleum and will certainly outlast petroleum reserves by a few hundred years. We shall discuss it first, before we turn to the renewable sources.

Coal was important to the chemical industry in the nineteenth and early twentieth centuries. It provided calcium carbide and hence acetylene, synthesis gas and hence ammonia and methanol, petroleum-like fuels, and all the aromatic chemicals contained in coke oven distillate. This distillate still provides some chemicals, although the quantities ($\sim 1.5\%$, see note) pale by comparison with those from petrochemicals. The conversion of coal to synthesis gas (Section 10.4) was developed in the nineteenth century as a source of gaseous energy. It was mixed with oil gas (see note to Section 12.2) to increase its calorific content. That coal-based synthesis gas can still provide the basis for chemicals was demonstrated most recently by the Eastman acetic anhydride process (Section 10.5.2.3). Coal-based synthesis gas is also the basis for the Fischer–Tropsch reaction (Section 12.2) and for substitute natural gas (Section 12.4). The hydrogenation of coal is another approach to converting coal to liquid fuels and conceivably to raw materials for chemicals.

After World War II, enthusiasm for coal-based chemicals waned as the cost of petrochemicals dropped. Interest revived in the 1970s at the time of the oil shocks but waned again in the 1980s because the price of petroleum dropped and there were huge discoveries of natural gas around the world especially in the Soviet Union (now the CIS).

12.1 CHEMICALS FROM COKE OVEN DISTILLATE

When coal is heated in the absence of air to a temperature of about 1000°C, coke forms together with many liquid and gaseous decomposition products. It is this distillate, also called coal tar, that provided the aromatics and many other chemicals for the early chemical industry. Some of these are shown in Figure 12.1.

The coke is almost pure carbon and is used in steel manufacture which, as it becomes more efficient and processes change, requires less coke. Nonetheless, some coke will always be needed, and therefore the chemical industry will always have available the chemicals that volatilize from the coke ovens.

A typical coking operation produces 80% coke by weight, 12% coke oven gas, 3% tar, and 1% light oil consisting of crude benzene, toluene, and xylenes. The coal tar is distilled to give four fractions:

1. Light oils, boiling below 200°C. These fractions are called light oils because they float on water. They are crudely fractionated, then agitated with concentrated sulfuric acid to remove olefins. The hydrocarbons are washed with dilute sodium hydroxide and redistilled to give benzene, toluene, xylenes, and "solvent naphtha," a mixture of indene, coumarone, and their homologs. This is a powerful solvent especially for coatings containing coal tar and pitch. Treated with a Friedel–Crafts catalyst such as aluminum chloride, it gives coumarone–indene thermoplastic resins, used for cheap floor tiles, varnishes, and adhesives.

2. The middle oil boils between 200 and 250–270°C. The most abundant chemical is naphthalene and it occurs with phenols, cresols, and pyridines in the tar. It crystallizes when the middle distillate from the tar is allowed to cool and, even in this impure form, is suitable for phthalic anhydride manufacture (Section 9.1.1). Alternatively, it may be purified by sublimation, an unusual purification process. Extraction of the remaining tar with aqueous sodium hydroxide takes the acidic phenols and cresols into the aqueous layer as phenates and cresylates. They are regenerated with carbon dioxide. Subsequent extraction of the remaining oil layer with acid gives nitrogen-containing bases, primarily pyridines.

3. Heavy oil comes off between 250 and 300°C if anthracene oil is taken off as a separate fraction but sometimes they are combined. It is used for wood preservatives generally under the name of creosote or (if the fractions are not separated) anthracene oil.

4. Anthracene oil comes off between 250 and 350/400°C, or 300 and 350/400°C if taken off separately. It contains anthracene, phenanthrene, and carbazole. Anthracene may be obtained by crystallization or solvent extraction. It is oxidized with nitric acid to anthraquinone, a raw material for dyestuffs.

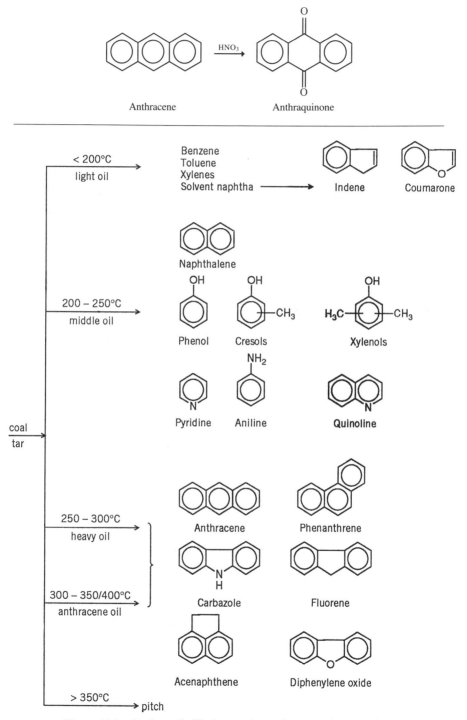

Figure 12.1 Coal tar distillation: major and some minor products.

A competing petrochemical process involves the reaction of benzene with phthalic anhydride in the presence of a Friedel–Crafts catalyst to give o-benzoyl-benzoic acid (Section 7.8), which is dehydrated to anthraquinone.

Some 60% of the tar remains as a residue called pitch. It is thermally polymerized for carbon electrodes, and the residual tar is used for road building and in paints, primarily to waterproof subsurface structures. These often comprise combinations of the tar with epoxy resins.

Benzene from coke oven distillate accounted for 8% of total benzene production in Europe in the early 1990s, but for only 1–2% in the United States (Section 7). The type and quantity of chemicals in coke oven distillate vary with the type of coal and method of coking. Typically, in the most volatile fraction of coke oven distillate, benzene comprises 70%, toluene 15%, and the xylenes 4%. The remainder consists of alicyclics and aliphatics plus some phenols and cresols.

More important in market terms is naphthalene, which is a raw material for phthalic anhydride in competition with o-xylene (Section 9.1.1). Until the early 1960s, coke oven distillate was the sole source of naphthalene, but it can now be obtained by catalytic reforming of heavier naphthas. Unfortunately, methylnaphthalenes form, and the methyl groups must be removed by hydrodealkylation if naphthalene itself is desired. This is the same reaction used to convert toluene to benzene (Section 8.1). Naphthalene production in the United States in the early 1990s was about 260 million lb. Twenty-seven percent of the production facilities were petroleum based. The remainder was from naphthalene from coke oven distillate.

The quantities of distillate available are limited by the demand for coke by the steel industry and could not be increased substantially to compensate for a shortage of petrochemical products.

12.2   THE FISCHER–TROPSCH REACTION

The Fischer–Tropsch reaction provides a route from coal to hydrocarbons. When synthesis gas (Section 10.4) at near atmospheric pressure is passed over an iron, nickel, or cobalt catalyst at 150–300°C, a mixture of alkanes and olefins with a broad range of molecular weights is formed. The olefins are formed first and may be reduced to alkanes. If hydrogen-rich synthesis gas is used, made from naphtha or methane instead of coal, alkanes may be the initial products (Section 10.4).

$$n\text{CO} + 2n\text{H}_2 \rightarrow \text{C}_n\text{H}_{2n} + n\text{H}_2\text{O}$$

$$2n\text{CO} + n\text{H}_2 \rightarrow \text{C}_n\text{H}_{2n} + n\text{CO}_2$$

$$n\text{CO} + (2n + 1)\text{H}_2 \rightarrow \text{C}_n\text{H}_{2n+2} + n\text{H}_2\text{O}$$

The hydrocarbons are predominantly C_5–C_{11} straight chain, although methane, ethylene, and propylene are also produced together with some higher molecular weight Fischer–Tropsch waxes used for candles. There are also oxygenated compounds such as alcohols and acids. The result is a petroleum-like mixture that can be used both as a fuel and a chemical feedstock. Ruthenium has been proposed as a catalyst for higher molecular weight hydrocarbons. Both fixed and fluidized-bed processes are operated; the products are shown in Table 12.1.

The mechanism of the Fischer–Tropsch reaction is complex and has not yet been completely elucidated. It is generally agreed that the above processes are basic. The water gas shift reaction also takes place but requires an iron catalyst.

$$CO + H_2O \rightleftharpoons CO_2 + H_2$$

Its equilibrium can be directed either way depending on temperature, pressure and reactant concentrations. High hydrogen concentrations lead to α-olefins and alkanes. Undesirable reactions include the formation of methane, the disproportionation of carbon monoxide to carbon dioxide and carbon, the reaction of hydrogen and carbon monoxide to give carbon and water, the decomposition of methane to hydrogen and carbon, and the oxidation of the metal catalyst. Formation of carbon fouls the catalyst. Addition of steam inhibits carbon formation and depresses slightly the yield of methane.

Since the hydrocarbons are straight chain, they have low octane numbers and must be isomerized for use as gasoline or in the alkylation reaction (Section 2.2). On the other hand, the straight-chain structure is ideal for steam cracking and catalytic reforming to provide the olefins and aromatics needed for chemical synthesis.

The process was subject to intensive development between the world wars and was operated successfully on a large scale in Germany during World War II.

TABLE 12.1   Fischer–Tropsch Products as Obtained at Sasol

Product	Fixed Bed at 220°C (wt %)	Fluidized Bed at 325°C (wt %)
CH_4	2.0	10
C_2H_4	0.1	4
C_2H_6	1.8	4
C_3H_6	2.7	12
C_3H_8	1.7	2
C_4H_8	3.1	9
C_4H_{10}	1.9	2
C_5–C_{11} (gasoline)	18.0	40
C_{12}–C_{18} (diesel fuel)	14.0	7
C_{19}–C_{23}	7.0	–
C_{24}–C_{35} (medium wax)	20.0	4
$C_{35}+$ (hard wax)	25.0	–
Water-soluble nonacid chemicals	3.0	5
Water-soluble acids	0.2	1

Currently, Fischer–Tropsch processing is underway in South Africa in three plants at Sasolburg. It was by this means that South Africa protected itself against the possibility that its racial policies might provoke a petroleum boycott. A change of government in the 1990s has obviated this need, and it has been announced that the plants will be used for chemical production. The first chemicals to be produced are α-olefins in the $C_5–C_6$ range, formed as indicated above and useful for linear low-density polyethylene (LLDPE) production (Section 3.1.4).

The Fischer–Tropsch process does not produce the same balance of products as an oil refinery. When it was operated on a scale to meet South Africa's gasoline needs, insufficient diesel fuel resulted. Accordingly, additional processing was required, which led to an excess of gasoline. In the early 1990s, this had to be exported at uneconomical international prices. A similar problem afflicted Brazil's gasohol program when the replacement of gasoline by ethanol led to a diesel shortage.

In spite of its superficial virtues, the Fischer–Tropsch process is costly and unreliable in operation. Coke is a solid; consequently reactors are bulky, mechanically complex, and therefore expensive. Utility and maintenance costs are high, and coal has to be dug laboriously out of the ground. The optimum route to gasoline from synthesis gas today is the MTG process described in Section 10.5.2.4. The Fischer–Tropsch products are created by growth of carbon chains on an iron-based catalyst and there is no restriction on their size. Thus a range of alkanes of varying molecular weights, many of them useless for gasoline, is produced. The Mobil process uses molecular sieve catalysts whose pore size is sufficiently small to control the molecular weight within the $C_5–C_{10}$ range.

We have described the Fischer–Tropsch reaction here rather than in Section 10.4 on synthesis gas because it does not seem sensible to make synthesis gas from petroleum and then reverse the process to make petroleum from synthesis gas. It is possible to imagine circumstances, however, under which that might be desirable. Petroleum-based synthesis gas is derived from methane or light naphtha. There are, however, many heavier petroleum fractions such as fuel oil and asphalt that are not used for gasoline or chemicals. If these could be converted cheaply to synthesis gas, which in turn could be reconverted to a lighter fraction suitable for gasoline or chemicals, the process might be worthwhile. Also there is a problem of how to capitalize on natural gas discovered in remote locations where its removal by pipeline is not practical. One suggestion is to produce synthesis gas on site and convert it to a more easily transportable liquid by the Fischer–Tropsch reaction.

12.3 SHELL MIDDLE DISTILLATE SYNTHESIS

A variation of the Fischer–Tropsch process, aimed at enhancing the middle distillate or diesel fuel fraction (Section 2.1), is Shell's Middle Distillate Synthesis

(SMDS). Natural gas rather than coal is used as the source of synthesis gas. A Fischer–Tropsch reaction with a zirconium-promoted cobalt catalyst promotes high selectivity to paraffin waxes. Table 12.1 indicates that wax production is high even in the normal reaction. Conversion is in excess of 80% in order to maximize the formation of the heavy paraffins. These waxes are subsequently hydrocracked (Section 2.2.6) to give a maximum yield of diesel fuel along with some naphtha and kerosene (Section 2.1). In addition, a solvent fraction and hydrocarbons for detergent feedstock are produced. These are dehydrogenated to olefins for alkylation of benzene. The alkylbenzenes are sulfonated to give detergents. The process is in operation in Malaysia.

12.4 COAL HYDROGENATION

Coal has the approximate empirical formula CH. To convert it to an aliphatic hydrocarbon mixture requires addition of hydrogen. In the Fischer–Tropsch process, this is derived ultimately from water. It is, however, possible to hydrogenate coal directly. This process, called the Bergius process, was operated in Germany in World War II. Coal, lignite, or coal tar was hydrogenated over an iron catalyst at 450°C and 700 bar, and 4 million tons of gasoline were synthesized in this way.

Less drastic conditions could be used for coal hydrogenation if a solvent could be found that would hydrogenate the coal in a liquid–solid phase process. In the hydrogen donor process, finely powdered coal reacts for 1–2 h with tetralin at about 200°C and 65 bar. The tetralin acts both as a solvent and a hydrogen donor, giving up four hydrogen atoms to form naphthalene.

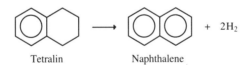

Tetralin   Naphthalene

The hydrogen presumably reacts with free radicals generated by the coal decomposition to stabilize them and prevent further cracking, which would lead to gas formation. Free radical coupling, which leads to coke formation, is also prevented. The naphthalene produced is rehydrogenated to tetralin. The same technique can be used on heavy crude oils and on residues from crude oil distillation. The cheap petroleum of the late 1980s argued against commercialization of this process.

There are thus three potential routes from coal to petroleum-like fuels. The Fischer–Tropsch process is more attractive than the Bergius process and the Mobil process is likely to prove the best of all. Whether or not coal ever becomes a raw material for petroleum-like fuels on a world scale depends on the severity of the petroleum shortage, the availability of the huge capital investment required, and the feasibility of various other possible sources of energy such as nuclear fission or nuclear fusion.

12.5   SUBSTITUTE NATURAL GAS

Methane made synthetically instead of being extracted from natural gas fields is known as substitute natural gas (SNG). It is described here because, in the long term, coal is its logical source. Today it is produced only in countries such as Japan where natural gas is not available and is made from petroleum fractions by a lower temperature variant of steam re-forming.

$$4C_nH_{2n+2} + (2n - 2)H_2O \rightarrow (3n + 1)CH_4 + (n - 1)CO_2$$

The Catalytic Rich Gas (CRG) process (450°C, potassium promoted nickel catalyst) and the Gas Recycle Hydrogenator (750°C, no catalyst) have both been in operation for many years. The former, together with a variant called the "double methanation process," requires a light naphtha feedstock whereas the gas recycle hydrogenator can use heavier hydrocarbons.

In the long term, however, the aim must be to obtain SNG from coal. Synthesis gas technology is relatively straightforward. The difficulty is one of solids handling. The various technologies involve fixed beds, fluid beds and entrained beds. The main processes are summarized in Table 12.2. We shall comment only on the Lurgi process, the slagging gasifier and the Texaco processes.

The supply of steam and oxygen in a molar ratio between 2:1 and 6:1 to a moving bed of coal (replenished at the top, with ash being withdrawn at the bottom, so that it is effectively a fixed-bed technology) leads to a mixture of hydrogen, carbon monoxide, carbon dioxide, and a little methane plus a mass of clinker (fused ash). The gases can be reformed, but the clinker is difficult to handle in a continuous plant. In the Lurgi process, developed in Germany in 1930s, a high ratio of steam to oxygen was used that kept the temperature down and produced a gas high in methane and low in carbon monoxide plus a fine ash that could be handled in an appropriate grate. The excess steam is expensive and leads to large amounts of dilute but corrosive effluent. It also leads to increased plant size for a given capacity and a low thermal efficiency.

A British Gas process—the Westfield Slagging Gasifier—involves a low steam/oxygen ratio that gives a high proportion of carbon monoxide in the gas plus a slag that, at the high temperatures achieved, can be drawn off as a liquid.

The Texaco process, used by Eastman, is a variant of the Koppers–Totzek process and involves an entrained bed. A stream of water carrying pulverized coal meets a stream of oxygen, creating a flame at 1400–1600°C, from which an SNG high in carbon monoxide and hydrogen and very low in methane and condensable hydrocarbons can be extracted. This is thermally inefficient because of the high temperature but the product is more suitable for chemicals than the moving-bed processes.

TABLE 12.2 Coal Gasification Processes

Process	Main Characteristics
Texaco	Pressurized entrained-bed process involving the use of a watery slurry of powdered coal. Less suitable for lignite. Product gas low in CH_4 and tar-free. Low H_2/CO ratio (~ 0.7).
Lurgi	Pressurized moving-bed process suitable for non-caking granular coal. Relatively high-steam consumption. Product gas is rich in methane, residual steam, and CO_2 and contains tar, H_2/CO ratio (~ 1.7).
Koppers–Totzek	Atmospheric entrained-bed process in which powdered coal is used as feed. Suitable for a wide range of coals. Relatively high oxygen consumption. Product gas low in methane, CO_2, and residual steam and free from tar, H_2/CO ratio (~ 0.5).
Winkler	Atmospheric stationary fluid-bed process suitable for reactive coals (lignite). Moderate O_2 and steam consumption. Moderately pure product gas. Moderate coal conversion. Operates at 800–1100°C
High-temperature Winkler (Rheinbraun)	Pressurized stationary fluid-bed process suitable for reactive coals (lignite). Higher gasification rates and more complete coal conversion than by Winkler process.
British Gas-Lurgi slagging gasifier	Pressurized moving-bed process primarily for non-caking granular coal. Less steam consumption, smaller reaction volumes and purer product gas than via Lurgi process.
Shell	Pressurized entrained-bed process in which a dry coal powder is used as feed. Suitable for wide range of coals including lignite. High thermal efficiency. Product gas of high purity, comparable to that obtained by Koppers–Totzek.

12.6 CALCIUM CARBIDE

The production of acetylene by way of calcium carbide and its uses were discussed in Section 10.3. The process uses energy extravagantly and, at 1993 US prices, the cost of electricity to make one pound of acetylene is greater than the total cost of a pound of ethylene. The use of carbide might have become attractive had cheap night-time electricity become available as a result of nuclear plants operating steadily around the clock. The question mark over the nuclear electricity program suggests that this is now only a remote possibility.

12.7 COAL AND THE ENVIRONMENT

Coal poses social and political as well as technical problems. The cheapest way to mine coal is by open pit or strip mining, which ruins large areas of country-side. On the other hand, traditional underground mining is a dangerous and expensive business that demeans the human spirit. Mechanization of mining and better safety standards have improved the miner's lot from what it was in the nineteenth century, but the situation is far from ideal.

The economic and moral dilemmas are underlined by the virtual closure of Britain's historic mining industry. British coal lies in narrow seams deeply buried. They are uncomfortable and dangerous to work, and the British miners saw themselves as the cream of the manual workers.

Meanwhile, demand for coal has declined. Oil and natural gas have replaced it in many applications, especially now that Britain has its own supplies from the North Sea. Coal is polluting and European Commission legislation on air pollution discriminates against relatively high sulfur British coal for electricity generation. The possibility of imports of cheap strip-mined coal (much of it produced, say the miners, by sweated child labor in Columbia) has made things worse. The collapse of the British coal industry has led to the disappearance of employment, the fragmentation of historic communities, and the depopulation of whole areas of the country.

Should one applaud the ending of an unpleasant and dangerous industry or mourn the disappearance of employment and community? Either way, the production of coal at a lower social cost may be a problem of crucial importance for future generations.

NOTES AND REFERENCES

The huge classic, *The Chemistry of Coal Utilization*, H. H. Lowry, Ed., originally published in 1945 was reprinted by Wiley, New York, in 1977. For a more succinct background to the industry, see G. J. Pitt and G. R. Millward, *Coal and Modern Coal Processing*, Academic, London, 1979. For more recent developments see N. R. Payne, *Chemicals from Coal: New Processes*, Critical Reports on Applied Chemistry, Wiley, Chichester, 1987.

The proportion of US organic chemicals coming from coal tar and crudes can be calculated from *Synthetic Organic Chemicals*, published by USITC and cited in Section 0.4.5. In 1990, 127 million tonnes of organic chemicals were derived from 52 million tonnes of primary products from petroleum and natural gas and 0.84 million tonnes of coal tar and crudes, from which it appears that coal made up 1.6% of feedstocks. Corresponding 1991 figures were 123, 54, and 0.76 million tonnes, respectively, so that the proportion dropped to 1.4%. The 1992 figures, however, are 138, 56, and 2.66 million tonnes suggesting a rise to 4.7%. This is highly unlikely and presumably represents a change of classification.

A major barrier to an extensive coal-based chemical industry would be the amount of coal needed. Bayer estimated at the end of the 1970s that to produce all the organic chemicals then made from petroleum would require an extra 250 million tonnes of coal/year in Western Europe. Total current production of hard coal, lignite and brown coal, not all suitable for chemicals, is only 400 million tonnes.

Section 12.2 Coal (town) gas, producer gas, water gas and synthesis gas are often confused. Coal or town gas is a coproduct of coke oven operation and is high in hydrogen and carbon monoxide. Producer gas was a low Btu gas with a high nitrogen content made by passage of air and steam over coke and is now obsolete. Water gas (or blue gas) is the gas obtained from the passage of steam over coke, when its composition is typically 40% CO, 50% H_2, 5% CO_2, and 5% N_2 and CH_4. A gas with higher H_2/CO ratio is obtained by steam reforming of methane and higher alkanes. For the manufacture of chemicals, a synthesis gas is required with the appropriate CO/H_2 ratio and, to the extent to which this is necessary, water gas is not the same as synthesis gas. Oil gas was obtained in the United States by thermally cracking petroleum distillates. It contained 20–25% ethylene, 13–16% propylene and some higher olefins and light paraffins. Oil gas was not available in the UK and synthesis gas was therefore rarely used as a fuel.

A proposed Fischer–Tropsch mechanism involving two simple reactions has been advanced by J. Falpe, *Chem. Eng. News*, October 26, 1981, p. 23.

The Fischer–Tropsch mechanism involving CO dissociation and CO hydrogenation has been elaborated by M. E. Dry, "The Sasol Fischer–Tropsch Process" *Applied Industrial Catalysis* Vol. 2, Academic, New York (1983).

Catalysts for the Fischer–Tropsch reaction are discussed by R. D. Srivastava, V. U. S. Rao, G. Cinquegrane, and C. J. Stiegel, *Hydrocarbon Processing*, February 1990, p. 59.

A good review of the modern conception of the Fischer–Tropsch reaction has been written by J. Haggin, *Chem. Eng. News*, August 23, 1990, p. 27.

The discovery of ZSM-5 led to a flurry of interest in methanol from coal, exemplified by several books: E. Supp, *How to Produce Methanol from Coal*, Springer-Verlag, 1990: A. Kasem, *Three Clean Fuels from Coal: Technology & Economics*, facsimile of 1979 edition, Bks demand; H. A. Dirksen, *Pipeline Gas from Coal by Methanation of Synthesis Gas*, Inst. Gas. Tech. 1963.

Section 12.5 An excellent review of coal gasification before everyone lost interest is by D. Rooke, *Chem. Eng.*, January 1978, p. 34.

Details of all the gasification technologies are to be found in H. D. Schilling, B. Bonn, and U. Krauss, *Coal Gasification: Existing Processes and New Developments*, Graham and Trotman, London, 1981 and in E. C. Mangold, *Coal Liquefaction and Gasification Technologies*, Ann Arbor, MI, 1979.

CHAPTER 13

FATS AND OILS

Naturally occurring triglycerides, that is esters of glycerol with saturated or unsaturated fatty acids, are called oils if they are liquid or fats if they are solid. They may be of animal or vegetable origin. They have the general formula: $ROOCCH_2$–$CH(OOCR')$–CH_2OOCR'', where R, R', and R'' are alkyl or alkenyl groups. Usually more than one fatty acid is present and the triglyceride is said to be "mixed." Some of the fatty acids most commonly found in triglycerides in fats and oils are shown in Table 13.1. They all have even numbers of carbon atoms.

Fats and oils are one of the three major groups of foodstuffs, the others being proteins and carbohydrates. Food is, however, surrounded with a host of

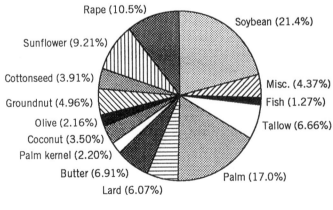

Total production = 85.67 million tonnes

Figure 13.1   World production of edible oils and fats, 1993.

cultural attitudes. In southern Europe, oil is widely used for cooking, whereas in northern Europe solid fats are preferred. In the Middle East, a market for butter scarcely exists, whereas in Europe and North America it is an important foodstuff. Therefore ways have been sought by which oils could be hardened to make them culturally acceptable to dwellers in northerly areas. The melting point of a fat or oil is related to the melting points of the fatty acids it contains, and these in turn depend on molecular weight (see the series lauric, myristic, palmitic, stearic), number of double bonds (see the series stearic, oleic, linoleic, linolenic), and the *cis* or *trans* configuration of the double bond (see the series linolenic, α-eleostearic, β-eleostearic). Crystalline structure of the fat also plays a part. To harden a fat or oil, therefore, it is hydrogenated and *cis–trans* isomerized over a nickel catalyst. The degree of hardness can be controlled, and fats of any desired consistency can be produced.

The role of fats and oils in cardiovascular disease has been a major area of concern in recent years. European and North American diets are said to be too rich generally in fats and oils. Hard fats have been associated with atheroma, and a switch to soft fats containing polyunsaturated side chains has been recommended. Thus a historic trend towards hard fats has been reversed. Opinion now seems to be stressing the virtues of monounsaturated fats, notably olive oil.

World production of fats and oils in 1993 was 85.67 million tonnes (189 billion lb). United States consumption was about 21.2 billion lb ($\sim$ 85 lb per head). This figure is similar to the production of propylene or benzene. The difference is that propylene and benzene are almost entirely used by the chemical industry, whereas 80% of oils and fats are eaten by humans and another 6% of poorer grade material is made into animal feed. The remaining 14% ($\sim$ 3 billion lb) is used by the oleochemical industry. That is about the same as butadiene production. In volume terms, oleochemicals are thus about as significant as a major component of the C_4 stream.

Figure 13.1 shows world production of the different oils and fats. Four oil crops—soybean, palm, rape, and sunflower—have grown rapidly over the past 30 years compared with the traditional oils and fats. They have gone from 26% of world production in 1958–1962 to 58% today. Animal fats have a relatively small share. Oil-producing countries specialize to some extent. Malaysia grows mainly palm oil and is responsible for 51% of world production. The Philippines grows coconut oil (43% of world production). China (27%), Europe (25%), India (19%), and Canada (15%) grow rapeseed. The CIS (25%), Europe (20%), and Argentina (16%) grow sunflower oils, while the United States has 56% of soybean oil production and Argentina another 22%. The United States, China, and the CIS dominate the cottonseed market.

Rapeseed oil containing 90% erucic acid is a new development as is a sunflower oil with 80% oleic acid. Rapeseed as a source of oil is important not only because the oil has a high concentration of unsaturated fatty acids, but also because it can be grown in colder climates than other oilseeds. Thus it is an important crop in Canada and northern Europe. Its importance in Europe, however, reflects the desire of the European Commission to support European farmers and replace imports rather than any economic advantages. At 1990 prices, the production cost of palm oil in Malaysia was $215/tonne. Soybean oil in the United States cost $315/tonne and rapeseed oil in Europe $750/tonne. The world market price was only $422/tonne.

The original rapeseed oil contained 40–50% erucic acid, a C_{22} fatty acid with a single double bond, together with so-called glucosinolates, which are toxic compounds inhibiting growth by blocking the capture of iodine by the thyroid gland. Animals (rats, guinea pigs, ducklings, and hamsters) fed on a high erucic acid diet concentrated fat in the heart muscle. While there is no evidence that erucic acid has a similar effect on humans, the oil market is geared to edible oils and it was thought prudent to develop a zero erucic acid rapeseed, now called canola oil, and to remove or breed out the glucosinolates. Some high erucic acid oil is still required for industrial needs, and thus a high erucic acid crop is also grown. It is important to prevent cross-pollination of one crop by the other. Rapeseed oil goes into lubricants for diesel engines and erucic acid amide is used as a lubricant in plastics extrusion.

The most important fat or oil in the United States is soybean oil, of which about 22 billion lb are produced yearly. Little of it is used by the chemical industry, which obtains most of its fatty acids (palmitic, stearic, and oleic) from tallow or other animal fats and tall oil from Kraft paper manufacture. The sources of fatty acids for the chemical industry are shown in Table 13.1.

Fats are extracted from the fatty animal tissues in slaughterhouse wastes by rendering. In "dry" rendering, heat alone is use to dry the material and liberate the fat. In "wet" rendering, the fat is liberated by hot water or steam, which is then separated by skimming or centrifuging.

Oils are extracted either by pressing (expression) or by solvent extraction. Expression, for example, of peanuts, used to leave about 6% of the oil in the so-called oil cake, but modern presses leave only 3–5%. Nonetheless, solvent extraction may be preferred because it leaves less than 1%. Soybean oil is invariably solvent extracted. The soybeans are heat treated and then mechanically pressed to flakes. The flakes are solvent extracted (leached) with hexane, which is subsequently stripped off and recycled.

The modern trend is the use of supercritical carbon dioxide for solvent extraction of food and personal products. It is used in the processing of high-cost products such as removal of caffeine from coffee and extraction of essential oils for the perfume industry. It was suggested as an attractive process for edible oils because it would avoid the traces of hexane contamination, but thus far the expense of the capital equipment has deterred the industry.

Once the oil has been extracted, it is subjected to a series of purification processes: degumming, bleaching, steam deodorization, hydrogenation, alkali refining, and possibly fractionation by melting point. In the degumming process, phospholipids or phosphatides are removed by washing the oil with hot water or dilute acid. They precipitate and are removed as a sludge and vacuum dried. The sludge contains four major constituents, which are esters of phosphatidic

TABLE 13.1 Common Fatty Acids

Formula	Trivial Name	Melting point (°C)	Double-Bond Position and Stereochemistry	Source
n-$C_{11}H_{23}COOH$	Lauric acid	44.2	—	Coconut oil, palm kernel oil
n-$C_{13}H_{27}COOH$	Myristic acid	53.9	—	Coconut oil, palm kernel oil
n-$C_{15}H_{31}COOH$	Palmitic acid	63.1	—	Most vegetable oils and animal fats
n-$C_{17}H_{35}COOH$	Stearic acid	69.6	—	Most vegetable oils and animal fats
n-$C_{17}H_{33}COOH$	Oleic acid[a]	16.0	cis-9	Most vegetable oils and animal fats (olives, nuts, beans, tall oil)
n-$C_{17}H_{31}COOH$	Linoleic acid[a]	−9.5	cis-9, cis-12	Tall oil, most vegetable oils (safflower, sunflower, soy)
n-$C_{17}H_{29}COOH$	α-Linolenic acid[a]	−11.3	cis-9, cis-12, cis-15	Linseed oil
n-$C_{17}H_{29}COOH$	γ-Linolenic acid		cis-6, cis-9, cis-12	Evening primrose oil
n-$C_{17}H_{29}COOH$	α-Eleostearic acid	48.5	cis-9, trans-11, trans-13	Tung oil
n-$C_{17}H_{29}COOH$	β-Eleostearic acid	71.5	trans-9, trans-11, trans-13	Tung oil
n-$C_{17}H_{32}(OH)COOH$	Ricinoleic acid[a]	5.0	cis-9	Castor oil
n-$C_{19}H_{29}COOH$	Eicosapentaenoic acid		cis-5, cis-8, cis-11, cis-14, cis-17	Fish oil
n-$C_{19}H_{29}COOH$	Arachidonic acid	−49.5	cis-5, cis-8, cis-11, cis-14	Animal fats and organs
n-$C_{21}H_{41}COOH$	Erucic acid	33.5	cis-13	Rapeseed (canola) oil, fish oil
n-$C_{21}H_{31}COOH$	Docosahexaenoic acid	22.6	cis-4, cis-7, cis-10, cis-13, cis-16, cis-19	

[a] Oleic acid: $CH_3(CH_2)_7CH=CH(CH_2)_7COOH$
Linoleic acid: $CH_3(CH_2)_4CH=CHCH_2CH=CH(CH_2)_7COOH$
Linolenic acid: $CH_3CH_2CH=CHCH_2CH=CHCH_2CH=CH(CH_2)_7COOH$
Ricinoleic acid: $CH_3(CH_2)_4CH_2CH(OH)CH_2CH=CH(CH_2)_7COOH$

acid. These are 1,2-diglycerides with a phosphoric acid group on the third carbon, and the phosphoric acid group is in turn esterified with choline, ethanolamine, serine, or inositol. If the structure is written

then if $R''=CH_2CH_2N^+(CH_3)_2$, the compound is called lecithin or phosphatidylcholine, $R''=CH_2CH_2NH_2$ is cephalin or phosphatidylethanolamine, and $R''=CH_2CH(NH_2)COOH$ is phosphatidylserine. The trade lecithin is a mixture of all three plus minor components. It is used primarily as a cationic surfactant in foods and specialty chemicals. For example, it is added to margarine to prevent spattering, and to chocolate to prevent "blooming."

Vegetable oils usually contain free fatty acids resulting from enzymatic decomposition. If the oils are to be used as foodstuffs, these are removed by alkali treatment in a so-called alkali refining process. The sodium salts of the fatty acids are separated. They are called soapstocks or "foots." The free fatty acids are regenerated by acidification.

The bleaching process decolorizes the oil by adsorption rather than by oxidation, which is the mode of action of conventional bleaches. The colored material is adsorbed on bentonite or montmorillonite clays. The oil may then be fractionated by melting point to extract a particular range of triglycerides, or it may be hydrogenated.

Hydrogenation is carried out over a nickel catalyst and "hardens" the oil, that is it raises the melting point. Margarine is a typical product. The hardening process involves reduction of double bonds, *cis–trans* isomerization, and shifting of double bonds along the chain.

Deodorization involves removal of the strong flavor associated with vegetable oils. It depends on a combination of high vacuum (0.004–0.008 bar) and high temperature (240–260°C) for 15–40 min and is a variant of vacuum steam distillation, which removes undesirable volatiles while not damaging the triglycerides. A distillate results (deodorizer distillate) containing odor bodies together with small amounts of two valuable products. One is tocopherol, the precursor of vitamin E and thus the source of natural vitamin E. The other is a sterol mixture containing stigmasterol and other sterols that are converted to cortisone (see note).

13.1 FATTY ACIDS

Oils and fats are saponified to glycerol and soap—the sodium salt of fatty acids—by treatment with alkali (Fig. 13.2) when soap is the desired end product.

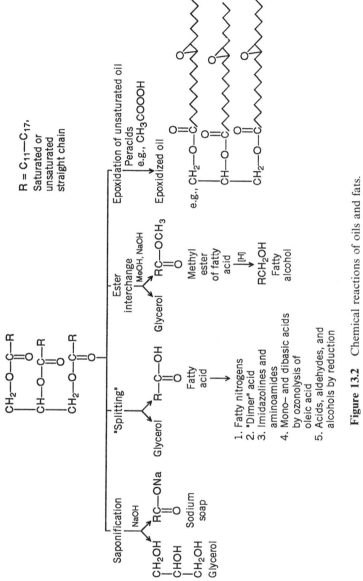

Figure 13.2 Chemical reactions of oils and fats.

More widely used in both the United States and Europe to obtain free fatty acids is "splitting"—continuous noncatalytic hydrolysis at high temperature and pressure. In smaller plants, continuous autoclave splitting is used in the presence of oxide catalysts such as zinc oxide, and very small operations may use batch processes with so-called Twitchell catalysts, which are combinations of sulfuric and sulfonic acids.

The separation of saturated from unsaturated acids, if required, is normally effected by cumbersome crystallizations. A somewhat less efficient but simpler process for separating fatty acids, primarily from tallow, is known as hydrophilization. An aqueous solution of a wetting agent is slurried with a crystallized fatty acid mixture. The higher melting stearic acid crystals are wetted preferentially and are sufficiently solubilized by the wetting agent to transfer into the aqueous phase. The aqueous layer is separated from the oleic acid by centrifugation. Heating melts the stearic acid, which separates from the water and is easily isolated.

The enzymatic hydrolysis of fats has been extensively studied but a practical process that proceeds at an acceptable rate has not yet been devised. Another source of the vegetable oil fatty acids used by the chemical industry is "foots" or soapstocks (see above).

The lowest molecular weight acids, lauric and myristic, come from coconut and palm kernel oils. Palmitic and stearic acids are found in most oils and fats, tallow being the most important. Oleic acid is a component of animal fats and a major constituent of many vegetable oils. Its ozonolysis gives azelaic and pelargonic acids (Section 13.5). 12-Hydroxyoleic (ricinoleic) acid with its OH is an oddity found in castor oil. Erucic acid (see above) is found in rapeseed oil from plants where it has not been bred out.

Linoleic acid is found in many vegetable oils, particularly linseed and safflower oil. Nonetheless, it is largely obtained from tall oil, which became important after World War II. Tall oil is not a triglyceride but a mixture of rosin and fatty acids. It is a byproduct of the pulping of southern pine for paper manufacture. Southern pine is converted to pulp by the Kraft process, in which sodium hydroxide is used to separate the desired cellulosic fibers from the undesired lignins (polymers of phenylpropane monomers containing OH and OCH_3 groups), rosins, fatty acids, and other materials in the wood. The fatty acids, mainly oleic, and linoleic, end up as their sodium salts in a smelly black liquid. Acidification gives rosin and fatty acids (mainly oleic and linoleic), which may be separated by distillation.

Rosin is a complex mixture of about 90% acids related to partially hydrogenated phenanthrene and 10% neutral matter. Of the resin acids, 90%,

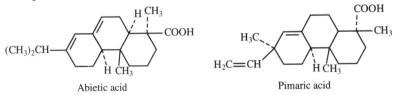

Abietic acid   Pimaric acid

including pimaric acid, are isomeric with abietic acid and the remainder with dihydro- and dehydroabietic acids in which one or both the double bonds have been reduced.

The rosin is famous for its use on violin bows, but far larger quantities are formulated into paints and varnishes. As its sodium salt, it is the important size in paper manufacture, and this is its major use.

The fatty acids with more than 18 carbon atoms are found in fish oils. They are of current interest because they are the precursors in the body of the prostaglandins, leukotrienes, thromboxanes, and prostacyclins. They cannot be synthesized by the body and must be obtained in diet.

Evening primrose oil is advertised as being rich in the important C_{18} fatty acid, γ-linolenic acid (GLA), an isomer of the α-linolenic acid from tall oil. The seeds of the evening primrose fall when ripe, that is over a long period. As a result, commercial harvesting gives only about 5% recovery, so that the oil is expensive. The active ingredient can be manufactured by fermentation using fungi of the class *Phycomycetes*. For example, species of *Mucor* can be grown on a glucose-containing medium in large fermenters ($\sim 220 \, M^3$), and an oil containing 7% GLA recovered from the fungal mycelium. Another useful species is *Phycomyces blakesleeanus*, its oil also containing large quantities of β-carotene.

In Russia, Japan, and China, impure petrochemical-based fatty acids with mixtures of odd and even numbers of carbon atoms have been produced by oxidation of petroleum wax. The products are generally regarded as inferior.

A large proportion of fatty acid production goes into soaps. Domestic soap is merely the sodium salt of the mixture of fatty acids obtained by hydrolysis of fats or oils. The major components are sodium stearate and palmitate. It is used also as a lubricant in rubber and polymer processing.

The decomposition of soap with hydrochloric acid into sodium chloride and a fatty acid mixture called stearine was discovered by Chevreul in 1823. In the nineteenth century, until the advent of cheap paraffin wax, it was the principal material used in household candles. Unlike the traditional tallow, it did not give acrolein, a powerful lachrymator, on combustion. It is still blended with paraffin wax in some candle formulations to improve the melting properties of the wax. Only in countries such as Denmark and Sweden, whose fishing industries produce a surplus of cheap stearine, is it still a major component of candles. Paraffin candles dominate the world market, which has passed its nadir but is again growing. This is partly due to fashion in developed countries but also to the frequent interruptions in electricity supply in developing countries, where the inhabitants are no longer content to sit in the dark. Paraffin wax comes from the heavier petroleum fractions, and ceresin or microcrystalline waxes from the same source are used in US candles. These are branched-chain paraffins, and consequently must be blended with an antioxidant to prevent oxidation at the tertiary carbon atoms. Fischer–Tropsch waxes are made in South Africa (Section 12.2). Those of higher molecular weight are used throughout the world for the fashionable long, thin candles, while the fraction equivalent to paraffin wax is used in the larger domestic market.

Salts of fatty acids are used as stabilizers for poly(vinyl chloride). These are the laurates or stearates of metals such as lead, barium, calcium, strontium, and zinc. Metal salts of cobalt, lead, manganese, calcium, and zirconium can be added to unsaturated oils to accelerate their oxidation and are thus used as driers in oil-based paints. Metal soaps, for example lithium, are also used to thicken lubricating oils to yield greases. Free fatty acids may be components of automobile lubricating oil formulations.

13.2   FATTY NITROGEN COMPOUNDS

Fatty acids may be converted into a large number of fatty nitrogen compounds of which the fatty amines, including quaternary amines, are the most important. Some of these are shown in Figure 13.3. They find many applications in industry as surface-active agents.

In Figure 13.3, the route to fatty nitrogen compounds starts with fatty acids in which the alkyl group may contain 7–17 carbon atoms. Normally it is saturated, although derivatives based on oleic acid are also produced. Treatment with ammonia converts the fatty acid to a nitrile through successive dehydration of the ammonium salt and the amide, neither of which need be isolated. This chemistry was observed earlier in the classic synthesis of hexamethylenediamine (Section 5.1.2). Amides, however, are articles of commerce and are used as a parting agent or slip agent to prevent plastic from adhering to the die during extrusion. The nitrile in turn may be reduced to a primary, secondary, or tertiary

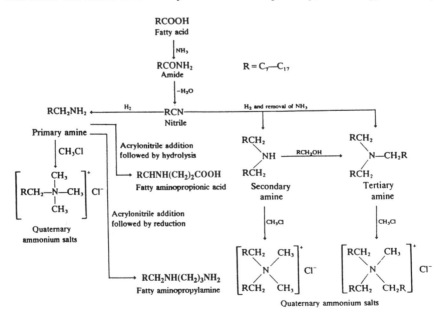

Figure 13.3   Fatty nitrogen chemistry.

amine. Conditions for primary amine formation require that ammonia be present to suppress secondary amine formation. Conversely, when the secondary amine is the desired product, ammonia must be removed continuously from the reaction mixture. The tertiary amine may also prepared by hydrogenation with removal of ammonia. It is better prepared, however, by the interaction of a di-(long-chain alkyl) amine with an alcohol. The primary, secondary and tertiary amines may be quaternized with methyl chloride or sulfate.

The most important use for distearyldimethylammonium chloride is as a textile softener for home laundering. It must be added during the last rinse, otherwise it will precipitate on contact with anionic detergents. Although still widely used, it is believed to present some ecological problems. Substitutes are certain ethoxylates (Section 3.7), which are nonionic and can be included as part of the detergent formulation.

Distearyldimethylammonium chloride and related quaternaries also react with bentonite clays to modify them, so that they are dispersible in organic solvents. As such they are the basis for high-performance greases, which result when the organoclay is dispersed in mineral oil. These organoclays are also widely used in oil-based paints to impart thixotropy.

An application of long-chain quaternary alkylammonium compounds is in phase-transfer catalysis (Section 16.10). A related application is liquid ion exchange used primarily in the extraction of uranium ore. Uranium is leached with sulfuric acid to give the complex anion $[UO_2(SO_4)_2]^{2-}$. A trifatty amine in which the fatty groups contain 8–10 carbon atoms is converted to a sulfate as shown in Eq. (a) and this organic-soluble salt, dissolved in kerosene, is mixed with the aqueous leach liquor containing the uranium anion. On vigorous stirring, liquid ion exchange takes place between the sulfate ion of the trifatty amine salt and the uranium-containing anion so that the latter is transferred to the organic solution whereas the sulfate is transferred to the aqueous layer (Eq. b). Thus the uranium, which was present in the aqueous phase at a concentration of considerably less than 1%, is not only concentrated in the organic phase but is separated from numerous impurities. It can be removed from the organic phase by stripping with an alkali (Eq. c), which converts it into so-called "yellow cake." The trifatty amine is recycled.

Salt formation:

$$2R_3N_{kerosene} + H_2SO_{4\,aqueous} \rightleftharpoons [(R_3NH)_2^+ SO_4^{2-}]_{kerosene} \qquad (a)$$

Liquid ion exchange:

$$[(R_3NH)_2^+ SO_4^{2-}]_{kerosene} + [UO_2(SO_4)_2]_{aqueous}^{2-} \rightleftharpoons$$
$$[(R_3NH)_2^+ [UO_2(SO_4)_2]^{2-}]_{kerosene} + SO_{4\,aqueous}^{2-} \qquad (b)$$

Stripping:

$$[(R_3NH)_2^+ [UO_2(SO_4)_2]^{2-}]_{kerosene} + Na_2CO_3 \rightleftharpoons 2R_3N_{kerosene} +$$
$$[UO_2(SO_4)_2]_{aqueous}^{2-} + 2Na^+ + H_2O + CO_2 \qquad (c)$$

Figure 13.3 shows another reaction sequence for primary amines. They may react with acrylonitrile to give a fatty aminopropionitrile. This in turn can be reduced to a fatty aminopropylamine or hydrolyzed to a fatty aminopropionic acid. Each of these compounds varies in degree of surface activity, and it is this variation that accounts for their specific applications.

In addition to the foregoing simple reactions, fatty acids undergo a number of more complicated reactions of chemical interest, although the tonnages involved are trivial compared with petroleum and natural gas. Many of the products, however, are not accessible from petrochemical sources. An example is "dimer acid," produced at a level of about 100 million pounds per year worldwide.

13.3 "DIMER" ACID

The dimerization of linoleic acid is shown in Figure 13.4a. Natural linoleic acid has double bonds in the 9,12 positions (Table 13.1) but, when heated, it isomerizes to the conjugated 10,12, or 9,11 structures. In Figure 13.4a, the 9,11 acid is shown. This diene may then undergo a Diels–Alder reaction with another molecule of the original 9,12 acid or either of the conjugated isomers. The figure shows the reaction with the 9,12 acid to give a typical Diels–Alder adduct—a cyclohexene with four side chains, two of which contain carboxyl groups.

The conjugation reaction can go in two ways, and the products can react with either of the double bonds in the 9,12; 10,12; or 9,11 acids. Furthermore, addition may be head-to-head or head-to-tail. The figure shows head-to-head addition. In all, 24 different products are possible. An added complication is that the double bonds can be *cis* or *trans*. Although most naturally occurring double bonds, including those in linoleic acid, are *cis*, some *trans* double bonds form at the temperature of the reaction. Whether they are *cis* or *trans*, however, markedly affects the kinetics of the reaction.

Linoleic acid is scarcer and more expensive than oleic acid and it was desirable to find a way to dimerize the cheaper oleic acid which, with its single double bond, will not undergo the Diels–Alder reaction. It will, however, dimerize over a natural acid clay catalyst, for example a montmorillonite, known as a "pillared clay." Such a clay consists of layers separated by ions, such as metal ions, which function as "pillars." Large organic molecules can undergo reactions in the spaces between the layers. Oleic acid dimerizes as shown in Figure 13.4b. The product is also known as "dimer acid," and many different chemical structures are present, although all contain 36 carbon atoms and two carboxyl groups. United States production in 1993 was 45 million lb. These dibasic acids are used in the production of specialty polyamide oligomers ('Versamids') with unusual adhesive and coating properties. For example, these are practically the only alcohol-soluble materials that adhere to polyethylene, and accordingly they are used as vehicles for printing inks for polyethylene film. The alcohol solubility is necessary since stronger solvents such as aromatic hydrocarbons dissolve the natural rubber printing rolls.

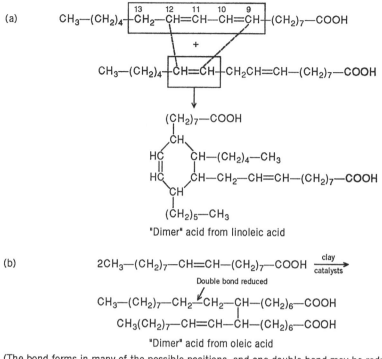

(a)

(b)

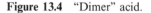

"Dimer" acid from oleic acid

(The bond forms in many of the possible positions, and one double bond may be reduced, as shown)

Figure 13.4 "Dimer" acid.

When dimer acid reacts with diethylenetriamine, triethylenetetramine, or higher polyalkylene amines, an amino-containing polyamide oligomer results. It is an important coreactant for epoxy resins for adhesives and particularly for maintenance paint for metals. The amino-containing resin imparts corrosion resistance by reacting chemically with the metal oxide surface (e.g., ferric oxide) that is invariably the top layer of iron exposed to air. The chemical reaction is salt formation that takes place between the acidic hydrated amine groups and the basic hydrated metal oxides.

An interesting specialty use involves the substitution of dimer acid esters, such as the di-(2-ethylhexyl) ester, for lubricating oil in two-cycle outboard engines in lakes in Switzerland that were being polluted by petroleum engine oil discharged from motor boats.

13.4 AMINOAMIDES AND IMIDAZOLINES

A small volume application of fatty acids involves their conversion to amino-amides and imidazolines. A fatty acid will react with a polyamine such as diethylenetriamine to give an aminoamide, a molecule of water being eliminated. Further dehydration leads to cyclization. Imidazolines have many of the properties of the fatty nitrogen compounds already described.

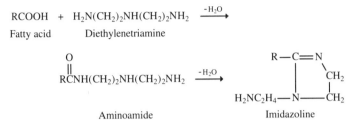

If quaternized with methyl chloride or sulfate, they are useful as textile softeners, such as the fatty quaternaries already discussed (Section 13.2). Their main use is as corrosion inhibitors, particularly in oil well applications. They are also used as asphalt emulsifiers and antistrippants. An asphalt emulsifier is an emulsifying agent that brings particles of liquid asphalt and water into close proximity to provide a convenient means for laying down a layer of asphalt on a road bed. An asphalt antistrippant causes asphalt to adhere to the rock or aggregate with which it is frequently mixed in road construction. The antistrippant is particularly useful if the road is wet. It functions by adsorbing onto the surface of the rock, which is usually silica. Adsorption is so strong that water on the surface of the rock is replaced. The fatty tails point away from the surface of the rock and are solvated by the asphalt. In this way, aminoamides and imidodazolines facilitate the bonding between the asphalt and the rock.

13.5 AZELAIC, PELARGONIC, AND PETROSELINIC ACIDS

Oleic acid may be cleaved at the double bond by treatment with ozone from an electrical discharge. The products are aldehydes that can be oxidized further to acids.

$$CH_3(CH_2)_7CH{=}CH(CH_2)_7COOH \xrightarrow{O_3} CH_3(CH_2)_7CHO + OHC(CH_2)_7CHO$$

Oleic acid   Pelargonic aldehyde   Azelaic aldehyde

$$\downarrow O_2 \qquad\qquad \downarrow O_2$$

$$CH_3(CH_2)_7COOH \qquad HOOC(CH_2)_7COOH$$

Pelargonic acid   Azelaic acid

A new pelargonic acid process has been developed by Celanese, and thus once more petrochemicals have encroached on natural products. The Celanese process involves the hydroformylation of 1-octene, followed by oxidation of the resulting aldehyde (Section 4.8.2).

Both azelaic and pelargonic acids are raw materials for specialty polymers and polyesters for synthetic lubricants. Thus a typical first generation synthetic automotive lubricant useful at very low temperatures is the diisodecyl or ditridecyl ester of azelaic acid. The pelargonic acid also finds its way into synthetic lubricants by way of pentaerythritol pelargonate (Section 3.11.3). However, most important today is the trimer of 1-decene (Section 3.3). A lubricant used in automobiles is actually a mixture of all three. In comparison with petroleum lubricants they are said to provide greater lubricity, better engine protection, and lower gasoline consumption. They are also easier to recycle. They may be blended with petroleum-derived hydrocarbons, which reduces the cost but increases the difficulty of recycling.

Odd number carbon atom acids are rare in nature and in industrial chemistry. Azelaic and pelargonic are the most accessible. It appears that odd number carbon acids are more surface active than are those with an even number.

In an effort to find natural products to compete with petrochemicals, some effort is being invested in the production of seed oils from *umbelliferae*, for example, coriander. Their seeds are rich in petroselinic acid, an isomer of oleic acid with a *cis*-6 double bond. Ozonolysis followed by oxidation gives adipic and lauric acids, products with larger markets than azelaic and pelargonic acids but adipic acid is already available from petrochemical sources.

$$CH_3(CH_2)_{10}CH{=}CH(CH_2)_4COOH \xrightarrow[(2)\,O_2]{(1)\,O_3}$$

Petroselenic acid

$$CH_3(CH_2)_{10}COOH + HOOC(CH_2)_4COOH$$

Lauric acid $\qquad\qquad$ Adipic acid

13.6 FATTY ALCOHOLS

Fats and oils may easily be "split" to fatty acids and glycerol (Section 13.1). They can also be converted to fatty alcohols and glycerol by hydrogenolysis (Figure 13.2). Because the fatty acid groups have mixed chain lengths, a mixture of fatty alcohols results. The hydrogenolysis was originally carried out with sodium in ethanol—the Bouveault–Blanc reaction—but hydrogen and a copper chromite catalyst at high pressures are now used. The products differ in that the latter method hydrogenates all the double bonds in the fatty alcohol whereas the Bouveault–Blanc procedure leaves them intact. A newer proprietary chromite catalyst, however, is said to preserve the double bonds.

In practice it is easier to convert the fatty acid in the triglyceride to its methyl ester by alcoholysis with methanol and then to subject the methyl ester to hydrogenolysis. Most vegetable oil-based alcohols are made in this way.

While esters undergo hydrogenolysis to fatty alcohols fairly easily, reduction of a fatty acid to a fatty alcohol is more difficult. A clever procedure that brings it about involves a copper chromite catalyst (as above) in a slurry in fatty alcohol. The small amount of alcohol esterifies some of the fatty acid, the hydrogenolysis catalyst serving also as an esterification catalyst. The ester undergoes hydrogenolysis (Fig. 13.5) to yield a fatty alcohol that esterifies more of the fatty acid to yield an ester for further hydrogenolysis. Thus, even though the feed is fatty acid, an ester forms in situ for the hydrogenolysis. Since this is a high molecular weight ester, it reacts more slowly than a methyl ester, but the process is still useful industrially.

The hydrogenolysis of the methyl esters or of fatty acids takes place at 250–300°C and 200–300 bar. Catalysts that cause lower molecular weight esters to undergo hydrogenolysis at considerably milder conditions such as 5–25 bar have been described recently (Section 3.7.2). Conceivably, modifications of these catalysts could be used for the hydrogenolysis of fatty acid esters under less strenuous conditions.

Fatty alcohols as well as α-olefins may also be obtained from ethylene oligomerization by use of aluminum trialkyls (Section 3.3.2) or by the SHOP process (Section 3.3.4). Here again petrochemicals have made an impact on traditional processes, but the vegetable oil-based routes, used primarily by Henkel in Germany and Proctor and Gamble in the United States, are competitive, at least when vegetable oil prices are low. Together, the two processes accounted for over 800 million lb of alcohols in the United States in 1993.

The possibility of using triglycerides directly in a hydrogenolysis reaction has long been discussed. The proposed catalyst is copper chromite in the form of lumps and the reaction is run at 180–250°C at up to 280 bar.

The special role now occupied by straight-chain primary alcohols in detergent technology results not only from their excellent detergent properties but also because products based on them biodegrade more quickly than compounds

$$C_{17}H_{35}COOH \; + \; 2 \, H_2 \; \xrightarrow[\text{slurried in stearyl}]{\text{Copper chromite cat.}} \; C_{17}H_{35}CH_2OH \; + \; H_2O$$

Stearic acid   alcohol, 300°C, 300 Bar   Stearyl alcohol

The stearyl alcohol esterifies some ot the stearic acid and this ester, stearyl stearate, on hydrogenolysis yields stearyl alcohol.

$$C_{17}H_{35}COOH \; + \; C_{17}H_{35}CH_2OH \longrightarrow C_{17}H_{35}COOCH_2C_{17}H_{35} \; + \; H_2O$$

Stearyl stearate

$\Big\downarrow H_2$

$2 \, C_{17}H_{35}CH_2OH$ (This esterifies more of the fatty acids so that the reaction can be repeated).

Figure 13.5   Hydrogenolysis of fatty acids.

containing a benzene ring. Furthermore, degradation of benzene ring-containing detergents are claimed eventually to lead to phenols, and these are toxic to fish. This has been hotly contested by the manufacturers of alkylbenzene sulfonates who advertise that their products are completely biodegradable.

13.7   EPOXIDIZED OILS

Unsaturated fats and oils can be epoxidized so that some of the double bonds are replaced by —C—C— groups. These compounds are added to poly(vinyl chloride) (PVC) often together with metal soaps to prevent degradation by light and heat.

Epoxidized oils are also secondary plasticizers, that is they add their softening power to that of any plasticizer that softens PVC when used on its own. Epoxidation of fatty acid esters such as butyl or hexyl oleate or "tallate" gives a primary PVC plasticizer (see note). Such materials are widely used in the United States where soy oil is produced on a large scale. In 1992, 60,000 tonnes of epoxidized soy oils were produced, representing about 12% of total plasticizer production. In Western Europe, on the other hand, epoxidized oils are normally used only for their stabilizing properties.

13.8   RICINOLEIC ACID

Ricinoleic acid, having a hydroxyl group, is a curiosity among naturally occurring fatty acids and is found as its triglyceride only in castor oil. Well over 100 million lb of castor oil are consumed yearly in the United States, the largest application being in paints and varnishes. Dehydration of the acid gives an isomer of linoleic acid that can be used in nonyellowing protective coating formulations. Castor oil itself may also be dehydrated to give a useful drying oil and it may be sulfated to give Turkey Red oil, a textile dye leveler. It is also used as a polyol (three hydroxyl groups per molecule) in polyurethane production.

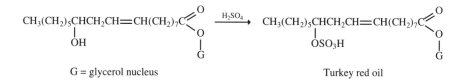

G = glycerol nucleus   Turkey red oil

High-temperature cleavage of ricinoleic acid at 275°C with concentrated aqueous sodium hydroxide yields 2-octanol and sodium sebacate.

$$CH_3(CH_2)_5CHOHCH_2CH=CH(CH_2)_7COOH \xrightarrow{\text{NaOH}}$$

Ricinoleic acid

$$CH_3(CH_2)_5\overset{..}{\underset{..}{C}}HOHCH_3 + NaOOC(CH_2)_8COONa$$

2-Octanol   Sodium sebacate

2-Octanol was an important foam depressor prior to the advent of the silicones. Sebacic acid condenses with hexamethylenediamine to give the specialty polyamide, nylon 6,10. Its dioctyl ester is an excellent PVC plasticizer, but its properties rarely justify its high price.

Thus far, castor oil has been the only source of sebacic acid, although a Japanese process has been described for the electrodimerization of 2 mol of adipic acid with elimination of two molecules of carbon dioxide.

Dry distillation of the sodium or calcium salt of ricinoleic acid at 500°C breaks the carbon chain between the eleventh and twelfth carbon atoms to yield *n*-heptaldehyde and undecylenic acid (Figure 13.6). The heptaldehyde may be reduced to *n*-heptanol, which is an acceptable plasticizer alcohol. The undecylenic acid is treated with hydrobromic acid in the presence of peroxide so that it will add "anti-Markovnikov." Replacement of the bromine with an amino group gives ω-aminoundecanoic acid (the real reason for the process), which can then be polymerized to nylon 11. Nylon 11 is used in highly specialized formulations and never became a tonnage product. Undecylenic acid, as its zinc salt, is a fungicide effective against *tinea pedis* or athlete's foot.

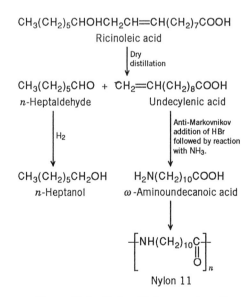

Figure 13.6 Nylon 11 from castor oil.

13.9 GLYCEROL

We have already described the production of glycerol from fats and oils (Section 13.1) and from propylene by way of allyl chloride (Section 4.10.2). Its major uses are in the formulation of cosmetics, toiletries, foods and beverages for its moisturizing, lubricating, and softening characteristics. It is also a humectant in tobacco and serves as a plasticizer for cellophane. World production in 1993 was almost 850 million lb.

Dynamite is glyceryl trinitrate (nitroglycerin) adsorbed on wood pulp. The compound is also used as a coronary vasodilator in cases of *angina pectoris*. It is delivered either as a sublingual tablet or a transdermal patch. Although it has been used for at least a century, its mode of action via the blood's nitric oxide pathways has only recently been discovered. Glycerol competes with other polyols such as ethylene glycol, pentaerythritol, and sorbitol as a raw material for polyethers for polyurethanes (Section 7.3.1).

Glycerol's most important chemical use is in alkyd resins (Section 9.1.1) and to a lesser extent in unsaturated polyester resins. Alkyds are condensates of glycerol or pentaerythritol (Section 3.11.3) with phthalic anhydride (Section 9.1.1) and fatty acids. They are the major vehicles for oil-based paints, but their use is declining because of the trend towards nonsolvent-based coatings. With propylene oxide, glycerol yields a co-reactant that gives polyurethane resins (Section 7.3.1) with isocyanates. Trimethylolpropane is usually preferred to glycerol in this application.

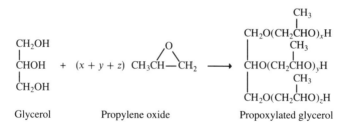

Glycerol monoesters, for example the monooleate and the monostearate, are used as nonionic surfactants especially in the food industry. Triacetin (glyceryl triacetate) is an antifungal used against athlete's foot, a fixative in perfumery and a plasticizer for the filter tips in cigarettes.

13.10 ALCOHOLYSIS OF FATS AND OILS

One of the most important commercial reactions of fats and oils that does not lead to foods or soaps is ester interchange or alcoholysis. If a triglyceride is heated with a polyol such as glycerol or pentaerythritol in the presence of a suitable catalyst such as sodium methoxide, mixed partial esters are obtained. The partial esters may then be reacted with a dibasic acid or its precursor to give an oil modified alkyd resin (see note). If the partial esters are reacted with toluene diisocyanate, oil modified urethanes are obtained.

13.10.1 Cocoa Butter

Transesterification of triglycerides has a significant application in the production of synthetic cocoa butter. As a general rule, triglycerides from animal sources (fats) are solid at room temperature, while fish and vegetable oil triglycerides are liquid. Cocoa butter is unique in being the only solid vegetable triglyceride. It is pressed from cocoa beans that have been harvested and fermented so that the pulp can be drained.

Cocoa butter consists of triglycerides that are mixed esters of palmitic acid $(C_{15}H_{31}COOH)$, oleic acid $(C_{17}H_{33}COOH)$, and stearic acid $(C_{17}H_{35}COOH)$. Only the oleic acid contains a double bond, hence the high melting point. If the alkyl residues are represented as P, Q, and S, respectively, then the main triglycerides in cocoa butter are (**I**) and (**II**) below.

$$
\begin{array}{ccc}
\text{CH}_2\text{OOCP} & \text{CH}_2\text{OOCS} & \text{CH}_2\text{OOCP} \\
| & | & | \\
\text{CHOOCQ} & \text{CHOOCQ} & \text{CHOOCQ} \\
| & | & | \\
\text{CH}_2\text{OOCS} & \text{CH}_2\text{OOCS} & \text{CH}_2\text{OOCP} \\
(\textbf{I}) & (\textbf{II}) & (\textbf{III}) \\
\end{array}
$$

Cocoa butter   Palm oil

Nature, in her wisdom, has not provided the correct ratio of cocoa-to-cocoa butter in the cocoa bean, for there is not enough butter to turn all the cocoa solids into chocolate. Surplus cocoa solids may be sold on their own as the base for a drink or blended with alternative fats to give a low-grade chocolate which, in some countries, cannot even be sold under the name of chocolate.

The substitute fats have two limitations. They lack the volatile flavor constituents of cocoa butter and they have the wrong melting characteristics. Cocoa butter melts over a range and becomes completely liquid just below the temperature of the mouth, hence a high-quality chocolate will "melt in the mouth" in a satisfying way.

Cocoa butter is expensive but palm oil (**III**), which resembles it in structure, is cheap and readily available. Unilever devised a process in which a mixture of palm oil and stearic acid was treated with a 1,3-specific lipase immobilized on kieselguhr. The solvent was n-hexane saturated with water. The tiny amount of water that dissolves in hexane means that appreciable overall hydrolysis does not take place. The processes of hydrolysis and reesterification take place at 313°C over a few hours.

Unfortunately, it was not possible to reproduce the flavor constituents of cocoa butter and this, together with labeling regulations, means that the process is currently being used only on a small scale. On the other hand, the potential market is huge. World production of cocoa butter is about 1.5 million tonnes compared with about 16 million tonnes of soybean oil and 6 million tonnes of butter. The British, in 1992, ate $2000 million of chocolate compared with a mere $500 million each of coffee and tea. If the flavor constituents of cocoa butter could be synthesized, the process would appear much more attractive.

13.11 THE FUTURE OF FAT AND OIL CHEMISTRY

Although fat and oil chemistry is mature, several recent developments have reawakened interest in it. On the pharmaceutical side, there is the physiological role played by fatty acids, and the use of fish and evening primrose oils as diet supplements (Section 13.1). Arachidonic acid, 5,8,11,14-eicosatetraenoic acid, has been shown to be the progenitor of the prostaglandins, remarkable chemicals with multiple physiological functions that occur in minute quantities in the body. Among other functions, good and bad, two of them cause inflammation. Their formation is inhibited by aspirin and this accounts for that drug's efficiency. In addition, there is interest in fat substitutes, and biodegradable fuels, surfactants, and lubricants.

13.11.1 Noncaloric Fat-Like Substances

There is a huge market for nonnutritive substances that function like fats for those who feel themselves to be overweight. Proctor and Gamble has developed over a period of many years a fully esterified fatty acid ester of sucrose called "Olestra." It is said to have the mouth feel, taste, and cooking and baking functionality of triglyceride fats. The product is made by an ester interchange reaction between sucrose, dissolved in water and soap, with fatty acid methyl esters. Apparently, esters from several oils are useful, but the patents stress the highly unsaturated safflower oil.

The methyl esters are prepared by an alcoholysis reaction between safflower oil and methanol. The coproduct is glycerol, and one estimate is that if the annual market for Olestra were one billion pounds, the world's glycerol supply would suffer from a 50% oversupply. This could provide impetus to find routes to convert glycerol to other useful chemicals. The status of the development has been summarized by Spinner (see note). By the mid-1990s the development had not been commercialized. A potential problem is that fat-soluble vitamins may dissolve in nonnutritive fats and be excreted.

13.11.2 Alkyl Polyglycosides

A new family of nonionic surfactants for detergents, commercialized by both Henkel and L'Air Liquide in the early 1990s, are called alkyl polyglycosides and are the first important fatty acid derivatives with an acetal linkage. They comprise acetals of a mixture of mono- and disaccharides of glucose, from which an acetal or glucoside has been formed with a fatty alcohol with 10–14 carbon atoms. The formula for a product from glucose and an alcohol where $n = 9$–13 is

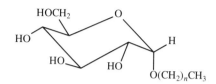

The use of the term glycoside indicates that sugars other than glucose may be involved. The preferred process for manufacture appears to involve two steps. In the first, the sugar reacts with n-butanol in the presence of weak acid to form an acetal. In the second, this intermediate reacts with a C_{10}–C_{14} fatty alcohol mixture to give the glucoside surfactant, with the release of the n-butanol for recycle.

The products are biodegradable and are said to be "green" since the sugar portion comes from corn starch and the fatty alcohols from coconut or palm kernel oil. The products are being targeted for shampoos and specialties such as cosmetics, as well as for laundry and dishwashing detergents.

These products are related to the sugar esters—C_{12} to C_{18} partial fatty acid esters of sucrose—which have been popular in Japan as nonionic rapidly biodegradable surfactants for many years. In 1995 Proctor and Gamble announced that it was replacing a portion of the alkylbenzene sulfonate in heavy duty detergents with a sugar-based surfactant.

13.11.3 Fatty Acid-Based Fuels and Lubricants

Concern about biodegradability, pollution, and depletion of nonrenewable resources has led to the development of fuels and lubricants based on fats and oils. Thus in the early 1990s "biodiesel" fuels emerged. These consisted of methyl esters of animal or vegetable oils such as rapeseed, sunflower, or tallow compounded with additives similar to those used for petroleum-based diesel fuel. The triglyceride-based products, unlike their petroleum-based counterparts, are free of sulfur. On combustion they are said to produce less smoke, less hydrocarbons, and less carbon monoxide. Nitrogen oxides emission, however, is not reduced.

Although biodegradable fuels that burn more cleanly are attractive, there are two questions. The first is how to dispose of coproduct glycerol, a problem also raised by nonnutritive fats (Section 13.10.1) particularly since it appears that methyl esters made by methanolysis of fats will be the primary materials. The second problem is one of supply. In the early 1990s, the world produced about 85 million tonnes/year of fats and oils, of which about 90% was used for food. Correspondingly, the United States alone used about 90 million tonnes of diesel fuel in 1992. It is therefore doubtful that oil and fat production can be increased to the point that will allow it to gain more than a niche market as a diesel fuel replacement. However, in areas in the world with a high ecological sensitivity, dimer acid esters (Section 13.3) have been used as fuel in two-cycle engines. Introduced in the early 1990s were oils for two-cycle engines based simply on esters of acids. These were intended as fuels for engines for lawnmowers, outboard motor boats and for engines for motorcycles, to replace the normal smog-producing motor oil–gasoline mixture. The problem with the latter fuel is well demonstrated in cities like Bangkok, where two-cycle engines on cab-equipped bicycles for human transport produce a serious smog problem. This is an example of a niche market for fatty-based fuels. An alternative fuel is a combination of gasoline with polybutenes (Section 5.2.3).

Of potential importance is the use of biodegradable fatty materials as lubricants. The lubricity of fatty materials is well established, and our ancestors used tallow on the moving parts of their chariots in 1400 BC. Esters with lubricant properties have been known since World War II, and synthetic lubricants for automobiles today may contain pentaerythritol tetraesters (Section 3.5) and esters of dibasic acids such as the di-(2-ethylhexyl) ester of azelaic acid (Section 13.5). Such compounds become much more biodegradable in the form of their fatty acid counterparts, and such materials are being proposed for use in automobiles, and as metal working lubricants, turbine oils, hydraulic fluids, and functional fluids generally.

NOTES AND REFERENCES

The classic work on the chemistry of fats and oils is *Bailey's Industrial Oil and Fat Products*, D. Swern, Ed., Wiley, New York, Vol. I (1979), Vol. II (1982), and Vol. III (1985). There is an excellent series of articles on oleochemistry in a special issue of the *Journal of the Oil Chemists' Society*, February, 1984. See also G. Hoffman, *Chemistry and Technology of Edible Oils and Fats and their High Fat Products*, Academic, New York, 1989, D. M. Small, *The Physical Chemistry of Lipids: Handbook of Lipid Research*, Plenum, New York, 1986, *Developments in Oils and Fats*, R. J. Hamilton, Ed., Blackie, London, 1994, and F. D. Gunstone and F. A. Norris, *Lipids in Foods: Chemistry, Biochemistry and Technology*, Pergamon, Oxford, UK, 1983, on which we have drawn. Developments in the oleochemical market are surveyed in F. D. Gunstone, *Oils and Fats go to Market, Chem. Br.* (1990) **26** 569. Figure 13.1 is derived from this source. B. Y. Tao reviews industrial products from soybeans in *Chem. Ind.* (1994) 906.

Other statistics on the world oil market come from *Fats and Oils, Oilcakes and Meals, 1994*, Basic Foodstuffs Service, Commodities and Trade Division, FAO, United Nations, Rome. Problems with erucic acid and glucosinolates are dealt with in P. N. Maheshwari, D. W. Stanley, and J. I. Gray, *J. Food Protection*, **44** 459 (1981) and J. M. Concon, *Food Toxicology*, Dekker, New York, 1987, Part A, pp. 352ff and 422ff.

H. B. Patterson has written three authoritative books on oil processing – *Hydrogenation of Fats and Oils*, Applied Science, New York, 1983, *Handling and Storage of Oilseeds, Oils, Fats and Meal*, Elsevier, Amsterdam, 1989, and *Bleaching and Purifying Fats and Oils*, American Oil Chemist's Society, Chicago, IL, 1993. Recent oilseed rape hybrids are reported in *Chem. Brit.* **31** 183 (1995).

Phosphatide chemistry and technology is detailed in an old but classic treatise, H. A. Wittcoff, *The Phosphatides*, Reinhold, NY, 1951.

Stigmasterol is one of the few naturally occurring sterols with an unsaturated side chain. The double bond makes possible side-chain cleavage and, with subsequent chemical reactions, transformation to progesterone. This in turn is converted to cortisone. A key step is the enzymatic insertion of oxygen in the 11 position. See *Pharmaceutical Chemicals in Perspective*, B. G. Reuben and H. A. Wittcoff, Wiley, New York, 1989, p. 325.

Section 13.1 A volume devoted to fatty acid chemistry, edited by E. H. Pryde, is *Fatty Acids*, American Oil Chemists' Society, Chicago, IL, 1979, reprinted 1985. See also the various articles in *Applewhite*, cited in the note to Section 13.11.

Information and statistics about fatty acids may be obtained from the Fatty Acids Producers' Council, 475 Park Ave. Co., New York, 10016.

The role of fats and oils in cardiovascular disease is discussed in innumerable articles including S. Sanders, *Chem. Ind.* 426 (1990) and J. Beare-Rogers, *Chem. Ind.* 131 (1995).

The extraction of oils with supercritical carbon dioxide is discussed by J. M. Snyder, J. P. Friedrich, and D. Christiansen, *J. Am. Oil Chem. Soc.* **61** 185, 1 (1984) and G. R. List and J. P. Friedrich, *J. Am. Oil Chem. Soc.* **66** 98, 1 (1989).

The hydrophilization process is described by Stein, *J. Am. Oil Chem. Soc.*, **45**, 471 (1968).

Section 13.2 The use of trifatty amines in solvent extraction operations has been described by J. E. House, *J. Am. Oil Chem. Soc.* **61**, 357 (1984).

Section 13.2 Organoclay technology has been described by W. S. Mardis, *J. Am. Oil Chem. Soc.* **61**, 382 (1984).

Developments such as a high erucic acid oil are described by L. H. Princen and J. A. Rothfus, *J. Am. Oil. Chem. Soc.* **61**, 281 (1984).

Section 13.6 The process for converting triglycerides to fatty alcohols by a one-step hydrogenolysis has been described in European Patent 334 118 (September 27, 1982) to Henkel.

Section 13.7 Isooctyl "tallate" is an isooctyl ester of tall oil fatty acids, predominantly a mixture of oleic and linoleic acids (Fig. 13.1). Plasticizers are discussed in detail in H. A. Wittcoff and B. G. Reuben, Part II, *loc. cit.* The epoxidized oils stabilize PVC by mopping up free chlorine radicals that would otherwise cause the PVC chain to "unzip" in a free radical chain reaction.

Section 13.10 Oil-modified alkyds are discussed in more detail in H. A. Wittcoff and B. G. Reuben, Part II, *loc. cit.* Section 5.6.

Section 13.10.1 The cocoa butter transesterification is described in *Catalysis at Surfaces*, I. M. Campbell, cited in Chapter 16. More details are given in A. R. McRae, *Biochem. Soc. Trans.* **17**, 1146 (1989). The field is reviewed in J. L. Harwood, Cocoa Butter. Food of the Gods?, *Chem. Ind.* October 21, 1991, p. 753.

Section 13.11 For a brief discussion of the arachidonic acid cascade and the functions of prostaglandins, see *Pharmaceutical Chemicals in Perspective*, cited above.

An excellent review of the use of enzymes for fatty chemical reactions has been authored by C. Wandrey, "*Biochemical Engineering for Oleochemicals*," in Proceedings, World Conference on "*Oleochemicals into the 21st Century*," T. H. Applewhite, Ed., American Oil Chemistry Society, Champaign, IL, 1991.

Section 13.11.1   J. Spinner's article, entitled " *Olestra Update*" is found in *Applewhite*, cited in the note for Section 13.11.

Section 13.11.2   Alkyl polyglycoside development is described in by P. Lorenz, *Problem Solving with New Fatty Alcohol Derivatives*, in *Applewhite*, cited in the note for Section 13.11.

Section 13.11.3   Biodegradable lubricants are the subject of articles by H. Kohashi, *Application of Fatty Acid Esters for Lubricating Oil* and L. Bogaerts, *Esters: Performance Oleochemicals for Food & Industrial Usage in Applewhite*, cited in the note for Section 13.11.

CHAPTER 14

CARBOHYDRATES

Carbohydrate sources for chemicals may be subdivided into four main groups: sugars, starch, cellulose, and the so-called carbohydrate gums. In addition, there are miscellaneous sources such as the pentosans found in agricultural wastes, from which furfural is made. We consider each of these as a source of chemicals. Fermentation processes are mainly carried out on carbohydrate substrates, so they too have been included briefly in this chapter.

14.1 SUGARS AND FURFURAL

For the chemist, the term sugar covers a multitude of mono-, di-, and trisaccharides composed of pentose and hexose units. In ordinary speech, however, it generally means sucrose. Sucrose is a major constituent of diet; the average citizen of a developed country consumes about 40 kg—about two-thirds of his/her body weight—annually. It is the purest crystalline organic substance to be sold in quantity to the general public, routinely reaching 99.96% purity on an anhydrous basis.

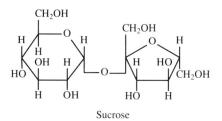

Sucrose

Sucrose is extracted from sugar cane, which is a member of the grass family, or sugar beet, which is a root crop. The cane is chopped and crushed and the juice extracted either with the aid of water or weak juice. The residue—bagasse—is a fairly pure cellulose and is either burned as fuel to make the sugar refinery self-sufficient for energy or converted to paper or hardboard.

The juice is purified, clarified, and concentrated from about 85 to 40% water in triple or quadruple effect evaporators to conserve energy. Vacuum evaporation supersaturates the liquid, and seeding precipitates sugar crystals, leaving black-strap molasses. These are used as cattle feed and as a substrate for citric acid, rum making, and other fermentations. In chemical terms, the extraction of sugar

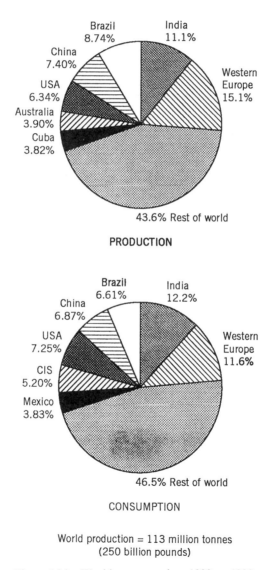

World production = 113 million tonnes
(250 billion pounds)

Figure 14.1 World sugar market, 1992 to 1993.

beet is similar, although both processes are far more complicated than is indicated above.

World production and consumption of sugar are indicated in Figure 14.1. Western Europe and Brazil are both the largest producers and consumers, the former growing beet and the latter cane. Western Europe, Cuba, Australia, Thailand, and Brazil are the major exporters, and the CIS, Western Europe, Japan, and the United States are the major importers.

There are few chemical uses for sucrose. Sucrose octaacetate is used as a denaturant in ethanol. The acetate isobutyrate and octabenzoate are used as plasticizers. The mono- and di-fatty acid esters are surfactants, and sucrose polyether polyols may be used in polyurethane formulations. The bacteria *Leuconostoc mesenteroides* and *Lactobacteriacae dextranicum* convert sucrose to dextran, a polysaccharide consisting of a backbone of D-glucose units with predominantly α-D(1 → 6) linkages. This is used as a plasma volume expander for transfusion in cases such as burns, where there is a drastic loss of body fluids. Dextran competes with "natural" materials such as fresh frozen blood plasma and albumen, and with degraded gelatin and hydroxyethylstarch. In the past, the "natural" products have been preferred where available, but with the current anxiety about infected blood, dextran is likely to benefit.

D-Glucose, known as dextrose, is manufactured on a large scale from corn-starch, which is hydrolyzed by a mixture of acid and the enzyme glucoamylase, producing a syrup of high nutritive value but not as sweet tasting as sucrose. Yields per hectare of corn are so much higher than those of sugar cane that sweeteners obtained in this way are in principle cheaper than sucrose.

D-Fructose is isomeric with D-glucose but is sweeter, and therefore offers the opportunity for a reduced calorie diet without sacrificing "sweetness." It is produced by the action of three enzymes on starch. The first is an amylase,

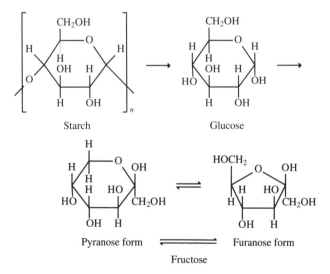

Starch   Glucose

Pyranose form ⇌ Furanose form

Fructose

which degrades the starch to lower molecular weight polymers; the second is an amyloglucosidase which converts these oligomers to glucose; and the third is an isomerase, which changes the glucose to fructose. The isomerase may be adsorbed on an insoluble carrier or fixed within the cellular microorganism in which it occurs.

Crystalline-D-fructose is more hygroscopic than dextrose and requires special packaging. Hence, it is always used as a syrup (high fructose corn syrup) in soft drinks, confectionery, and foodstuffs.

Sweeteners based on corn make up about 40% of the US sweetener market. High-fructose syrup production started in 1967 and by 1980 had overtaken the per capita consumption of sucrose. In Western Europe, pressure from the sugar farmers has prevented the wide use of high-fructose syrup within the European Community.

Lactose is the only other sugar available in large quantities and is extracted from waste skim milk. It is used as an acidulant in foods, in cheese production, in printing and dyeing, and in leather production. Fermentation of whey, lactose, sucrose, or glucose with *Bacillus acid lacti* or related *Lactobacilli* gives lactic acid:

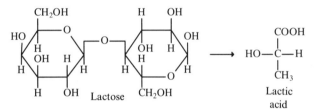

Lactic acid is currently of research interest because a polymer, polylactate, is a biodegradable plastic. It is now used for surgical sutures but, if it were cheap enough, it could have wider applications.

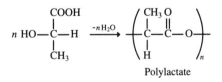

Polylactate

Glucose may be oxidized to gluconic acid, reduced to sorbitol, and esterified to α-methylglucoside as shown in Figure 14.2. Gluconic acid is primarily used as a food additive, and α-methylglucoside has been of some interest in alkyd resins. A new glucoside surfactant based on glucose and fatty alcohols is described in Section 13.6.

Sorbitol is the starting material for the classic synthesis of vitamin C (ascorbic acid) shown in Figure 14.3. The crucial stage is the second, and it is an early example of the use of biotechnology to supplement the synthetic skills of

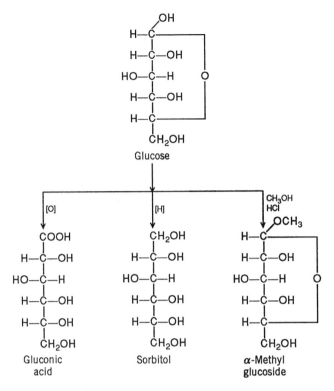

Figure 14.2   Reactions of glucose.

the chemist. Conversion of D-sorbitol to L-sorbose would present a formidable problem in organic chemistry. The appropriate bacterium, *Acetobacter suboxydans*, selectively oxidizes the hydroxyl group on the C2 of sorbitol, making feasible the remainder of the synthesis.

Sorbitol is also condensed with ethylene oxide to give a range of sorbitol–ethylene oxide emulsifiers that are used, for example, to stabilize "synthetic" whipped cream. It is used as a diluent in nonnutritive sweeteners, for example, mixed with aspartame (Section 14.5.1) to provide bulk, so that consumers can sprinkle a teaspoonful of sweetener on their breakfast cereal instead of having to measure out a tiny quantity of pure aspartame.

14.1.1   Furfural

Oat hulls, corn cobs, sugar cane stalks, wood, and many other vegetable wastes contain polymers (pentosans) of pentose sugars such as arabinose. On dehydration with hydrochloric or sulfuric acid, furfural is produced (see below). Furfural is used as a selective solvent in petroleum refining and has been used in the extractive distillation of butadiene to separate it from other C_4 olefins

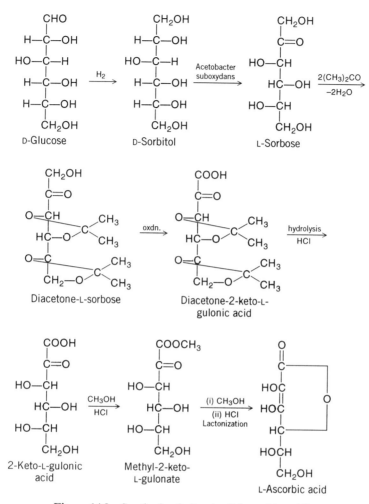

Figure 14.3 Synthesis of vitamin C (ascorbic acid).

(Section 5.1). With phenol it gives phenol–furfuraldehyde resins, which are used to impregnate abrasive wheel and brake linings.

Reduction of furfural gives furfuryl alcohol and tetrahydrofurfuryl alcohol. The latter can be dehydrated and its ring expanded by passage over an alumina catalyst at 270°C. The product is 2,3-dihydropyran. The largest use for furfural in the United States is for conversion to furfuryl alcohol, which in turn undergoes an acid-catalyzed condensation polymerization to yield resins useful as binders in the preparation of foundry cores for molding.

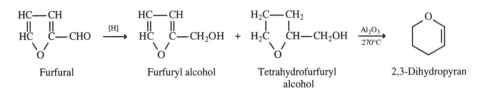

Furfural   Furfuryl alcohol   Tetrahydrofurfuryl   2,3-Dihydropyran
                                                 alcohol

Tetrahydrofuran is made by the decarbonylation of furfural with a zinc–chromium–molybdenum catalyst followed by hydrogenation. It may also be made by the dehydration of 1,4-butanediol (Section 10.3.1) and by the acid hydrogenolysis of maleic anhydride or maleic esters. These are by far the largest sources, and the furfural route accounts for less than 10% of US production.

Like ethylene and propylene oxides, tetrahydrofuran can be oligomerized to a polyether with hydroxyl end groups. The polymer is called poly(tetramethylene glycol) and is a component of the elastomeric fiber, spandex (Section 5.4).

$$n \quad \underset{O}{\bigcirc} \quad \xrightarrow{\text{cat.}/H_2O} \quad HO(C_4H_8O)_nH$$

Poly(tetra-
methylene glycol)

There is only one producer of furfural in the United States, Great Lakes Chemical, who acquired the technology from the Quaker Oats Company, who pioneered it. About 6 tonnes of corncobs or other vegetable wastes are required for every tonne of liquid product, so collection of feedstock involves high labor costs.

Another development that bears on furfural is the interest in the sugar alcohol, xylitol, as an ingredient in chewing gum, candy, and sweet cereals to prevent dental caries. Xylitol, like furfural, is derived from pentosans. It is produced in a plant in Finland, but a similar plant that was planned for the United States was never built.

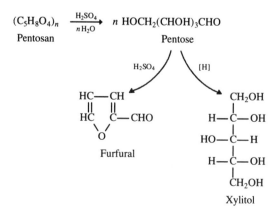

14.2 STARCH

Starch is one of the most important chemical products of the vegetable kingdom and is found in practically all plant tissues, especially in seeds (e.g., wheat and rice) and tubers (e.g., potatoes). Commercial starch in the United States is extracted chiefly from corn and to a lesser degree from wheat. In Europe potatoes are an important source. About one-half of the starch produced in the United States is hydrolyzed with hydrochloric acid to glucose or partially hydrolyzed to starch–glucose syrups that can be sold as such or isomerized to fructose (Section 14.1).

Chemically, starch comprises two distinct polymers of α-D-glucopyranoside. The linear polymer, amylose, is composed of several hundred glucose units connected by α-D-(1 → 4) glucosidic linkages (Fig. 14.4a). The branched-chain polymer, amylopectin (Fig. 14.4b), is of much higher molecular weight and contains between 10,000 and 100,000 glucose units. The segments between the branched points contains about 25 glucose units joined as in amylose, while the branched points are linked by α-D-(1 → 6) bonds. Most cereal starches contain about 75% amylopectin and 25% amylose.

Because amylose has straight chains, its molecules can approach each other within molecular range and form hydrogen bonds. This interaction is so strong that amylose can scarcely be dispersed in water. Amylopectin, on the other hand, has branched chains that have more difficulty in approaching each other, and it is therefore readily dispersed.

If starch, which is an amylose–amylopectin mixture, is dispersed in water, a gel results that forms a "skin" on standing. The skin is a result of hydrogen bonding. So-called waxy starches, such as waxy maize starch, which arose from genetic experimentation, are very high in amylopectin, and consequently they form gels that do not "skin."

Amylose is prevented from skinning if it is converted into a derivative such as the phosphate. Phosphorus oxychloride or sodium trimetaphosphate are the reagents. Only a small number of phosphate groups per polymer–molecule are required to interfere with the hydrogen bonding. Consequently, starch phosphates are widely used as thickeners in the food industry and in other applications where a stable, high-viscosity starch paste is required.

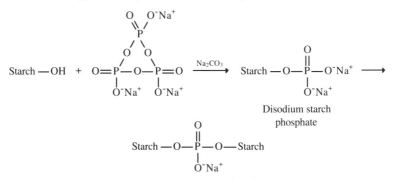

Disodium starch phosphate

Starch dihydrogen pyrophosphate

(a)

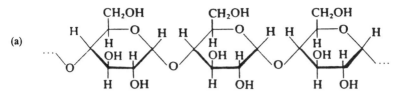

Part of amylose molecule showing α-D-(1→4)-glucosidic linkages

(b)

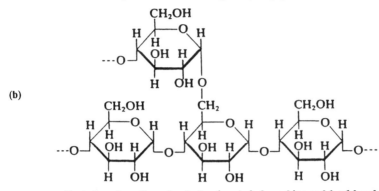

Part of amylopectin molecule showing chain branching and 1→6 bonds

(c)

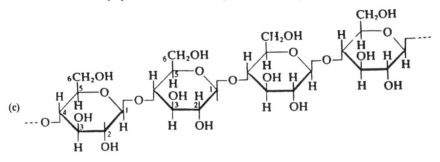

Part of cellulose molecule showing β-D-(1→4)-glucosidic linkages

(d)

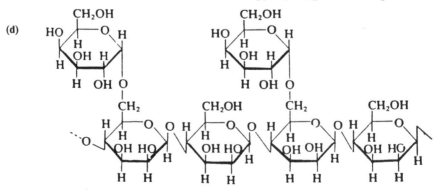

Part of guar gum molecule showing chain of mannose units with pendant galactose units

Figure 14.4   Starch, cellulose and guar structures.

Other cross-linking agents for starch are also approved by the FDA, the most widely used being epichlorohydrin and linear mixed anhydrides of acetic and di- or tribasic carboxylic acids.

Starch is used in adhesives and as a size in textiles and paper manufacture. Sizing is a method for altering the surface properties of paper fibers. For example, many grades of paper are coated with an aqueous suspension of pigments (such as clay) in adhesive (such as starch) to provide a smoother surface, control the penetration of inks, and generally improve the paper's appearance. Dextrinized or degraded starch is used as the adhesive.

Starch acetates are made from starch and acetic anhydride, the extent of acetylation being indicated by the degree of substitution (DS). This is the number of hydroxyl groups per glucose unit that are esterfied. Total substitution there- fore represents a DS of 3. Acetyl starches are used in the food, paper, and textile industries.

Hydroxyethylstarch, made by the action of ethylene oxide on starch, is used in paper coating and sizing, because it disperses more easily than starch itself and provides a dispersion of better clarity. It competes with dextran (Section 14.1) as a plasma volume expander. Starch may also be reacted with cationic reagents such as dimethylaminoethyl chloride to give "cationic starch," which is used as a flocculant and to impart greater strength to paper products. Acrylonitrile may be polymerized onto a starch polymer with the aid of a ceric sulfate catalyst to give a graft copolymer which, on partial hydrolysis of the nitrile groups, is capable of absorbing large amounts of water (Section 4.3).

There are many other chemically modified starches that find specialized uses in the food industry but which are outside the scope of this book. The reader is referred to sources quoted at the end of this chapter.

14.3 CELLULOSE

Cellulose is the primary substance of which the walls of vegetable cells are constructed. It occurs in plants, wood, and natural fibers, usually combined with other substances such as lignin, hemicellulose, pectin, fatty acids, and rosin. It accounts for about 30% of all vegetable matter. It may be represented by the formula $(C_6H_{10}O_5)_n$.

Cellulose is a linear polymer composed largely of glucose residues in the form of anhydroglucopyranose joined together by β-glucosidic linkages (Fig. 14.4c). The β-linkages make the cellulose molecule very stiff compared with amylose, which has α-glucosidic linkages. It also makes cellulose more difficult to hy- drolyze than starch. It is this difficulty that makes cellulose indigestible by humans. Accordingly, one of its uses is in diet foods, providing bulk and satiety but not calories. Had nature not insisted on β linkages, a vast source of nutrition would be available to humans. On the other hand, cellulose with α linkages would not be stiff and would not possess the structural properties it has in wood or the strength it has in textiles. Cows and other ruminants have digestive systems containing organisms with enzymes that hydrolyze cellulose, and they are therefore able to use grass and similar materials for food.

Cellulose may be obtained from wood or derived in very high purity from cotton linters. The majority of production is used in paper manufacture. In the manufacture of pulp used in cheap paper, the wood fibers are simply pulped mechanically. In the premium chemical pulp, lignins, fatty acids, and rosins are first removed by sodium hydroxide treatment.

Cellulose also has a range of chemical uses. Cellulose derivatives (Fig. 14.5) may be water-soluble like methylcellulose, hydroxyethylcellulose, or sodium carboxymethylcellulose, or water-insoluble like the cellulose esters and ethylcellulose.

Methylcellulose, sodium carboxymethylcellulose, and hydroxyethylcellulose are all thickening agents and form protective colloids. For example, they are used in water-based latex paints to produce desirable flow and viscosity and to help stabilize the emulsion and pigment dispersions. Protective colloids affect the washability, brushability, rheological properties, and color acceptance of a paint. The same properties make them useful in foods such as ice cream and in inks and adhesives.

Methylcellulose is also used as a base for paper coatings since it is a good film former. Hydroxyethylcellulose is an adhesive and a binder in nonwoven fabrics. Sodium carboxymethylcellulose (CMC) has a major use as a soil suspending agent in detergents, for if it were not present the dirt would tend to redeposit on the articles being washed. Instead the CMC forms a protective colloid and holds it in suspension. Carboxymethylcellulose is used in textile sizing, paper coating, and in oil well drilling mud, a material that helps to bring to the surface the dirt and rock particles dislodged by the drill. Diethylaminoethylcellulose is a cationic material useful in cotton finishing.

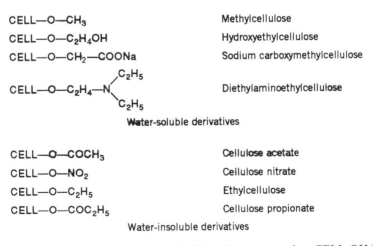

Figure 14.5   Cellulose derivatives (Cellulose is represented as CELL-OH.)

Sodium carboxymethylcellulose is made by spraying powdered cellulose first with sodium hydroxide solution and then with chloroacetic acid. The product contains some sodium chloride, which is removed by washing if food grade material is required. The commonest grade has a degree of substitution of 0.7 (0.7 hydroxyl groups substituted per glucose unit).

$$\text{Cell—OH} \ + \ \text{NaOH} \ \longrightarrow \ \text{Cell—ONa}$$
Cellulose

$$\text{Cell—ONa} \ + \ \text{ClCH}_2\text{COONa} \ \longrightarrow \ \text{Cell—OCH}_2\text{COONa}$$
Sodium   Carboxymethyl cellulose
chloroacetate

Cellulose triacetate is made by the action of equal quantities of acetic anhydride and glacial acetic acid on "chemical cotton," a form of cellulose obtained by purification and conversion of cotton linters. All three hydroxyl groups in each glucose unit are acetylated.

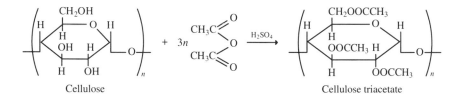

Cellulose   Cellulose triacetate

Cellulose acetate is obtained by partial hydrolysis of cellulose triacetate with water and a small amount of acetic and sulfuric acids. Addition of a large excess of water at the appropriate stage stops the hydrolysis and precipitates the acetate (with an average of two acetyl groups per glucose unit) as flakes. Cellulose acetate and triacetate may be used as plastics molding materials or spun into fibers and used for textiles. Cellulose triacetate is more difficult to process than the acetate but can be heat-set to give wash-and-wear fabrics. The largest use, however, for cellulose acetate is for cigarette filters. Cellulose propionate, acetate-propionate, and acetate-butyrate are also used as plastics materials and for films and lacquers. Ethylcellulose is useful in coatings and some plastics applications.

Cellulose nitrate or nitrocellulose was an early explosive (guncotton), plastic (celluloid), and surface coating (lacquer) both for wood and metal. Thus it was the first material used for coating automobiles by assembly-line procedures, its rapid drying making semiautomated production possible. It is obtained by nitration of cellulose, about two nitro groups per glucose unit being introduced. It is also used for fabric coating, as the basis for lacquers for furniture (where it is being displaced by "melamine precatalyst" formulations) and in plastic moldings and film.

In addition to chemical conversion, cellulose may be altered physically or "regenerated." Two processes are used, both of which start with highly purified cellulose from wood pulp, and the products are known as viscose rayon and cuprammonium rayon. Viscose rayon is produced by conversion into the soluble xanthate by "ripening" with concentrated sodium hydroxide followed by treatment with carbon disulfide:

0.5–0.6 Xanthate groups per glucose unit are inserted. The solution is known as "viscose." It is extruded through a spinerette—a metal disk with many tiny holes—into an acid coagulating bath that regenerates the cellulose as a fiber. The fibers are washed and spun. The spinning orients them by stretching. The carbon disulfide is regenerated and recycled. The xanthation serves to solubilize the cellulose into a form from which it can readily be regenerated. During the regeneration, additional hydrogen bonding is created, thus increasing strength.

The premium cuprammonium rayon is made by dissolution of cotton linters or wood pulp in ammoniacal copper oxide. The solution is extruded through spinnerettes into an acid bath in the same way as viscose. Cuprammonium rayon is chemically similar to viscose rayon but gives a finer yarn that is used in sheer fabrics.

Less vigorous treatment at the pulping stage leads to a higher molecular weight cellulose, which yields so-called polynosic rayon with longer, finer fibers and high wet-strength.

Cellophane film is also made from viscose solution, but the cellulose is regenerated in sheet form instead of fibers. It has a high moisture transmission and is lacquered with a waterproofing agent, traditionally cellulose nitrate, which also imparts heat sealing capability. Poly(vinylidene chloride) is now widely used for this purpose. It is still a major packaging film for decorative items such as flowers and candy boxes despite the advent of plastic films such as polyethylene and Saran.

Cellulose sponges are also made from xanthate, which is cast into blocks together with sodium sulfate crystals of various sizes. The xanthate is decomposed to regenerate cellulose, and when the sulfate is washed out it leaves holes in the sponge. Recovery of sodium sulfate is a major part of the operation.

The strength properties of cellulose, and to a degree its ability to hydrogen bond, make it useful for the formation of paper and nonwoven fabrics. For paper, cellulose is "beaten" or "pulped" until very finely divided particles result. Glassine is prepared from such finely divided particles that it is translucent. Opaque papers result from less finely divided particles. The wet and dry strength, bulk density, water and oil resistance, and gas permeability of paper are properties than can be modified either by adding chemicals to the pulp or by coating the finished sheet. Urea–formaldehyde and melamine–formaldehyde resins increase wet strength. Starch and rosin soap are widely used for sizing.

Nonwoven fabrics result when rayon is chopped into very fine particles known as fibrils and converted to sheets by various processes including (with the aid of a bonding agent) the paper-making process.

14.4 GUMS

Gums, like starch and cellulose, are carbohydrate polymers. They differ from them in that the monomeric unit may be a sugar other than glucose, and the chemical configuration and the way in which the units are joined may be different.

The molecular weight of gums is usually between 200,000 and 300,000, which is about 1500 monomer units. Guar (Fig. 14.4d) is a typical gum. It consists of a chain of mannose units joined by $1 \rightarrow 4$-glycosidic linkages, and attached to every other mannose unit is a pendant galactose unit.

The main gums and their origins are shown in Table 14.1. Each gum has characteristic properties slightly different from other gums. Frequently, the commercially important differences lie in the rheological properties of the dispersions of the gums in water. Gums may be chemically modified just as cellulose is, and the most useful derivatives are carboxymethyl, hydroxypropyl, and dimethylaminoethyl gums.

The applications of gums are wide. Guar gum is the most important and may be considered typical. Guar has many times the thickening power of starches and may be used in combination with them. Its derivatives are useful as flocculents for precipitating mineral slimes and as a suspending agent for ammonium nitrate that not only leads to a much cheaper explosive than dynamite or nitroglycerin, but also to one that is more effective, because it assumes the shape of the cavity where the blast is to start.

Carboxymethylguar gum is an anionic material useful as a print gum paste. This means that it serves as a binder for pigment used to impart color and design to cloth. In contrast, diethylaminoethylguar gum is cationic and is used in paper

TABLE 14.1 Natural Gums

Source	Examples
Plant seeds	Guar gum, locust bean gum
Seaweed extracts	Alginates, carrageenan, agar
Tree exudates	Gum Arabic, karaya gum, gum tragacanth
Citrus fruits	Pectin
Animal skin and bones	Gelatin
Fermentation	Xanthan gum

manufacture. It is a particularly effective flocculent of "fines," the minute cellulose particles in the paper matrix, onto which it adsorbs. By helping to retain the "fines," it increases the yield of product and, since fewer "fines" are in the water that drains from the machine, the pollution problem is diminished. Guar itself strengthens the paper by hydrogen bonding to the fibers and helping them achieve a linear rather than a random configuration.

Xanthan is an unusual gum, in that it is obtained not from animals or plants but by fermentation of carbohydrates with a bacterium, *Xanthamonus campestris*. The gum is a complex glucose polymer, and its aqueous solutions are unusually stable, showing unchanged viscosity over broad temperature, salt concentration, and pH ranges. The product is therefore used to thicken oven cleaners based on strong alkali as well as the acid solutions used as metal cleaners. Its largest use is as a component of oil well drilling mud that must contain saline water. Its largest potential use is in so-called enhanced oil recovery where it thickens the water used to "push" oil, unobtainable in any other way, through the dense, oil-bearing rock formation. Unlike other gums such as guar, xanthan does not adhere to surfaces. If it did, it would deposit as a film on the rock surface and lose its thickening power.

14.5   FERMENTATION AND BIOTECHNOLOGY

When supplied with suitable nutrients, single-cell microorganisms—yeasts, molds, fungi, algae, bacteria including the important antibiotic-producing *Actinomycetes*—thrive and multiply. As they do so, various waste products of their metabolisms accumulate. The microorganisms can tolerate only low concentrations of their own wastes. Nonetheless, under certain circumstances, these wastes, which can be either intra- or extracellular, can be concentrated and used. The process is known as microbial conversion or fermentation, and it occurs as a result of the catalytic action of various enzymes produced by the microorganisms on the nutrient or substrate. Thus fermentation can also be brought about by pure enzymes or portions of cells that contain enzymes such as mitochondria.

The substrate is usually but not always a carbohydrate. In the past, there was interest in the use of gas oil (Section 2.1) or other petroleum hydrocarbon substrates for the production of single-cell protein. Much of the development work was done in the days of seemingly abundant petroleum, and all plants are thought to have closed. An ICI methanol-based plant, opened in 1979, survives. In the presence of aqueous nutrient salt solutions containing inorganic sulfur, phosphorus, and nitrogen, a variety of *Pseudomonas* will grow to give a bacterial culture, where the dried cells contain up to 81% protein. This is supplemented by various essential amino acids that are underrepresented in single-cell protein and used as animal feed.

ICI operates a plant for production of mycoprotein—a relation of the mushroom—which can be processed to yield a proteinaceous food with the fibrous texture of meat. The product is known as "Quorn." ICI markets it especially to vegetarians and announced in 1992 that it was increasing productive capacity. A joint venture of Phillips Petroleum and Petrofina also produces speciality proteins for human consumption.

In the United States, inexpensive soy protein is available, and there is little incentive for the development of other sources. Soybean production is also being expanded in Brazil, and there should be adequate supplies of vegetable protein in the Western Hemisphere for the foreseeable future. On the other hand, there are many protein-short countries in the world where fermentation protein might be helpful, and there is believed to be a plant in the CIS.

Fermentation is used in the chemical industry only when an economical chemical process is not available. Its largest volume application is in sewage treatment, where obnoxious amines and sulfur compounds in the sewage are oxidized to nitrates and sulfates. Other wastes are anaerobically digested to methane or aerobically oxidized. The next largest application is the production of alcoholic beverages, where the product can either be consumed in dilute form (beer and wine) or concentrated by distillation (whisky, brandy, gin, or vodka).

The fermentation reaction is usually efficient and inexpensive, and the conditions required are mild. "Waste heat" from power stations, refineries, and factories could easily be used as an energy source. On the other hand, nutrients tend to be expensive, reactions are slow, and the product is so dilute that huge tank capacities are required. Unless the product precipitates (e.g., single-cell protein), its isolation can be tedious and expensive. If the fermentation is aerobic, mass transfer of oxygen to the required site demands intricate engineering. Contrary to popular opinion, fermentation processes are not ecologically friendly. In general, they produce large volumes of waste water and mycelium with a high biochemical oxygen demand, which must be treated before discharge into a waterway. Nonetheless, there are products for which fermentation methods are uniquely suitable.

Antibiotics are an example. Practically all antibiotics—penicillins, cephalosporins, macrolides, tetracyclines—are made by fermentation. Penicillin is made by fermentation of a corn-steep liquor substrate (a cheap form of sucrose) with *Penicillium chrysogenum* (see note). For the production of semisynthetic penicillins, such as amoxycillin, the penicillin from this first stage is then cleaved to 6-aminopenicillanic acid with an immobilized amidase (Section 16.8).

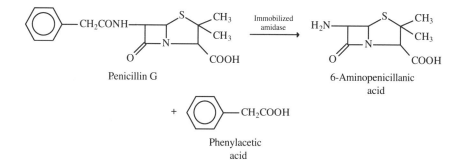

Penicillin G

6-Aminopenicillanic acid

Phenylacetic acid

Penicillin has been synthesized in the laboratory, but the process would be hopelessly uneconomic compared with the biotechnological route.

Lactic (Section 14.1) and citric acids are made by fermentation. Citric acid has a structure that makes economical chemical synthesis difficult. It is obtained by growing *Aspergillus niger* on a molasses, starch, or hydrocarbon substrate.

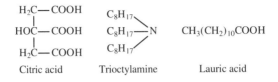

Citric acid   Trioctylamine   Lauric acid

One method for separation of the acid from the fermentation broth is unique. The mixture is shaken with a trioctylamine-lauric acid "couple" in an organic solvent. The citric acid displaces the lauric acid from the acid–base couple and enters the organic phase leaving the impurities behind. After phase separation, the organic phase is shaken with hot water. The entropy of the phase transfer is such that, at the higher temperature, the process is reversed and the citric acid returns to the aqueous phase from which it can simply be extracted (see note).

Other noteworthy fermentation processes include the Weizmann process, which gives acetone and *n*-butanol by fermentation of corn with *Bacillus clostridium acetobutylicum*, which was important in World War I but is now obsolete; the production of evening primrose oil (Section 13.1); the use of an immobilized cell to hydrolyze acrylonitrile to acrylamide (Section 4.10.3); the production of vitamin C (Section 14.1); the 11-hydroxylation of progesterone in the synthesis of cortisone (see note); and the production of synthetic cocoa butter (Section 13.10.1). A route to hydroquinone from glucose has been reported but not yet commercialized (see note).

14.5.1 Amino Acids

The synthesis of L-amino acids by fermentation has been pioneered largely in Japan. Production of every essential amino acid by fermentation is now possible. Because demand is small, many of them are still being made by chemical methods followed by resolution of the resulting DL-racemate. The most important amino acids made by fermentation are L-glutamic acid, L-lysine, and L-aspartic acid.

L-glutamic acid is produced by fermentation of glucose or sucrose from molasses or other sugar refinery wastes. The bacterium is *Micrococcus glutamicus*, and nitrogen is supplied in the form of ammonia. Monosodium glutamate is used as a flavor enhancer in prepared foods, such as packaged soups. Only the L form enhances flavor. Monosodium glutamate is also widely used in Southeast Asia and in Chinese restaurants throughout the world. Excessive doses may lead to *Kwok's disease*, the so-called *Chinese restaurant syndrome.*

$$^-OOCCH_2CH_2CHCOO^-Na^+$$
$$|$$
$$NH_3^+$$

Monosodium glutamate

Lysine was originally manufactured by Du Pont, who used a conventional synthetic route followed by classical resolution of the D and L isomers. It was also isolated from blood meal by General Mills Chemicals. The L form resulted, obviating the need for the tedious resolution. These methods were displaced by fermentation routes based on *Clostridium glutamicum* and ammonium acetate or carbohydrate–ammonia substrates. Toray then developed a new route based on the nitrosyl chloride chemistry and the chlorocyclohexane byproduct from its caprolactam synthesis (Section 7.2.2).

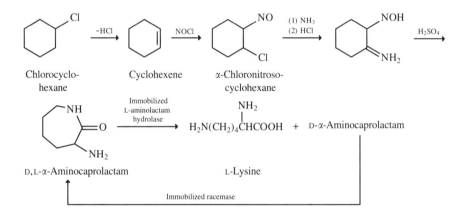

Chlorocyclo-   Cyclohexene   α-Chloronitroso-
hexane   cyclohexane

D,L-α-Aminocaprolactam   L-Lysine

Only the L-α-aminocaprolactam is attacked by the L-hydrolase so that only L-lysine is produced. The D-α-aminocaprolactam is racemized with the aid of a second immobilized enzyme to a D,L-mixture for reuse.

L-Aspartic acid is an amino acid with wide uses in the food and pharmaceutical industries. Its significance on a tonnage scale arises because it is a starting material for the nonnutritive sweetener, aspartame.

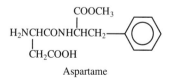

Aspartame

The enzyme *aspartase* promotes the addition of ammonia across the double bond of fumaric acid to give the L form of aspartic acid:

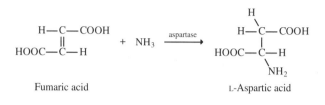

Fumaric acid   L-Aspartic acid

Microbial strains of *Escherichia coli* with high *aspartase* activity are immobilized in a κ-carrageenan gel cross-linked with glutaraldehyde and hexamethylenediamine and operate at a temperature of 38°C.

DL-amino acid mixtures in general may be resolved by use of immobilized enzymes. If a racemic mixture is acetylated and passed over an immobilized L-acylase, the acetyl group is hydrolyzed from the L-amino acid but the D-amino acid remains unchanged. The mixture may then be separated by crystallization.

14.5.2 Polymers

A fermentation route to nylon 6,6 is described in Section 7.2.1. The production of xanthan gum (Section 14.4), which has a molecular weight of more than 1 million, is an example of the production of polymers by fermentation, but the product does not have the physical properties generally associated with polymers. ICI has developed a route to a biodegradable polyester copolymer of γ-hydroxybutyric and γ-hydroxyvaleric acids by fermentation of glucose with *Alcaligenes eutrophus*.

$$HOCH_2CH_2CH_2COOH \qquad HOCH_2CH_2CH_2CH_2COOH$$
γ-Hydroxybutyric acid   γ-Hydroxyvaleric acid

The polymer is stereoregular and has a molecular weight of 100,000 to 400,000. After separation from microbial cells and purification, it can be formed by conventional polymer processing techniques. The stereospecificity and high molecular weight, which give it strength, are unique for step-growth polymerization which, when performed conventionally, gives molecular weights as low as 10,000. The polymer, *polyhydroxybutyrate*, is expensive but its biodegradability makes it attractive and it is finding a markets in the blow-molding of bottles and other packages, particularly for "green" cosmetics.

Proposed uses include medical applications such as sutures, staples, ocular inserts, cardiovascular grafts, bone screws, and related surgical items. Because the polymer is biodegradable, postsurgical removal is not required. Also because of its biodegradability, the polymer might be useful in controlled release of drugs, applicators for feminine hygiene products, for the encapsulation of pheromones, and for agricultural film in which fertilizers and insecticides are imbedded. A decreased price would make it a candidate for general purpose biodegradable film. In 1992 ICI announced that manufacture was planned in the

United States, the material having already been manufactured in the United Kingdom for several years. An alternative polymer appears to be *polylactate*, a polyester from lactic acid.

ICI has also developed a fermentation route to poly-*p*-phenylene. This polymer has excellent high-temperature resistance and hydrolytic stability but is highly intractable with no known solvent, and it can be molded only at high temperatures and pressures. ICI found a strain of *Pseudomonas putida* that would convert benzene to benzene *cis*-glycol when supplied with benzene and ethanol in the absence of oxygen. Esterification with acetic acid leads to a diester, which is polymerized with a free radical initiator. The intermediate polymer is not aromatic. It can be solubilized and is converted into fibers and supported or unsupported film. Heating volatilizes the solvent and completes the polymerization to polyphenylene in situ. The acetic acid is regenerated for recycle.

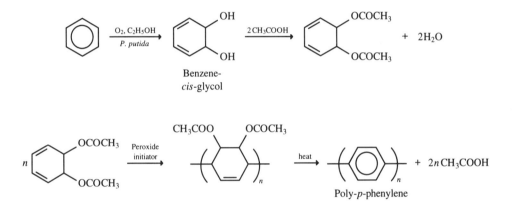

14.5.3 Proteins by Recombinant DNA Technology

The revolution in biotechnology over the past decade has made possible the manufacture of almost any protein by recombinant DNA technology. Large scale single-cell protein production has already been mentioned in Section 14.5. The new technique, by which protein producing genes are spliced into single-cell organisms, is altogether remarkable. Insulin, interferon, a range of blood clotting factors for hemophiliacs, human growth hormone, bovine growth hormone, and a range of vaccines are already on the market. Organisms have already been cloned to produce others should it become economic. We can scarcely conclude this section on fermentation without mention of this development.

Implications for the heavy chemical industry, however, are remote. The largest volume genetically engineered product on the market is insulin, which sells at about \$50–60 g (\$23,000–\$27,000/lb). Annual production is measured in pounds rather than tonnes. The adaptation of these methods to bulk chemicals present a whole series of new problems and will not happen for a long time.

14.5.4   A Fermentation Scenario

Discussion of fermentation, however, raises the question as to whether the chemical industry could survive were supplies of oil and natural gas depleted. In terms of technology, the answer is encouraging. In between the world wars, fermentation was a major route to organic chemicals, providing ethanol, *n*-butanol and acetone as feedstocks. Production was expensive and tonnages were low. Today reaction pathways exist by which most modern organic chemicals and polymers could be made from these feedstocks more efficiently than they were 60–80 years ago. Thus ethanol can be dehydrated to ethylene. Ethylene can be dimerized to 2-butene, and the metathesis reaction (Section 2.2.9) then permits the production of two molecules of propylene from one each of ethylene and 2-butene.

Ethanol is also valuable as a fuel, and gasohol is used in several countries including the United States and Brazil. It appears, however, that production requires more energy than the ethanol provides when burned (Section 12). Furthermore, if all the corn, wheat and other crops grown by world farmers were converted to ethanol, it would still give only 6–7% of the energy equivalent of present world oil production.

Methane is produced by anaerobic fermentation of sewage sludge and of organic wastes generally. About 75% of the calorific value of sludge can be recovered in this way, which sounds impressive but in fact serves mainly to make sewage plants independent of outside sources of power. In India and China, biomass fermentation processes are well developed and are used in rural areas to generate heating and cooking gas from accumulated biomass wastes. The so-called biogas makes an important contribution to an improved lifestyle in these regions.

It is estimated that biomass (which includes residues of the forest industry, corncobs, oat hulls, and various plants) could supply $5–10 \times 10^{15}$ Btu ($\sim 5000–10,000$ terajoules) of fuel and chemicals by the year 2020. This would amount to 5–10% of present US energy consumption.

In summary, therefore, fermentation is currently valuable to do what the chemist cannot do, such as antibiotic and L-amino acid production, and steps in vitamin C and cortisone synthesis. It will provide energy via methane food protein and nutrition via alcoholic beverages. At a cost, it can provide ethanol, which can be dehydrated to ethylene, on which a lion's share of the chemical industry is based. How much energy, food, or ethylene will be produced by fermentation, however, depends on economics which, in spite of some exceptions, tend to favor petrochemicals. Wider use of fermentation will depend on the seriousness of shortages and the alternative routes devised to alleviate them.

NOTES AND REFERENCES

Section 14.1 Statistics on sugar come from *World Sugar Situation and Outlook*, US Dept. of Agriculture, Foreign Agricultural Service, Circular Series FS-2-93, Washington, DC, 1993.

Section 14.2 R. L. Whistler, J. N. BeMiller, and E. F. Paschall, *Starch: Chemistry and Technology*, 2nd ed., Academic, Orlando FL, 1984, is the classic book on starch, the earlier edition having appeared in 1965. The immobilization of enzymes for high-fructose syrup is dealt with in this volume by N. E. Lloyd and W. J. Nelson, *Glucose and Fructose Containing Sweeteners from Starch*, p. 635. See also T. Galliard, *Starch: Properties and Potential*, SCI Critical Reports on Applied Chemistry, Vol. 13, Wiley, New York, 1987.

Section 14.3 Cellulose makes up about 50% of wood together with 25% of lignin (phenylpropane polymers) and 25% of hemicelluloses (carbohydrate polymers built up from molecules of simple sugars). Wood is thus a source of such polymers. In prehistoric times it was a precursor of coal and oil. In principle, therefore, it should be possible to obtain chemicals from wood. In practice, it has proved much more difficult than getting them from oil. Two rather old articles describing what can be done are by I. S. Goldstein, *AIChemE Symposium Series* **74**, 11 (1978), and in a paper *Chemicals from Wood: Outlook for the Future* to the 8th World Forestry Congress, Jakarta, Indonesia, October, 1978.

Section 14.4 A third edition of R. L. Whistler and James N. BeMiller, *Industrial Gums, Polysaccharides and their Derivatives*, Academic, San Diego, CA, 1993 has recently appeared and will become the standard work on the subject. Y. Paik and G. Swift review polysaccharides as raw materials for the detergent industry in *Chem. Brit.* **31** 55 (1995).

Section 14.5 A recent account of the Weizmann process together with a scholarly account of developments in biotechnology to the present day appears in Robert Bud, *The Uses of Life: A History of Biotechnology*, Cambridge University Press, Cambridge, 1993.

The production of single cell protein is described in N. Calder, Food from Gas Oil, *New Sci.*, **36**, 468 (1967). ICI and BP both produce educational publications on the topic of single-cell protein.

The pharmaceutical processes mentioned here are described in more detail in our book *Pharmaceutical Chemicals in Perspective* cited in Section 0.4.5. Some of the immobilized enzyme processes are taken from Ian Campbell's book on catalysis cited in Chapter 16.

The Miles citric acid process is described by A. M. Baniel, European Patent 49 429 (1982), A. M. Baniel, R. Blumberg, and K. Hajdu, US Patent, 4 275 234 (1981) and J. E. Alter and R. Blumberg, US Patent 4 251 671 (1981).

The biotechnological route to hydroquinone and benzoquinone was reported by S. Borman, *Chem. Eng. News*, 14 December 1992, and J. W. Frost and K. M. Draths, *Chem. Brit.* **31** 206 (1995), where other glucose-based chemicals are also discussed.

Section 14.5.2 The poly-*p*-phenylene process is described in European Patent Application 0 076 606A1 (April 13, 1983) to ICI. See also S. C. Taylor, Paper presented at BIOTECH'86, Online Publications Pinner, UK, 1986.

CHAPTER 15

HOW POLYMERS ARE MADE

The polymer industry stands out above all others as a consumer of heavy organic chemicals, and it converts these to the products called plastics, fibers, elastomers, adhesives, and surface coatings. The terms polymer and resin are used synonymously in the chemical industry, but the terms plastics, elastomers, and fibers have specific meanings. Moreover, it is incorrect to refer to all synthetic polymers as plastics. A plastic is a material that is formed or fabricated from a polymer, usually by causing it to flow under pressure. Thus if a polymer is molded, extruded, cast, machined, or foamed to a particular shape, which may include both supported and unsupported film, the polymer can be described as a plastic. Often a plastic contains pigments and additives such as antioxidants, plasticizers, and stabilizers. Fibers and elastomers are defined in Section 15.6.

This chapter includes some of the chemistry of individual polymer manufacture but is intended more as a broad description of how to synthesize a polymer, how to influence its properties, and how these properties relate to end uses that affect everyone's daily lives.

In 1995 the US polymer industry is expected to produce about 87 billion lb of polymers (Table 15.1) compared with the figure of about 450 billion lb (Section 1.5) for the output of the US organic chemical and polymer industry.

The latter figure refers to chemicals actually isolated before being subjected to another reaction. Thus we might conclude that the polymer industry consumes only about one-fifth of the chemical industry output. Such a conclusion ignores an important element of double counting implicit in the statistics. If 1 billion lb of ethylene and 3 billion lb of benzene are produced from naphtha and thus react together to give about 4 billion lb of ethylbenzene, which is then dehydrogenated to a like amount of styrene, which in turn is polymerized to almost 4 billion lb of polystyrene, then the production statistics will record 16 billion lb of chemicals. It is difficult to eliminate this element of double counting al-

TABLE 15.1   United States Polymer Production, 1995 (estimated)

Polymer Use	(billion lb)
Plastics	62.0
Fibers	10.0
Elastomers	6.0
Coatings	5.0
Adhesives	4.0
Total	87.0

together, but one estimate is that the 87 billion lb of polymers consume about 216 billion lb of chemicals—about 48% of the total of all chemicals produced. To that should be added such materials as solvents for surface coatings, plasticizers for poly(vinyl chloride) (PVC), as well as many compounding and processing aids. Thus we can say with confidence that the polymer industry consumes well over half the tonnage output of the organic chemical industry.

Polymers may be subdivided into two categories, thermoplastic and thermosetting. Thermoplastics soften or melt when heated and will dissolve in suitable solvents. They consist of long-chain molecules often without any branching (e.g., high-density polyethylene). Even if there is branching (e.g., low-density polyethylene) the polymer may still be two dimensional. Thermoplastics may be used in the five main applications of polymers—plastics, fibers, elastomers, coatings, and adhesives—as shown in Table 15.2. These are discussed further in Section 15.6.

Thermosets decompose on pyrolysis and are infusible and insoluble. They have elaborately cross-linked three-dimensional structures and are used for plastics, elastomers (lightly cross-linked), coatings, and adhesives but not fibers, which require unbranched linear molecules that can be suitably oriented by stretching during the spinning and drawing processes.

Table 15.3 shows the estimated sales of polymers for plastics applications in the United States in 1993. The largest volume plastic is low-density polyethylene, and if low- and high-density polyethylene are added together, then polyethylene emerges as the most important plastic worldwide.

Sales of thermoplastics for plastics applications are about nine times those of thermosets in spite of the fact that thermosets, especially urea–formaldehyde and phenol–formaldehyde resins, have been produced commercially for much longer than any of the thermoplastics. The thermosets have been unable to share more extensively in the phenomenal growth of plastics because they are difficult to process and do not lend themselves to the high production speeds that can be achieved, for example, by modern injection molding machines. Methods of processing plastics are summarized in the notes.

TABLE 15.2   Major Applications of Polymers

Plastics

Extruded products
  Low-density polyethylene
  Poly(vinyl chloride)
  Polystyrene and styrene copolymers
  High-density polyethylene
  Poly(ethylene terephthalate)
  Polypropylene
  Acrylonitrile–butadiene–styrene copolymers
  Cellulose acetate
  Cellulose acetate butyrate

Molded products
  Polystyrene and styrene copolymers
  High-density polyethylene
  Polypropylene
  Low-density polyethylene
  Poly(vinyl chloride)
  Phenolics
  Urea–formaldehyde
  Melamine–formaldehyde
  Acrylics
  Cellulose acetate
  Cellulose acetate butyrate

Film and Sheet
  Low-density polyethylene
  Poly(vinyl chloride)
  Regenerated cellulose (cellophane)
  Acrylics
  Poly(ethylene terephthalate)
  Polypropylene
  High-density polyethylene

Foams
  Polyurethane
  Polystyrene
  Many others

Fibers
  Poly(ethylene terephthalate)
  Nylon (polyamides)
  Aramids
  Polyacrylonitrile copolymers
  Polypropylene
  Rayon
  Cellulose acetate
  Glass

TABLE 15.2   Continued

Elastomers

Styrene–butadiene rubber
Polyisoprene
Ethylene–propylene and ethylene–propylene-diene-
    monomer terpolymers
Polybutadiene
Butadiene–acrylonitrile copolymers
Silicone
Sulfochlorinated polyethylene
Styrene–butadiene–styrene thermoplastic rubber

Coatings

Paper and Textile applications
    Low-density polyethylene
    Polystyrene and styrene copolymers
    Poly(vinyl chloride)
    Poly(vinyl acetate)
    Urea–formaldehyde
    Melamine–formaldehyde

Conventional coatings
    Alkyds
    Oils
    Acrylics
    Poly(vinyl acetate)
    Poly(vinyl chloride)
    Epoxy
    Cellulose acetate
    Cellulose acetate butyrate
    Urea–formaldehyde
    Urethanes
    Polystyrene and styrene copolymers
    Unsaturated polyesters

Adhesives

Laminating
    Phenol–formaldehyde
    Urea–formaldehyde
    Melamine–formaldehyde

Conventional
    Phenol–formaldehyde
    Urea–formaldehyde
    Melamine–formaldehyde
    Poly(vinyl acetate)
    Epoxy
    Cyanoacrylate

TABLE 15.3   Production of Selected Polymers (United States 1993)

Polymer	(million lb)
Thermoplastic resins	**53,541**
Low-density and linear low-density polyethylene	12,044
High-density polyethylene	9,912
Polypropylene	8,614
Polystyrene	5,367
Acrylonitrile–butadiene–styrene, styrene butadiene copolymers, and other styrene copolymers	3,025
Nylon-type polyamide plastics	768
Poly(vinyl chloride) and copolymers	10,257
Polyester thermoplastics, mainly poly(ethylene terephthalate)	2,536
Polycarbonate	617
Polyphenylene-based alloys	219
Polyacetal	181
Thermosetting resins	**10,341**
Phenol–formaldehyde	3,078
Urea–formaldehyde	1,743
Melamine–formaldehyde	270
Polyurethanes	3,476
Unsaturated polyesters	1,264
Epoxy	510
Synthetic fibers	**9,297**
Acrylics and modified acrylics	433
Nylon	2,664
Polyolefin (mainly polypropylene)	2,138
Polyester, mainly poly(ethylene terephthalate)	3,557
Rayon and cellulose acetate (excluding filter tips, including di- and tri-acetate)	505
Elastomers	**1,993**
Styrene–butadiene rubber	817
Polybutadiene	473
Ethylene–propylene and EPDM terpolymers	227
Nitrile rubber (solid)	78
Polychloroprene	72
Other (butyl, styrene–butadiene latex, nitrile latex, polyisoprene and miscellaneous)	328

Source: *Chem. & Eng. News*, 4 July 1994, p. 34; *Modern Plastics*, January 1995, p. 63.

15.1   POLYMERIZATION

Before considering polymer properties, we shall describe how molecules link together to form polymers. These are two types of polymerization: addition or chain growth (also called simply chain) polymerization and condensation or step growth (also called simply step) polymerization. The terms chain and step are more accurate than the older terms, addition and condensation. Chain growth polymerization often involves monomers containing a carbon–carbon double bond, although cyclic ethers, such as ethylene and propylene oxides, and aldehydes, such as formaldehyde, polymerize in this way. Chain growth polymerization is characterized by the fact that the intermediates in the process—free radicals, ions, or metal complexes—are transient and cannot be isolated.

Step growth polymerization occurs because of reactions between molecules containing functional groups, for example, the reaction between a glycol and a dibasic acid to give a polyester.

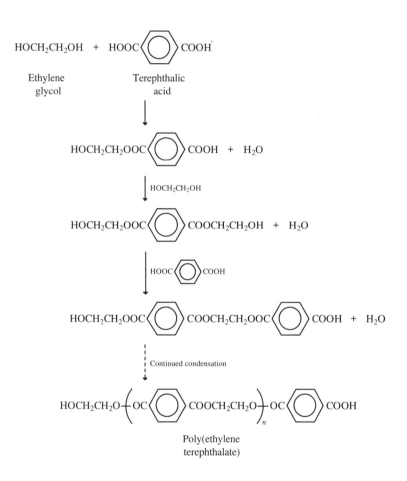

The low molecular weight intermediates are called oligomers, a term also used for the low molecular weight products obtained by chain growth polymerization. In the polyesterification shown here, an oligomer can have two terminal hydroxyl groups, two terminal carboxyls, or one of each. The hydroxyls can react further with terephthalic acid and the carboxyls further with ethylene glycol. Alternatively, two oligomers can condense. The continuation of these reactions, familiar from simple esterfication chemistry, leads to the final polymer. The step growth or condensation reactions can be stopped at any time, and low molecular weight polyesters (terminated by hydroxyl or carboxyl groups) isolated. Step growth polymerization, as opposed to chain growth polymerization, is therefore defined as a polymerization in which the intermediates can be isolated.

Usually a small molecule such as water is given off, but this is not always so. In the polymerization of the cyclic monomer caprolactam, for example, both functional groups are in the same molecule, and there is no byproduct. Indeed, 1 mol of water is needed to start the polymerization by hydrolyzing the caprolactam to 6-aminocaproic acid. Each molecule of the latter that self-condenses does indeed give off a molecule of water, but it is needed to hydrolyze more caprolactam to 6-aminocaproic acid. Because no small molecule is given off, the reaction reaches an equilibrium in which about 10% of the caprolactam remains unreacted. The monomer and oligomers (see below) that are always present must be removed by washing with water. This polymerization is carried out under the same conditions as are used to produce nylon from two bifunctional reagents and is clearly a step growth reaction.

$$n\,\overline{\mathrm{HNCH_2CH_2CH_2CH_2CH_2CO}} \xrightarrow{\mathrm{H_2O}} n\,\mathrm{H_2NCH_2CH_2CH_2CH_2CH_2COOH}$$

Caprolactam   6-Aminocaproic acid

$$\xrightarrow{\mathrm{-H_2O}} \mathrm{H-\!\!\left(\!NHCH_2CH_2CH_2CH_2CH_2CO\!\right)_{\!\!n}\!\!-OH}$$

Nylon 6

Caprolactam can also be polymerized by a chain mechanism using ionic initiators. Figure 15.1 demonstrates the ionic polymerization of caprolactam with sodium methoxide as the initiator. Ionic polymerization is discussed in greater detail in Section 15.3.6.

15.2 FUNCTIONALITY

Functionality is a measure of the number of linkages one monomer may form with another. A monomer that, when polymerized, may join with two other monomers is termed bifunctional. If it may join with three or more molecules, it is tri- or polyfunctional. Glycols and dibasic acids are clearly bifunctional. Similarly, 6-aminocaproic acid, the reaction product of water and caprolactam, is bifunctional because it contains a carboxyl and an amino group. The functionality rules state that if bifunctional molecules react, only a linear polymer will result (Fig. 15.2a). If a trifunctional monomer (Y in Fig. 15.2b) is added, chain branching can occur (see note). If there is sufficient quantity of it, an elaborate three-dimensional network can result. If some of the chains attached to the

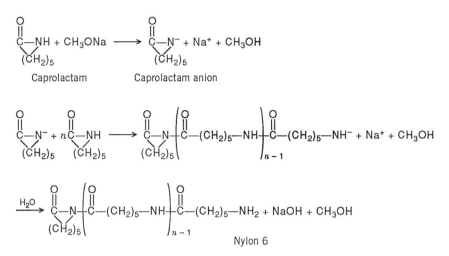

Figure 15.1 Ionic polymerization of caprolactam.

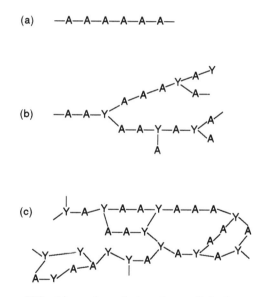

Figure 15.2 Linear, branched, and cross-linked structures.

Y groups in Figure 15.2c are thought of as coming off at right angles to the plane of the paper, some concept of the three-dimensional structure can be gained.

Glycerol has a functionality of three and if condensed with a dibasic acid can give the multifunctional oligomers shown below. Cross-linked polymers soon become insoluble and infusible as their molecular weight builds up. The cross-linking may lead to excellent strength characteristics, but the infusibility and insolubility means that such polymers are difficult to convert into shapes. The chemistry of thermoset polymers produced by step growth polymerization is further discussed in Section 15.4.

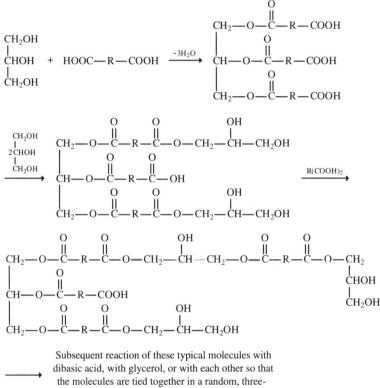

Subsequent reaction of these typical molecules with dibasic acid, with glycerol, or with each other so that the molecules are tied together in a random, three-dimensional network

The ethylenic double bond so important in chain growth polymerization has a functionality of two even though the organic chemist would regard the double bond as a single functional group. However, the ability of the extra electron pair of the ethylenic linkage to enter into the formation of two bonds makes it bifunctional. Thus ethylene polymerizes to form a linear thermoplastic polymer. The double bond in propylene contributes a functionality of two, but propylene also possesses allylic hydrogen atoms that are activated by peroxide initiators so that a cross-linked structure results. Thus propylene has a functionality greater than two toward peroxide catalysts. On the other hand, Ziegler–Natta catalysts do not activate the allylic hydrogen atoms. Propylene shows a functionality of two towards them, and a linear polymer results. This chemistry is harnessed in some ethylene–propylene rubbers. A linear ethylene–propylene copolymer is made with a Ziegler–Natta catalyst. It may then be cross-linked with a peroxide catalyst.

In the production of unsaturated polyester resins, a linear liquid oligomer is made by step growth polymerization, typically of propylene glycol with phthalic and maleic anhydrides. Each of these reagents exhibits functionalities of two in an esterification reaction. The maleic anhydride, however, has a double bond that can undergo chain growth polymerization. Thus subsequent treatment of the unsaturated polyester with styrene and a peroxide catalyst leads to a solid, infusible thermoset copolymer in which polyester chains are cross-linked by polystyrene chains. The maleic anhydride has a functionality of two in both the step growth and chain growth polymerizations. In chain growth polymerization, cross-linking results because the unsaturated polyester with its multiple double bonds has a functionality much greater than two.

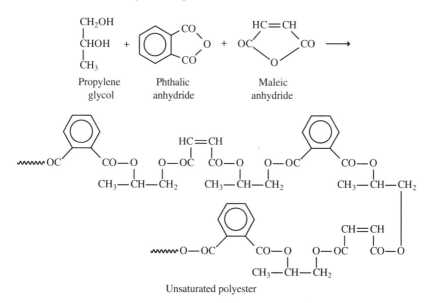

Unsaturated polyester

The fact that monomers exhibit different functionalities toward different reagents and polymerization techniques provides a means by which an initial polymerization can give a linear polymer than can subsequently be cross-linked by a different technique. Further examples of this are given in Section 15.4.

Conjugated structures, such as those in butadiene, are considered to have a functionality of only two. When they polymerize, linear polymers are formed that still contain double bonds.

$$n\text{H}_2\text{C}{=}\text{CH}{-}\text{CH}{=}\text{CH}_2 \longrightarrow -(\text{CH}_2{-}\text{CH}{=}\text{CH}{-}\text{CH}_2)_n \quad \text{or} \quad -(\text{CH}_2{-}\text{CH})_n$$
$$\qquad\qquad\qquad\qquad\qquad\qquad\qquad\qquad\qquad\qquad\qquad\qquad\qquad\qquad\qquad\qquad\qquad\qquad\qquad\text{CH}{=}\text{CH}_2$$

Butadiene　　　　　　　　　1,4-Polybutadiene　　　　　　　　1,2-Polybutadiene

These can subsequently react to form a cross-linked polymer but, in the initial polymerization, butadiene is bifunctional.

Pyromellitic dianhydride has a functionality of two when reacted with diamines, and linear polyimides result as shown in Figure 15.3 (see page 422). On the other hand, this anhydride reacts with water to form a tetracarboxylic acid that has a functionality of four when reacted with compounds containing hydroxyl groups.

The functionality of a molecule is not always obvious. The situation with a double bond has already been discussed. Formaldehyde has a functionality of two and will polymerize in concentrated solution to the common laboratory reagent, paraformaldehyde:

$$n\text{HCHO} \rightarrow \{\text{OCH}_2\}_n$$

A similar polymer, a polyacetal, may be formed by anionic or cationic polymerization of formaldehyde or trioxane. Although very high molecular weights can be obtained, the product is of little commercial value because it "unzips" easily to regenerate formaldehyde. This can be prevented and practical polymers obtained by acetylation of the hydroxyl end groups. Alternatively, a formaldehyde–ethylene glycol copolymer is made and is then subjected to conditions that would normally degrade it. The formaldehyde groups at the end of the chains "peel off" until an ethylene glycol unit is encountered. Depolymerization ceases, and a stable polymer molecule with hydroxyethyl end groups is left with the following structure:

$$\text{HOCH}_2\text{–CH}_2\text{–O}(\text{CH}_2\text{O})_n\text{CH}_2\text{–CH}_2\text{OH}$$

Ethylene glycol end-capped polyacetal

Aldehydes in general can be polymerized anionically or cationically (Section 15.3.6) to give polymers with a ⌁C–O–C–O–C–O⌁ backbone.

15.3   STEP GROWTH AND CHAIN GROWTH POLYMERIZATIONS

Step growth polymerization can be described as a simple chemical reaction carried out repeatedly. Polyesterification, for example, is brought about by the same catalysts as esterification reactions, and the equilibrium is pushed to the ester side of the equation by removal of the byproduct water either by simple distillation or as an azeotrope. A major difference between a simple condensation reaction and a polycondensation is that the high molecular weight of the polymer product increases the viscosity of the reaction mixture if the polymer is soluble in it; if not, it precipitates. To solve the viscosity problem, the reaction may be carried out in a solvent, a technique that is particularly useful if the polymer is to be used in a surface coating that requires solvent. More often the engineer is called on to devise equipment with powerful stirrers that can accommodate viscous masses.

In the production of many thermoset polymers, polymerization is interrupted at an early stage before cross-linking starts. The product is still fusible and soluble and is known as a "B-stage" polymer. In situ curing, usually with the aid of heat and a catalyst, is relied upon to build up molecular weight and achieve the cross-linked state. Phenolics are often used as B-stage polymers (Section 15.4). In polyimide formation, an intermediate chemical species, an "amic" acid, is formed (Fig. 15.3). This is soluble, albeit in very strong solvents such as N,N-dimethylformamide. The solution can, however, be laid down as a film and then heated further to achieve polyimide formation. Polyimides are linear as Fig. 15.3 indicates but are almost intractable because of their insolubility and rigidity.

The molecular weight and tendency to gel of a polycondensation polymer may be controlled by addition of a monofunctional compound known as a "chain stopper." In the production of polymeric plasticizers such as poly-(ethylene glycol adipate), for example, butanol is used as a chain stopper. Chain stoppers are also important in the production of alkyds (Section 9.1.1).

Chain growth polymerization proceeds rapidly by way of transient intermediates to give the final polymer. We can write an overall equation,

$$n\text{CH}_2=\text{CHX} \quad \rightarrow \quad \text{+(CH}_2=\text{CHX)}_n$$

but it provides no indication of the reaction mechanism. Polymerization is started by a chain initiator that converts a molecule of monomer into a free radical or an ion or else by a catalyst that converts the monomer to a metal complex. The free radicals or ions then undergo so-called propagation reactions that build up the polymer chain. In the case of metal complex catalysis, especially Ziegler–Natta catalysis, the propagation takes place on the surface of the metal complex catalyst. Finally, there must be a chain termination step in which the transient intermediate, now a polymer chain, is stabilized.

In chain growth polymerization, repeating units are added one at a time, as opposed to step growth polymerization where oligomers may condense with one another. Propagation and termination steps are very rapid. Once a chain is initiated, monomer units add on to the growing chain quickly, and the molecular weight of that unit builds up in a fraction of a second. Consequently, the monomer concentration decreases steadily throughout the reaction. Prolonged reaction time has little effect on molecular weight but does provide higher yields. At any given time, the reaction mixture contains unchanged reactant and "fully grown" polymer chains but a very low concentration of growing chains. The growing chains cannot readily be separated from the reaction mixture.

In step growth polymerization (Section 15.1) the monomer does not decrease steadily in concentration; rather it disappears early in the reaction because of the ready formation of low molecular weight oligomers. The molecular weight of a given polymer chain increases continually throughout the reaction, and thus long reaction times build up the molecular weight. After the early stages of the reaction, there is neither much reactant nor a great deal of "fully grown" polymer present. Instead there is a wide distribution of slowly growing oligomers. If desired this distribution can be calculated and the separate oligomers isolated from the reaction mixture.

The remainder of this section will be devoted to chain growth polymerization and polymerization procedures. Step growth polymers will be discussed further in Section 15.4.

15.3.1 Free Radical Polymerization

Free radical polymerization is initiated by free radicals from compounds such as benzoyl peroxide which, on heating, decomposes to give benzoylperoxy radicals, some of which eliminate carbon dioxide to give phenyl radicals. One of the free radicals (R· in the diagram) then adds on to a molecule of monomer such as ethylene, vinyl chloride, or styrene to convert that monomer to a radical. Initiation is now complete, and the initiating free radical is incorporated into one end of what will become the polymer chain. Its concentration in a high molecular weight polymer is so small that it does not affect final properties. The radical now reacts with another molecule of monomer to give a larger free radical, and this chain propagation process continues until the chain is terminated.

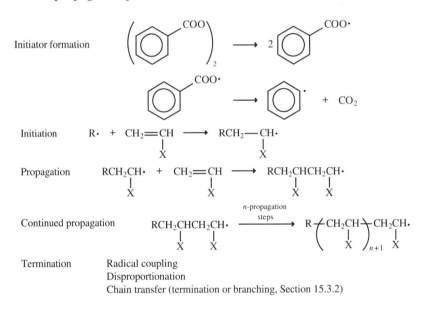

The most important free radical initiators are benzoyl peroxide, dicumyl peroxide, dialkyl peroxides (especially methyl ethyl ketone peroxide, used for unsaturated polyester resins, and di-*tert*-butyl peroxide), and peroxyesters of general formula R–CO–O–O–R.

What can stop the chain? The three possible processes are called coupling, disproportionation, and chain transfer. Coupling occurs when two growing free radicals collide head to head to form a single stable molecule with a molecular weight equal to the sum of the individual molecular weights. In the disproportionation reaction two radicals again meet, but this time a proton transfers from one to the other to give two stable molecules, one saturated and the other with a terminal double bond. Above 60°C polystyrene terminates predominantly by coupling whereas poly(methyl methacrylate) terminates entirely by disproportionation. At lower temperatures both processes occur.

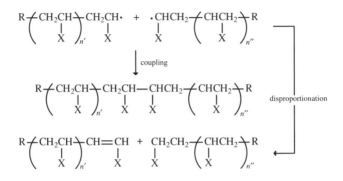

Chain transfer can cause either termination or branching. It is discussed in Section 15.3.2.

Because propagation reactions in chain growth polymerizations are very fast, polymerizations can and sometimes do become explosive. Termination steps occur rarely relative to the propagation reaction, not because they are slow but because the concentrations of free radicals are normally so low that encounters between them are rare.

In the polymerization of ethylene it makes no difference which end of the molecule is attacked by the free radical. With unsymmetrical monomers such as vinyl chloride or styrene, however, there could be head-to-tail propagation (Eq. 1), head-to-head propagation, or completely random addition. In the head-to-head case, the side chain or heteroatom X would sometimes occur on adjacent carbon atoms (Eq. 2).

$$\text{wwCH}_2\text{—CH·} + \text{CH}_2\text{=CH} \longrightarrow \text{wwCH}_2\text{—CH—CH}_2\text{—CH·} \longrightarrow \text{etc.} \qquad (1)$$
$$| | | |$$
$$X X X X$$

$$\text{wwCH}_2\text{—CH·} + \text{CH}\text{=CH}_2 \longrightarrow \text{R—CH}_2\text{—CH—CH—CH}_2\text{·} \longrightarrow \text{etc.} \qquad (2)$$
$$| | | |$$
$$X X X X$$

Head-to-head propagation rarely occurs because the unpaired electron in the free radical prefers to locate itself on the –CHX end of the monomer molecule where it has a better opportunity to delocalize. The free radical is thus more stable. Head-to-tail polymerization is the norm, and only an occasional monomer molecule slips in the "wrong way." Termination by coupling, of course, creates a head-to-head structure.

The relative rates of the initiation, propagation, and termination processes are reflected in the key property of molecular weight on which many of the other properties of the polymer depend. If the rate of initiation is high, for example, then the concentration of free radicals at any given moment will be high, and they will stand a good chance of colliding and coupling or disproportionating. A high initiation rate will therefore lead to a low molecular weight polymer.

For a high molecular weight polymer, a low initiation rate is required together with a high propagation rate. We might also say a low termination rate, but, because termination steps have no activation energy, they occur on every collision and are diffusion controlled. The termination rate is decreased by increase in viscosity or decrease of concentration in a system. If propagation and termination steps have comparable rates, a polymer will not result. A propagation rate thousands of times the termination rate is required, and the molecular weight is, in fact, a function of the ratio of propagation to termination rates.

15.3.2   Chain Transfer

Another factor that affects molecular weight is chain transfer. A growing polymer radical may extract a hydrogen atom from a finished polymer chain. This "finished" polymer chain now becomes a radical and starts to grow again. If the hydrogen atom is extracted from the end of the chain, the new chain simply continues to grow linearly. But if, as is more probable statistically, the hydrogen atom is extracted from the body of the chain, then further propagation occurs at right angles to the original polymer chain and a branch forms.

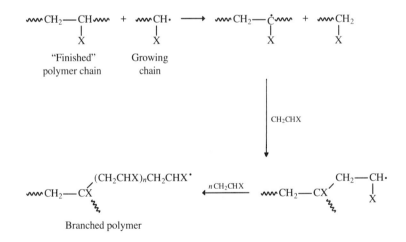

Branching can have a marked effect on polymer properties. It can keep polymer molecules from achieving molecule nearness and hence reduce cohesive forces between them. Correspondingly, branching makes it harder for polymer crystals to form (Section 15.5.1). The following equation shows how branching takes place in polyethylene.

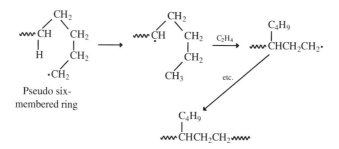

A growing polymer molecule with a free radical end can bend to form a pseudo six-membered ring that facilitates the transfer of the free radical site from the end of the chain to a carbon atom within the chain. The chain then starts to grow from this new site with the net result that the branch has four carbon atoms. Low-density polyethylene is indeed characterized by C_4 branches.

Chain transfer can occur not only to another polymer chain but also to a molecule of monomer. The new radical will then propagate in the usual way. If this happens often, a low molecular weight polymer will form.

Chain transfer is undesirable except when it is used intentionally to limit molecular weight. It can be controlled by addition of chain-transfer agents. These are materials from which hydrogen atoms can readily be abstracted. If a growing radical is liable to extract a hydrogen atom, it will do so preferentially from the chain-transfer agent rather than another polymer molecule. The problem of branching will thus be avoided but not that of reduced molecular weight. Dodecyl mercaptan is used as a chain-transfer agent in low-density polyethylene and rubber polymerizations. When it loses a hydrogen atom, a stable disulfide forms. The formation of a stable compound after loss of a proton is a key characteristic of a chain-transfer agent.

$$C_{12}H_{25}SH \rightarrow 2H + C_{12}H_{25}S\text{–}SC_{12}H_{25}$$

$$\underset{\text{mercaptan}}{\underset{\text{Dodecyl}}{\phantom{C_{12}H_{25}SH}}} \qquad\qquad \underset{\text{Dodecyl disulfide}}{\phantom{C_{12}H_{25}S\text{–}SC_{12}H_{25}}}$$

Phenols may be used similarly because they give up their phenolic hydrogen readily and the resulting phenoxy radical is relatively stable and does not add to monomer.

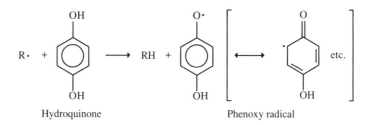

During storage, monomers are sometimes stabilized with dihydric phenols (e.g., hydroquinone and *tert*-butylcatechol) or aromatic amines (e.g., methylene blue) so that they do not polymerize spontaneously.

Methylene blue

Hydroquinone is ineffective in the absence of oxygen, so its mode of action is probably more complicated than suggested here. If a free radical should appear in the monomer, it immediately accepts a proton from the inhibitor and is "squelched." When it is desired to polymerize the monomer, the inhibitor must be removed, usually by distillation.

15.3.3 Copolymerization

Discussion of propagation rates leads to the topic of copolymerization. Copolymers are polymers made from two or more monomers and are of four kinds: regular, random, block, and graft. In regular copolymers, the monomer units alternate in the chain (–A–B–A–B–A–B–A–B–); in random copolymers they follow each other indiscriminately (–A–A–B–A–B–B–B–A–B–). Block copolymers (Section 15.3.8) consist of a group of one polymerized monomer followed by a group of the other (–A–A–A–A–B–B–B–B–), and graft copolymers result when a polymer chain of one monomer is grafted on to an existing polymer backbone (Section 15.3.9):

$$\begin{array}{c} \text{B—B—B—B—B—B—} \\ | \\ \text{—A—A—A—A—A—A—A—A—A—A} \end{array}$$

Many step growth copolymers are regular. An example of a regular chain growth copolymer is one based on maleic anhydride and styrene. Maleic anhydride will not homopolymerize but reacts rapidly in a polymerization reaction with another monomer. Thus one styrene monomer will react with a maleic anhydride monomer. The resulting free radical,

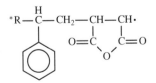

*R = Initiator residue

has a choice of reacting further with another styrene molecule or with another maleic anhydride. It chooses the styrene because the maleic anhydride will not react with itself. The free radical

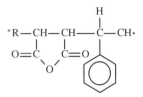

chooses a maleic anhydride monomer because the reaction rate between styrene and maleic anhydride is much greater than between styrene–styrene. In this way a completely regular copolymer, ∿ S–MA–S–MA–S∿, is obtained.

In a random copolymer, the monomer units are not in an orderly sequence. To form a random copolymer, the two monomers must react with themselves at a rate comparable to that at which they react with each other. If the propagation rates differ widely, the first polymer molecules to be formed will consist almost entirely of the fast reacting monomer and, when all of it is used up, the slow reacting material will polymerize to give a polymer consisting almost entirely of the slow reacting material. The possible propagation reactions are as follows. It is the relative rates of these processes that decide whether a random copolymer, two homopolymers, or something in between is obtained.

$$\sim\!A\cdot \ + \ A \ \longrightarrow \ \sim\!A\!-\!A\cdot$$
$$\sim\!A\cdot \ + \ B \ \longrightarrow \ \sim\!A\!-\!B\cdot$$
$$\sim\!B\cdot \ + \ B \ \longrightarrow \ \sim\!B\!-\!B\cdot$$
$$\sim\!B\cdot \ + \ A \ \longrightarrow \ \sim\!B\!-\!A\cdot$$

Copolymerization serves several functions. First, a copolymerizing monomer may be included to plasticize the polymer, that is to make it softer. Because vinyl acetate gives too brittle a film for water-borne paints, it may be copolymerized with 2-ethylhexyl acrylate. Second, the copolymerizing monomer may insert functional groups. In unsaturated polyesters (Section 4.2) the maleic anhydride provides double bonds that may subsequently be cross-linked by chain growth polymerization. In elastomers, a comonomer with two double bonds is almost always used. One double bond engages in chain growth polymerization, and the other remains intact on each recurring unit so that sites for "vulcanization" or cross-linking are present. Thus butyl rubber is a copolymer of isobutene with a small amount of isoprene which provides the residual double bonds.

Finally, copolymerization can be used to reduce crystallinity (Section 15.5.1). Low-density polyethylene is about 50% crystalline. By making a copolymer with propylene, this crystallinity is destroyed, and a polymer results that becomes an elastomer on cross-linking.

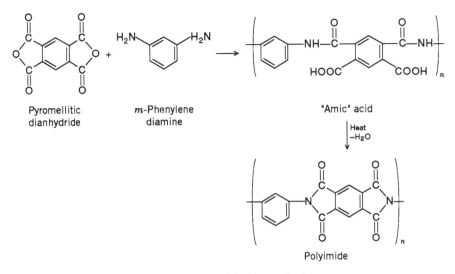

| Pyromellitic | m-Phenylene | "Amic" acid |
| dianhydride | diamine | |

Polyimide

Figure 15.3 Polyimide synthesis.

Copolymerization plays an important role in the synthesis of linear low-density polyethylene (LLDPE). High-density polyethylene (HDPE) (Section 3.1.3) requires mild conditions for manufacture, whereas low-density polyethylene (LDPE) requires very severe conditions such as 1200 bar and 200°C. Chemists learned how to make a polymer whose properties approximate those of LDPE simply by making a copolymer of HDPE, under the mild conditions HDPE requires, using the comonomers 1-butene, 1-hexene, or 1-octene in concentrations of 6–8%. These destroy some of the crystallinity of HDPE, which is above 90%, lowering it to about 50%, which is the crystallinity of LDPE. Linear low-density polyethylene has greater tensile strength than LDPE because its branches (cf. Fig. 15.3b) are all the same length—two carbon atoms if 1-butene is used, four-carbon atoms if 1-hexene is used, and six-carbon atoms if 1-octene is used. Because of its lesser energy requirements, LLDPE production has grown rapidly, although this economic advantage is partially offset because the C_6 and C_8 comonomers are more expensive than ethylene.

15.3.4 Molecular Weight

We have referred several times to the molecular weight of a polymer. This is not as simple a concept as it sounds. Since the chains in a sample of polymer do not all have the same number of recurring units, the molecular weight of a polymer is always an average. A broad molecular weight distribution is often desirable, for oligomers may serve as lubricants during processing and as plasticizers thereafter.

Molecular weight of polymers is commonly expressed in two ways: by number average $\bar{M}_n$ and by weight average $\bar{M}_w$. The number average is obtained by adding the molecular weights of all the molecules and dividing by the number of molecules. If we have n_1 molecules of molecular weight M_1, n_2 of molecular weight M_2, and n_x of molecular weight M_x then

$$\bar{M}_n = \frac{n_1 M_1 + n_2 M_2 + \cdots n_x M_x + \cdots}{n_1 + n_2 + \cdots n_x \cdots}$$

The weight average, on the other hand, is calculated according to the weight of all the molecules at each molecular weight. Let w_1 be the weight (in molecular weight units) of molecules of molecular weight M_1, w_2 the weight of molecules of molecular weight M_2, and so on, then

$$\bar{M}_w = \frac{w_1 M_1 + w_2 M_2 + \cdots w_x M_x + \cdots}{w_1 + w_2 + w_3}$$

But the total weight of all molecules with molecular weight w_1 is $M_1 n_1$, so we can substitute $w_1 = M_1 n_1$, $w_2 = M_2 n_2$, $w_x = M_x n_x$, and so on in the above equation, whence

$$\bar{M}_w = \frac{n_1 M_1^2 + n_2 M_2^2 + \cdots n_x M_x^2 + \cdots}{n_1 M_1 + n_2 M_2 + \cdots n_x M_x \cdots}$$

$\bar{M}_n$ tells us where most of the polymer molecules are relative to the molecular weight distribution. The parameter $\bar{M}_w$, on the other hand, tells us where most of the weight is regardless of the molecular weight distribution. Because $\bar{M}_w$ is biased towards molecules with higher molecular weight, it will be larger than $\bar{M}_n$.

As an example consider three persons, two weighing 100 lb and one weighing 200 lb. Their number average weight is $(100 + 100 + 200)/3 = 133\frac{1}{3}$ lb, but their weight average is $(100^2 + 100^2 + 200^2)/(100 + 100 + 200) = 150$ lb. In the first instance, we can consider that a person was selected at random; in the second, that a pound of weight was selected at random. The second selection (weight average) will naturally lead to a higher result because the pound of weight will tend to be selected from the heavier persons.

The parameters $\bar{M}_w$ and $\bar{M}_n$ both provide a narrow view of molecular weight, but their ratio, $\bar{M}_w/\bar{M}_n$, will tell us something about the molecular weight distribution. If $\bar{M}_w/\bar{M}_n = 1$ then all the molecules have the same molecular weight, and as the distribution of molecular weights becomes wider this ratio increases.

Boiling point elevation, freezing point depression, osmotic pressure, and end group analysis give number average molecular weight; light scattering and sedimentation methods give weight averages. Viscosity measurements give a value somewhere between the two.

The molecular weight profile of a polymer can be determined only by fractionation. Cumbersome solvent precipitation techniques give numerous fractions, and the molecular weight of each is determined. The fractions must be so narrow that for each of them $\bar{M}_w/\bar{M}_n$ is effectively unity

15.3.5 Polymerization Procedures

Chain growth polymerizations, whether initiated by free radicals as we have already described, or by ions or metal complexes as we describe later, are carried out by four different procedures—bulk, solution, suspension, and emulsion polymerizations.

In bulk polymerization, the monomer and the initiator are combined in a vessel and heated to the proper temperature. This procedure, although the simplest, is not always the best. The polymer that forms may dissolve in the monomer to give a viscous mass, and heat transfer becomes difficult. Heat cannot escape, and the polymer may char or develop voids. If the exotherm gets out of hand, the system may explode.

Even so the polymerization of ethylene by the high-pressure method is a bulk polymerization and is one of the polymerizations carried out on the largest scale. Fortunately, the polymer does not dissolve in the monomer. Instead it collects in the bottom of the reactor and is drawn off. The exotherm still presents a problem, and the strictest possible control of temperature and heat transfer is necessary. The polymerization of methyl methacrylate to "Lucite" (Plexiglas, Perspex) is also carried out in bulk.

Fluid-bed processes are essentially bulk polymerizations and represent one way of handling the exotherm. They have become popular because they provide an economical way to make many grades of high-density polyethylene, LLDPE (Section 3.1.4), and polypropylene. The fluid bed comprises small particles of the preformed resin, fluidized by inert gas. The gaseous monomer and catalyst are injected into the fluid bed and the polymer forms around the nuclei provided by the particles of preformed resin. Conversion is only 2% but ethylene is easily recycled.

The other polymerization procedures are all designed to solve the problem of heat transfer. In solution polymerization, the reaction is carried out in a solvent that acts as a heat sink and also reduces the viscosity of the reaction mixture. The snags with solution polymerization are first, it is frequently difficult to remove the last traces of solvent from the polymer, and second, the solvent participates in chain-transfer reactions so that low molecular weight polymers result. Solution polymerization is useful if the product is to be used in solvent. Solvent-based poly(vinyl acetate) adhesive is an example. Slurry polymerization is a variant of solvent polymerization and is used for the important polymerizations of ethylene to high-density (low-pressure) polyethylene in one version of the Phillips process (Section 15.3.11), and propylene to polypropylene. A small amount of solvent is combined with the monomer and catalyst in the reactor. The solvent forms a slurry with the catalyst and aids in its distribution throughout the reaction mixture. At the same time it helps to remove exotherm. Initially it was necessary to separate the catalyst by a cumbersome process. This has led to the development of catalysts that separate more readily. Also, catalysts are now available that give such high yields, and thus are present in such low concentrations, that they can be left in the polymer without affecting its properties.

In suspension polymerization, the monomer and catalyst are suspended as droplets in a continuous phase such as water. These droplets have a high surface/volume ratio so heat transfer to the water is rapid. The droplets are maintained in suspension by continuous agitation and also, if necessary, by addition of a water-soluble polymer such as methylcellulose that increases viscosity of the water. Finely divided inorganic materials such as clay, talc, aluminum oxide, and magnesium carbonate have a similar stabilizing effect on the suspension. The need to remove these materials is one of the disadvantages associated with their use. Poly(vinyl chloride) is frequently made by suspension polymerization.

The final procedure is emulsion polymerization, a technique that was developed as part of the synthetic rubber program during World War II. The products are particularly useful for the formulation of water-based paints. As its name implies, it uses an emulsifying agent, usually various kinds of soap. In solution this forms micelles in which the nonpolar hydrophobic ends of the soap molecules point inward, and the polar hydrophilic groups point outward and interact with the water. If monomer is added, it is absorbed into the micelle to give a stable emulsion particle. If more monomer is added than can be absorbed in the micelles, a separate monomer droplet phase may form that is also stabilized by the soap molecules, the droplets being a micron or more in diameter.

A water-soluble composite initiator called a "redox" catalyst is then added. This consists of a mixture of a reducing agent and an oxidizing agent. An example is ferrous ammonium sulfate and hydrogen peroxide. In the absence of monomer, the former would reduce the latter in a two-stage process:

$$Fe^{2+} + H_2O_2 \rightarrow Fe^{3+} + OH^{\cdot} + OH^-$$
$$Fe^{2+} + OH^{\cdot} \rightarrow Fe^{3+} + OH^-$$

If monomer is present, however, the hydroxyl free radical can initiate polymerization. Other redox systems include benzoyl peroxide–ferrous ammonium sulfate, hydrogen peroxide–dodecyl mercaptan, and potassium persulfate–potassium thiosulfate, which gives radical ions:

$$S_2O_8^{2-} + S_2O_3^{2-} \rightarrow SO_4^{2-} + SO_4^{\cdot} + S_2O_3^{\cdot}$$

Persulfate   Thiosulfate
   ion   ion

These polymerizations must be carried out with rigorous exclusion of oxygen, which is an inhibitor in this case, although an initiator in low-density polyethylene production.

The free radicals diffuse into the micelles, and polymerization takes place within them. Diffusion into the droplets also occurs but, since they have a far lower surface/volume ratio than the micelles, virtually none of the polymerization takes place within them. As polymer is formed, the micelles grow by diffusion of monomer from the droplets into the micelles. Rather than providing a site for polymerization, the droplets serve as reservoirs for monomer that will later react in the micelles.

Polymerization within a micelle may take as long as 10 s. Very high molecular weights are produced, higher than by any of the three other procedures. The product is a latex, a dispersion of solid particles in water, which is frequently a desirable form for a polymer. For example, poly(vinyl acetate) or polyacrylate latices are used as such for "emulsion" paints. On the other hand, if solid polymer is required, the dispersion must be broken and the polymer precipitated.

There are two important differences between emulsion and suspension polymerization. In emulsion polymerization, the catalyst or initiator is in the aqueous phase, not dissolved in the monomer. Also, particles produced are at least an order of magnitude smaller than those obtained from suspension polymerization.

15.3.6 Ionic Polymerization

Free radical initiation is the most widely used way to produce polymers (Fig. 15.4). A second method involves initiation by ions, either anions or cations. Table 15.4 provides a list of ionic initiators useful for polymerization. In Table 15.5 there is a list of monomers and an indication as to whether they can be

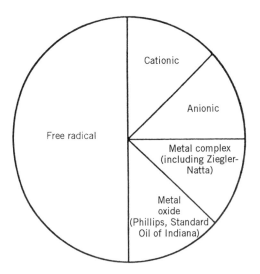

Figure 15.4 Methods of polymerization: use distribution.

TABLE 15.4   Ionic Initiators

Initiators	Sample Monomers[a]
Cationic Types	
Lewis acids	
BF_3 (with H_2O, ROH, ROR)	1, 2
$AlCl_3$, $AlBr_3$ (with H_2O, ROH, RX)	2, 3
$SnCl_4$ (with H_2O)	3
$TiCl_4$	4
$FeCl_3$ (with HCl)	3
I_2 (with Zn halides)	10
Brønsted acids	
H_2SO_4	3 (low mol wt), 4
$KHSO_4$	
HF	3
$HClO_4$	
Cl_3COOH	5
Active salts	
$(C_6H_5)_3C^+BF_4^-$; $(C_6H_5)_3C^+SbCl_6^-$	3
$C_2H_5O^+BF_4^-$	3, 4
$Ti(OR)_4$	
Anionic Types	
Free metals[b] ⎱ in toluene, naphthalene, liquid ammonia etc. Na ⎰ K	3, 6, 7
Bases and salts	
KNH_2, $NaNH_2$	3, 6
$Ar_2N^-K^+$	8
NaCN	
$NaOCH_3$	3
RLi, RK, RNa R may be $C_4H_9^-$, $(C_6H_5)_3C^-$, $C_6H_5CH_2^-$, $C_6H_5-\overset{CH_2^-}{\underset{CH_2}{\diagup\!\!\!\diagdown}}$	9

[a] 1 = 2-butene; 2 = isobutene; 3 = styrene; 4 = propylene; 5 = isopropenylbenzene; 6 = butadiene; 7 = stilbene; 8 = 2-cyano-1,3-butadiene; 9 = acrylonitrile; 10 = vinyl ethers.
[b] Operate by way of production of radical anions and subsequent reaction of these to actual initiating anions.

TABLE 15.5 Methods of Polymerizing Monomers

Monomer		Polymerization Mechanism			
		Anionic	Cationic	Free Radical	Metal Oxide or Coordination Catalyst
$CH_2=CH_2$	Ethylene		+	+	+
$CH_2=CHCH_3$	Propylene		+		+
$CH_2=C(CH_3)_2$	Isobutene		+		
$CH_2=CH-CH=CH_2$	Butadiene	+		+	+
$CH_2=C(CH_3)CH=CH_2$	Isoprene	+		+	+
$CH_2=CHC_6H_5$	Styrene	+	+	+	+
$CH_2=CHNO_2$	Nitroethylene	+			
$CH_2=CHOR$	Vinyl ethers		+		+
$CH_2=CH-N\diagdown\genfrac{}{}{0pt}{}{CO-CH_2}{CH_2-CH_2}$	Vinylpyrrolidone		+	+	
$CH_2=C(CH_3)COOCH_3$	Methyl methacrylate	+		+	+
$CH_2=C(CN)COOCH_3$	Methyl α-cyanoacrylate	+		+	
$CH_2=CHCN$	Acrylonitrile	+		+	

polymerized ionically or cationically. Many of them can be polymerized also by free radicals and by the metal complex catalysts discussed later. Ethylene may by polymerized cationically and also with the aid of free radicals. Propylene, however, has allylic hydrogen atoms on the methyl group, and any attempt at free radical polymerization leads to low molecular weight cross-linked structures.

As a general rule, monomers containing electron-withdrawing groups are more easily polymerized anionically, whereas those with electron-donating groups are more easily polymerized cationically. Nonetheless styrene, which contains the electron-withdrawing phenyl group, may be polymerized both anionically and cationically and, for that matter, by free radicals. Cationic polymerization of styrene, however, yields low molecular weight polymers.

Ionic polymerization is usually unsuitable for the preparation of copolymers. This is because the differences in the stabilities of organic ions are much greater than those between the corresponding radicals. It represents a serious limitation to ionic polymerization. An exception is block copolymers (Section 15.3.8), which may be prepared by ionic polymerization because the monomers are added successively not simultaneously.

The initiation step in anionic polymerization is the production of an anion from the monomer by the strong base. This is shown in the equation that follows where butyl lithium is the initiator. Butyl lithium and other anionic initiators such as sodium or potassium amides in liquid ammonia, or sodium cyanide in dimethylformamide are expensive and not recoverable. Consequently, this procedure is used only where there is no cheaper method of polymerization available and when the value of the product justifies the high initiator cost.

$$CH_2{=}CH + C_4H_9Li \rightarrow C_4H_9CH_2CH^-Li^+ \quad \text{Initiation}$$
$$\underset{X}{|} \qquad\qquad\qquad\qquad \underset{X}{|}$$

Butyl lithium

$$C_4H_9CH_2CH^-Li^+ + CH_2{=}CH \longrightarrow C_4H_9CH_2CHCH_2CH^-Li^+ \quad \text{Propagation}$$
$$\underset{X}{|} \qquad\qquad\quad \underset{X}{|} \qquad\qquad\qquad \underset{X}{|}\ \underset{X}{|}$$

The propagation step in anionic polymerization is superficially similar to that in free radical polymerization (Section 15.3.1), but actually there are differences. Ions "like" to be solvated, and the solvating power of the polymerization medium may affect the propagation rate. Also, an ion is always associated with a counterion of opposite charge, which in the foregoing equation is Li^+. This counterion may be completely dissociated from the negative ion or it may be associated with it as an ion pair, and this too can affect the course of propagation.

The use of anionic initiation leads to a radical ion that can propagate at both ends of the polymer chain. If styrene is treated with sodium in naphthalene, the sodium first transfers an electron to the naphthalene, which in turn transfers it to the styrene. The styrene has become an anion with an odd number of electrons, that is, it is also a free radical and is called a radical ion. It will combine with more monomer to give a chain with an anionic end and a free radical end.

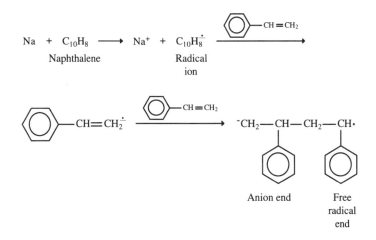

This species could conceivably add monomer from the two ends by different mechanisms. More likely, because the charge is sufficiently delocalized, two of the radical ends couple to give a divalent anion that propagates from both ends by an ionic mechanism:

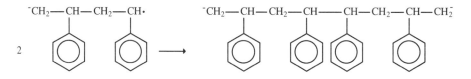

Chain termination is more complicated than in free radical polymerization, where it takes place by way of coupling and disproportionation (Section 15.3). Neither of these is possible because two negative ions cannot easily come together. The reluctance of ionic chains to terminate leads to the so-called "living" polymers (Section 15.3.7). Termination may result because of proton transfer from solvent, weak acid, polymer, or monomer. Thus water will quench an anionically initiated polymer. Proton transfer is not true destruction of a transient species, and termination only occurs if the new species is too weak to propagate.

Termination

The recombination of a chain with its counterion or the transfer of a hydrogen to give terminal unsaturation, frequent in cationic polymerization (see below), is unlikely in anionic systems. For example, if the counterion is Na^+ the transfer to it of H^- is improbable.

Termination can be brought about by a cation-generating small molecule such as silicon tetrachloride. Four chains can terminate at the silicon atom, so the molecular weight of the polymer has been quadrupled.

$$4 \text{ww} CH_2-CH^-Li^+ \xrightarrow{\text{SiCl}_4} \left(\text{ww} CH_2-CH \right)_4 Si \; + \; 4LiCl$$

$$\underset{X}{|} \qquad \qquad \underset{X}{|}$$

Termination

For a "three-armed star-shaped" polymer, a terminating agent such as 1,3,5-tris(chloromethyl)benzene may be used. This is a unique aspect of ionic termination that has no counterpart with free radicals. The radial block polymer is much less viscous than a linear polymer of similar molecular weight, and it is more soluble simply because its shape provides more opportunities for solvation. Thus it couples the benefits of very high molecular weight with easier handling properties.

An alternative procedure for preparing star-shaped polymer is to start with multifunctional initiators such as $C(CH_2C_6H_4Li)_4$. An initiator for a three-armed star-shaped polymer is the alkoxide of triethanolamine, $N(CH_2CH_2ONa)_3$.

Ionic polymerization may also be cationic. Table 15.5 shows which monomers may be polymerized cationically. Initiation is by proton donors such as conventional acids and Lewis acids, and these give rise to carbenium ions. Boron trifluoride in water is typical of a Lewis acid.

$$BF_3 \; + \; H_2O \longrightarrow H^+(BF_3OH)^-$$

$$H^+(BF_3OH)^- \; + \; CH_2{=}CH \longrightarrow CH_3CH^+(BF_3OH)^-$$

$$\underset{X}{|} \qquad\qquad \underset{X}{|}$$

Initiation

$$\text{ww} CH_2CH^+(BF_3OH)^- \; + \; CH_2{=}CH \longrightarrow \text{ww} CH_2CHCH_2CH^+(BF_3OH)^-$$

$$\underset{X}{|} \qquad\qquad \underset{X}{|} \qquad\qquad \underset{X}{|} \; \underset{X}{|}$$

Propagation

$$\text{ww} CH_2CH^+(BF_3OH)^- \longrightarrow \text{ww} CH{=}CH \; + \; H^+(BF_3OH)^-$$

$$\underset{X}{|} \qquad\qquad\qquad \underset{X}{|}$$

$$\text{ww} CH_2-CH^+(BF_3OH)^- \xrightarrow{\text{NH}_3} \text{ww} CH_2-CHNH_2 \; + \; H^+(BF_3OH)^-$$

$$\underset{X}{|} \qquad\qquad\qquad \underset{X}{|}$$

Termination

Propagation occurs as in anionic polymerization, and termination occurs when a proton is transferred back to the counterion leaving a polymer molecule with terminal unsaturation. Unlike anionic polymerization, the initiator is regenerated and can go on to generate other chains or even to attack the solvent. Termination can also be brought about by addition of a small molecule such as ammonia, and a polymer with an amine end group is formed. Again the initiator is regenerated.

It is only by way of ionic polymerization that functional end groups can be attached to polymer molecules. With anionic polymer molecules that grow in two directions, as described, CO_2 yields carboxyl groups; ammonia will provide amine end groups; potassium isocyanate, isocyanate end groups; and HCl, chlorine end groups. If the molecular weight of a polymer is very high, the effect of these end groups is negligible. On the other hand, if the ratio of initiator to monomer is such that the number of recurring units is low, an oligomer is obtained. This termination procedure provides an elegant method for the manufacture of difunctional compounds such as dibasic acids and diisocyanates. But, as so often happens, elegance and expense go together. One mole of expensive initiator is required for every mole of difunctional compound produced.

15.3.7 Living Polymers

Living polymers are an important ramification of ionic polymerization. The polymer theory that has been outlined so far was developed between 1935 and 1950. The importance of initiation and propagation reactions was recognized, but no one worried very much about termination. It was understood that termination was more difficult in ionic than free radical polymerization because the charges on the growing chains repelled one another, but it was assumed that termination would come about somehow or other.

Eventually, it was realized that it did not have to occur at all. If styrene is polymerized ionically with sodium naphthalide in tetrahydrofuran solution, and care is taken not to introduce agents that terminate chains, a polymer is formed whose chain length can be estimated from the viscosity of the polystyrene solution. The ends of the chains are unterminated, and the polymer is described as "living." If further styrene is added, weeks or even months later, there will be a marked increase in viscosity showing that the polymer chains have started to grow again once they are supplied with fresh monomer.

15.3.8 Block Copolymers

Instead of styrene, in the above example, some other monomer such as isoprene may be added to the living polymer, and a copolymer results. Copolymers can be achieved by other means (Section 15.3.3), but they are usually random. With the living polymer technique, the copolymer is ordered, consisting of a chain of X molecules followed by a chain of Y molecules. If desired a further set of X molecules or of any other monomer can be added. Such materials are called block copolymers.

Block copolymers can also be made by condensation techniques. Spandex is an elastomeric fiber whose use in the 1990s increased at a rapid rate. One form of spandex comprises a block polymer in which there is a flexible block comprising poly(tetramethylene ether) made by oligomerizing tetrahydrofuran to obtain a hydroxyl-terminated polyether.

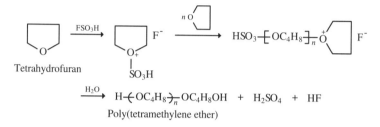

Tetrahydrofuran

Poly(tetramethylene ether)

Reaction of 2 mol of the polyether with 1 mol of methylene diphenylene diisocyanate (Section 7.3.1) yields an isocyanate-terminated prepolymer. The fluorosulfonic acid, because it presents ecological problems, is being replaced by solid acid catalysts.

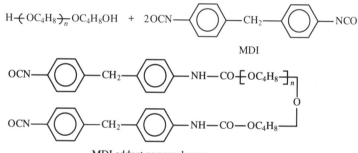

MDI

MDI adduct or prepolymer

The prepolymer, on reaction with ethylene diamine, yields a copolymer with a hard polyurea block adjacent to the soft polyether block. Elasticity is contributed by crimping.

$$OCN-R'-NCO \ + \ H_2NC_2H_4NH_2 \ \longrightarrow \ \left(\begin{matrix} O \\ \| \\ CNH-R'-NHCNHCH_2CH_2NH \end{matrix}\right)_m$$

Prepolymer   Spandex

Condensation reactions can be used to prepare block polymers only if the reactions take place at low temperatures. At high temperatures, existing bonds in many polymers break and reform to provide random distribution and thus a random copolymer.

Block copolymers have industrial applications, especially in the formation of so-called thermoplastic elastomers. The two blocks may be insoluble in one another and thus tend to repel each other. They will, however, tend to associate with similar blocks in other polymer molecules. This is illustrated in Figure 15.5 for a styrene–butadiene–styrene block copolymer. It is quite different from the common free radical polymerized styrene–butadiene rubber, which is a random copolymer. Polybutadiene is a flexible rubbery material; polystyrene is hard and brittle. Furthermore, polystyrene is highly insoluble in polybutadiene. The polystyrene blocks therefore associate with other polystyrene blocks with which they are more compatible, and physical bonding, that is, van der Waals forces, results. Although these forces are not very strong they nonetheless give an element of cross-linking so that the polymer at room temperature has many of the properties of a cross-linked material. At higher temperatures, however, the weakness of the forces dissociates the "cross-links" so that the polymer can be processed as if it were a simple thermoplastic.

The useful range of temperature for a block copolymer is determined by the glass transition temperatures (T_g) of the blocks that constitute it. The glass transition temperature is a property of amorphous polymers and is discussed in more detail in Section 15.5.2. At this stage it is sufficient to say that, if an amorphous polymer is melted and then allowed to cool, it will at its precise glass transition temperature cease to be soft, pliable, flexible, and plasticizing and become hard, rigid, and glassy.

If the styrene–butadiene–styrene block copolymer (e.g., Shell's "Kraton") is to be an elastomer, it must be used above the T_g for polybutadiene. Equally, if it is to retain physical cross-linking it must be used below the T_g for polystyrene. By varying the flexible and rigid blocks, other thermoplastic elastomers result, some of which are now articles of commerce. Two are of interest because of their high service temperatures. One is a block copolymer of poly(tetramethylene ether)

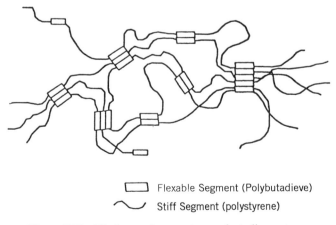

☐ Flexable Segment (Polybutadieve)

⌣ Stiff Segment (polystyrene)

Figure 15.5 Block copolymer: styrene–butadiene–styrene.

with hydroxyl end groups, obtained from tetrahydrofuran, and poly(butylene terephthalate) from terephthalic acid and 1,4-butanediol (DuPont's "Hytrel"). Its service temperature ranges from -50 to $150°C$ as opposed to -70 to $80°C$ for the styrene–butadiene–styrene copolymer. The other is a block copolymer of polypropylene as the soft segment with a hard segment comprising fully cured ethylene-propylene-diene-monomer (EPDM) rubber (Section 5.1). This is Monosanto's "Santoprene" and has a service temperature of -50 to $135°C$. High-temperature properties are also obtained by Shell in a hydrogenated version of styrene–isoprene–styrene.

Polymers of this sort have found application in rubber footwear, in rubber soles for shoes, in asphalt modifiers and in both solvent-based and hot melt adhesives.

15.3.9 Graft Copolymers

Chains are usually grafted onto a polymer backbone by creation of a free radical site along the backbone, which initiates growth of a polymer chain. Less often the backbone possesses functional groups, and chains can be condensed onto it.

An example of graft copolymerization is the production of high-impact polystyrene. Polystyrene is a useful low-cost plastic. Unfortunately it is brittle, and under stress it tends to craze or stress crack. These defects are alleviated by graft copolymerization, although the grafted polymer is no longer transparent. Polybutadiene is dissolved to the extent of 5–10% in monomeric styrene and an initiator added. Because polybutadiene readily undergoes chain transfer, polystyrene chains grow on the polybutadiene backbone, and an impact resistant graft copolymer results.

Polyacrylonitrile chains can be grafted onto a starch backbone with the aid of a ceric sulfate initiator or ionizing radiation from a cobalt 60 source. Typically three chains of acrylonitrile, each with a molecular weight of about 800,000, graft onto each starch molecule. The graft copolymer has markedly different properties from starch itself, able to absorb as much as 1000 times its own weight of water (Section 4.3).

15.3.10 Metal Complex Catalysts

The third method to bring about chain growth polymerization is by the use of metal complex catalysts. Karl Ziegler, who spent World War II at the Kaiser Wilhelm Institute in Germany trying to find ways to oligomerize small molecules into gasoline, found that titanium tetrachloride or titanium trichloride combined with an alkyl aluminum catalyzes the polymerization of ethylene. The two components of the catalyst form a solid complex, a proposed structure for which is shown on p 438. Ziegler found that his catalyst produced a high molecular weight linear crystalline polyethylene without any of the chain branching or oxygen bridges obtained in the high-pressure free radical polymerization. It was stronger and denser than the conventional material. The conditions required—about atmospheric pressure and 60°C—were astonishingly mild. Ziegler offered his discovery to Imperial Chemical Industries in the United Kingdom at a remarkably low price, but the latter were heavily committed to their own high-pressure process and were not interested. Other companies did license his process throughout the world, but a competing process developed by Phillips Petroleum in the United States proved to have advantages (Section 15.3.11) and initially was more widely used. Thus Ziegler's contribution, which attracted well-earned attention throughout the scientific world, did not find its greatest application in polyethylene manufacture.

In 1955, about 3 years after Ziegler's breakthrough, the Italian chemist, Giulio Natta, who was working for the Italian chemical giant, Montecatini, tried the new catalyst system on propylene. It does not take great scientific intuition to realize that if a catalyst works on ethylene it might also work on propylene. This, however, had not been the case with free radical polymerization. The allylic hydrogen atoms on propylene are labile and easily displaced, so that several free radical sites developed on the monomer and growing polymer. A useless, low molecular weight, cross-linked polymer was obtained. With Ziegler catalysts, however, the propylene polymerized smoothly to a high molecular weight linear polymer, and in addition—and this was the dramatic thing that won Ziegler and Natta a Nobel prize—the polymer was stereoregular. Few discoveries in organic chemistry have created as much interest or excitement.

A stereoregular polymer may be defined as one in which the substituent groups are oriented regularly in space. The structure **I** in Figure 15.6 shows such a polymer with a substituent, CH_3, in a regular formation. The carbon atom on which the substituent occurs is asymmetric, that is, four different groups are attached to it. It had long been recognized that polymers could contain asymmetric carbon atoms, but Natta was the first person to synthesize such a polymer in which all the asymmetric carbons atoms had the same orientation. This polymer is said to be isotactic because all the substituents are similarly placed. Another type of stereoregularity is shown as **II** in Figure 15.6 in which the substituents point alternately forward and back. Such polymers are said to be syndiotactic. Low temperature tends to favor formation of syndiotactic structures. The conventional nonstereoregular polymers **III** (Fig. 15.6) have their substituents placed randomly and are said to be atactic.

The discovery of Ziegler–Natta catalysis meant that almost overnight a procedure had become available for polymerization of unsaturated compounds that could not be polymerized by way of free radicals. The method was versatile and offered scope for further research. There are many transition metal salts besides the titanium trichloride and tetrachloride that Ziegler used and an equally large number of organometallic compounds with which to combine them. The adding of ligands increased the possibilities for influencing results. Since Ziegler's discovery, hundreds of combinations and thousands of ratios of constituents have been evaluated. Polymer chemists today have power to achieve practically any molecular configuration they think will give them the properties they are seeking. It is this versatility that makes Ziegler–Natta catalysis such a powerful

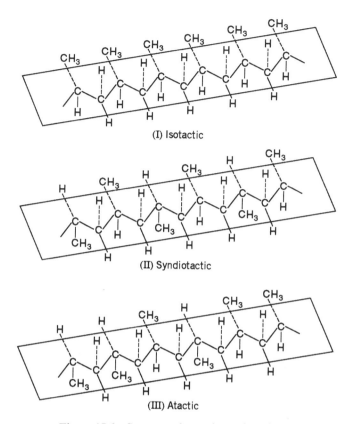

Figure 15.6 Stereoregular and atactic polymers.

tool. Metal oxide catalysis (Section 15.3.11), so important for ethylene polymerization, does not have this versatility.

The versatility is illustrated by the polymerization of butadiene. Polybutadiene may have either a 1,2 or a 1,4 configuration (Fig. 15.7). The 1,4 polymer has a double bond, which can be *cis* or *trans*. The 1,2 polymer has vinyl side chains, and these can be arranged in atactic, isotactic, or syndiotactic configurations. Thus five different polybutadienes exist, and four of them (not the atactic form) have been synthesized with the aid of Ziegler–Natta catalysts. The structures are shown in Figure 15.7.

The stereospecificity of Ziegler catalysts has allowed chemists to do what nature can do with its highly specific enzymes. Nature is able to synthesize optically active compounds, sterically complex antibiotic molecules, and also steroregular polymers. Natural *hevea* rubber is *cis*-1,4-polyisoprene while *trans*-1,4-polyisoprene is the nonelastomeric *balata* or *gutta percha*. In the synthesis of *hevea* rubber in nature, every step is catalyzed by an enzyme. The starting material is acetic acid, a material manipulated with particular ease by nature as is illustrated by its various metabolic pathways in the body. The final step is the polymerization of the monomer, isopentenyl pyrophosphate, which nature prefers over isoprene, by a polymerase present in the rubber plant.

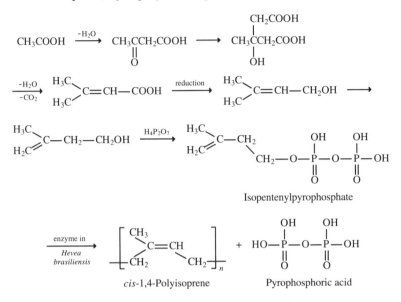

Isopentenylpyrophosphate

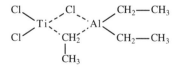

cis-1,4-Polyisoprene   Pyrophosphoric acid

With Ziegler catalysis, scientists can duplicate nature's precision and produce materials that are similar to either *hevea* rubber or *gutta percha*. Thus the chemical industry can and does produce "synthetic natural rubber." It is not quite the same as natural rubber, because end groups and molecular weight distribution differ. Hence, it is almost but not quite as resilient. Equally significant on the plane of ideas is that Ziegler–Natta catalysis enables chemists to mimic nature by making steroregular polymers. It is satisfying too that Ziegler catalysis works, as does nature, at moderate temperatures and pressures compared with, for example, the formidable 1200 bar and 200°C required for traditional low-density polyethylene synthesis.

What causes stereospecificity? How do Ziegler catalysts work? The mechanism is not fully understood, but it is known that polymerization takes place at active sites on the catalyst surface. The catalyst is an electron-deficient solid complex of an aluminum alkyl and a titanium halide, the alkyl group on the titanium atom coming from the alkyl aluminum portion of the catalyst. This is one of several possible structures.

The electron deficiency occurs between the titanium–carbon and carbon–aluminum bonds. Titanium has an octahedral configuration with one ligand vacancy,

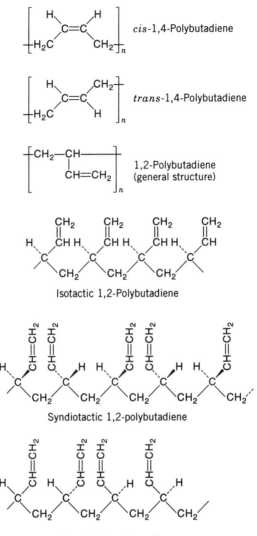

Figure 15.7 The isomeric polybutadienes.

as shown in Figure 15.8, and a monomer, for example, propylene, may become π bonded to the titanium. After this insertion, new bonds form and the old ones break. The net result is that the propylene molecule inserts itself between the titanium atom and the alkyl group. The ligand vacancy again exists so that the same progression can happen all over again, and another propylene molecule can be incorporated into the alkyl chain. This is the propagation step, and the polymer chain grows by successive insertion of monomer units at the surface of the titanium complex catalyst.

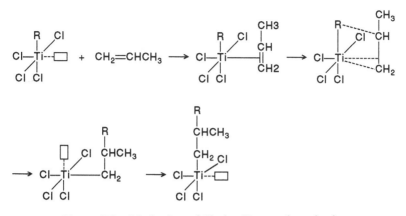

Figure 15.8 Mechanism of Ziegler–Natta polymerization.

The system is heterogeneous, and the catalyst is insoluble in the monomer and in the solvent. The insertion of monomer molecules takes place at the solid–liquid interface, and the polymer chain grows from the insoluble catalyst into the solvent. It is the solvating effect of the solvent on the polymer that attracts the chain away from the catalyst surface and into the solvent and allows further monomer to have access to the titanium atom. Some homogeneous metal-catalyzed polymerizations can also be carried out.

The mild conditions in the use of Ziegler–Natta catalysis are not only a bonus but also a prerequisite for it. At higher temperatures bonds around the catalyst would tend to break and reform, and stereospecificity would be lost. The mild conditions also insure linear polymers and eliminate the branching that is characteristic of free radical initiation. Furthermore, the linear chains can get very near to each other, which gives them high cohesive strength and crystallinity and confers certain desirable properties discussed in Section 15.5.1.

15.3.11 Metal Oxide Catalysts

Before Ziegler discovered his catalyst, studies on supported metal oxide catalysts were underway. Researchers at Standard Oil of Indiana developed a molybdenum oxide catalyst supported on silica or alumina that gives high-density polyethylene. Their discovery predated Ziegler's but they did not exploit it because a consultant's evaluation was negative. The consultant, interested in making film, could see no virtue in a stiff, structural-like polymer. The conventional wisdom associated with this story is that it is not enough to invent. One must also recognize the importance of the invention.

Another oxide catalyst system, chromic oxide on silica or alumina, was developed almost concurrently with Ziegler's catalyst by Phillips Petroleum. With it polymers can be obtained of higher molecular weight than those obtained by the Ziegler method, and these tend to be intractable. For example, they are difficult to remove from the kettle. More tractable polymers result when hydrogen and about 1% of a comonomer, 1-butene, is included. The hydrogen controls molecular weight by serving as a chain stopper, which can in this case be regarded as a chain-transfer agent (Section 15.3.2). The comonomer controls density. The reaction takes place in a hydrocarbon solvent at 100°C and 40 bar.

Chromium oxide is the most active catalyst, although oxides of Ti, Zr, Ge and Th are also effective. The best supports are silica or aluminosilicates with low alumina contents. It is important that they have low mechanical strength to permit the breakup of the catalyst particles during polymerization.

X-ray work suggests that the catalytically active species is Cr(II) and that the polymerization centers contain isolated chromium moieties. Polymerization is initiated by the formation of chromium–carbon bonds on the surface of the catalyst to give a chromium alkyl. The oxidative addition of ethylene to a divalent chromium ion with vacant coordination positions may take place by one of the following mechanisms:

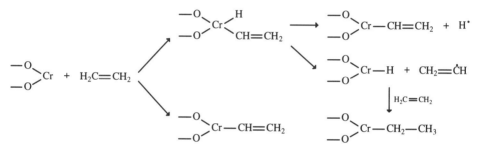

Propagation probably occurs as in Ziegler–Natta catalysis by the sequential insertion of ethylene molecules into the chromium–carbon bond.

The mechanism of metal oxide catalysis remains uncertain. Theoretical interest, perhaps unjustifiably, is less because neither the Standard Oil of Indiana nor the Phillips process can be used to make polypropylene (see note).

15.3.12 Metallocene Catalysts

Metallocene-based single site and constrained geometry catalysts are the latest development in the design of "tailored" polymers. In the 1930s, ethylene was first polymerized industrially to low-density polyethylene at high pressures with free radical initiators, then in the 1950s to high-density polyethylene at much lower pressures with metal oxide and Ziegler–Natta catalysts. In the 1980s, addition of α-olefin comonomers to Ziegler polyethylene provided linear low-density polyethylene at low pressures, which inhibited growth of the high-pressure process. It might have seemed as if the possibilities had been exhausted, but technology moves inexorably. In 1991, Exxon started a plant for production of polyethylenes based on so-called metallocene single-site catalysts, and Dow started a plant based on constrained geometry single-site catalysts in 1993. These innovations have caused great excitement.

Metallocenes are organometallic coordination compounds in which transition metals are sandwiched between cyclopentadienyl rings. The first metallocene was ferrocene **IV**. Its rings are capable of free rotation and are parallel to one another. Compounds of catalytic interest are discussed below.

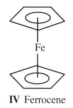

IV Ferrocene

The key difference between metallocene catalysts and the conventional Ziegler–Natta catalysts is that Ziegler–Natta catalysts are heterogeneous and have many nonidentical active sites, only some of which are stereospecific. Metallocene catalysts, on the other hand, are homogeneous or supported and have a single active polymerization site. As a result, polymers made with them are effectively a single molecular species, having a very narrow molecular weight distribution. Comonomers are taken up in a uniform manner and their distribution within the polymer chains will be similar. This narrow distribution provides polymers with lower crystallinity, greater clarity, lower heat seal temperatures, and better resistance to extraction, hence lower taste and odor.

The drawback of the narrow distribution is that ease of processing is diminished. Lower molecular weight polymers act as plasticizers for the processing of the higher molecular weight material. This problem is overcome in two ways. Addition of conventional Ziegler–Natta catalysts to metallocenes or use of two different metallocene catalysts can give broad or bimodal molecular weight distributions. Alternatively, a controlled amount of long-chain branching can be incorporated to give narrow molecular weight distribution branched polymers that are more readily processed.

The metallocenes of catalytic interest are those of the Group IVB metals, titanium, zirconium, and hafnium. Compound **V** is a typical zirconocene. Metallocene rings are not necessarily parallel and they may be substituted. The cyclopentadienyl ring can be part of a condensed ring system such as fluorene or indene (**VI**), and the cyclopentadienyl rings may be joined by silyl or alkylene

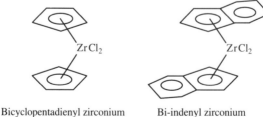

Bicyclopentadienyl zirconium   Bi-indenyl zirconium
dichloride **V**   dichloride **VI**

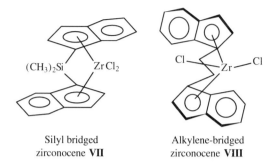

Silyl bridged
zirconocene **VII**

Alkylene-bridged
zirconocene **VIII**

bridges (**VII** and **VIII**). Such metallocenes do not allow ring rotation and restrict
access to the metal. By definition, metallocenes contain two cyclopentadienyl
rings but monocyclopentadienyl compounds of transition metals are sometimes
also included in the term.

Metallocene catalysts are of low activity unless used with cocatalysts, usually
a methylaluminoxane but sometimes an ionic activator. Methylaluminoxanes,
in turn, are linear or cyclic polymers formed when trimethylaluminum reacts
with water:

$$(CH_3)_3Al \; + \; H_2O \; \longrightarrow \; \begin{matrix} H_3C \\ \diagdown \\ \diagup \\ H_3C \end{matrix} Al{-}O{-}\left(\begin{matrix} CH_3 \\ | \\ Al{-}O \end{matrix}\right)_m Al\begin{matrix} \diagup CH_3 \\ \diagdown CH_3 \end{matrix} \; + \; \left(\begin{matrix} CH_3 \\ | \\ Al{-}O \end{matrix}\right)_m$$

Linear   Cyclic

$m = 4\text{--}20$

An example of a cyclic methylaluminoxane, where $m = 5$ is

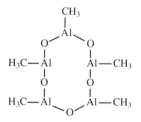

A simple example of a single-site catalyst suitable for polymerization of ethylene
is the reaction product of the zirconocene (**V**) and a methylaluminoxane.
Compound **VIII** is also used with methylaluminoxane and (**V**) is used with the
ionic activator $C_6H_5N^+(CH_3)_2B^-(C_6H_5)_4$. The mechanism of the polymerization process is similar to that of conventional Ziegler–Natta polymerization.
The monomer π-bonds to the metal atom and then inserts into the growing
polymer chain, as was shown in Section 3.1.3. The activities of the metallocenes
are said to be high. One gram of a proprietary metallocene is claimed to generate
4000 lb of polypropylene, a ratio of 1.8 million : 1.

The most advanced group of metallocene catalysts are the constrained geometry catalysts. These are based on a constrained geometry ligand attached to a transition metal catalyst center. A Group IV transition metal is bonded to a cyclopentadiene ring, and the ring and a hetero atom are bonded to both by a suitable bridge. An example suitable for ethylene copolymers is **IX**.

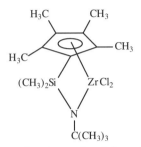

(*t*-Butylamino)dimethyl(tetramethyl-n^5-cyclopentadienyl)silane-zirconium dichloride (**IX**)

The cyclopentadienyl–zirconium–nitrogen bond angle is said to be less than 115°C and the metal is therefore more open for monomer and comonomer insertion. It permits the use of higher olefins such as octene, vinylcyclohexene and even styrene as comonomers and also allows polymer molecules with terminal unsaturation to act as comonomers to give long-chain branching. The branching permits easier processing, as mentioned above.

The ethylene–styrene copolymers prepared via constrained geometry catalysts are "pseudorandom". After incorporation of one bulky styrene unit, the active site becomes crowded, and at least one unit of ethylene must be added before there is again room for a styrene. Thus the polymer is not iso- or syndiotactic, but all the phenyl groups in the polymer are separated by at least two methylene groups. The product is amorphous (unlike conventional Ziegler–Natta polymers) and elastomeric. The ethylene-1-octene copolymer is also elastomeric.

Commercial exploitation of metallocene-based products is still at an early stage. Syndiotactic polypropylene is available in quantity for the first time. Atactic polypropylene for such applications as adhesives and bitumens is said to be planned. Exxon is offering a range of ethylene copolymers with propylene, butene and hexene. Dow is offering a range of ethylene-1-octene copolymers. The methylaluminoxane cocatalyst is particularly expensive and metallocene-based products sell at a premium into the food packaging, personal care and medical markets. They are specialty products and are unlikely to make serious inroads into the markets for the large-tonnage commodity polymers in the near future. It is evident, nonetheless, that the fine tuning of polymer properties by variation of polymerization technique offers almost limitless possibilities. The discovery of Ziegler–Natta catalysis was merely the first step along the road.

15.4 EXAMPLES OF STEP POLYMERIZATION

15.4.1 Phenoplasts and Aminoplasts

There are two types of phenolic resins (phenoplasts), known as novolacs, and resoles. Both are made by step polymerization. Novolacs are made by condensing excess phenol with formaldehyde in the presence of an acid catalyst. Fusible polymers result with the phenol rings joined by methylene bridges but no free methylol groups. They are linear thermoplastic resins that may be stored or sold in that form. Cross-linking via the formation and condensation of free methylol groups is brought about by heat in the presence of a curing agent that provides more formaldehyde under alkaline conditions. An example is hexamethylene tetramine **X**, a condensate of formaldehyde and ammonia, the latter providing the alkaline conditions.

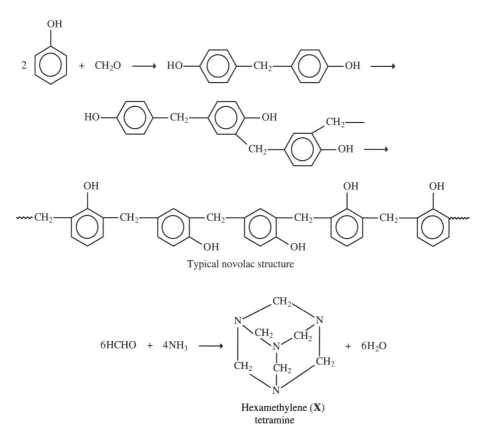

Typical novolac structure

Hexamethylene (**X**)
tetramine

Resoles are made with an alkaline catalyst and sufficient formaldehyde to allow for cross-linking. The process gives a cross-linked structure in one operation (a "one-stage" resin). However, the reaction is stopped at a so-called B-stage before cross-linking occurs. At that stage the polymer is linear, soluble and fusible but contains free methylol groups. The methylolphenols condense to give low molecular weight linear polymers called resoles, which contain occasional oxygen bridges. Typical structures are

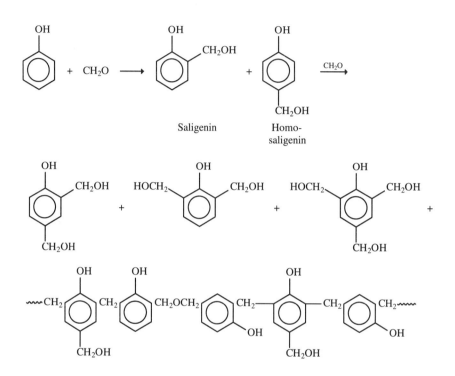

Saligenin

Homo-saligenin

The resin has sufficient shelf life for use in compounds for automotive and electrical applications. Cross-linking is brought about by further heating and pressure, which causes the free methylol groups to condense.

In these reactions the formaldehyde shows a functionality of two while the phenol has three active sites—two positions *ortho* and one *para* to the hydroxyl group. Only two of them are used in novolac formation because of the scarcity of formaldehyde. Thermoplastic phenolics useful for the preparation of varnishes result from the condensation of formaldehyde with *para* substituted phenols.

Urea will also give cross-linked resins with formaldehyde (aminoplasts) under slightly alkaline conditions (Section 10.5.1.1). Methylolureas are formed first. There follows a series of condensations that include ring formation, and the product is a complex thermoset polymer of poorly defined structure of which the following may be typical.

Urea has a functionality of four in urea–formaldehyde resins, corresponding to the four labile hydrogen atoms. Melamine (Section 10.5.1.1) with three amino

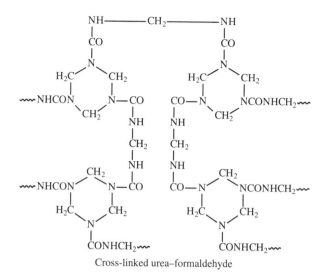

Cross-linked urea–formaldehyde

groups and six labile hydrogens has a functionality of six, and it too will form thermoset resins with formaldehyde—the so-called melamine–formaldehyde.

15.4.2 Polyurethanes

Polyurethanes result from the reaction of di- or polyisocyanates with di- or polyols. They can be manufactured by a one-step process in which the main reactants are mixed with catalysts, fillers, reinforcing and coloring agents, blowing agents, and minor constituents and placed in a mold to give the final product. Alternatively, a two-step process is employed in which a relatively high molecular weight prepolymer is assembled first. It is then reacted with a low molecular weight diamine or diol to give two or three-dimensional higher molecular weight polyurethanes. An example is spandex from MDI prepolymer, as described in Section 15.3.8. The two-step process is the basis of RIM technology (Section 7.3.1) and also avoids the handling of toxic isocyanates during shipment and final processing.

If a diol undergoes step polymerization with a diisocyanate, a linear thermoplastic polyurethane is obtained because both monomers are bifunctional.

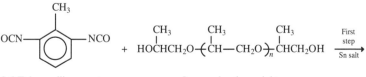

2,6-Toluene diisocyanate
(A mixture of 2,4 and 2,6
isomers is normally used)

Low molecular weight
polypropylene glycol

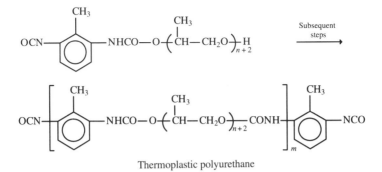

Thermoplastic polyurethane

Poly(propylene glycol) is a bifunctional oligomer synthesized from propylene oxide (Section 4.7). To obtain the more useful cross-linked polyurethanes, trifunctional reagents are required. Sometimes toluene diisocyanate (TDI) is converted to a trifunctional reactant by reaction with trimethylolpropane. The new reagent has the advantage of being considerably less toxic than TDI because of its lower vapor pressure. Alternatively, trifunctional hydroxyl compounds may be made by reaction of propylene oxide with glycerol.

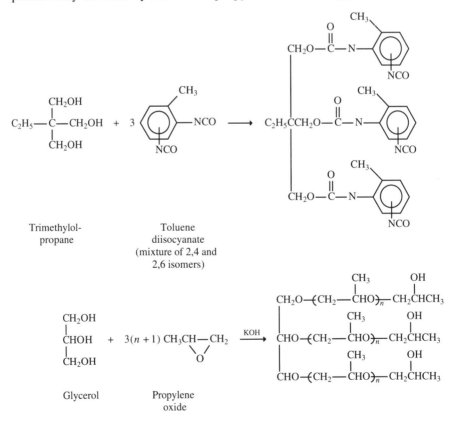

Castor oil (Section 13.1) is a naturally occurring triglyceride that contains three hydroxyl groups and is another useful starting material.

Toluene diisocyanate is the raw material for about 60% of polyurethanes and MDI, which is 4,4′-diphenylenemethane diisocyanate and oligomers of it, is also important. The presence of the trimer and tetramer in the product mixture means that the product has a functionality greater than two. The aliphatic isocyanate HMDI, hexamethylene diisocyanate, $OCN(CH_2)_6NCO$, leads to coatings with good color retention and weathering properties. It is toxic and is used in the form of a trimer with a biuret structure. A biuret forms from the interaction of an isocyanate group with a urea. It exemplifies a second type of linkage found in polyurethanes. The HMDI biuret trimer forms as follows:

$$OCN(CH_2)_6NCO \xrightarrow{H_2O} OCN(CH_2)_6NHCOOH \xrightarrow{-CO_2}$$

$$OCN(CH_2)_6NH_2 \xrightarrow{OCN(CH_2)_6NCO}$$

$$\underset{\text{Urea derivative}}{OCN(CH_2)_6NH\overset{\overset{O}{\|}}{C}NH(CH_2)_6NCO} \xrightarrow{O CN(CH_2)_6NCO}$$

$$OCN(CH_2)_6\overset{\overset{O}{\|}}{N}CNH(CH_2)_6NCO$$
$$\underset{\underset{\text{Biuret trimer}}{HN(CH_2)_6NCO}}{\overset{|}{C}=O}$$

A third linkage found in polyurethane foams is the allophonate group, which forms from the interaction of an isocyanate with a urethane linkage.

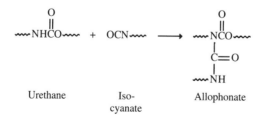

| Urethane | Iso-
cyanate | | Allophonate |

Still another form in which isocyanates are used is as isocyanurates. These are isocyanate trimers and have an advantage over biurets in being more stable. A typical isocyanurate is the trimer of 3-isocyanatomethyl-3,5′,5-trimethyl-cyclohexyl isocyanate, trivially known as "isophorone diisocyanate" **XI** (Section 4.6.2). The conversion to trimer (**XII**) takes place in the presence of a basic catalyst.

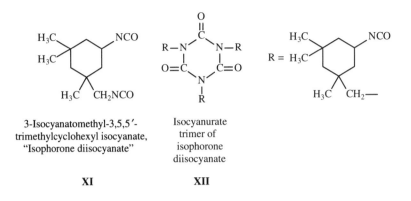

3-Isocyanatomethyl-3,5,5'-
trimethylcyclohexyl isocyanate,
"Isophorone diisocyanate"

Isocyanurate
trimer of
isophorone
diisocyanate

XI   **XII**

In the formation of polyurethane foams, carbon dioxide for foaming may be produced by the addition of water, which gives a carbamate, which in turn decomposes to an amine and CO_2.

$$\text{\mmNCO} + H_2O \xrightarrow[\text{amine}]{\text{Tertiary}} \text{\mm[NHCOOH]} \longrightarrow \text{\mmNH}_2 + CO_2\uparrow$$

Carbamate   Amine   Carbon
                                           dioxide

This, however, is an expensive way to obtain a gas for foaming, and specialized foaming agents are used. When water is added, urea linkages form by the condensation of the amine with more isocyanate:

$$\text{\mmNCO} + NH_2\text{\mm} \longrightarrow \text{\mmNHC—NH\mm}$$
$$\underset{O}{\overset{\|}{}}$$

Polyurethane foams contain both these and biuret linkages.

15.4.3 Epoxy Resins

The curing of epoxy resins is interesting because functionality is generated in the course of the reaction.

Epoxy resins are typically condensates of bisphenol A with epichlorohydrin. If a large excess of epichlorohydrin is used, a simple molecule results from the condensation of 2 mol of epichlorohydrin with 1 mol of bisphenol A.

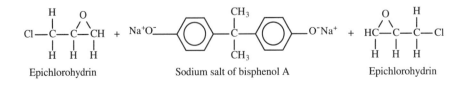

Epichlorohydrin   Sodium salt of bisphenol A   Epichlorohydrin

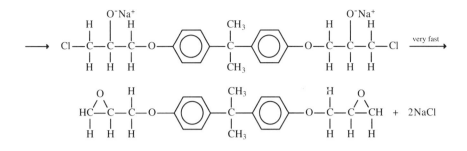

If the reactants are close to equimolar, on the other hand, a low molecular weight polymer **XIII** is formed where n is between 1 and 4. In either case the terminal groups are epoxy groups.

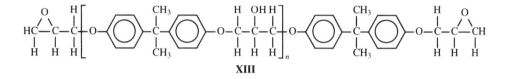

XIII

The epoxy groups will react with a multifunctional amine such as ethylene diamine to give a cross-linked resin. When one of the primary amine groups reacts with an epoxy group, a hydroxyl group and a secondary amine group are generated (**XIV**). Both of these groups can react further with epoxy groups in principle, although the hydroxyl group will react only at high temperatures. The aliphatic amine groups react at room temperature and, if two molecules of ethylene diamine react with one polymer molecule of epoxy resin, a molecule (**XV**) is generated with four amine groups: two primary and two secondary. The

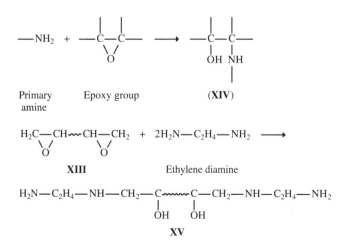

amine groups have six active hydrogens between them, each of which can react with more epoxy resin, and thus the conditions for cross-linking have been established.

The interaction of the secondary amine group with an epoxy group generates a tertiary amine. Tertiary amines are catalysts for the self-polymerization of epoxy groups to polyethers, so yet another polymerization mechanism has been introduced and is shown in the following equation. This polymerization is chain growth, whereas the polymer formation resulting from the condensation of amine and epoxy groups is step growth.

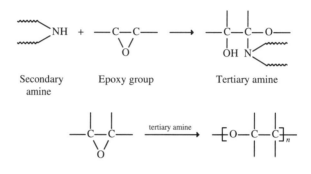

Secondary   Epoxy group   Tertiary amine
amine

15.5   POLYMER PROPERTIES

In Section 15.4.3 we described how monomers were converted to polymers, bifunctional monomers leading to linear thermoplastic resins and polyfunctional monomers providing cross-linked thermosetting resins. Only one property, average molecular weight, has been mentioned. Molecular weight and mechanical strength are related since strength increases rapidly between 50 and 500 monomer units. Further increases in molecular weight have little effect. In this section we describe additional properties of polymers and discuss the factors that make them useful: high viscosity, tensile strength, and toughness.

15.5.1   Crystallinity

Crystallinity is the key factor governing polymer properties. An easy way to visualize it is to regard crystallinity as a situation in which the polymer chains fit into an imaginary pipe. That is, they align themselves in bundles with a high lateral order, and the chains lie side by side. To do this they must be linear, not coiled, and there must not be bulky groups or branching to prevent the polymer chains from achieving molecular nearness.

There is an analogy, although not an exact one, with the crystallization of nonpolymeric materials like sodium chloride or n-hexane, where the ions or molecules must fit into a crystal lattice. With polymers, however, it is long-chain molecules, not ions or small molecules, that must fit, and the structure into which they fit is not a lattice but an imaginary cylindrical tube. Figure 15.9 attempts to illustrate this concept, and shows how the chains may line up in an ordered fashion. It is, however, unlikely that all the chains in a polymer or even all of a single chain will be able to enter into the ordered structure of complete crystallinity, although nylon and high-density polyethylene come close. Usually, the ordered regions are small—microcrystalline—and are scattered through the polymer, which is otherwise amorphous. Where the polymer is crystalline, it is platelike and of uniform thickness. Emanating from these crystalline regions are the sections of the chains that are not incorporated into the crystal lattice. They form the amorphous part of the polymer and may actually coil back over the crystalline platelets. Thus one can legitimately talk of the degree of crystallinity of a polymer. Some polymers are almost totally crystalline, others almost totally amorphous. Even a single polymeric substance can exist with a range of crystallinities depending on how it was made and processed.

Two factors govern the tendency of a polymer to crystallize. One is the ease with which the polymer chains will pack into a "crystal," and the other is the

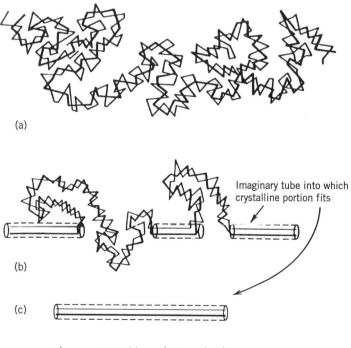

Figure 15.9 Polymer crystallization: (a) Noncrystalline; (b) partially crystalline; and (c) completely crystalline.

magnitude of the attractive forces between neighboring molecules of the polymer. The first of these means that crystalline polymers are more likely to form from chains that do not have bulky substituents, and where there is not a great deal of branching. Poly(ethylene terephthalate) is without bulky side chains and is crystalline after orientation, that is after the fibers are stretched or drawn. The rigid benzene ring makes the polymer chains stiff and unwilling to coil.

Isotactic polymers made by Ziegler–Natta polymerization are also highly crystalline unless, as in *para*-substituted polystyrenes, a bulky group keeps the polymer chains apart. Polystyrene, polymerized as it normally is by peroxide catalysis, is an example of an amorphous polymer. Isotactic polystyrene made by Ziegler–Natta catalysis, however, is a highly crystalline material with properties very different from those of the amorphous product. It commercialization was inhibited because it crystallizes very slowly and therefore changes its properties after processing. However, metallocene catalysts (Section 15.3.12) seem to have made possible a syndiotactic form that is scheduled for commercialization in the late 1990s. It has a high-melting point and is scheduled to function as an engineering polymer, that is one that will replace metals.

Other amorphous polymers are exemplified by poly(methyl methacrylate) and polycarbonates, which are condensates of bisphenol A and phosgene (Section 7.1.2.2). The pendant groups on the methacrylate polymer are bulky, and the two phenyl groups in the polycarbonate are not coplanar. Thus crystal formation is obstructed. The benzene rings of the polycarbonate participate in so-called crystallites, which form by their random association. These, plus the polymer's stiffness, contribute to its strength. Most copolymers have little crystallinity because their structures are nonlinear.

In general, crystalline polymers are opaque because light is reflected or scattered at the boundaries between the microcrystalline and amorphous regions. Amorphous polymers are transparent and glass like. Two exceptions should be noted. If a crystalline polymer is biaxially oriented, that is stretched simultaneously in two directions, as in drawn poly(ethylene terephthalate) (Mylar) sheet, then the whole sheet is in effect a single crystal and is transparent. Furthermore, in a few polymers, the most important of which is poly (4-methyl-1-pentene), the refractive index of the crystal is identical with that of the amorphous region. No light scattering occurs at boundaries, and the crystalline polymer is clear and transparent.

The second factor leading to crystallization is the forces of attraction between neighboring molecules. These comprise hydrogen bonding, which is the strongest, and the various kinds of van der Waals forces, dipole–dipole forces, dipole–induced dipole forces, and London dispersion forces. They vary in strength from about 1–2 kcal/mol (5–10 kJ/mol) per unit of polymer chain in elastomers to 5–10 kcal/mol (20–40 kJ/mol) in fibers. London forces are weak whereas the others can be quite strong. Cellulose and nylon provide examples of hydrogen bonding, PVC and polyacrylonitrile of dipole–dipole interaction, and polyethylene of London forces.

We can extend the analogy between crystallization of *n*-hexane and polyethylene. As *n*-hexane is cooled, the thermal motion of the molecules decreases until it can no longer overcome the forces of attraction between them. Accordingly, the molecules pack into the orientation of lowest energy, that is, the crystal lattice, and the sample solidifies or crystallizes.

The molecules of polyethylene are hundreds of times larger than those of hexane. Although the intermolecular forces between the $-CH_2-$ units are about the same in the two molecules, the total force per molecule in the polymer will be much higher. Furthermore, the polyethylene chains will not have the freedom of movement of hexane molecules. Instead they will be wriggling and coiling. There is little chance that one of them will ever be fully extended. The chains will also become entangled with one another, which will hinder molecular motion. Molten polymers are viscous both because of chain entanglement and intermolecular forces. The latter factor is negligible at high temperatures, and viscosity under these conditions is due largely to chain entanglement.

When the temperature of molten polyethylene is reduced, molecular motion diminishes as it does in hexane. Eventually, there will be a tendency for the chains to pack into a crystal lattice. In order to do this they will need to be extended and not coiled, but the chance of a chain being fully extended is small. There will, however, be large portions of polymer chains that are extended. These will pack into an ordered crystal lattice and provide the microcrystalline regions, while the tangled, coiled portions will form the disordered amorphous regions seen in Figure 15.9.

In block copolymers (Section 15.3.8) it is possible to create a block that is highly crystalline together with one that is amorphous to obtain a final copolymer with special properties.

Crystalline polymers tend to have greater mechanical strength and higher melting point than amorphous polymers. Because the chains in the crystalline regions are closely packed, they would also be expected to have higher densities, and this too is observed. For example, low-density polyethylene has a tensile strength (Section 15.5.4) of 2000–2500 psi (140–175 kg/cm^2), a softening point of 85–87°C, and a specific gravity of 0.91–0.93. High-density polyethylene, on the other hand, has a tensile strength of 3500–5500 psi (245–385 kg/cm^2), a softening point of 127°C, and a density of 0.94–0.97 g/cm^3.

The degree of crystallinity of a polymer is measured by X-ray diffraction by the same technique used for single crystals. High-density polyethylene may have as much as 90% crystallinity; the low-density material has only about 55%, which is still high considering the extent of chain branching in the polymer. It occurs because the lengths of polymer chain between the branches are capable of getting close enough to other chains for crystallization to take place.

Crystallinity is also related to orientation. If one slowly flexes a wide rubber band, an appreciable amount of heat is generated, which can be felt if the band is touched to the lips. This results from the friction of one polymer molecule rubbing against another as the stretching action causes them to align. Before it is

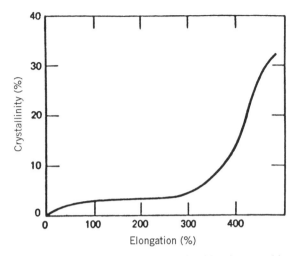

Figure 15.10 Crystallization of cured rubber by stretching.

stretched, the rubber band is largely amorphous. The alignment on stretching is tantamount to crystallization, which also causes the translucent rubber band to become opaque. Orientation of the polymer has caused it to crystallize. The heat generated can therefore also be thought of as heat of crystallization. The degree of crystallinity that can be induced in an elastomer by stretching can reach 30% (Fig. 15.10).

The orientation or crystallization of polymer molecules by stretching or drawing is an important step in the processing of polymers for use as textile fibers. When nylon and polyesters are manufactured they have low crystallinity. The stretching or drawing of the fibers causes the polymer molecules to line up or crystallize to give the longitudinal strength required in fibers.

A difference between fibers and rubbers is that the latter have much greater "elastic memory" and do not remain stretched and crystalline. In other words, it is hard to stretch them beyond their limit of crystallinity. The reason is that rubber is lightly cross-linked, and these cross-links tend to pull the rubber molecules back to their original disordered configuration. In addition, since stretching is a disorder–order transition, the entropy decreases. The return to the unstretched state is favored by the corresponding entropy increase. In nylon the "cross-linking" takes the form of hydrogen bonds that break and reform as the fiber is stretched. Indeed, stretching orients the molecule so that more hydrogen bonds are formed than are broken. Consequently, the elastic memory is quickly exceeded on stretching.

15.5.2 Glass Transition Temperature, Crystalline Melting Point, and Softening Temperature

If an amorphous material such as polystyrene is melted and then allowed to cool, it does not solidify sharply. First it goes from a viscous liquid to a rubbery solid, then to a leathery solid. Finally, when all of the molecules have lost their thermal "wriggling" motions, it becomes a glassy solid, recognizably polystyrene. The last change is a sharp one, and the temperature at which it occurs is called the glass transition temperature of the polymer, T_g. Obviously, it has a bearing on the properties of the polymer at the service temperature at which it is to be used. A polymer that is a soft, leathery material above its T_g may be a hard, brittle, amorphous one below the T_g.

The glass transition temperature is associated primarily with amorphous polymers, although crystalline polymers also have a glass transition temperature, because all polymers have amorphous regions between the microcrystalline regions.

If a crystalline polymer is used above the T_g of the amorphous regions, the latter will be flexible, and the polymer will be tough. If the temperature is below the T_g, however, the amorphous regions will be glassy and the polymer brittle. A similar situation for block copolymers is described in Section 15.3.8.

The temperature at which a molten polymer changes from a viscous liquid to a microcrystalline solid is called the crystalline melting point, T_m, of the polymer. If the solid polymer is partly crystalline, the change is accompanied by sudden changes in density, refractive index, heat capacity, transparency, and similar properties. It is analogous to the melting point of a nonpolymeric chemical compound but is not as sharp, and melting and solidification take place over a small range. The value of T_m depends on chain structure, intermolecular forces, and chain entanglement.

The softening point is an arbitrary measure of the temperature at which a polymer reaches a certain specified softness. It is of great importance as the upper service temperature of a polymer but has little significance on the molecular level.

Table 15.6 indicates typical values for T_g and T_m for various polymers. The glass transition temperature is about one-half to two-thirds of T_m for most polymers if the temperatures are in degrees absolute. Deviations from this may be due to unusual molecular weight distributions, chain stiffness, and symmetry. These values, unlike the melting points of pure organic compounds, can be considered only "typical." The glass transition temperature may vary with the molecular weight of the polymer, its method of preparation, its end group distribution, and with the degree of crystallinity in a given polymer sample. For a completely unoriented material the glass transition temperature will be very low. When the material is oriented or converted to a crystalline state, the glass transition temperature increases. The glass transition temperature for poly(ethylene terephthalate) may vary from -80 to $180°C$.

15.5.3 Molecular Cohesion

Molecular cohesion is the average force between the repeating units of a polymer chain and its neighbors. The forces are van der Waal's forces or hydrogen bonds. Their magnitude can be calculated from cohesive energy density (see note).

TABLE 15.6 Typical T_g and T_m Values for Polymers

	Temperature (°C)	
	T_g	T_m
cis-Polybutadiene	− 101	4
cis-Polyisoprene	− 73	29
trans-Polyisoprene	− 58	70
Linear polyethylene	− 70 to − 20	132
Polypropylene	− 16	170
trans-1,4-Polybutadiene	− 9	139
Nylon 6,6	47	235
Poly(methyl methacrylate)	49	155
Poly(vinyl chloride)	70	140
Polystyrene	94	227
Polycarbonate	152	267
Cellulose triacetate	111	300
Poly(tetrafluoroethylene)	135	327

Hydrogen bonds contribute most to molecular cohesion. Dipole–dipole forces contribute less, and London dispersion forces the least. The strength of the molecular interactions diminishes rapidly, actually with the sixth power of the distance between the molecules. Thus bulky amorphous polymers have relatively low intermolecular forces whereas those in crystalline polymers are much higher because the molecules are much closer together.

15.5.4 Stress–Strain Diagrams

Many of the quoted properties of polymers are derived from stress–strain diagrams. These are graphs of the deformation in a polymer sample (expressed as percent of elongation) produced by a particular applied stress (a tension expressed as pounds per square inch, kilograms per square centimeter, or meganewtons per square meter). Such diagrams can be generated quickly from a given polymer sample of controlled size in a testing laboratory. For reproducibility a standard sample and rate of extension must be used. An example (for a hard, tough plastic) is shown in Figure 15.11.

In the initial stages of the extension, the stress–strain diagram is linear. That is, the material obeys Hooke's law, and stress is proportional to strain. The gradient or slope of this straight section is called the initial modulus of elasticity and is a measure of the stiffness of the material (Young's modulus). If the applied stress is removed at this stage, the polymer will return to its initial length. After a certain stress the graph ceases to be linear, and the extension is nonreversible; that is, a permanent deformation is produced. The yield point is defined as the maximum in the stress–strain curve, as shown in the diagram, and has an

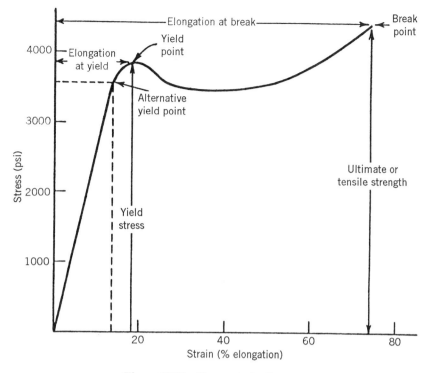

Figure 15.11   Stress–strain diagram.

elongation at yield and a yield stress associated with it. Some authorities define the yield point as the point at which deformation becomes irreversible, that is at which Hooke's law ceases to be obeyed. Because many polymers do not obey Hooke's law, the modulus is frequently expressed as pounds per square inch of tensile strength at a given degree of elongation such as 2%. This is called the 2% modulus. Some elastomers are better described by a 100 or 300% modulus.

After the yield point the polymer stretches relatively easily and, after stiffening a little toward the end, it breaks, and the curve comes to an abrupt end. The break point has associated with it an elongation at break or upper limit of extensibility and an ultimate tensile strength. The area under the curve up to the break point is called the work-to-break. The tensile strength is the strength required to pull the polymer apart, and the work-to-break is the work required to do it and is a measure of the ability of the polymer to resist not only tension or "pulling apart" but also other stresses such as bending, compression, impact, and twisting.

Figure 15.12 shows the type of stress–strain diagrams obtained for fibers, thermoplastics, and elastomers. Fibers, because they have been oriented, have high tensile strength and modulus and resist elongation. The work-to-break is small, although the polymer is strong and resists "pulling apart." The elastomer,

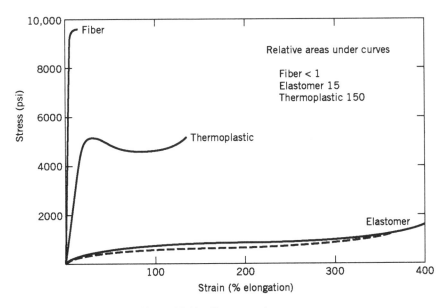

Figure 15.12 Stress–strain curves.

on the other hand, has high elongation but low tensile strength and modulus. Thus it has little resistance to deforming forces other than elongation. As its extension is reversible, a return stress–strain curve can be obtained if the stress is reduced before the sample breaks. This is indicated by the dotted line. The area within this hysteresis loop is the work dissipated as heat during the cycle.

The thermoplastic shown has quite a high modulus and tensile strength and a fairly high elongation. The area under the curve is large; hence this type of thermoplastic would be used if a wide variety of deforming forces were expected, whereas a fiber would have best resistance to simple tension. Elastomers are useful because they "bounce back" and can absorb energy by hysteresis.

15.6 CLASSES OF POLYMERS

Polymers are used in five main ways: as elastomers, plastics, fibers, surface coatings, and adhesives. Table 15.7 shows the combination of polymer properties required for the first three applications in terms of the four major properties.

Elastomers have a low modulus. Modulus is a measure of stiffness, and rubbers are not stiff. On the other hand, they need to be highly extensible, and elongations of 1000% are common. The crystallinity of an elastomer is low when the material is unstressed, but stretching leads to a higher degree of crystallization (Section 15.5.1). Molecular cohesion in elastomers must be low because otherwise the polymer chains will not easily slip over one another when the elastomer is stretched or slide back reversibly when the tension is released.

TABLE 15.7 Polymer Properties

	Elastomers	Plastics	Fibers
Modulus (psi)	15–150	1500–200,000	150,000–1,500,000
(kg cm^{-2})	1–10		10,000–100,000
Extensibility, upper limit (%)	100–1000	20–100	< 10
Crystallization tendency	Low when unstressed	Moderate to high	Very high
Molecular cohesion (cal/monomer unit)	1000–2000	2000–5000	5000–10,000
Examples	Natural rubber, polychloroprene, polybutadiene	Polyethylene, polypropylene, polystyrene, poly(vinyl chloride), poly(vinyl acetate)	Silk, cellulose, nylon, polyester, acrylics, polypropylene

The second class of polymers is plastics. These are defined as polymers or resins that have been made into shapes, usually under pressure. Shaping may be done by processes such as molding, casting, extrusion, calendering, laminating, foaming, blowing, and many others. They are reviewed briefly in the notes to this section. Nonetheless, the term plastic has a specific meaning in this context and should not be used to refer to polymers in general. The moduli, as well as other properties of plastics, vary widely according to their applications. For example, molding, calendering, or extrusion of a thin film or sheet requires a material with a low modulus so that the sheet is flexible. On the other hand, manufacture of a bleach bottle demands a stiff polymer—the stiffer the better so long as it is not brittle—so that the walls can be made thinner and material saved.

The extensibility demanded in elastomers is not needed in plastics, although a degree of extensibility is important so that the work-to-break is high enough for the plastic to resist twisting and impact. Generally, the properties of plastics are intermediate between those of elastomers and fibers.

Fibers, if they are going to be knitted or woven into dimensionally stable garments, require unrelenting properties. They should have high modulus and low extensibility. The fibers must be strong so that a single thread will not "pull apart," and this requires a high modulus and molecular cohesion, properties closely related to high tensile strength and crystallinity.

It is more difficult to generalize about surface coatings and adhesives than about the other groups. A coating may require high extensibility and low modulus if it is to be applied to a soft rubber surface. On the other hand, a coating for a baked phenolic sheet may require low extensibility and high modulus. The property that both coatings and adhesives require is high adhesion, and the problem is to achieve this while maintaining a reasonable level of the diametrically opposed property of cohesion. In addition, a coating will require resistance to abrasion.

In general, coatings and adhesives tend to have low moduli, somewhere between those of elastomers and plastics. They must have some extensibility, particularly if the material is to be used on a dimensionally unstable substrate such as wood. It follows that coatings and adhesives have low crystallinity.

NOTES AND REFERENCES

The *Encyclopedia of Polymer Science and Engineering* and *The Encyclopedia of Chemical Technology* cited in Section 0.4.1 are major sources of information about polymers. P. T. Flory, *Principles of Polymer Chemistry*, Cornell University Press, Ithaca, NY, 1953, although dated, has hardly been superseded for many of its discussions of basic polymer chemistry. A standard work, F. W. Billmeyer, Jr., *Textbook of Polymer Science*, 2nd ed., Wiley, New York, 1971 is also classical. Sound on polymer science and well referenced, it is nonetheless somewhat difficult to use.

Polymer research is summarized annually in H. J. Canton, Ed., Advances in Polymer Science, Springer, Berlin, 1958–●● Fourteen volumes are now available.

A recent user-friendly book is I. M. Campbell, *Introduction to Synthetic Polymers*, OUP, Oxford, UK, 1994.

Section 15.2 Functionality has been described by Flory, *op. cit.*, p. 31 ff.

Section 15.3.3 Reactivity ratios of monomer pairs is described in detail in *Copolymerization*, G. E. Ham, Ed. High Polymers Series No 18, Wiley, New York, 1964.

Section 15.3.4 An excellent discussion of molecular weight determination is contained in Flory, *op. cit.*, p. 266 ff.

Section 15.3.6 The May 1978 issue of *J. Phys. Chem.* is devoted entirely to articles on radical ions. The issue is in honor of Professor M. Szwarc who did the early work in this field. P. Plesch (private communication) has proved that the first step in cationic polymerization by aluminum chloride is the disproportionation:

$$2AlCl_3 \rightleftharpoons AlCl_2^+ + AlCl_4^-$$

Section 15.3.8 For a discussion of block polymers see D. C. Allport and H. Janes, *Block Copolymers*, Wiley-Interscience, New York, 1973.

Section 15.3.10 A celebration of Ziegler–Natta catalyst science and technology is contained in a book entitled *A Memorial to Karl Ziegler, Coordination Polymerization*, J. C. Chien, Ed., Proceedings of the American Chemical Society Symp., UCLA, Los Angeles, CA, April 1974, Academic, New York, 1975.

Section 15.3.11 Several companies, including Du Pont, Hercules Phillips, Montecatini and Standard Oil of Indiana, filed patents almost simultaneously on polypropylene production. A bitter patent suit of almost 20-years duration resulted and Phillips Petroleum Co. emerged the victor with a composition of matter patent on crystalline polypropylene. The decision was based on the fact that the inventors, P. Hogan and R. Banks, had early on tried a supported chromia–silica–alumina catalyst on propylene in an attempt to prepare C_6 and C_8 oligomers in the gasoline range. They obtained mostly a liquid polymer with about 10% of a solid material. The solid turned out to be identical with Natta's polypropylene. In further work, Hogan and Banks decided that the chromia catalyst was not adequate for polypropylene production and turned to ethylene where results were far better and Phillips "Marlex" catalyst was born. Phillips patent applies only to the United States.

Section 15.3.12 The pioneering work on metallocenes was reported by H. Sinn, W. Kaminsky et al. *Angew. Chem.* **69**, 686 (1957). Because of their commercial significance, few of the modern developments have been published in the scientific literature, but there is an excellent review by K. Sinclair and R. Wilson, Metallocene catalysts: a revolution in olefin polymerization, *Chem. Ind.* 857 (1994). The claims for high catalyst activity appeared in *Chemical Marketing Reporter*, June 13, 1994, p. 1. Dow's plans were discussed in *Chem. Week*, February 2, 1994, p. 10 and January 5/12, 1994, pp. 27–31. The Exxon–Dow competition is reviewed in *Chem. Week* September 15, 1993, p. 5. Hundreds of patents have been filed. Exxon's basic patents are European Patent Applications 0 206 794 A1 (December 30, 1986) and 0 128 045 A1 (December 12, 1984); World Patents 8 702 991 A1 (May 21, 1987) and 9 200 333 A2 (January 9, 1992), and US Patent 5 026 798 (March 17, 1992). The most important constrained geometry catalyst patent is Dow's European Patent Application 0 416 815 A2 (March 13, 1991).

Section 15.4.2 Chlorofluorocarbons used to be the favorite foaming agent for polyurethanes but are being phased out. For other agents, see H. A. Wittcoff and B. G. Reuben, Part 2, Section 2.7, cited in the main bibliography.

Section 15.5.3 The cohesive energy density δ^2 can be calculated from the latent heat of vaporization, ΔH_v cal/mol, or the surface tension, ρ dyn/cm, by the relationships:

$$\delta^2 = \frac{\Delta H_v - RT}{M/D} = \frac{14\rho}{(M/D)^{1/3}}$$

where R is the gas constant (1.987 cal/mol K; 8.314 J/mol K), T is the absolute temperature, M is the molecular weight, and D is the density in g/cm^3. The units of δ^2 are calories per cubic centimeter and those of δ are cal$^{1/2}$ cm$^{-3/2}$. The latter unit is called the hildebrand.

Section 15.6 The following is a brief review of plastics fabrication techniques, adapted with permission from *The Chemical Economy*, cited in the general bibliography.

Compression molding (Fig. 15.13) used for thermosetting polymers is practically the oldest method of fabricating plastics and is still widely used. The polymer, which is either linear or only partially cross-linked, is placed in one half (the "female" half) of a mold and the second or "male" half compresses it to a pressure of about 1 tonne per square inch (1.55×10^6 kg/m^2). The powder is simultaneously heated, which causes the resin to cross-link. Transfer molding is a cross between this and injection molding.

Casting was also used before World War II. In the sheet casting of poly(methyl methacrylate), monomer is partly polymerized and the viscous liquid then poured into a cell made up of sheets of glass separated by a flexible gasket that allows the cell to contract as the casting shrinks.

Injection molding (Fig. 15.14). In injection molding, polymer is softened in a heated volume and then forced under pressure into a cooled mold where it is allowed to harden. Pressure is released, the mold is parted, the molding is expelled, and the cycle is repeated. Injection molding is a versatile technique and can be used for bottle manufacture by a method identical with blow molding except that the initial "bubble" is injected rather than extruded.

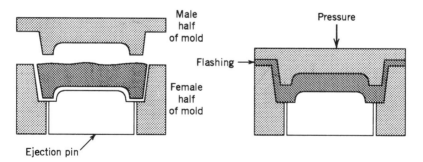

Compression molding

Figure 15.13 Compression molding.

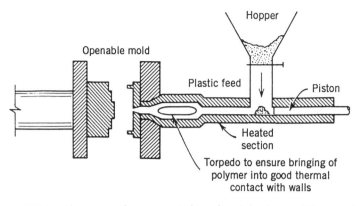

Figure 15.14   Diagrammatic representation of an injection molding machine.

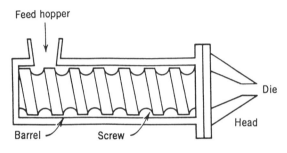

Figure 15.15   Extrusion.

Extrusion (Fig. 15.15) is a method of producing lengths of plastics materials of uniform cross section. The extruder is similar to a domestic mincing machine with the added facility that it can be heated or cooled. The pellets enter the screw section via the hopper, are melted, and then pass through the breaker plate into the die. The molten plastic is forced out of the die with its cross section determined by the shape of the die, but not identical with it because of stresses induced by the extrusion process. Extrusion can be used for coating electrical wiring by means of cross-head die.

Blow molding Blow extrusion (Fig. 15.16), in which the initial lump of polymer is formed by an extrusion process, is the commonest form of blow molding. A short length of plastic tubing is extruded through a crossed die and the end is sealed by the closing of the mold. Compressed air is passed into the tube and the "bubble" blown out to fill the mold. Suitable control of this process can lead to the biaxial orientation of the polymer as in poly(ethylene terephthalate) soft drink bottles.

A variation of the process is used in the manufacture of thin film (Fig. 15.17). A tube of plastic is continuously extruded and expanded by being blown to a large volume, and consequently a small wall thickness. The enormous bubble of plastic is cooled by air jets and taken up onto rollers. This material can either be slit down the side to give thin film or turned into plastic bags by heat sealing the bottom of the tube.

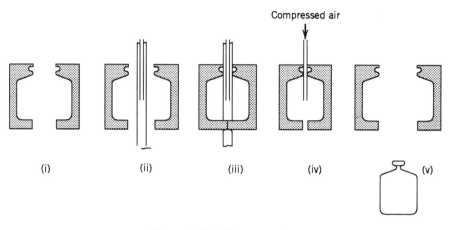

Figure 15.16 Blow extrusion.

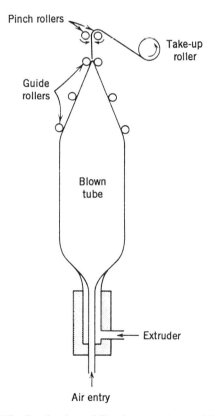

Figure 15.17 Production of film by extrusion and blowing.

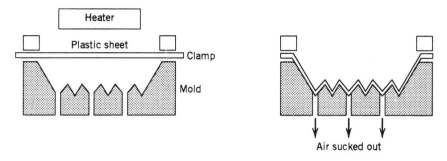

Figure 15.18   Vacuum forming.

Vacuum forming (Fig. 15.18) is the opposite of blow molding. A sheet of heat-softened plastic is placed over a mold and the air sucked out. The plastic is drawn down and conforms to the shape of the mold.

In **calendering** a preheated polymer mix is turned into a continuous sheet by passage between two or more heated rolls that squeeze it to the appropriate thickness. If fabric or paper is fed through the final rolls, the plastic can be pressed on it to give a plastic coated material. If the rolls are patterned, a "sculpted" appearance can be obtained.

Laminating The best known laminates are the Formica-type decorative laminates for household use. Brown paper is impregnated with an alcoholic solution of a resole and is then cut up and arranged in piles with a suitably printed melamine–formaldehyde impregnated decorative sheet on top. The whole is pressed at about 150–180°C and 3 tonnes/in.[2] (4.65×10^6 kg/m^2) to give the finished laminate.

CHAPTER 16

INDUSTRIAL CATALYSIS

Many of the organic chemicals discussed in the previous chapters were already being made before World War II. They were made by fermentation or from coal by "traditional" organic chemistry in batch processes. The advent of cheap olefinic feedstocks derived from oil and natural gas led to a switch to "hot tube" industrial organic chemistry, that is, to continuous processes that produced hundreds of thousands of tons of product per year. The basis for this new organic chemistry is frequently the "appropriate" catalyst, whose exact nature and formulation is often a closely guarded industrial secret.

In 1966, when catalyst theory was in its infancy, it was estimated that 70% of industrial processes involved catalysis. The percentage is higher now. According to the Pimental report (see note) 20% of the US gross national product is generated with the help of catalysis. This chapter is therefore devoted to a brief review of catalyst technology with emphasis on ways in which the concept of a catalyst has changed and on recent trends in the industry. First, we shall discuss the questions of catalyst choice and then some of the chemical engineering aspects of catalsyt use. Next, comes the markets for catalysts and finally a discussion of the different types of catalysts encountered in industry. Table 16.1 shows a summary of catalytic reactions.

16.1 CATALYST CHOICE

Unlike feedstocks and raw materials, the overriding factor for industrial catalysts is not their cost. As long as catalysts can be reused, cost is not crucial. The outstanding feature of modern industrial catalysts is their selectivity, that is the extent to which they make a reaction go in a desired direction. Next there is the access they provide to unusual "nonequilibrium" products. Then there is the

TABLE 16.1   A Summary of Catalytic Reactions[a]

Class	Functions	Examples
Transition metals	Hydrogenation Dehydrogenation Hydrogenolysis (oxidation)	Fe, Ni, Pd, Pt, Ag
Semiconductors and transition metal oxides	Oxidation Dehydrogenation Desulfurization (hydrogenation)	NiO, ZnO, MnO_2, Cr_2O_3, Bi_2O_3/MoO_3, WS_2, SnO_2, Fe_2O_3
Insulator oxides (main group oxides)	Dehydration (hydration)	Al_2O_3, SiO_2, MgO
Acids and acidic mixed oxides	Polymerization Isomerization Cracking Alkylation (hydrolysis) Esterification	H_3PO_4, H_2SO_4 BF_3, SiO_2/Al_2O_3, zeolites V_2O_5/Al_2O_3
Bases	Polymerization (esterification)	Na/NH_3
Transition metal complexes	Hydroformylation Polymerization Oxidation Metathesis	$[Co_2(CO)_8]$, $RhCl[(C_6H_5)_3P]_3$ $TiCl_4/Al(C_2H_5)_3$ $CuCl_2/PtCl_2$ WO_3, WCl_6
Dual function catalysis	Isomerization plus hydrogenation/ dehydrogenation	Pt on SiO_2/Al_2O_3
Enzymes	Varied	Amylase, urease, proteinases
Reactions between immiscible reactants	Varied	Quaternary ammonium salts

[a] Data partly based on G. C. Bond, *Heterogeneous Catalysis: Principles and Applications*, Clarendon Press, Oxford, 1974 (see notes). Less important functions are given in parenthesis.

conventional role of catalysts in increasing the rate of reaction. Fourth is the problem of recovering catalysts for reuse. In choosing a catalyst, there is also the question of homogeneous versus heterogeneous catalysis. We shall discuss these in turn.

16.1.1   Reaction Velocity and Selectivity

Catalysts are defined as materials that increase reaction velocity without themselves being consumed. That is only part of the story. The effect of a catalyst can be so large that a quantitative difference becomes qualitative. Thus platinum catalyzes the combination of hydrogen and oxygen. In the absence of a catalyst, the reactants can apparently remain for hundreds of years without reaction. It would appear that the platinum did not accelerate the reaction but rather that it brought it about. The same applies to many industrial catalysts. Silver "catalyzes" the oxygen–ethylene reaction to ethylene oxide, but yields of ethylene oxide in its absence are zero.

Selectivity is crucial. A material that does not react at a measurable rate in the absence of a catalyst may react to give quite different products in the presence of different catalysts and at different temperatures and pressures. Ethanol passed over copper can give either acetaldehyde or ethyl acetate depending on conditions. Passage over alumina gives ethylene or diethyl ether.

$$C_2H_5OH \xrightarrow{\text{Cu}} CH_3CHO + H_2$$

$$2C_2H_5OH \xrightarrow{\text{Cu}} CH_3COOC_2H_5 + 2H_2$$

$$C_2H_5OH \xrightarrow{\text{Al}_2O_3} CH_2{=}CH_2 + H_2O$$

$$2C_2H_5OH \xrightarrow{\text{Al}_2O_3} C_2H_5OC_2H_5 + H_2O$$

Thus catalysis often provides a reaction route that proceeds in parallel with other existing thermal or catalytic routes. Useful catalysts cause the desired reaction to be favored overwhelmingly. Increasing the selectivity of catalysts is an important activity of development chemists. A high selectivity not only brings the financial reward of increased yield, it also reduces the problems of byproduct or effluent disposal.

A catalyst often permits a reaction to take place at measurable speed under milder conditions than would otherwise be possible. That, in turn, may lead to a different equilibrium and a different major product. In the production of ethylene oxide (Section 3.7), no reaction would take place at 275°C in the absence of a catalyst. At a temperature at which a reaction *would* take place, only carbon dioxide and water would be formed. The silver catalyst enables the temperature to be kept sufficiently low for the ethylene oxide-forming reaction to proceed at a rate higher than the overall combustion reaction. It also means that the sequential oxidation of ethylene oxide proceeds slowly enough for the ethylene oxide to be isolated.

This is illustrated in a semiquantitative way in Figure 16.1. Ethylene and oxygen exist in a potential energy well, as shown, and there are other potential energy wells corresponding to ethylene oxide, acetaldehyde, peracetic acid, and

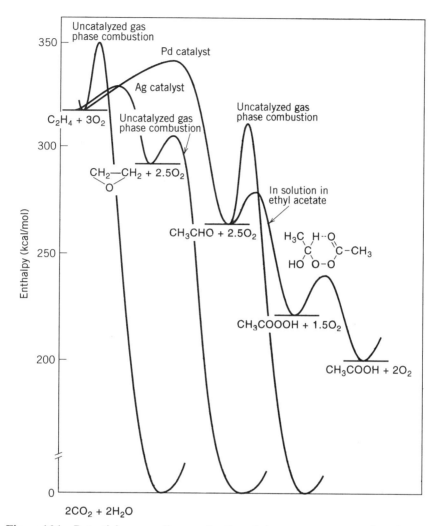

Figure 16.1 Potential energy diagram for the ethylene–oxygen system (based on enthalpy of $2H_2O + 2CO_2$ as zero).

carbon dioxide–water as indicated. The maxima are the activation energies required for the various possible catalyzed and uncatalyzed reactions.

In the absence of a catalyst, ethylene and oxygen combust directly to carbon dioxide and water. In the presence of a silver catalyst, however, the initial activation barrier is reduced and the ethylene oxide minimum becomes accessible. In the presence of a palladium catalyst, a different potential energy surface becomes accessible with acetaldehyde and peracetic acid as local minima.

The commercial development of the ethylene oxide process has concentrated on these points. The silver catalyst consists of about 15% silver as a finely divided layer on a support that acts as a heat sink to prevent undue temperature rise of the catalyst by the reaction exotherm. Furthermore, an inhibitor—several parts per million of dichloroethane—is often added to the system. Inhibitors are "anticatalysts." This one decomposes to ethylene and chlorine. The chlorine adsorbs on the silver surface, influences the chemisorption of oxygen, and hence inhibits the combustion reactions.

Other measures to increase selectivity have included the promotion of the catalyst with 3.5×10^{-4} to 3×10^{-3} equivalents of potassium, rubidium, or cesium ions per kilogram of catalyst (measured as silver oxide). The effect is specific—sodium and lithium have no effect, and quantities of the active metals outside the above range do not seem to improve selectivity.

In Figure 16.1, the effect of the catalyst is always to reduce the activation enthalpy (energy) of a reaction. If the rate constant k of a reaction is represented by the Arrhenius equation $Ae^{-E/RT}$, where A is the preexponential factor, E is the activation enthalpy (energy), R is the gas constant, and T is the absolute temperature, then the implication is that E is always reduced and A remains more or less unchanged. Indeed many textbooks state that catalysts always act in this way.

In fact, catalysts act to reduce the *free energy of activation*, so that A varies as well as E. In some examples, a catalyst raises E but also raises A by such a large amount that it still acts as a catalyst. For different catalysts of the same reaction, it often happens that a graph of $\log A$ against E gives a straight line. This phenomenon is called the compensation effect and variations in A of up to 10^{14} have been reported (see note).

16.1.2 Recovery of Unchanged Catalyst

If catalyst cost is to be unimportant, catalysts must be cheap or have exceptionally high activity or be recoverable in high yield. Many catalysts, especially in homogeneous reactions, cannot be recovered unchanged at the end of a reaction.

When sulfuric acid catalyzes the esterification of ethanol with acetic acid, the reaction products contain materials such as ethyl hydrogen sulfate. The sulfuric acid has also been solvated by the byproduct water. This dilution reaction is highly exothermic. Because the catalyst alters the overall chemistry of the reaction, it alters the position of equilibrium and the free energy change, and fails to increase the backward and forward rates proportionately. The sulfuric acid is not present in its original form at the end of the reaction. Its recovery from the reaction products requires a further process involving expenditure of chemical energy and reversal of the hydration and sulfate-forming reactions. In practice, sulfuric acid is so cheap that, after use as an esterification catalyst, it would probably be discarded.

An example of a high-activity catalyst is the fourth generation Ziegler–Natta catalyst for polypropylene, which produces 25,000 lb of product for only 1 lb of catalyst. Although the catalyst is expensive, its activity is so high that only a tiny quantity is required, and its recovery is not worthwhile.

Expensive catalysts include the precious metals used in automobile emission control catalysts. They contribute about 75% of the cost of such catalysts. The North American consumption of platinum in this end-use in 1994 was 7.16 tonnes, of palladium 2.49 tonnes and of rhodium 1.18 tonnes. At an average price of $12.5 per g for precious metals, this amounts to about $135 million, and the recovery and recycling are worthwhile.

16.1.3 Catalyst Deactivation

Homogeneous catalysts are obviously more difficult to recover than heterogeneous catalysts which, in principle, can simply be removed by filtration. With heterogeneous catalysts, however, the difficulty is not so much in the "recovery" part of the definition but in the "unchanged" part. Three routes to deactivation of heterogeneous catalysts are recognized: sintering, poisoning, and fouling. Sintering covers changes in surface area and structure and is irreversible. Poisoning involves chemisorption on active sites of the catalyst and is more likely to be reversible. Fouling involves physical blockage of active sites and is usually reversible.

All these processes may result from chemical reactions. Practically all industrial solid catalysts must be removed from the reactor at intervals, varying from a few months to a year, for regeneration. In the catalytic cracking of hydrocarbon fractions to gasoline (Section 2.2.2), a second reactor is included in the plant for continuous regeneration of catalyst, and it is only this that makes the process feasible. This is also the basis for the Du Pont transport bed process for maleic anhydride production (Section 5.4).

In the manufacture of substitute natural gas by the Catalytic Rich Gas Process (Section 12.5), the catalyst is nickel on high surface area γ-alumina promoted with potassium. The alumina has a cubic close-packed structure. When not used as a catalyst, γ-alumina is stable up to 1100°C. In plant use, however, at a temperature as low as 400°C, it undergoes an irreversible phase change to a hexagonal close-packed structure, α-alumina or corundum. The collapse of the fine pore structure of γ-alumina and subsequent formation of corundum also leads to agglomeration of the nickel crystallites and loss of metal surface area. Hence, there is a drastic reduction in catalyst activity. The lifetime of the catalyst, which cannot be recovered unchanged, is governed by the rate of sintering of alumina and nickel.

A final example is the gas-phase coking of hydrocarbons at high temperatures. It is catalyzed by surfaces, especially those containing metal particles. Nickel is especially active. Electron micrographs show that the carbon deposits as spiral fibers growing from the surface and topped by a metal crystallite, which promotes further growth. The etching of the surface is thus an essential part of the catalytic process. As in the previous examples, the textbook requirement that a catalyst be recoverable "unchanged" at the end of a reaction is not strictly met.

16.1.4 Access to Nonequilibrium Products

Thermodynamicists stress the importance of equilibrium in chemical reactions. If hydrocarbons were to react to equilibrium, however, the products would be carbon and hydrogen. If sufficient oxygen were added, furthermore, the equilibrium products would be carbon dioxide and water. Most useful chemical reactions do not go fully to equilibrium. They are stopped at some local minimum in the free energy surface as in the case of ethylene oxide discussed in Section 16.1.1. Selective catalysts direct reacting systems towards these local minima.

Catalysts are said not to affect the position of equilibrium in chemical reactions but may do so in a number of ways. In the esterification reaction described in Section 16.1.2, the amount of sulfuric acid added pushes the equilibrium to the right because of its effect on the overall free energy change accompanying the reaction.

A more modern example of a catalytic system where the catalyst appears to affect the position of equilibrium is the disproportionation of toluene (Section 8.1). The zeolite catalyst makes possible much greater than thermodynamic yields of p-xylene. The desired isomer, the p-xylene, is able to diffuse away from the catalytic site much more quickly than the undesired isomers. Thus, as p-xylene is removed, a fresh equilibrium is continually being established. This is described further in Section 16.9.

16.2 HOMOGENEOUS AND HETEROGENEOUS CATALYSIS

Catalysts are designated homogeneous or heterogeneous depending on whether they function in a single phase or at a phase boundary. This distinction appears to have little theoretical significance. A homogeneous acid catalyst such as sulfuric acid acts in the same way as a heterogeneous acidic catalyst such as silica–alumina. In certain examples of the Friedel–Crafts reaction, it is uncertain whether catalysis is homogeneous or heterogeneous. In some industrial processes, there is the possibility of operating in either mode.

Homogeneous catalysts can be added easily to a reaction system but may be difficult to remove from the products. They are usually used in the liquid phase, although some reactants may be introduced as gases or solids. There is only a handful of homogeneous gas-phase catalyzed reactions. One example is the air oxidation of sulfur dioxide to sulfur trioxide catalyzed by nitric oxide—the old lead chamber process for sulfuric acid. Another is the cracking of acetic acid to ketene at about 700°C, catalyzed by diethyl phosphate (Section 10.5.2.3).

For most but not all homogeneous catalysts, the rate increase is proportional to the amount of catalyst added. Some catalytic effects saturate, and a few appear in the rate equation with a power other than one. Homogeneous catalysts are readily reproducible and give high selectivities. For this reason they are the preferred systems for academic study and have led to most of the insights into the mechanisms of catalytic reactions.

In heterogeneous catalysis, the mass of the catalyst is much less important than its surface area. Preparation and pretreatment of solid catalysts is an art and the activity of a catalyst depends on its previous history. Catalysts achieve high-surface areas because of their microporosity. Areas of 1000 m^2/g are not uncommon. Diffusion of reactants to a surface and of products away from it may present problems, as may heat transfer into or out of catalyst particles. On the other hand, heterogeneous catalysts have the great advantage of being easily recoverable. They can be used in static or fluidized beds in continuous processes and in systems where thermodynamic constraints demand high temperatures, and where solution methods would therefore be impossible. They are therefore more widely used in industry.

16.2.1 Reactors for Heterogeneous Catalysts

The mixing of homogeneous catalysts with reactants presents few problems. In polymerization reactions there may be difficulties if the mixture becomes viscous, but in general stirring is adequate. It is sometimes possible to stir a heterogeneous catalyst into a liquid reaction mixture in the same way, but gas-phase reactions occurring on solid catalysts are usually carried out either in fixed or fluidized beds.

In fixed beds (Fig. 16.2a), the catalyst is compressed into pellets of approximately 1 cm diameter, and these are packed into long tubes, which are contained in a shell through which a heating or cooling medium can be circulated. The

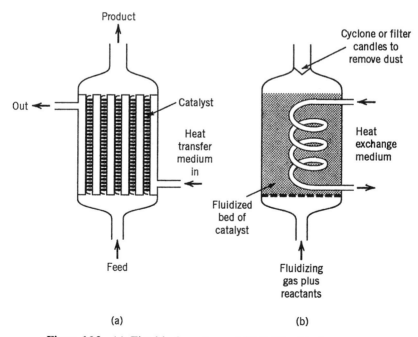

(a)   (b)

Figure 16.2 (a) Fixed-bed reactor and (b) fluidized-bed reactor.

advantage of the system is its reliability. It works for varying flows of gases and for catalysts that are "sticky" or liable to agglomerate. There are three main drawbacks. First, the tubes are difficult to remove and repack when the catalyst needs to be regenerated. Second, there are problems with diffusion of reactants into and products out of the compressed pellets. Third, there are problems with heat transfer. Even if tubes as narrow as 3 cm in diameter are used in the reactor, it is still difficult to maintain a constant temperature. Active sites in the middle of catalyst pellets may achieve much higher temperatures than the rest of the system. Thermal shock may even fragment the pellets leaving dust, which may block the tube. In endothermic reactions, local cooling may bring the reaction to a halt.

Fluidized beds (Fig. 16.2b) contain catalyst in small particles like grains of sand. The particles are suspended in a rapid flow of gas, which may be the reactant or an inert fluidizing gas. If the flow is accurately maintained, the bed behaves like a homogeneous fluid. It can be drawn off into other vessels for regeneration (Section 2.2.2). Heat can be added or removed by coils in the bed. Temperature equilibration is rapid. Problems of diffusion are much reduced because the catalyst particles are smaller. One drawback of the system is its intolerance to variations in gas flow. Fluidization of the bed may also be hindered by attrition, channelling, and slugging. Attrition means the wearing away of the catalyst particles as they rub against each other in the bed. A fine dust is continuously carried out of the bed and must be removed by cyclones at its exit. Channelling occurs when the gases manage to find a channel through the bed, perhaps near the wall, and travel through without adequate contact with the catalyst. Slugging is a similar phenomenon, where the catalyst particles, agglomerating into large lumps, do not fluidize.

Fluidized beds are the more advanced technology and are used whenever the nature of the catalyst and the reaction conditions permit it. They are especially valuable when accurate control of temperature is important as in highly exo- or endothermic reactions.

16.2.2   Immobilization of Homogeneous Catalysts

The ease of separating heterogeneous catalysts from reaction mixtures means that there is sometimes a need to turn a homogeneous reaction into a heterogeneous one. An early example is the proton-catalyzed stoichiometric hydration of ethylene (Section 3.9). This was originally carried out indirectly by passage of ethylene into sulfuric acid at 55–80°C and 10–35 bar, followed by hydrolysis of the ethyl hydrogen sulfate and regeneration of the sulfuric acid. In the newer catalytic process, ethylene is passed over phosphoric acid absorbed into celite—a porous diatomaceous earth that serves as a support—and this provides a heterogeneous catalyst. The lower effectiveness of phosphoric acid as a protonating agent means that more drastic conditions are required—300°C and 70 bar—but the phosphoric acid remains physically absorbed onto the porous interior of the celite and there is still pore space to allow access of the ethylene.

An example of the anchoring of a more sophisticated catalyst to a solid surface is the binding of the so-called Wilkinson complex (Section 16.7.2) to a polystyrene support. The polystyrene is first converted (functionalized) with a diphenylphosphino group to a polymer that may be represented as $[PS]-P(Ph)_2$. This then reacts with the Wilkinson complex:

$$[PS]-P(Ph)_2 + RhCl[P(Ph)_3]_3 \rightarrow [PS]-P(Ph)_2RhCl[P(Ph)_3]_2 + P(Ph)_3$$

The immobilized complex preserves the ability of the Wilkinson complex to hydrogenate olefins, although the reaction may be slower.

Polystyrene is not the only support for the anchoring of catalysts. Inorganic materials are also useful. Silica is an example, and the catalyst combines with the hydroxyl groups on its surface.

Immobilization of man-made catalysts generally leads to deactivation, hence there are only a few useful examples of it. Highly active "natural" catalysts, such as enzymes, and complete living cells, such as bacteria, on the other hand, can be attached to solid substrates without serious loss of catalytic activity. They can usefully be employed as heterogeneous catalysts. Immobilized enzymes and methods of immobilization will be discussed further in Section 16.8.

16.3 CATALYST MARKETS

The world market for catalysts in 1990 was about $6 billion and had risen to just under $8 billion by 1993. The US share of the 1990 market was $2.11 billion (Fig. 16.3a). The US market divided almost evenly between petroleum refining catalysts (35%), chemical processing (31%) and environmental applications (automobile emissions 30.2% together with industrial pollution 3.8%). This is shown in Fig. 16.3b. The similar sales hide large differences in tonnage.

The total US market for catalysts for chemical processing in 1990 amounted to $660 million by value and about 200 million lb by weight, an average of $3.30 /lb. Catalysts are expensive and are truly specialty chemicals. They are typically more expensive than the bulk pharmaceutical aspirin ($\sim$ $1/lb) and about the same price as vitamin C ($\sim$ $3 4 /lb). Tonnages are also of the same order of magnitude as for pharmaceuticals.

Furthermore, of the 650 billion lb of chemicals produced annually in the United States, about 60% are produced with the aid of catalysts. Thus 1 lb of an average catalyst produces about 1500 lb of chemicals (allowing for the catalyst volume being somewhat out of date). The polypropylene example in Section 16.1.2, where 1 lb gives 25,000 lb, is exceptional for a catalyst although typical for olefin polymerization.

Petroleum refining catalysts (Fig. 16.3c) on average are the cheapest and cost about one-half as much per pound as chemicals catalysts. They are dominated by the cheap acid catalysts used for alkylation, which account for 90% by weight but only about 32% by value in this sector. The new solid acid catalysts are

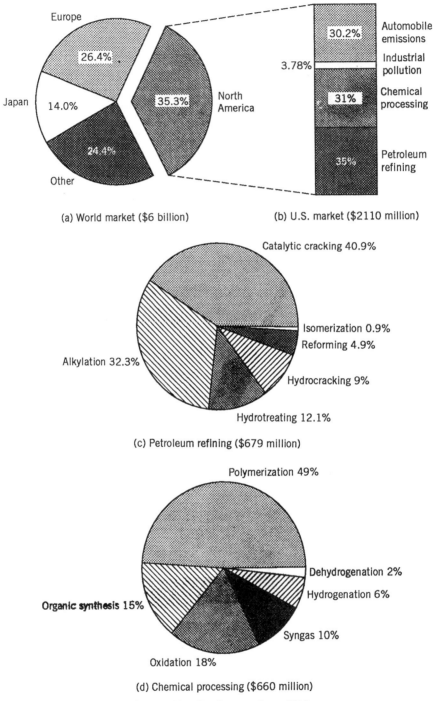

(a) World market ($6 billion)

(b) U.S. market ($2110 million)

(c) Petroleum refining ($679 million)

(d) Chemical processing ($660 million)

Figure 16.3   Catalyst markets, 1990.

more expensive. The high activity zeolite catalysts used for catalytic cracking are also more expensive and account for about one-tenth of the tonnage but about 40% of sales value. The other petroleum refining catalysts—for hydrotreating, hydrocracking, reforming and isomerization—are of less significance. In the chemicals sector (Fig. 16.3d), the largest group is described as polymerization catalysts. Materials that promote polymerization may be divided into true catalysts such as metal complexes, metal oxides, and anionic and cationic catalysts (Chapter 15), and initiators, which appear as end groups in the final polymer. Although the formal equations depicting the polymerization reactions are similar, compounds that appear in the final product differ fundamentally from those that are regenerated and can in principle be separated and reused. Nonetheless, this distinction is not always recognized in the commercial world, and the figure given above is inflated by the inclusion of free radical initiators. The most expensive polymerization catalysts are the Ziegler catalysts used for polypropylene and polyethylene, and the dibutyl tin and triethylene diamine (DABCO or diazabicyclooctane) (Section 3.11.5) for polyurethanes.

The second group in the chemicals list is organic synthesis. It includes a variety of catalysts for esterification, hydrolysis, alkylation (cumene and ethylbenzene dominate) and halogenation.

Oxidation is third in importance and about one-half of this market by value is the silver catalyst used to make ethylene oxide. In terms of tonnage, the relatively cheap catalysts for the oxychlorination of ethylene (Section 3.4) make up about one-third by weight but only about 5% by value. Another expensive oxidation catalyst is the manganese and cobalt salts in acetic acid-plus-bromine promoter for oxidation of p-toluic acid to terephthalic acid (Section 9.3.1). Palladium on charcoal is also used in this process to catalyze hydrogenolysis of byproduct p-carboxybenzaldehyde to p-toluic acid (Section 9.3.1). Added together, these produce a process that is expensive in terms of catalyst. Further cost is added by the decomposition of the acetic acid, only part of which can be recovered for reuse.

The fourth category in Figure 16.3d is the iron-based catalysts for ammonia and chromium-based catalysts for methanol from synthesis gas. The cost is relatively low but the tonnage for ammonia alone is about one-fifth of total catalysts for chemicals.

The main hydrogenation catalysts are

- Raney nickel for margarine and related processes.
- Nickel and to a lesser extent palladium or platinum on lithium oxide for hydrogenation of benzene to cyclohexane.
- Silver gauze for dehydrogenation or oxidative dehydrogenation of methanol to formaldehyde.
- Cobalt and rhodium catalysts for the oxo process.

The main dehydrogenation process is the conversion of ethylbenzene to styrene.

Catalysts for automobile emission control are based on platinum, palladium, and rhodium and account for about 33% of world and 58% of North American demand for these metals (Section 16.1.2). In 1994, for emission catalysts worldwide, 17% of platinum, 13% of palladium, and 10% of rhodium was recycled material. North American figures were 29, 16, and 28% respectively. Platinum is used for catalytic reforming and rhodium for the oxo reaction (Sections 2.2.3 and 4.8) but recovery in the chemical industry is close to 100%. World consumption of platinum for chemical uses in 1994 was only 5.9 tonnes (North America 2.0 tonnes) and of rhodium 312 kg.

Thus the automobile market is dominated by precious metals, but the petroleum and chemicals markets are dominated in terms of tonnage by simple acidic or metal oxide catalysts. Nonetheless, there is a range of relatively modern high technology catalysts, which repay their higher prices. They are often sold together with "know-how" and are steadily improved. Examples are the zeolites, and the catalysts for ethylene oxide production and Ziegler polymerization.

Environmental applications are seen as the growth market for catalysts, while petroleum refining catalysts are expected to be static. The 1993 world market was estimated at $3 billion environmental, $3.1 billion chemical processing, and $1.8 billion petroleum refining. Corresponding 1998 figures are predicted as $5.1 billion, $3.6 billion, and $2 billion, implying growth rates of 11, 3, and 2%, respectively.

16.4 CATALYSIS BY ACIDS AND BASES

Acid catalysis is the most widely used form of catalysis in the chemical and refining industries. Alkylation was described in Section 2.2.5 and is brought about by cheap acidic catalysts. Sulfuric acid was used in the past but hydrogen fluoride has replaced it in most modern refineries. Solid catalysts, more acceptable ecologically, are on the horizon. The mechanism is a typical acid catalysis in which the catalyst donates a proton to the reactant to give an ion, in this case a carbenium ion, that reacts as shown below for the alkylation of isobutane by butene:

$$(CH_3)_2C{=}CH_2 \ + \ HF \ \longrightarrow \ (CH_3)_2\overset{+}{C}{-}CH_3 \ + \ F^-$$

$$(CH_3)_2\overset{+}{C}{-}CH_3 \ + \ (CH_3)_2CHCH_3 \ \longrightarrow \ (CH_3)_3C{-}CH_2{-}CH(CH_3)_2 \ + \ H^+$$

Sulfuric acid is also the usual catalyst in esterification reactions, which proceed by a similar mechanism:

$$\underset{\text{Reagent acid}}{R{-}\overset{\overset{\displaystyle O}{\|}}{C}{-}OH} \ + \ H_3O^+ \ \xrightarrow{\text{fast}} \ R{-}\overset{\overset{\displaystyle OH}{\|}}{C}{=}\overset{+}{O}H \ + \ H_2O$$

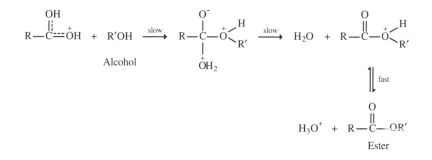

Brønsted and Lewis acids also act as catalysts by proton donation. Thus HCl/AlCl₃ catalyzes the Friedel–Crafts reaction between benzene and propylene (Section 4.5) by the following mechanism, which has been confirmed by isotopic labeling.

$$AlCl_3 + HCl \rightleftharpoons H^+[AlCl_4]^-$$

$$CH_3CH{=}CH_2 + H^+ \rightleftharpoons CH_3\overset{+}{C}H{-}CH_3$$

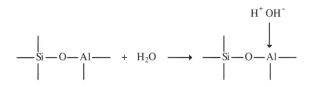

Catalytic cracking (Section 2.2.2) resembles alkylation and provides another example of acid catalysis. It differs in that a solid surface provides the acidity, and it acts either by accepting an electron pair from another species (Lewis acid) or donating a proton (Brønsted acid). The original catalytic cracking catalyst was silica–alumina, developed by Houdry in the 1930s. The modern catalyst is the rare earth form of a zeolite structure and it largely suppresses the reactions leading to the excessively volatile C_3–C_4 products.

Zeolites are aluminosilicates and are the basis for shape-selective catalysis (Section 16.9). Their structure is dominated by the fourfold coordination of silicon, each atom being bonded to four tetrahedrally arranged oxygen atoms. Substitution of a silicon atom in such a structure by aluminum, with its valency of three, leaves a vacant tetrahedral position in the lattice at which a pair of electrons can readily be accepted to complete the valency octet. Thus the species can function as a Lewis acid. Alternatively, interaction with water can lead to a Brønsted acid site with an available proton:

The cracking of a larger to a smaller alkane on such an acid site may be represented as in Figure 16.4.

Catalysts for catalytic cracking are the most valuable market in the field of petroleum refining. They have to be tailored to meet changing demands for the balance and specification of products. In the past, gasoline was required to meet only a few specifications; now it must meet strict composition standards. In the past, catalytic cracking produced 30–35% of the gasoline pool. Now, it is seen as a source of feedstocks for other applications such as increased yields of C_4 and C_5 olefins for the production of octane-building oxygenates and alkylates.

Basic catalysis is rarer in industry than acidic catalysis. Examples include one-stage phenol–formaldehyde resins (Section 15.4.1) and isocyanate formation. The opening of an epoxide ring frequently depends on a basic catalyst. An example is the reaction of ethylene oxide with acrylic acid to give hydroxyethyl

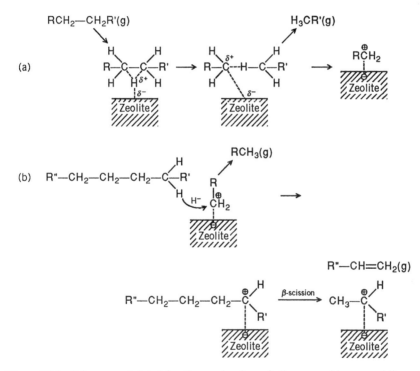

Figure 16.4 Scheme postulated for the mechanism of alkane cracking on acidic zeolite catalysts. (a) Protolytic route to formation of chemisorbed carbenium ion *via* cracking of larger alkane (RCH_2—CH_2R'). (b) Route for replacement of carbenium ion *via* hydride transfer from a gas phase alkane (R''—CH_2—CH_2—CH_2—CH_2—R'), showing the β-scission process leading to the release of an alkene (R''—CH=CH_2) to the gas phase. After a scheme given by A. Corma, J. Planelles, J. Sáuchez–Marîn and F. Tomás (1985) *Journal of Catalysis*, **93**, 30–37. Reproduced with permission from Catalysis at Surfaces, I. M. Campbell, Chapman and Hall, 1988.

acrylate, a trifunctional monomer used in baking enamels. The catalyst is a tertiary amine or quaternary ammonium salt.

$$H_2C = CHCOOH + H_2C - CH_2 \longrightarrow H_2C = COOCH_2CH_2OH$$

Acrylic acid   Hydroxyethyl acrylate

Whether a catalyst is acidic or basic can alter the reaction products. Alkylation of toluene with propylene in the presence of a Friedel–Crafts (acidic) catalyst leads to ring alkylation to give *o*- and *p*-isopropyltoluene. A potassium catalyst, however, promotes alkylation of the methyl group to give isobutylbenzene, an intermediate for the manufacture of the nonsteroidal antiinflammatory agent, ibuprofen.

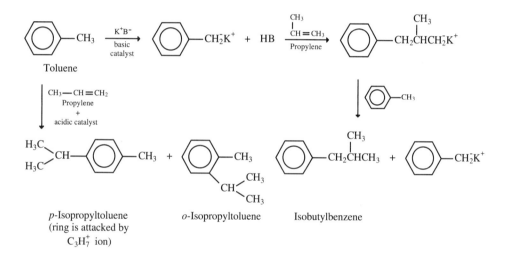

p-Isopropyltoluene
(ring is attacked by
$C_3H_7^+$ ion)

o-Isopropyltoluene   Isobutylbenzene

Similarly, methanol and carbon monoxide give acetic acid in the presence of an acid catalyst and methyl formate in the presence of a base.

$$CH_3OH + CO \xrightarrow{\text{acid}} CH_3COOH$$

$$\xrightarrow{\text{base}} CH_3 - O - \overset{\overset{\displaystyle O}{\|}}{C} - H$$

16.5   DUAL FUNCTION CATALYSIS

Catalytic reforming (platforming, Section 2.2.3) is brought about by a dual function catalyst, a mixture of two catalysts each of which performs differently. Reforming catalysts are based on mixtures of an acidic isomerization catalyst (SiO_2/Al_2O_3) with a hydrogenation–dehydrogenation catalyst (Pt). The modern form is a platinum–rhenium alloy supported on alumina with sufficient sulfur added to poison the metal surface partially. The combined action of rhenium and sulfur inhibits coking.

Typical feedstocks are cyclohexane and methylcyclopentane. A dehydrogenation catalyst can convert hexane to cyclohexane to cyclohexene and then to benzene, and an acidic catalyst can convert methylcyclopentene to cyclohexene. Only a dual function catalyst can convert methylcyclopentane to benzene. For this transformation, the two types of catalytic sites are required with transfer of an olefinic intermediate between the sites. The same effect could not in many cases be achieved by successive beds of the two catalysts. In a reaction of the type A $\rightleftharpoons$ B $\rightarrow$ C, in which the equilibrium of the first reaction lies to the left, two beds of single function catalyst would not bring about reaction because not much of B would be formed in the first bed. On a dual function catalyst, however, the small quantity of B is removed as soon as it is formed. More A consequently changes to B, and the reaction to give C is accomplished. The distances between the different catalytic sites govern the effectiveness of a dual function catalyst. n-Heptane will reform over a mixture of catalysts of particle size 1.0–10 μm but not 100–1000 μm.

Hydrotreating catalysts are cobalt-molybdenum or nickel–molybdenum on alumina or possibly silica. These are typical hydrogenation–dehydrogenation catalysts based on transition metals. Hydrocracking catalysts are palladium or nickel-tungsten on zeolites. Increasingly strict regulations on sulfur content of refinery products mean that there is pressure on catalyst producers to upgrade their products, and there is a question as to whether the more expensive but more efficient noble metal catalysts might replace transition metals.

16.6   CATALYSIS BY METALS, SEMICONDUCTORS, AND INSULATORS

The majority of heterogeneous catalysts are metals and metal oxides. Pure metals are rarely used in industrial processes, except for silver for ethylene oxide, and noble metals for hydrogenation–dehydrogenation and hydrogenolysis.

Metals and metal oxides may alternatively be classified as p- and n-type semiconductors and insulators. At the surfaces of these materials, reactants can adsorb. Physical adsorption is weak ($\Delta H \approx -40$ kJ/mol) and does not lead to catalytic activity. Chemisorption (dissociative adsorption) on the other hand is strong ($\Delta H \approx -400$ kJ/mol) and the adsorbents themselves dissociate and form chemical bonds with the surface.

In the Haber process, for example, the reactants both adsorb and dissociate on iron. Hydrogen dissociates freely even at liquid air temperatures, but nitrogen does not do so until about 450°C, and this is the rate-determining step. Once the nitrogen molecules have dissociated (with adsorbed atoms written cat≡N where cat is the catalyst) the atoms can react readily with neighboring hydrogen atoms to give cat=NH, cat–NH$_2$, and finally cat$\cdots$NH$_3$ from which the NH$_3$ is easily desorbed.

Table 16.2 lists the heats of adsorption of nitrogen on various surfaces. On glass and aluminum only physical adsorption occurs, and these materials do not catalyze ammonia production. Iron, tungsten, and tantalum all give dissociative adsorption, but the preferred catalyst is iron because it has the smallest heat of adsorption, and therefore the products are most easily desorbed.

The theory of catalysis by way of chemisorption provides a reasonable explanation for catalysis by metals. The mode of action of "pure" metal oxides and nonstoichiometric metal oxides is more complicated, and an explanation had to await a quantum mechanical theory of solids and the application of this to heterogeneous catalysts. Together with the crystal and molecular orbital ligand field thoeries, these now provide substantial theoretical underpinning for the catalytic effects of semiconductive metal oxides.

The large majority of catalytically active metals belong to the transition series. Consequently, it appears that catalytic activity is related to the state of the d bands, corresponding to the assembly of d orbitals at the catalyst surface. The crystal surface of a semiconductor may be thought of as having a supply of electrons and a supply of "holes" where electrons can locate themselves. These either donate electrons to adsorbed molecules or draw them out. Thus they participate in reactions as free valences, so that the addition of a heterogeneous catalyst to a reactant system is in some ways like the addition of free radicals. In general, oxidation reactions are catalyzed by p-type semiconductors that have surplus "holes," whereas hydrogenations are brought about by n-type semiconductors that have excess electrons. This fits in with a definition of oxidations as reactions in which electrons are lost and reductions as reactions in which they are gained. Insulators are effective for dehydration.

Another way of looking at this is to think of catalysts as weakening chemical bonds either by feeding electrons into antibonding orbitals on adsorbed molecules or by withdrawing them from bonding orbitals.

The theory of heterogeneous catalysis by semiconductors is complicated and is still developing. Fundamental knowledge about the mechanism of heterogeneous catalysis may make it possible one day to tailor-make a catalyst that functions as efficiently and selectively as the semiconductor devices used in

TABLE 16.2 Heats of Adsorption of Nitrogen on Various Surfaces

Surface	Approximate ΔH (kJ/mol)
Glass	-7
Aluminum	-42
Iron	-293
Tungsten	-397
Tantalum	-585

modern electronics. It may also be possible to minimize problems of catalyst poisoning either by modification of catalyst structure or by the admixture of antidotes to the feedstock or intermittently to the catalyst.

16.6.1 Catalysts for Automobile Emission Control

The most widespread application of metal catalysts is in automobile emission control. The exhaust gases from an internal combustion engine running on unleaded gasoline contain nitrogen, water, oxygen, and carbon dioxide, all of which are harmless, if one discounts the possible contribution of carbon dioxide to the greenhouse effect. They also contain unburned and cracked hydrocarbons, carbon monoxide, and oxides of nitrogen NO and NO_2 (known together as NO_x). These are the major pollutants.

A catalytic converter must oxidize the hydrocarbon and the carbon monoxide while reducing the NO_x to nitrogen. It must perform at a low temperature, because unburned hydrocarbons are worst during start-up, but must also operate as high as 600–700°C. Contact times are likely to be between 100 and 400 ms, so efficient catalysis is essential. To ensure adequate contact between gases and catalyst, the catalyst is supported on a monolith with narrow channels.

The demands on the catalyst are so severe that the base metals proved inadequate and noble metals on alumina supports must be used. At one point, it was thought that a two-stage conversion would be necessary. First, the engine would run with excess fuel to permit easy reduction of NO_x in an oxygen-poor environment. Air would then be added to the exhaust gases, and the hydrogen and carbon monoxide would be oxidized in an oxygen-rich environment. The addition of rhodium to the platinum catalyst made this unnecessary. If the engine is controlled to operate at the stoichiometric air/fuel ratio of 14.6:1, the oxidation and reduction reactions proceed simultaneously. Addition of palladium and cerium to the catalysts can also provide benefits.

16.7 COORDINATION CATALYSIS

Coordination catalysis uses transition metals or their compounds bound to ligands. As noted in Section 16.6, transition metals have long been used as catalysts. Their d-orbitals can activate organic molecules to undergo otherwise inaccessible reactions. The addition of ligands, moreover, produces a highly organized environment in the coordination sphere. Such an environment permits stereochemical control of reactions leading to the synthesis of pure optical isomers or *cis* or *trans* products.

Coordination catalysis has provided many of the dramatic synthetic advances since World War II. The first industrial example was the oxo reaction in which an α-olefin was treated with CO and H_2 at 150°C and 200 bar in the presence of a cobalt catalyst to give the linear and branched-chain aldehydes with one more carbon atom. The mechanism goes via the hydrocarbon soluble intermediate $Co_2(CO)_8$, dicobalt octacarbonyl, and is shown in Section 4.8.

Carbon monoxide insertion or alkyl transfer to carbon monoxide occurs widely in organometallic chemistry (Section 10.5.2.2). In the simple case, the usually less desirable branched-chain product predominates because of the greater stability of the secondary organometallic intermediate. The use of tri-phenylphosphine rhodium hydrocarbonyl with its bulky ligands permits milder conditions and a preponderance of linear product (Section 4.8).

The Wacker process (Section 3.5) proceeds by way of a coordination complex as does the carbonylation of methanol to acetic acid (Section 10.5.2.2), which is believed to involve a methyl iodide intermediate and a rhodium–iodine–carbon monoxide complex:

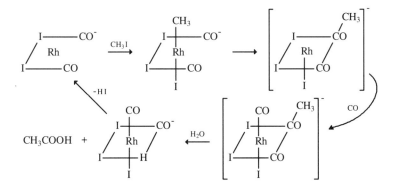

Similar intermediates are probably involved in the Halcon–Eastman route to acetic anhydride.

The Dupont synthesis of adiponitrile from butadiene and hydrocyanic acid (Section 5.1.2) involves a coordination catalyst with a ligand to give anti-Markovnikov addition.

Coordination catalysis is usually performed homogeneously in solution but sometimes there is a choice. In the metathesis of olefins, for example (Section 2.2.9), WO_3 will act as a heterogeneous catalyst, whereas WO_3 plus $C_2H_5AlCl_2$ in ethanol will work in solution. Similarly, in the Wacker process for vinyl acetate (Section 3.6), the mechanism was worked out for the homogeneous liquid-phase reaction, but the system proved too corrosive, and the workable industrial process finally involved a gas-phase reaction with a heterogeneous catalyst.

16.7.1 Catalysis to Stereoregular Products

Coordination catalysis can lead to stereoregular and regioselective products. An example is Ziegler–Natta catalysis, which gives stereoregular polymers (Section

4.1). A further example is *trans*-1,4-hexadiene, which is used as the diene in ethylene–propylene–diene monomer elastomers (Section 5.1.3.6). *trans*-1,4-Hexadiene has been made by Du Pont since 1963 and has the distinction of being the first chemical to be made by a rhodium-catalyzed process. The US production approaches 10,000 tonnes/year.

The catalyst is a dilute solution of rhodium chloride in ethanolic hydrogen chloride, which gives $RhCl_3(H_2O)_n$. This rhodium compound is reduced to a rhodium(I) complex, which may be in equilibrium with a rhodium(III) hydride. Ethylene and butadiene are passed into the mixture. Butadiene reacts with the rhodium complex to give an η-crotyl complex, which in turn couples with ethylene to give *trans*-1,4-hexadiene:

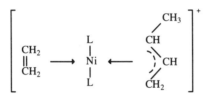

Ziegler catalysts based on nickel, cobalt, and iron salts also bring about the reaction. Cobalt and iron give the *cis* isomer, while nickel gives the industrially important *trans* isomer. The butadiene–nickel–ligand–ethylene complex, where L = ligand, has the approximate structure:

$$
\left[
\begin{array}{c}
\underset{\displaystyle CH_2}{\overset{\displaystyle CH_2}{\|}}
\end{array}
\longrightarrow
\begin{array}{c}
L \\ | \\ Ni \\ | \\ L
\end{array}
\longleftarrow
\begin{array}{c}
CH_3 \\ / \\ CH \\ \diagdown \\ CH \\ / \\ CH_2
\end{array}
\right]^{+}
$$

in which the two hydrocarbons are bonded to the nickel atom, as ligands, through their double bonds.

A third example is the synthesis of β-formylcrotyl acetate, shown in Figure 16.5. Beta-formylcrotyl acetate (**I**) is an important intermediate in the synthesis of vitamin A (see note). Its precursor is 1,4-diacetoxy-2-butene (**II**). The precursor is made as a *cis–trans* mixture from acetylene and formaldehyde via butynediol (**III**) and butenediol (**IV**) or, in a newer process, from butadiene (**V**) and acetic acid (**VI**). The latter process also gives 1,2-diacetoxy-3-butene (Section 10.3.1).

Two routes to β-formylcrotyl acetate have been developed. In the BASF process, the mixture of diacetoxybutenes is heated with platinum(IV) chloride in a stream of oxygen and chlorine. The lowest boiling isomer (**VII**) distils out in high yield and is hydroformylated with a rhodium catalyst. Usually, in such reactions, a ligand is added to give a linear aldehyde but, in this case, it is omitted and the desired branched product (**VIII**) predominates in the mixture of products (**VIII**) and (**IX**). Also, the rhodium, unlike cobalt, does not promote

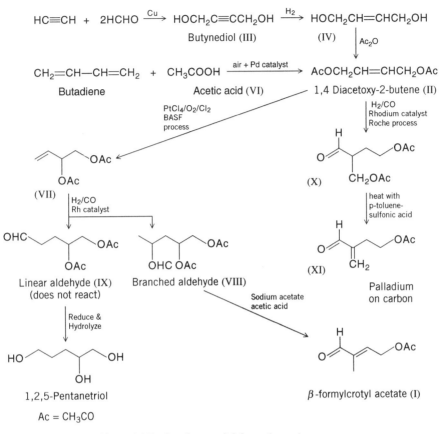

Figure 16.5 Syntheses of β-formylcrotyl acetate.

the migration of the double bond that would be required for the formation of a linear product.

Treatment with sodium acetate in acetic acid selectively eliminates acetic acid from the branched product to give the desired β-formylcrotyl acetate. The linear aldehyde does not react and is separated, reduced and hydrolyzed to 1,2,5-pentanetriol, used as an intermediate in the manufacture of synthetic lubricants.

The Roche process uses rhodium in a different way. It starts with the isomer **(II)** in Figure 16.5, which is hydroformylated with a conventional catalyst $RhH(CO)[P(C_6H_5)_3]_3$ that has been pretreated with sodium borohydride. Because the original double bond was internal, the aldehyde that results is still branched but has the structure **(X)**. Heating with p-toluenesulfonic acid selectively eliminates acetic acid to give **(XI)**, and palladium on carbon isomerizes the double bond to β-formylcrotyl acetate.

The latest catalysts for stereoregular polymer synthesis are the metallocenes, and these are discussed in Section 15.3.12.

16.7.2 Asymmetric Synthesis

Monsanto's asymmetric synthesis of levodopa (see note) was a landmark not only in industrial chemistry but in organic chemistry generally. The key step was the hydrogenation of the olefin (**XII**) to a specific optical isomer of dihydroxyphenylalanine (**XIII**).

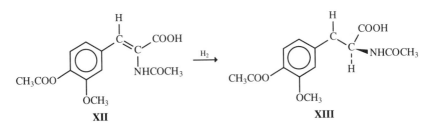

The hydrogenation was carried out in the presence of a soluble rhodium catalyst bearing a chelating biphosphine ligand, DIPAMP (**XIV**):

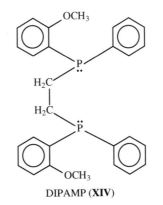

DIPAMP (**XIV**)

The ligand is itself asymmetric and creates an unsymmetrical environment about the rhodium atom that leads to selective hydrogenation on the back face of the C=C bond.

Most of the early stereospecific syntheses involved hydrogenation. Different stereospecific catalysts turned out to be valuable in different hydrogenations. The Wilkinson catalyst $RhCl[P(C_6H_5)_3]_3$ is selective for unhindered C=C double bonds and causes little isomerization in the substrate, that is the organic reagent. The catalyst $[Ir(cod)(PCy_3)py]^+$ is active for very hindered C=C groups and, if a functional group such as –OH is present in the olefin, the catalyst will bind to it and then add hydrogen to the double bond. [The symbols (cod), (PCy$_3$), and py, respectively, represent the ligands cycloocta-1,5-diene, tricyclohexylphosphine, and pyridine.]

In addition to hydrogenation, a reaction of potential importance is stereoselective epoxidation, which adds an oxygen atom to one face of an olefinic double bond. The olefin is treated with *tert*-butyl hydroperoxide in the presence of a titanium complex of diethyl tartrate. The (+) and (−) tartrate isomers produce different optical isomers of the epoxide. For example, allyl alcohol reacts to give either (2R)- or (2S)-glycidol. Either isomer can be reacted with 1-naphthol to give an epoxide which, with isopropylamine, gives the active (2S) form of the antiangina drug propranolol (see note). The naphthol is so large, it sterically prevents (R) formation.

There is every indication that chiral syntheses will be important in the pharmaceutical and fine chemical industries in the future. Indeed it is likely that the US Food and Drug Administration (FDA) will insist that new drugs are marketed as the single pharmacologically active enantiomer rather than as a racemic mixture. Asymmetric synthesis will be of less importance in the bulk organic chemicals industry.

16.8 ENZYMES

Enzymes are the oldest industrial catalysts. In certain respects they are also among the newest. Enzymes are biological catalysts, and many show high specificity. For example urease will only catalyze urea hydrolysis.

$$OC(NH_2)_2 + H_2O \rightarrow CO_2 + 2NH_3$$

Some enzymes will attack certain chemical groups wherever they occur (group specificity). For example, proteolytic enzymes will split the peptide linkage. Proteolytic enzymes will only attack peptides made up either from L-amino acids or from D-amino acids, and this is called stereochemical specificity.

Enzymes are proteins but may be associated with nonproteins (cofactors) essential to their activity. Cofactors may either be simple metal ions like Zn^{2+}, Mg^{2+}, Fe^{2+}, Fe^{3+}, or Cu^{2+}, or organic molecules. In the latter case they are called coenzymes or prosthetic groups. Enzyme activity is usually related to a small region of the molecule referred to as the active center. At low concentrations of substrate, the rate of enzyme action is directly proportional to both enzyme and substrate concentration. If the concentration of substrate is raised, however, the rate ceases to increase and becomes independent of substrate concentration. Thus enzymes are only efficient in dilute solution. Furthermore, they operate only under a limited range of pH (rarely < 4) and temperatures (usually < 50°C), and reactions become very slow as 0°C is approached. Nonetheless, molecule for molecule, enzymes are much more effective than nonbiological catalysts.

The mechanism of their action varies widely. The push–pull mechanism of acid–base catalysis is one route. The enzyme forms a complex with a molecule of reactant. The latter is bound to two active sites on the enzyme molecule and one of these "pushes" electrons while the other "pulls" them, to give a concerted action. Another mechanism is transition state compression to facilitate simultaneous bond making and breaking. Hydrogen bonding stabilizes intermediates.

Enzymes are used in isolated applications such as the addition of proteolytic enzymes to detergents. Fermentation is, of course, based on enzymes (Section 14.5). Although that was an important route to chemicals between World Wars I and II, enzymes as opposed to fermentation have not recently found application in large tonnage chemical production except for high-fructose syrups (Section 14.1) and 6-aminopenicillanic acid. That and various lower tonnage enzyme processes are discussed in Section 14.5.1.

One problem of enzyme technology was solved when it was realized that enzymes retain their activity when attached to an insoluble matrix by a chemical linking agent. In all likelihood, the enzyme is not truly "fixed" and retains some ability to accommodate itself to the shape and conformation of the molecules whose chemical reactions it is catalyzing. Similarly, the substrate to which the enzyme attaches must be able to accommodate to the shape presented by the enzyme. The "immobilization" means that enzymes can be reused may times in continuous reactors. Enzymes are manufactured in bulk by conventional fermentation techniques. They are immobilized in various ways. For intracellular enzymes, either the cells are stabilized by entrapment in an aqueous gel or attached to the surface of spherical particles, or they are homogenized and cross-linked with glutaraldehyde, $OHCCH_2CH_2CH_2CHO$, to form an insoluble yet penetrable matrix.

Extracellular microbial enzymes are immobilized in the form of proteins purified to varying degrees. This usually involves the coupling of the free amino groups on lysine $[H_2N(CH_2)_4CH(NH_2)COOH]$ residues, again with glutaraldehyde, to give an insoluble, active, stabilized enzyme.

There are more sophisticated, less widely used methods. Thus one may start with polystyrene lightly cross-linked with divinylbenzene to give a gel. Treatment with formaldehyde, hydrochloric acid, and zinc chloride gives a chloromethylated polystyrene, in which some of the benzene rings contain $-CH_2Cl$ groups and these will react with amino groups on the enzyme to bind it to the polymer.

Another method uses trialkoxysilanes as coupling agents. A range of agents is available and they couple with enzymes to give compounds of the type $[enzyme]-CH_2CH_2CH_2-Si(OR)_3$. The three carbon bridge is sufficient to ensure that the active site of the enzyme is not obstructed on immobilization. The trialkoxysilicon group will bind to a glass surface to immobilize the enzyme:

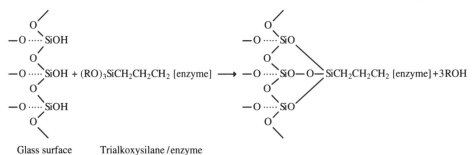

Glass surface   Trialkoxysilane /enzyme

16.8.1 Catalytic Antibodies

Catalytic antibodies are an exciting area of catalyst development in the pharmaceutical industry. Antibodies are chemicals produced by the body as part of its defense against antigens, that is infecting organisms (pathogens) or their toxic products. Cells may be stimulated by infection to produce these antibodies. A person infected with measles, for example, develops the measles antibody and does not get the disease a second time. Immunization with attenuated measles virus produces the same effect. An alternative method, known as passive immunization (e.g., for diphtheria) is to inject the antibodies (immunoglobulins) from one person into another. Once an antibody has been identified, it is also possible by genetic engineering techniques to develop cells that will yield larger quantities. The technique is an elegant way to produce enzymes to order. The interaction between an antigen and its antibody is chemical and specific. A known antigen can be used to identify an unknown antibody and vice versa. Antibodies can be isolated.

An example of the sort of reaction that can be carried out with the aid of catalytic antibodies is the selective catalysis of an unfavorable reaction pathway. In the normal way, the hydroxy–epoxide (**XV**) (Fig. 16.6) cyclizes spontaneously to the substituted tetrahydrofuran (THF) (**XVI**). The aim of the researchers was to direct the reaction towards the disfavored tetrahydropyran (**XVII**). They first synthesized a hapten. A hapten is a small molecule that reacts with proteins, polypeptides or other carrier substances to give an antigen. In this case, an *N*-oxide hapten (**XVIII**) was synthesized, which resembled the transition state of the disfavored reaction pathway. It was injected into an animal and induced a range of antibodies against an antigen that had the structural features of the transition state. The researchers extracted and purified 26 antibodies and evaluated their catalytic properties. Two of them were regioselective for the formation of the desired product and one was highly stereoselective, so that the desired tetrahydropyran resulted.

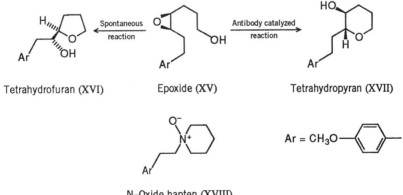

Tetrahydrofuran (XVI) Epoxide (XV) Tetrahydropyran (XVII)

N–Oxide hapten (XVIII)

Figure 16.6 Antibody-catalyzed reactions.

Catalytic antibodies have recently been reported that catalyze peptide bond formation without the need for multiple protecting groups and without residual hydrolytic side reactions. Peptide synthesis is potentially an important branch of biotechnology.

16.9   SHAPE-SELECTIVE CATALYSTS

Many naturally occurring clays and minerals are microporous and may be used as adsorbents (e.g., fuller's earth as cat litter), drying agents, and support for catalysts. Synthetic zeolites with uniform and structured porosities are available. These act as specific catalysts that not only bring about selective chemical reactions but can even generate greater than equilibrium concentrations of products.

Zeolites are aluminosilicate solids with a three-dimensional polymeric framework. Their basic building block is a tetrahedral unit with a silicon or aluminum atom (the so-called T atom) at the center and oxygen atoms at the corners (Fig. 16.7a). Each oxygen atom is bonded to a further T atom, that is, it is shared with another tetrahedral unit. The T atoms can unite via T–O–T linkages to give squares (Fig. 16.7b) or hexagons (Fig. 16.7c) of T atoms. Combination of these square and hexagonal units gives the sodalite cage (Fig. 16.7d), a basic substructural unit. The sodalite cage has oxygen atoms with free bonds able to link to other sodalite cages.

If the sodalite cages unite via their hexagonal faces, an X- or Y-type zeolite is obtained (Fig. 16.7e). These have voids in the center, called supercages, with a diameter of 1.3 nm, and pores or apertures leading to them with a diameter of 0.74 nm. If the sodalite cages unite via their square faces, an A-type zeolite (Fig. 16.7f) is obtained with supercage diameter 1.1 nm and pore aperture 0.42 nm.

Thirty-six zeolite structures are recorded in the literature. The size of voids and apertures can nonetheless be fine-tuned to an even greater extent than this suggests by chemical and physical treatments. In a particularly significant development, Mobil included tetrapropylammonium ions instead of sodium ions in the parent solutions. Between 40 and 90% (typically 70%) of sodium ions were replaced in this way. The $(n\text{-}C_3H_7)_4N^+$ ions are much bigger than Na^+ ions and the resulting zeolite had a structure intermediate between those of the A- and X- or Y-type zeolites. It was called ZSM-5. Incorporation of tetramethylammonium ions as a template led to offretite, and tetrabutylammonium ions to ZSM-11. Mobil discovered ZSM-5 in 1968 but failed to recognize its activity for 5 years. They then found it would convert methanol by way of dimethyl ether to a gasoline containing aliphatic compounds and mixed aromatics up to durene (1,2,4,5-tetramethylbenzene). It also provided aromatics from alkanes, olefins, and oxygenated material and can thus be used to increase the octane number of gasoline. Fischer–Tropsch hydrocarbons (Section 12.2) have an octane number of about 60 and passage over ZSM-5 can raise this to 90.

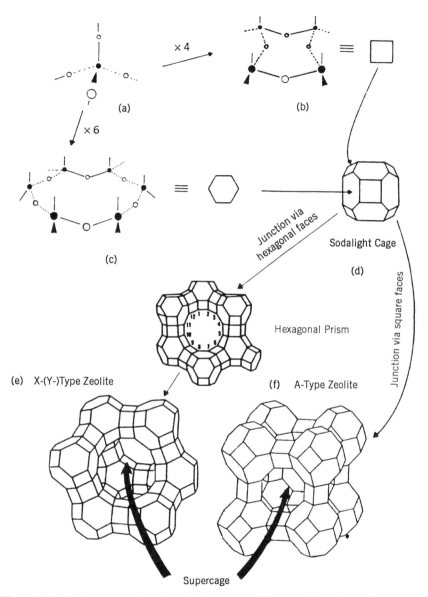

Figure 16.7 Basic structural units and the modes of their combination in the common zeolites. T atoms (Si or Al) are denoted as ● in the simple structures and lie at the intersections of lines in the complex structures. Oxygen atoms are shown in the simple structures when they are represented as ○: in the complex structures an oxygen atom lies at the midpoint of each line. Reproduced from *Biomass, Catalysts and Liquid Fuels*, I. M. Campbell (1983) Holt. Rinehart and Winston Ltd, Eastbourne, p. 116, with permission.

The major use of ZSM-5 is for xylene isomerization, which is discussed below. It catalyzes the disproportionation of toluene to benzene and *p*-xylene (Section 8.1) and is important in one process for the condensation of benzene and ethylene to ethylbenzene (Section 3.8).

Zeolites are normally made from sodium silicate and aluminate. Hence, the finished structures contain sodium ions. If the zeolite is required for catalyst use, it is usual to replace the sodium by ammonium in an ion-exchange process and then to heat the ammonium zeolite so that ammonia is given off and the zeolite is left containing strongly acid H^+ sites. Such zeolites have names preceded by an H, that is, ZSM-5 becomes HZSM-5.

Pores and voids make up about one-half of the volume of zeolites but these "holes" are not accessible to all molecules. Small molecules can diffuse in and out with ease, but larger ones cannot. A-type zeolite admits *n*-alkanes but not branched-chain alkanes. Benzene is also rejected. There is thus a *molecular sieve* action. Sometimes, this provides a method of separation, such as straight-chain from branched-chain hydrocarbons. In other cases, molecules are excluded from catalytic sites on account of their size. Molecules that do gain access to the voids, however, are subject to strong localized electrostatic fields. In most current industrial applications, this leads to acid catalysis.

The shape selectivity of a zeolite catalyst may apply to the reactants, the transition state or the products. An example of reactant selectivity used to raise the octane number of gasoline is the cracking of straight-chain alkanes over A-type zeolites. Branched-chain materials are excluded from the acid sites. Over an amorphous silica–alumina, the branched-chain alkanes would be preferentially cracked or isomerized.

An example of product selectivity is the transalkylation of toluene over HZSM-5 modified with magnesium and phosphorus. The products are benzene and *p*-xylene (Section 8.1). Benzene diffuses very rapidly out of the pores. *p*-Xylene diffuses moderately fast, while the other two isomers diffuse at 0.1% of the rate of the *para* isomer. Thus they have a high chance of isomerizing again before they escape. At operating temperatures, the thermodynamic distribution of xylenes is about 25% each of *ortho* and *para* and 50% of *meta*; zeolites are claimed to give 88–95% *para*.

The most important example of transition state selectivity is the absence of coking in such processes as the formation of gasoline from methanol over HZSM-5 in the Mobil process. Coking proceeds via polynuclear aromatic hydrocarbon intermediates, while the zeolite discriminates against molecules with more than 10 carbon atoms. Hence, coking is avoided and the spread of molecular weights is narrow compared to products from the Fischer–Tropsch or similar processes.

The Mobil process still appears to be the cheapest route to gasoline from methanol but the drop in crude oil prices in the 1980s has reduced interest in it. Toluene disproportionation, however, has been widely commercialized.

Related to work on zeolites is research on clay minerals and particularly smectite clays. Like zeolites these are aluminosilicates and consist of layers, the separation of which can be varied and between which catalysts can lodge. They are known as intercalated catalysts. The collapse of the layered structure at temperatures above 200°C may be avoided by incorporation of "spacers" such as aluminum or zirconium hydroxides to give so-called pillared clays. Currently, the only commercial process based on smectite clays is the dimerization of oleic acid (Section 13.4).

The number of commercialized processes involving synthetic zeolites is also small. Nonetheless, there is extensive research in progress on microporous aluminophosphates, showing that silica is no longer necessary in the formation of a molecular sieve. There is also interest in extra-large pore molecular sieves and in zeolites where catalytically active metal atoms are deposited in the pores to give "uniform" heterogeneous catalysts, in which active sites are distributed in a spatially uniform way throughout the bulk of the solid. Among other things, this raises the possibility of zeolite-based industrial processes other than acid catalysis. Shape selective catalysis is intriguing and holds great potential.

16.10 PHASE-TRANSFER CATALYSIS

Phase-transfer catalysis is used for the preparation of polycarbonates (Section 7.1.2.2), estradiol carbamate and various fine chemicals. It is not likely to be involved in the manufacture of large tonnage heavy organic chemicals but is an unusual and elegant catalytic technique that is energy sparing and gives high yields at low residence times under mild conditions. It is therefore typical of the methods that will be attractive in the future. It finds application where reactants are immiscible. An example is the production of penicillin esters (see note) in which the carboxylate group of an aminopenicillanic acid is esterified with a labile group that will hydrolyze in the gut. The desired reaction is

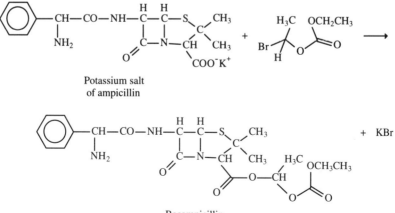

Potassium salt
of ampicillin

Bacampicillin

but the ampicillin salt is water soluble, whereas the acid bromide dissolves in organic solvents and would be hydrolyzed by water. Mild conditions are essential to avoid decomposition of the acid bromide and the lactam ring of the ampicillin. The acid bromide is therefore dissolved in an organic solvent such as dichloromethane or chloroform and brought into contact with an aqueous solution of the ampicillin salt at 25°C. A phase-transfer catalyst, such as tetrabutylammonium chloride, is added to the aqueous phase.

A simple interpretation of what occurs is that the tetrabutylammonium cation is lipophilic and migrates into the organic layer as an ion pair carrying with it the ampicillin anion, which is less hydrophilic than the other anions present. Esterification of the ampicillin anion then takes place in the organic layer, and the tetrabutylammonium ion pairs with the bromide ion that is generated. The latter is so hydrophilic that it carries the tetrabutylammonium cation back into the aqueous phase and the procedure is repeated until esterification is complete. Kinetic measurements indicate that the mechanism is more complicated and may involve inverse micelle formation and interfacial reactions.

An opportunity for phase-transfer catalysis to be used in a large-tonnage process was frustrated by environmental considerations. Ethylene dibromide is at present made by bromination of ethylene (Section 3.11.9). The elemental bromine is obtained from bromide-containing brines from which it is displaced by chlorine and steam in an expensive energy-intensive process. Dead Sea bromine, in Israel, developed a process by which ethylene dichloride was mixed with Dead Sea brines in the presence of a phase-transfer catalyst. Bromide is more lipophilic than chloride, hence the bromide ions were carried to the organic layer to yield the dibromide. The main use for ethylene dibromide, however, is as a lead scavenger in leaded gasoline. Its market is therefore declining and it was never worthwhile to commercialize the new process.

16.11 CATALYSTS OF THE FUTURE

The objectives of catalyst research in the future have been listed as follows:

1. To advance catalyst theory to the point where catalysts can be designed. Remarkable progress has been made in the past few years in understanding catalyst action. The use of computer modeling is now routine in the development of new pharmaceuticals but is only just beginning to be used in the design of catalysts. Nonetheless computer techniques can aid catalyst design not only at the molecular and electronic level but also in the modeling of transport phenomena. There is scope for a breakthrough.

2. To achieve higher selectivities in known reactions. This might involve improvement of existing catalysts or development of new ones. Most of the catalysts discussed above were primarily developed to give higher selectivities, and we may expect to see advances in, for example, catalysis by zeolites.

Ammoxidation of propylene to acrylonitrile (Section 4.4) shows what is possible. The earliest catalysts (in the laboratory) gave 6% yields. Bismuth phosphomolybdate raised this to 65% and a plant was built in 1959. An antimony–uranium oxide catalyst gave laboratory yields of 80% in 1966 (72.5% in the plant). In 1972, promoted bismuth–molybdenum oxides on silica raised plant yields to 77%. A fourth generation catalyst composed of an oxide complex of iron, selenium, and tellurium appears to give about 83% and a series of patents in the early 1980s claimed 87% for a related complex. In addition, the newer catalysts have reduced byproduct acetonitrile production to the point that there is a shortage for the small volume uses that do exist.

On the other hand, there is enormous scope for improvement in the classic iron catalysts for the Haber process. Current yields are between 15 and 17% per pass. Higher yields with present catalysts would necessitate higher pressures and the use of expensive reciprocating rather than centrifugal compressors. Hence, there is great scope for improved catalysts but there are formidable difficulties in finding them.

Since the above was written, however, Kellogg and BP reported a new catalyst with 10–20 times greater activity. It consists of promoted ruthenium on a proprietary support. The greater activity means that the process can be operated at lower temperatures (350–490°C) and pressures (70–105 bar) while maintaining reasonable conversions per pass. It is more tolerant to variations in the N_2/H_2 ratio. The saving in energy is estimated at 21 kJ/mol and the saving in cost at \$2–6/tonne of ammonia.

3. To synthesize catalysts with greater activity (e.g., Ziegler catalysts as in Section 16.1.2). Greater activity improves economics. Homogeneous catalysts, with high activity could be converted to heterogeneous catalysts that still retain activity. Enzymes may be immobilized without substantial loss of activity but, in general, only a highly active homogeneous catalyst can be successfully immobilized.

There has been considerable interest in the use of clusters of metal atoms bound to carbon monoxide as highly active catalysts. Many catalytically active metals such as palladium, rhodium, platinum, osmium, rhenium, and ruthenium form clusters. The compound $Rh_6(CO)_{16}$ is active in the methanol–carbon monoxide route to acetic acid, but appears to cleave during the reaction. In spite of much research, cluster catalysts have not yet been commercialized.

Clusters of metal atoms of a size where they are just beginning to develop the properties of a solid metal are also very active as are intermetallic compounds. Many of these are known, one component being a rare earth. Lanthanum pentanickel $LaNi_5$, for example, absorbs large quantities of hydrogen reversibly and may be used in low-temperature hydrogenations. Currently $Rh_6(CO)_{16}$ is used in acetic acid production. Sometimes the two metals may be bonded either directly or via a ligand. One metal activates a hydrogen atom and the other the molecule.

There is also interest in the catalytic properties of heteropolyacids. An example that has been commercialized is the production of *tert*-butanol from the isobutene in mixed butenes over the heteropolyacid $H_3PMo_{12}O_{40}$.

4. To devise catalysts to solve pollution problems. The breakthrough in this area was the development of the platinum–palladium–rhodium catalysts that oxidize unburned hydrocarbons and carbon monoxide in the catalytic converter of automobiles. All the same, a cheaper catalyst is a desirable aim, as is a catalyst that would promote decomposition of nitrogen oxides to molecular nitrogen and oxygen in the presence of water and carbon dioxide. Other aims are catalysts to desulfurize flue gases and to remove organochlorine compounds from water. The latter might involve hydrogen or hydrogen donors such as hydroaromatics, and a catalyst, based perhaps on iron, cobalt, or ruthenium. There is a host of other environmental problems that could be helped by novel catalysts.

5. To achieve catalysts for entirely new reactions. A major research aim today is the functionalization of methane, ethane, propane, and butane to replace more expensive olefins as petrochemical precursors. Selective catalysts for low-temperature liquid-phase oxidation of methane to methanol and for oxidative coupling of methanol to ethylene would be especially valuable (Section 11.1). Among the few successes so far in the functionalization of alkanes is the conversion of *n*-butane to maleic anhydride (Section 5.4). A process for conversion of propane to acrylonitrile is slated for commercialization in the late 1990s (Section 11.2.1) and uncommercialized processes exist for the conversion of ethane to vinyl chloride. A laboratory process has also been reported for direct conversion of methane to acetic acid in aqueous solution at 100°C. The oxidants are carbon monoxide and oxygen and the catalyst rhodium trichloride plus a source of iodide ions or 5% palladium on carbon.

6. To develop catalysts that mimic natural catalysts. Nature manages to produce complex organic molecules in a single stereochemical form under mild conditions. The synthesis of stereoregular and optically active compounds is already possible in some cases (Sections 16.7.1 and 16.7.2). Some natural enzymes may be isolated, immobilized, and used as catalysts to give "natural" products. Most remarkable of all is recombinant DNA engineering or gene splicing, which makes possible the synthesis of proteins and peptides.

16.11.1 Artificial Photosynthesis

Tradition demands that scientific books should finish with some reference to the ultimate problems facing the world (or at any rate the developed world), and how science might help solve them. The problem might be anything from the rise in the level of the oceans to the rise in Islamic fundamentalism. We have no wish to depart from this tradition and our choice of problem, less fashionable than it was in the 1970s perhaps, is the energy crisis.

The point is that the world energy crisis has not been abolished by the discoveries of oil and natural gas in the 1980s; it has merely been postponed. The world is still reliant on fossil fuels and, apart from coal, these will inevitably be depleted, if not in the next 50 years, then in the next 100.

Various sources of renewable nonpolluting energy have been suggested. Nuclear fission has been a disappointment. Nuclear fusion is a hope for the future but the problems of getting energy into and out of a fusion reactor have not yet been solved. Much more attractive environmentally and technically is the development of some method of harnessing the sun's energy. In a sunny country, solar energy provides about 800 W/m^2. Solar energy equivalent to the total world fossil fuel energy reserves falls on the earth in less than 14 days and the annual world's energy requirement within 26 min. The problem is to concentrate the energy.

Solar furnaces, in which mirrors concentrate sunlight onto a furnace, have been built in Sicily (Eulios) and California (SEGS I + II). They have an efficiency of 10% but a high capital cost. Solar ponds, pioneered in Israel, produce a temperature difference of 60–70°C, which can be used to drive generators, but are hampered by capital cost and low thermodynamic efficiency. Photovoltaic cells produce electricity at efficiencies between 10 and 20% but widespread use is inhibited by the high cost of the very pure silicon, gallium arsenide or gallium antimonide required. A plant at Kibbutz Samar north of Elath, Israel, is expected to provide energy at 20 ¢/kWh compared with coal-fired power stations at 8 ¢/kWh.

Nature's answer has been the development of the remarkable catalyst, chlorophyll, which catalyzes the photochemical reaction of carbon dioxide and water to give carbohydrates. Sugar cane is said to be the most efficient photosynthesizer but only achieves an efficiency of about 1%. The Brazilian gasohol experiment has produced ample data on culture of sugar cane as an energy source and the process is labor rather than capital intensive. Is it possible to develop a catalyst with a higher efficiency than chlorophyll that would harness solar energy?

The feedstock would have to be carbon dioxide or water or both. Water requires only 238 kJ/mol to split it into hydrogen and oxygen; carbon dioxide requires 283 kJ/mol to give carbon monoxide and oxygen. Carbon dioxide and water will give methanol at a cost of 703 kJ/mol. These energies correspond to wavelengths of light of 502, 422, and 170 nm, respectively. The first falls acceptably in the green region of the spectrum, the second is in the blue but a high proportion of visible light will be useless, and the third is in the vacuum ultraviolet (UV) and could not be brought about by solar energy in a single step. Thus the majority of work on artificial photosynthesis has concentrated on splitting of water.

No reproducible cyclic system for artificial photosynthesis has yet been found, but the system shown in Figure 16.8a will work in principle. Ruthenium–trisbipyridine (**XIX**) (called the sensitizer) absorbs visible light and is promoted to an excited state. The excited state passes on its energy to methylviologen (**XX**) (called the quencher) and is oxidized to the trivalent state. This interacts with water and ruthenium(IV) oxide to generate oxygen and regenerate the bivalent compound. Meanwhile, the methylviologen, better known as the herbicide paraquat, is reduced to the monovalent state. In contact with a platinum/titanium dioxide catalyst, it splits water to hydrogen and a hydroxyl ion and reverts to the bivalent state. The titanium dioxide serves as a support for the platinum, thus stabilizing it in a suspension. The MV^{2+}/Pt electron transfer can then occur on the surface.

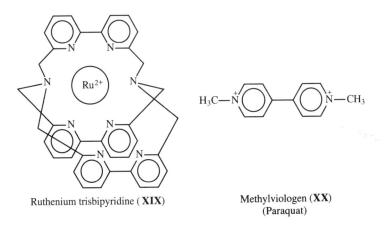

Ruthenium trisbipyridine (**XIX**)

Methylviologen (**XX**)
(Paraquat)

This system is not reproducible in a very clear way. In particular, the oxygen-carrying step requires high purity and selectivity for the requisite catalytic step. Recent work has focused on improvements in the separate halves of this system, with sacrificial compounds to transfer energy to or from the ruthenium–tris-bipyridine. Figure 16.8b shows a system for the generation of hydrogen, with triethanolamine being sacrificially oxidized. Figure 16.8c shows the generation of oxygen with a cobalt complex in place of the methylviologen quencher accepting energy from the ruthenium trisbipyridine. This half-cycle is the more difficult and requires rigorous exclusion of atmospheric oxygen.

Systems of this kind are capable of wide variation with changes in sensitizer, quencher, and catalysts. There is great interest in light-harvesting polymers in which the sensitizer and quencher exist in a single molecule and their interaction occurs by energy transfer along a series of covalent bonds (see note).

The difficulties with these artificial photosynthetic systems are that the absorbed energy can be degraded in a number of ways, only one of which leads to the desired products. Efficiencies of 2.8% have been achieved, and it is thought that 10% is an achievable target with the present technology. The advantage of these systems is that they are largely homogeneous. Capital costs would be relatively low. The product hydrogen is a useful, nonpolluting source of energy and can be transported over distances more cheaply and efficiently than electricity.

Artificial photosynthesis is still some way from commercialization but is attractive in the sense that it avoids the high temperatures of nuclear fusion and resembles rather the mild conditions that characterize natural processes. On the other hand, the efficiencies of current photosynthetic processes are simply not good enough. Can they be improved? In principle, there is no reason why not.

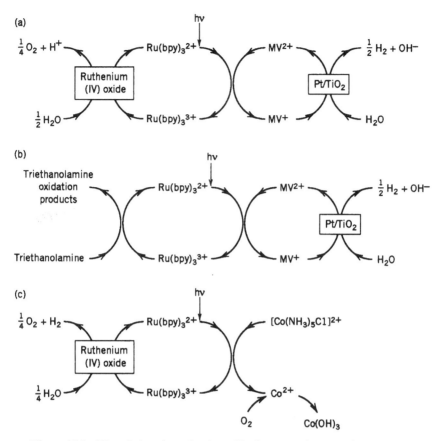

Figure 16.8 Photoinduced production of hydrogen and oxygen from water.

On the other hand, nature has had millions of years to get it right, and the best it has been able to do is chlorophyll. Unlike plant species, of which there are millions, only the four variants of chlorophyll have evolved in global history. Why are there no evolutionary half-way points, no somewhat modified chlorophylls, not quite as good but capable of occupying a niche? Is there really something better? Here is the ultimate challenge to the catalyst chemist, indeed to the chemistry community as a whole.

NOTES AND REFERENCES

An excellent general account of industrial catalysis is given in the new fourth edition of J. I. Kroschwitz and M. Howe-Grant, Eds, *Kirk–Othmer Encyclopedia of Chemical Technology*, New York: Wiley, 1993, Vol. 5, pp. 320–460, and the chemical engineering aspect is summarized in R. H. Perry, C. H. Chilton, and D. W. Green, *The Chemical Engineer's Handbook*, 6th ed., McGraw-Hill, New York, 1984.

The Royal Society of Chemistry, London, produces *Catalysis: A Review of Chemical Literature*, Ed, J. J. Spivey et al. By 1994, the series had reached Vol. 10. There are two authoritative continuing multivolume series: Frankenburg, W. G. Eley, et al., Eds., *Advances in Catalysis*, Academic, New York, 1948– and Anderson, J. R. and Boudart, M., *Catalysis—Science and Technology*, (7 vols.) Springer-Verlag, Berlin, 1981– .

Classic books on heterogeneous catalysis include G. C. Bond, *Catalysis by Metals*, Academic, New York, 1962; J. M. Thomas and W. J. Thomas, *Heterogeneous Catalysis*, Academic, London, 1967, and C. N. Satterfield, *Mass Transfer in Heterogeneous Catalysis*, Cambridge, MA: MIT, 1970. Table 5.1 is largely drawn from G. C. Bond, *Heterogeneous Catalysis: Principles and Applications*, Clarendon Press, Oxford, UK, 1974. More recent are B. C. Gates, *Catalytic Chemistry*, Wiley, New York, 1992 and an exceptionally useful paperback, I. M. Campbell, *Catalysis at Surfaces*, Chapman & Hall, London, 1988. A complete issue of *Chemical Reviews* (May 1995, Vol. 95) was devoted to state-of-art reviews of topics in heterogeneous catalysis.

Homogeneous catalysis is tackled in the review series Ugo, R. (Ed.), *Aspects of Homogeneous Catalysis*, Dordrecht, Holland: Kluwer, 1970– . Both types of catalysis are covered in B. Delmon and G. Jannes, Eds., *Catalysis: Heterogeneous and Homogeneous*, Elsevier, Barking, Essex, UK, 1975.

M. V. Twigg, Catalysis doyen of ICI Agricultural Division, has produced a second edition of his *Catalyst Handbook*, Wolfe, London, 1989. Other publications with a practical slant include a chapter: J. Pennington, Catalysts and Catalysis, in *Introduction to Industrial Chemistry*, 2nd ed., C. A. Heaton, Ed, Glasgow, 1991; a book: R. Pearce and W. R. Patterson, Eds., *Catalysis and Chemical Processes*, Leonard Hill, London 1981 (we liked this one particularly) and a 3-volume *magnum opus*: Leach, B. E., *Applied Industrial Catalysis*, Academic, New York, 1984.

Plenum Press, New York, is publishing a series under the general titile *Fundamental and Applied Catalysis*. The first—J. T. Richardson, *Principles of Catalyst Development*, 1989—is a crisp and clear introduction to heterogeneous catalysis. Other volumes are V. P. Zhdanov, *Elementary Physicochemical Processes on Solid Surfaces*, 1991; J. R. Jennings, *Catalytic Ammonia Synthesis: Fundamentals and Practice*, 1991; Kenzi Famaru, *Dynamic Processes on Solid Surfaces*, 1993, and Jacques Vedrine, *Catalyst Characterisation*, 1994. J. R. Jennings has also produced a volume in the *Critical Reviews of Applied Chemistry Series* entitled *Selected Developments in Catalysis*, Elsevier, Barking, Essex, UK, 1991.

Section 16 The Pimental report is entitled *Opportunities in Chemistry*, NAS/NRC/02-86, National Academy Press, Washington, DC, 1985.

P. N. Rylander, H. Greenfield, and R. L. Augustine, *Catalysis of Organic Reactions*, Dekker, New York 1993 is primarily an academic text but two of the authors have been in industry, and it provides an up-to-date survey of the issues of concern in catalyst research.

Section 16.1 For a not too technical introduction to the subject, see V. Haensel (the discoverer of catalytic reforming) and R. L. Burwell. Jr. "Catalysis" *Sci. Am.*, **225** No. 6, 46 (1971).

Section 16.1.1 Figure 16.1 is based mainly on standard thermodynamic data together with activation energies from G. J. Minkoff and C. F. H. Tipper, *Chemistry of Combustion Reactions*, Butterworth, London, 1962; C. Bamford and C. F. H. Tipper, *Comprehensive Chem. Kinetics*, **5**, Elsevier, Amsterdam, 1972; and K. Weissermel and H. J. Arpe, *Industrial Organic Chemistry*, Verlag Chemie, Weinheim, 1978. The last of these shows that the peracetic acid → acetic acid reaction goes via an α-hydroxyethyl peracetate transition state formed from peracetic acid and acetaldehyde. The peracetic acid and transition state enthalpies were deduced by the methods in S. W. Benson, *Thermochemical Kinetics*, 2nd ed., Wiley, New York, 1976.

The spread of *A* values in the compensation effect is reported by F. S. Feates, P. S. Harris, and B. G. Reuben, *J. Chem. Soc. Faraday Trans.* **70**, 2011 (1974).

Section 16.1.2 The irreversible degradation of CRG catalysts is documented in A. Williams, G. A. Butler, and J. Hammonds, *J. Cat.*, **24**, 352 (1974); W. H. Gitzen, *Alumina as a Ceramic Material*, p. 35. *Special Publication* #4, American Ceramic Soc. Columbus, OH (1970); and G. Yamaguchi and H. Yanagida, *Bull. Chem. Soc. Jpn.*, **36**, 1155 (1963).

Section 16.1.3 See Butt, J. B. and Petersen, E. E., *Activation, deactivation and poisoning of catalysts*, Academic, San Diego, CA, 1988.

Section 16.3 This section is based on A. M. Thayer, *Chem. Eng. News*, 11 July 1994; J. Haggin, *Chem. Eng. News*, 25 July 1994; I. Young and A. Wood, Catalysts—Reacting to Changing Markets, *Chem. Week*, June 26, 1991; S. C. Stinson, Catalysts: A Chemical Market Poised for Growth, *Chem. Eng. News*, December 5, 1983, p. 19; and J. M. Winton, Catalysts '88, Restructuring for Technical Clout, *Chem. Week*, June 29, 1988, p. 20. Noble Metal Statistics come from *Platinum 1995*, Johnson Matthey, London.

Section 16.4 The classic book on acid–base catalysis is R. P. Bell, *The Proton in Chemistry*, Cornell University Press, Ithaca, NY, 1959. The neglected area of basic catalysis was covered in H. Pines (the inventor of alkylate for gasoline) and W. M. Stalick, *Base-Catalyzed Reactions of Hydrocarbons and Related Compounds*, Wiley, New York, 1977.

Section 16.5 Physical techniques for investigation of the properties of adsorbed layers are summarized by J. T. Yates, *Chem. Eng. News*, August 26, 1974, p. 9.

Section 16.5.1 Catalytic converters are described in K. C. Taylor, Catalysts in Cars *CHEMTECH* **20**, 551 (1990).

Section 16.6 Catalysis by chemisorption is relatively well understood and is discussed by R. P. H. Gasser, *An Introduction to Chemisorption and Catalysis by Metals*, OUP, Oxford, UK, 1985. It is possible that catalysis at surfaces can be understood non-mathematically on the basis of organometallic chemistry. Readers unhappy with the quantum theory of solids should try Basset, J-M, Ed., *Surface Organometallic Chemistry: Molecular Approaches to Surface Catalysis*, Dordrecht, Holland, Kluwer, 1988. The mechanism for the carbonylation of methanol appeared in D. Forster, *J. Am. Chem. Soc.* **98**, 846 (1976). A recent very short and inexpensive book (85 pages, $4.99) is R. A. Henderson, *The Mechanism of Reactions at Transition Sites*, OUP, Oxford, UK, 1994.

Section 16.6.1 See Catalysts and Automotive Pollution Control, A. Crucq Ed., Elsevier, Barking, 1991, especially M. Bowker et al., *ibid.* p. 409.

Section 16.7.1 The Dupont hexadiene synthesis is described by C. A. Tolman, *J. Am. Chem. Soc.* **92**, 6777 (1977). See also "New Perspectives in Surface Chemistry and Catalysis," *Chem. Soc. Rev.* **6**, 373 (1977). The synthesis of vitamin A is described by H. A. Wittcoff and B. G. Reuben, *Industrial Organic Chemicals in Perspective, Part II*: *Technology, Formulation and Use*, Wiley, New York, 1980, p. 373.

Section 16.7.2 For the synthesis of levodopa, see H. A. Wittcoff and B. G. Reuben, Part II, *op. cit.*, p. 268. There is a continuing series on asymmetric synthesis: Morrison, J. D.

et al., Eds, *Asymmetric Synthesis*, Wiley, New York. Three seminal articles appeared in the late 1980s by G. W. Parshall and W. A. Nugent. They were: Making Pharmaceuticals via Homogeneous Catalysis, *CHEMTECH* **18**, 184 (1987), Functional Chemicals via Homogeneous Catalysis, *ibid.* p. 314; and Homogeneous Catalysis for Agrochemicals, Flavors and Fragrances, *ibid.* p. 376. The accumulated wisdom has appeared in book form: G. W. Parshall and S. D. Ittel, *Homogeneous Catalysis*, Wiley, New York, 1992.

There are many other books on what is a fashionable topic. For the industrially oriented reader we suggest R. A. Aitken and N. Kilenyi, *Asymmetric Synthesis*, Chapman & Hall, London, 1992; A. N. Collins, G. N. Sheldrake, and J. Crosby, *Chirality in Industry—the Commercial Manufacture and Applications of Optically Active Compounds*, Wiley, Chicheser, UK, 1992; R. A. Sheldon, *Chirotechnology: Industrial Synthesis of Optically Active Compounds*, Dekker, New York, 1993 and R. Noyori, *Asymmetric Synthesis in Organic Catalysis*, Chichester, UK, Wiley, 1994.

Use of the chiral auxiliary $[(C_2H_5)Fe(CO)(PPh_3)]$ for asymmetric synthesis is discussed by S. G. Davies, *Aldrichimica Acta* **23**, 31 (1990).

The FDA views on chiral drugs at the time of writing are explained in Fox, J., *Chem. Ind.* 270 (1993).

Section 16.8   A sound introduction to enzyme catalysis is provided by T. Palmer, *Understanding Enzymes*, Ellis Horwood, New York, 1991. See also J. E. Bailey and O. F. Ollis, *Biochemical Engineering Fundamentals*, 2nd ed. McGraw Hill, New York, 1986 and A. Fersht, *Enzyme Structure and Mechanism*, 2nd ed., W. H. Freeman, San Francisco, 1988. Immobilization techniques are discussed in K. Musbach, Ed., *Methods in Enzymology*, Vol. **44**, Academic, New York, 1976. Binding of enzymes to polystyrene is discussed in H. H. Szmant, *Industrial Utilization of Renewable Resources*, Technomic, Lancaster PA, 1986, and by silanes in E. P. Plueddemann, *Silane Coupling Agents*, Plenum, New York, 1982.

Lysine synthesis is described in H. A. Wittcoff and B. G. Reuben, Part II, *op. cit.* The 6-aminopencillanic acid synthesis is described by B. G. Reuben and H. A. Wittcoff, *Pharmaceutical Chemicals in Perspective*, Wiley, New York, 1989, p. 132, and in E. J. Vandamme (Ed.), *Biotechnology of Industrial Antibiotics*, Dekker, New York, 1984.

Section 16.8.1   We have drawn on two secondary sources: R. M. Baum, *Chem. Eng. News*, April 19, 1993, p. 33, and G. M. Blackburn and P. Wentworth, *Chem. Ind.* 338 (1994), together with J. K. Janda, C. G. Shevlin, and A. Lerner, *Science*, **259**, 490 (1993). Peptide synthesis is reported by R. F. Hirschmann and A. B. Smith, *Science* **265**, 234 (1994) and reviewed in *Chem. Eng. News*, July 11, 1994.

Section 16.9   J. D. Idol, "Horizons in Catalysis," *Chem. Ind.* 272 (1979) is an excellent general article on the development of molecular sieve catalysis. Zeolites for Industry—a full issue of *Chem. Ind.* April 2, 1984—has articles by J. Dwyer, A. Dyer, R. P. Townsend, L. V. C. Rees, and S. M. Csicsery. An article on Shape Selective Catalysis in Zeolites appears in *Chem. Brit.* **21**, 473 (1985). The New Zealand methanol-to-gasoline process is described by C. J. Maiden, Catalysis Research Bears Fruit, *CHEMTECH*, **18**, 32, (1988). A rare article on clays is E. Chynoweth, Shaping Environmental Solutions with Clay-Based Catalysts, *Chem. Week*, June 26, 1991, p. 55. Extra-large pore molecular sieves are described in M. E. Davies, *Chem. Ind.* 137 (1992), and "uniform" heterogeneous catalysts in J. M. Thomas, J. Chen, and A. George, *Chem. Brit.* **28**, 991 (1992).

The idea of encapsulating metal complexes in zeolite to give biomimetic catalysts is discussed by P. C. H. Mitchell, *Chem. Ind.* 308 (1991). Computer-aided design of zeolites is the theme of *Modelling of Structure and Reactivity in Zeolites*, C. R. A. Catlow, Ed., Academic, London, 1992.

Section 16.10 Phase transfer catalysis is discussed exhaustively in a new edition of the classic, E. V. Dehmlov, and S. S. Dehmlov, *Phase Transfer Catalysis*, 3rd ed., VCH Verlag, Weinheim, 1993. Ch. M. Starks, another pioneer of the technique, has edited *Phase Transfer Catalysis*, No. 326 in the ACS Symposium series, American Chemical Society, Washington DC, 1987. Still useful is W. E. Keller, *Phase Transfer Reactions—Fluka Compendium*, 2 vols. Stuttgart: G. Thieme Verlag, 1986/1987. Industrial aspects are covered by B. G. Reuben and K. Sjøberg, *CHEMTECH* **11**, 315 (1981). The penicillin esters are discussed in B. G. Reuben and H. A. Wittcoff, *Pharmaceutical Chemicals in Perspective*, Wiley, New York, 1989, p. 134.

Section 16.11 The future is overt in *Perspectives in Catalysis: A Chemistry for the 21st Century Monograph*, J. M. Thomas and I. Zamaraev, Eds. Blackwell Scientific, Oxford, UK, 1992. See also Ten Challenges for Catalysis, *Chem. Eng. News*, 31 May 1993, p. 27.

A pioneering book is E. R. Becker and C. J. Pereira, *Computer-Aided Design of Catalysts*, Decker, New York, 1993. See also C. R. A. Catlow, note to Section 16.9.

Methane conversion is discussed by J. Haggin in several articles in *Chem. Eng. News*. January 22, 1990, pp. 20–26. A methane-to-acetic acid process with a peroxysulfate catalyst has been known for some time but the new route uses simpler reagents. It is reported by M. Lin and A. Sen, *Nature (London)*, **368**, 613 (1994) and in *Chem. Brit.* **30**, 624 (1994).

Section 16.11.1 For a review of artificial photosynthesis see H. Dürr, Artifizielle Photo-Synthese, *Magazin Forschung*, **1**, 61 (1989). More recent work includes H. Dürr, S. Bossmann, R. Schwarz, M. Kropf, R. Hayo, and N. J. Turro, *J. Photochem. Photobiol. A: Chem.* **80**, 341 (1994) and H. Dürr, S. Bossmann, G. Heppe, R. Schwarz, U. Thiery, and H.-P. Trierweiler, *Proc. Indian Acad. Sci. (Chem. Sci.)* **105**, 435 (1993). Work on the production of molecules with sensitizer and quencher in the same molecule is reported in M. A. Fox, W. E. Jones Jr., and D. M. Watkins, "Light-Harvesting Polymer Systems," *Chem. Eng. News*, March 15, 1993, p. 38, which also provides a detailed reading list.

INDEX